5th AIChE Southwest Process Technology Conference 2013

Galveston, Texas, USA
3-4 October 2013

ISBN: 978-1-5108-1651-0

Printed from e-media with permission by:

Curran Associates, Inc.
57 Morehouse Lane
Red Hook, NY 12571

Some format issues inherent in the e-media version may also appear in this print version.

Printed by Curran Associates, Inc. (2016)

AIChE
120 Wall Street, FL 23
New York, NY 10005-4020

Phone: (800) 242-4363
Fax: (203) 775-5177

www.aiche.org

Additional copies of this publication are available from:

Curran Associates, Inc.
57 Morehouse Lane
Red Hook, NY 12571 USA
Phone: 845-758-0400
Fax: 845-758-2634
Email: curran@proceedings.com
Web: www.proceedings.com

TABLE OF CONTENTS

DISTILLATION & SEPARATION

UPSTREAM OIL & GAS

STS SOCIAL HOUR & DINNER PROGRAM

PROFESSIONAL ETHICS

PROCESS CONTROL & OPTIMIZATION

CHEMICAL PROCESS TECHNOLOGY

ExxonMobil
Taking on the world's toughest energy challenges.™

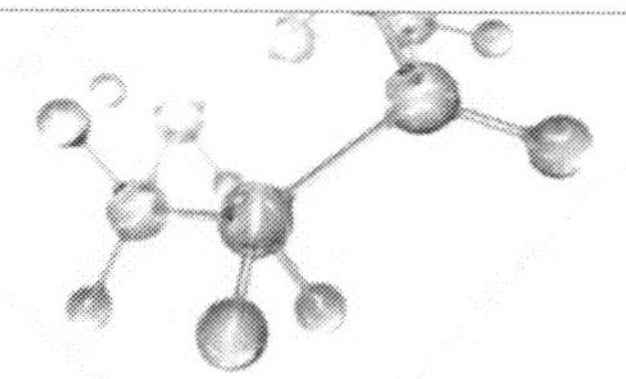

U.S. Chemical Industry Revitalization

AIChE Southwest Process Technology Conference
October 3, 2013
Lynne Lachenmyer
Senior Vice President, ExxonMobil Chemical Company

Chemical Industry Vital to U.S. Economy

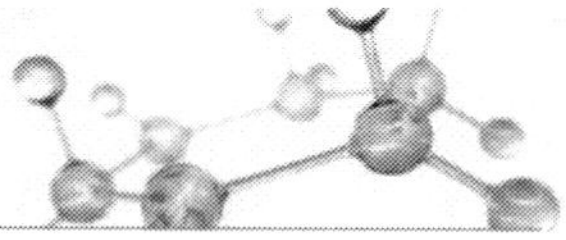

$770 billion	Annual sales – one of America's largest industries
12	Percent of U.S. exports
$39 billion	Annual capital investment
800,000	Industry employees
25	Percent of U.S. GDP supported

Source: American Chemistry Council, June 2013

ExxonMobil
Taking on the world's toughest energy challenges.™

Chemicals: Fastest Growing Sector

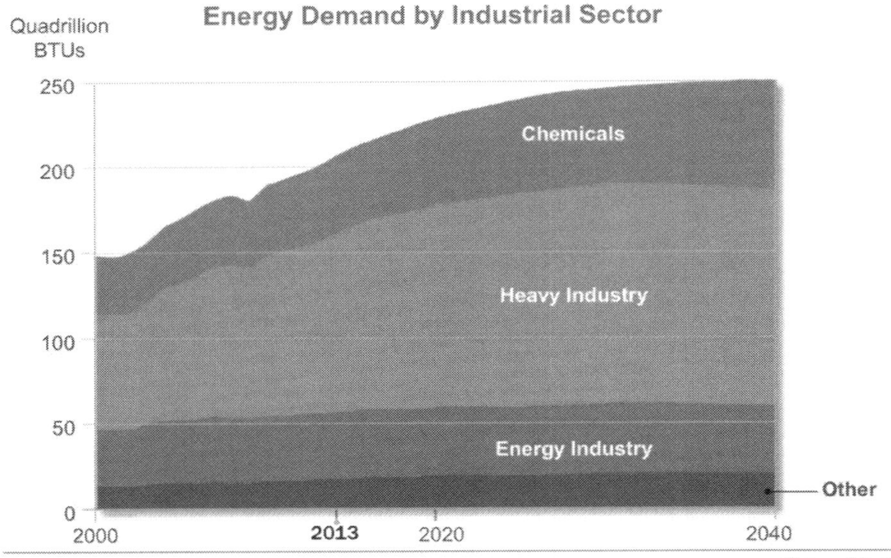

ExxonMobil 2013 Outlook for Energy

3

ExxonMobil
Taking on the world's toughest energy challenges.

Population Drives Global Energy Demand

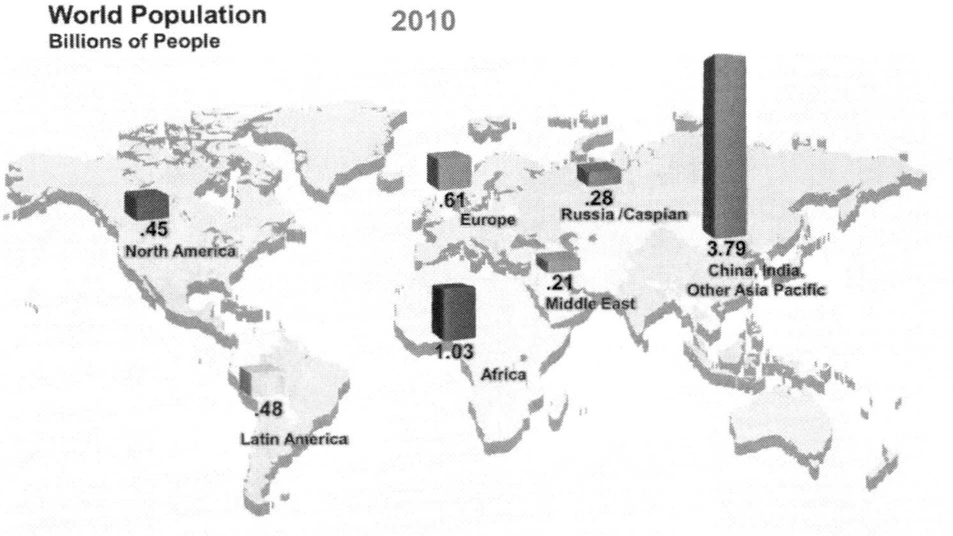

ExxonMobil 2013 Outlook for Energy

4

ExxonMobil
Taking on the world's toughest energy challenges.

Population Drives Global Energy Demand

World Population
Billions of People

2010 / 2040

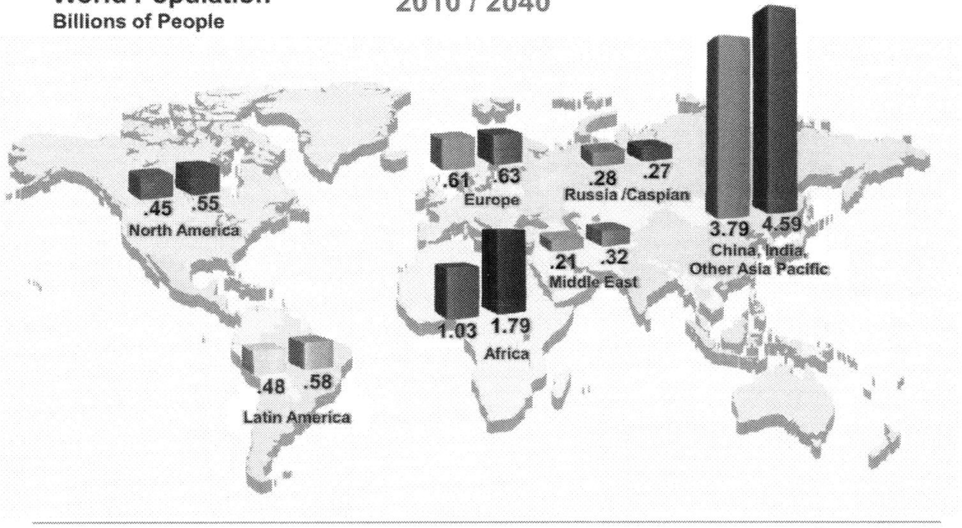

Energy Mix Continues to Evolve . . .

Global Energy Mix Quadrillion BTUs

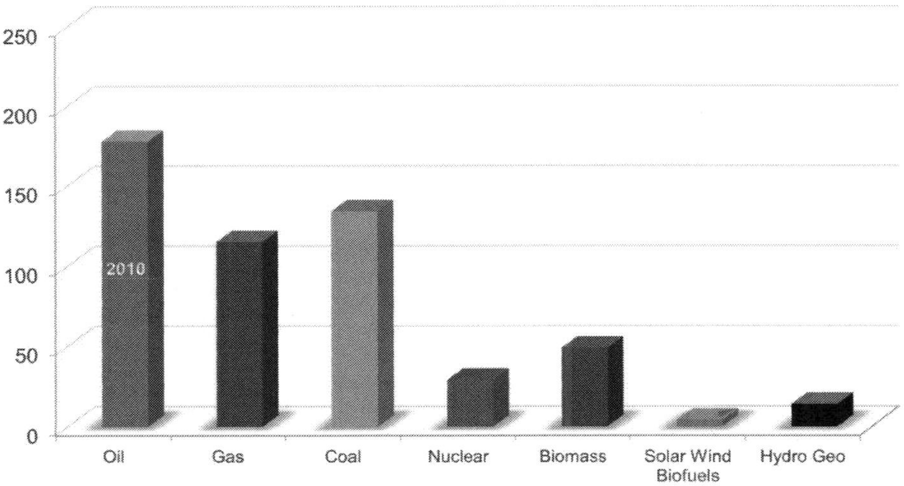

Natural Gas Leads the Growth

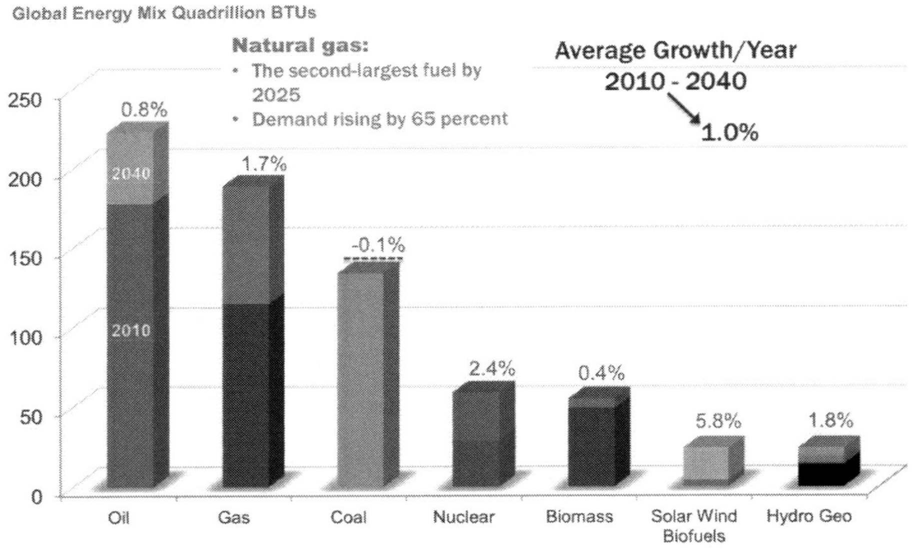

ExxonMobil 2013 Outlook for Energy 7

ExxonMobil
Taking on the world's toughest energy challenges.

Natural Gas Abundant, Widespread

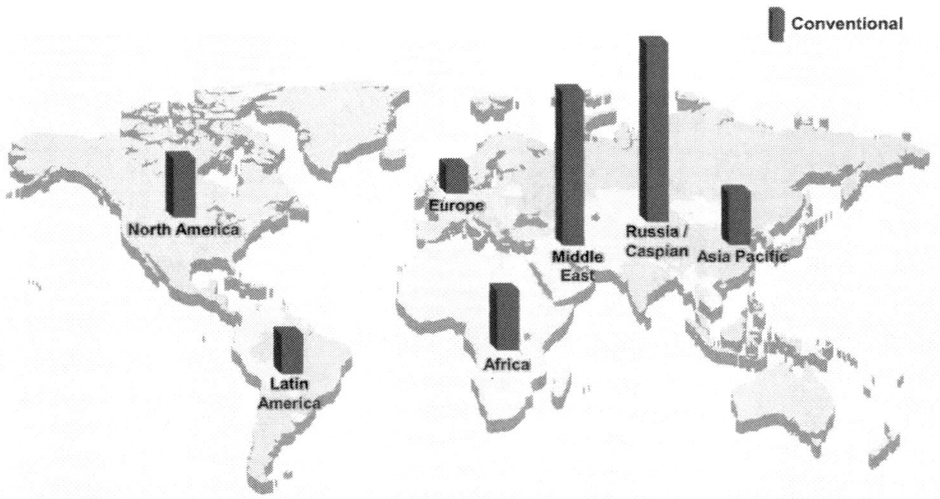

Source: IEA

ExxonMobil 2013 Outlook for Energy 8

ExxonMobil
Taking on the world's toughest energy challenges.

Natural Gas Abundant, Widespread

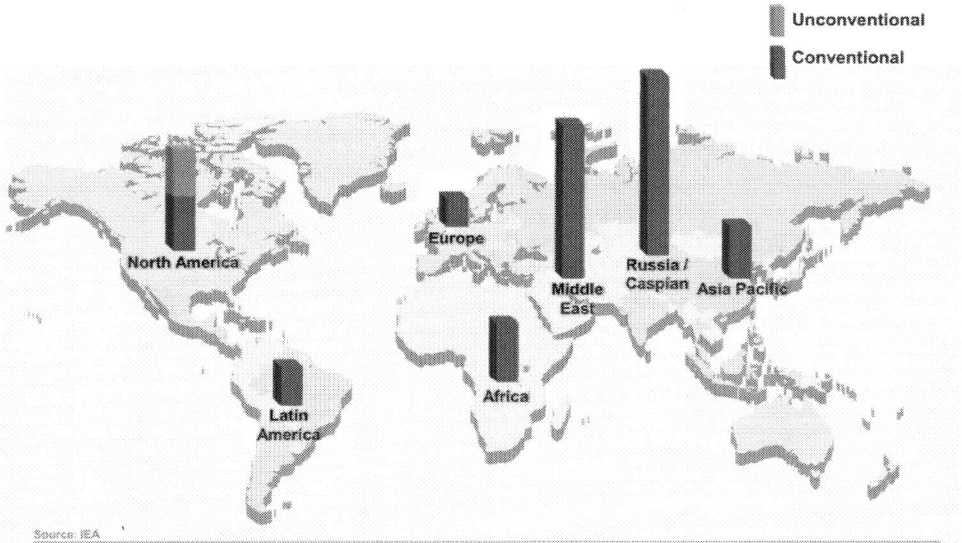

Source: IEA

ExxonMobil 2013 Outlook for Energy

9

ExxonMobil
Taking on the world's toughest energy challenges.

Natural Gas Abundant, Widespread

Thousand trillion cubic feet

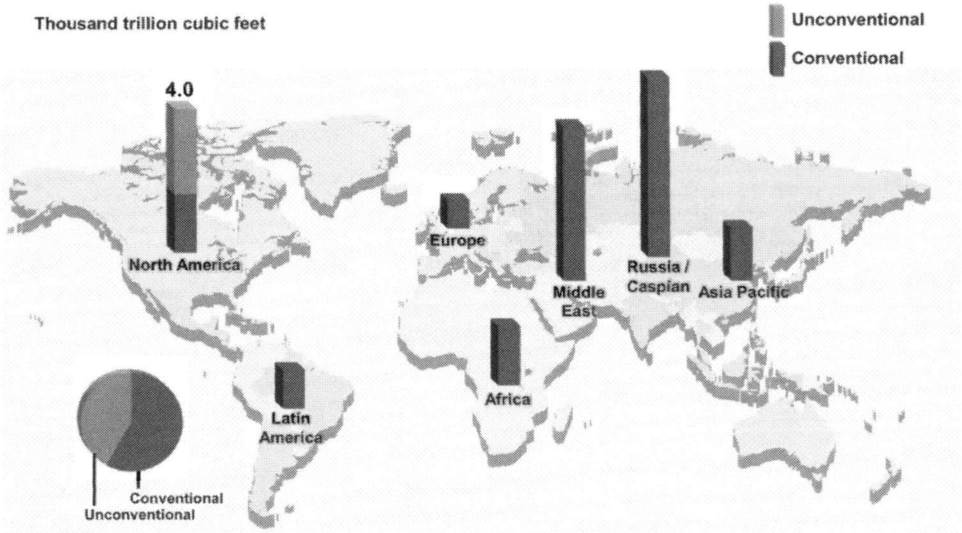

Source: IEA

ExxonMobil 2013 Outlook for Energy

10

ExxonMobil
Taking on the world's toughest energy challenges.

Natural Gas Abundant, Widespread

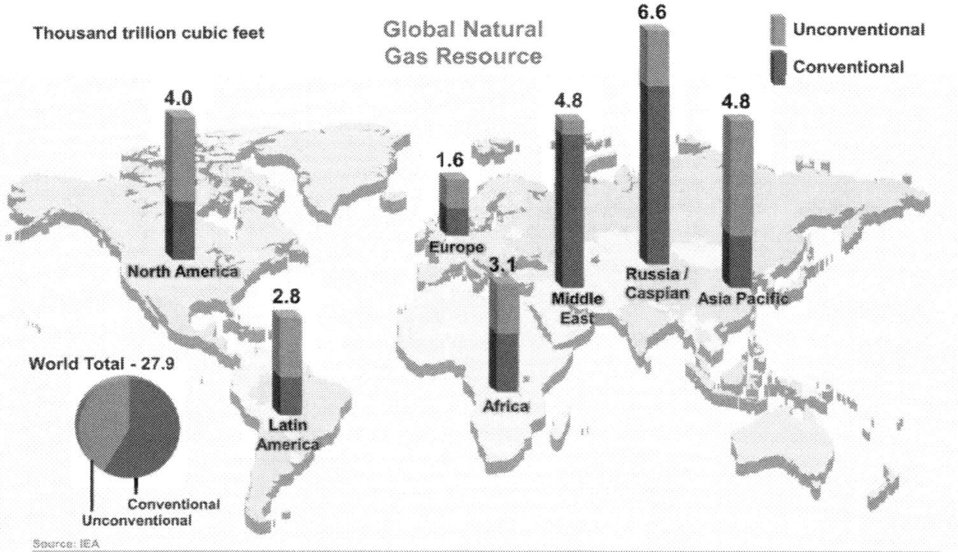

Technology Enables Energy Supplies

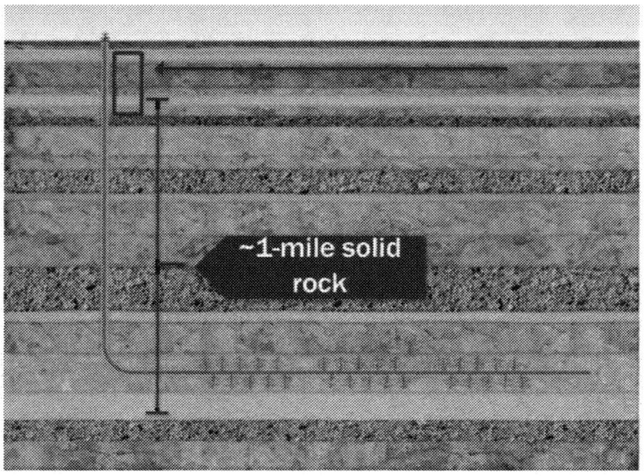

- A combination of horizontal drilling and hydraulic fracturing offers economic access
- The industry has used hydraulic fracturing on more than 1 million wells over 60 years
- Every step of this process can be done safely and responsibly.

Technology Enables Energy Supplies

13

ExxonMobil
Taking on the world's toughest energy challenges.

Technology Enables Energy Supplies

14

ExxonMobil
Taking on the world's toughest energy challenges.

Technology Enables Energy Supplies

15

ExxonMobil
Taking on the world's toughest energy challenges.™

Technology Enables Energy Supplies

16

ExxonMobil
Taking on the world's toughest energy challenges.™

Technology Enables Energy Supplies

17

ExxonMobil
Taking on the world's toughest energy challenges.

Technology Enables Energy Supplies

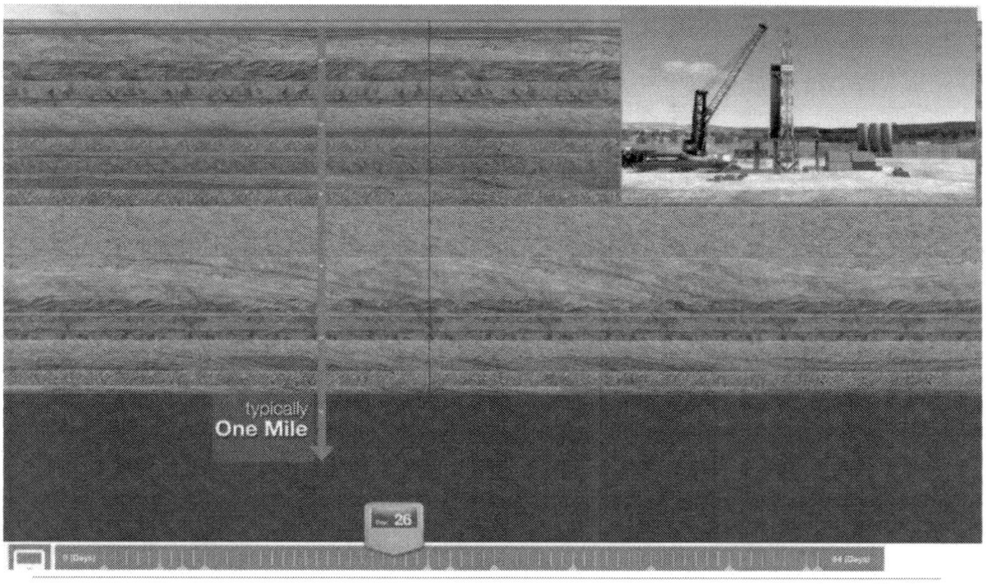

18

ExxonMobil
Taking on the world's toughest energy challenges.

Technology Enables Energy Supplies

19

ExxonMobil
Taking on the world's toughest energy challenges.

Technology Enables Energy Supplies

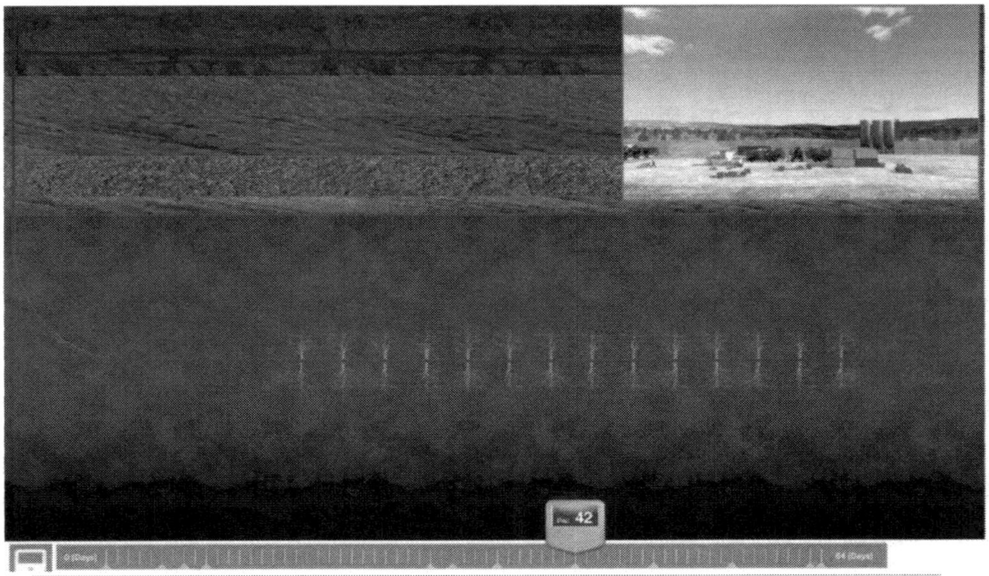

20

ExxonMobil
Taking on the world's toughest energy challenges.

Technology Enables Energy Supplies

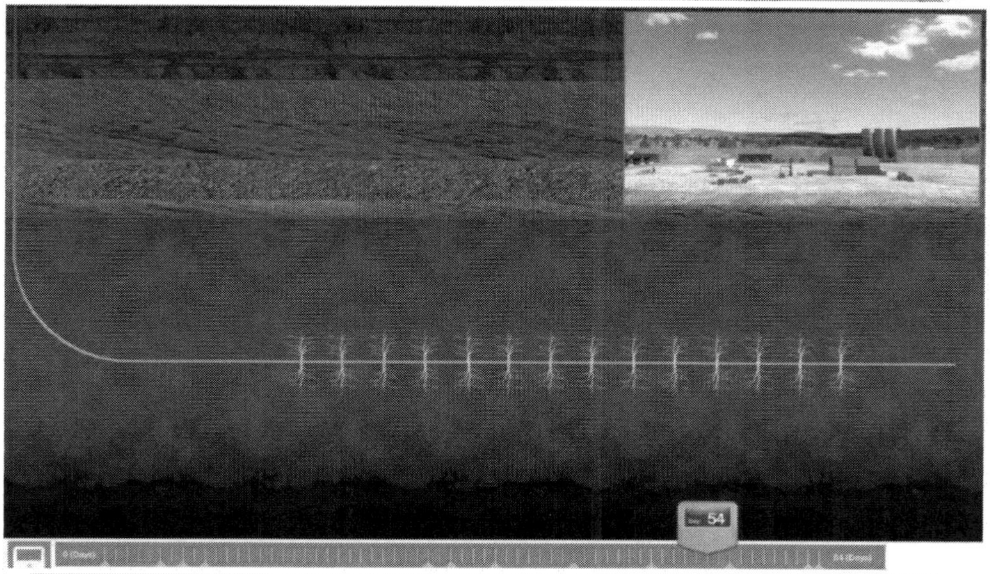

ExxonMobil
Taking on the world's toughest energy challenges.™

Technology Enables Energy Supplies

ExxonMobil
Taking on the world's toughest energy challenges.™

Technology Enables Energy Supplies

ExxonMobil
Taking on the world's toughest energy challenges.

Technology Enables Energy Supplies

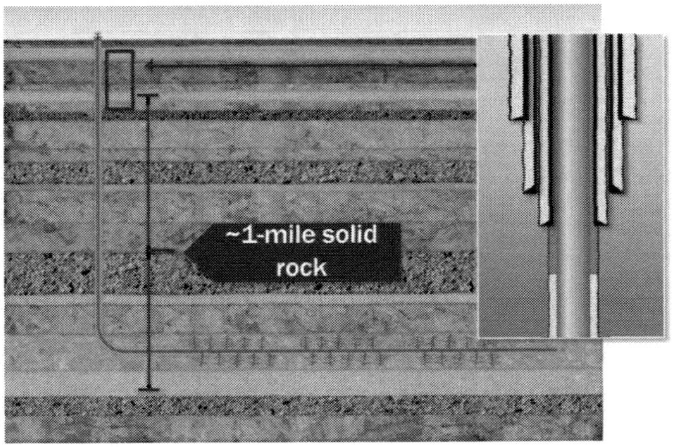

- A combination of horizontal drilling and hydraulic fracturing offers economic access
- The industry has used hydraulic fracturing on more than 1 million wells over 60 years
- Every step of this process can be done safely and responsibly.

ExxonMobil
Taking on the world's toughest energy challenges.

Major Change in U.S. Feedstocks

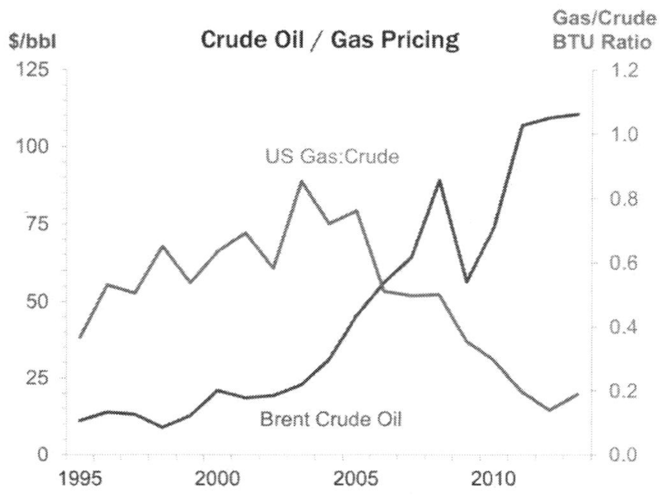

- **Major shift in energy market in mid-2000s**
- **Crude driven by emerging market growth / global events**
- **Shale gas supply drives the shift in relative economics**

Increased gas production provides advantaged ethylene feedstocks

Source: ExxonMobil Estimates

NA Polyethylene Industry Advantage

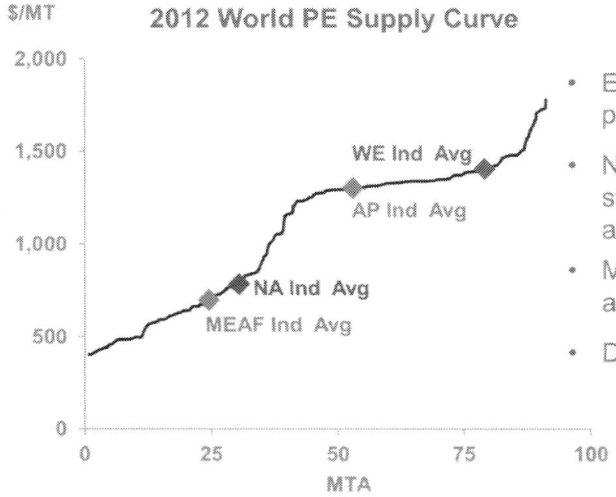

- Ethane and naphtha-based production historically in parity
- North America advantage from strong crude and ethane availability
- Multiple NA steam cracker announcements
- Derivatives for export markets

Shale gas has transformed competitiveness of NA polyethylene industry

Source: ExxonMobil Estimates

Breadth of Industry

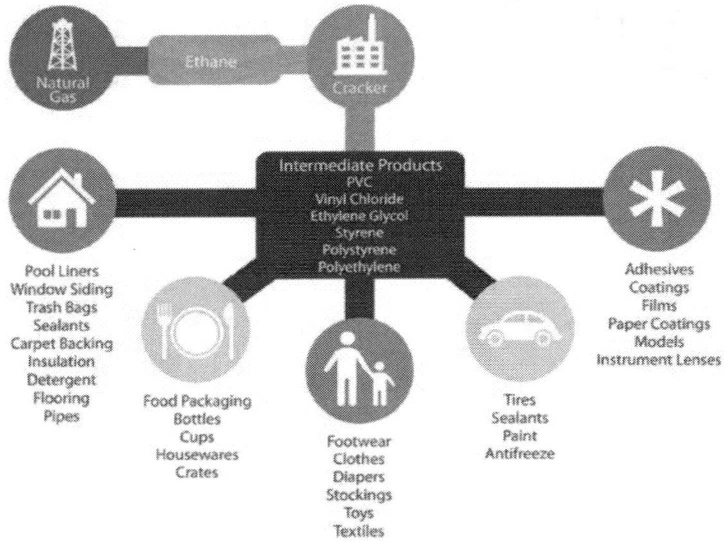

27

ExxonMobil
Taking on the world's toughest energy challenges.

Exxon Mobil Corporation

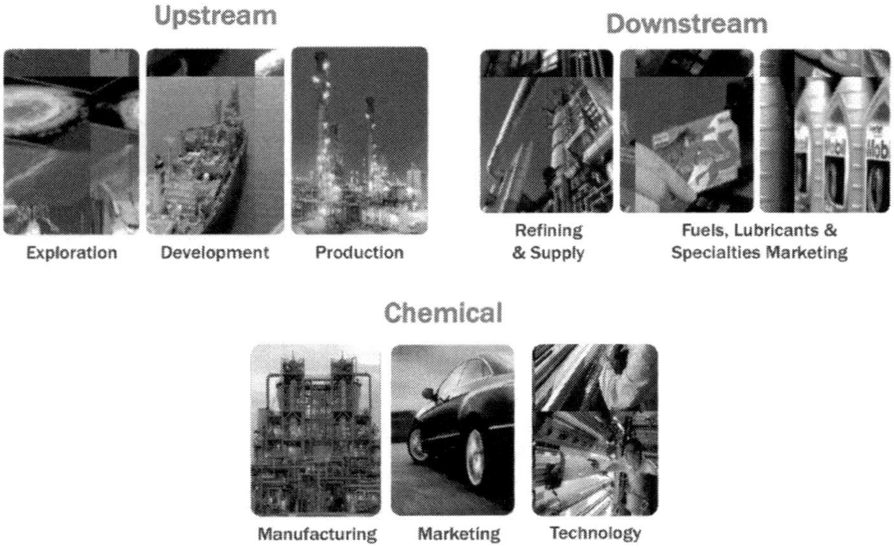

28

ExxonMobil
Taking on the world's toughest energy challenges.

Shale Gas Fuels Manufacturing Expansion

Economic, Jobs Growth from Shale

ExxonMobil TX Project

- 10,000 construction jobs
- 3,800 other jobs
- $870 million annually to regional economy
- $90 million annual added tax revenues

[illegible] U.S. Industry Projects

- $84 billion investment
- 1.2 million construction jobs
- [illegible]46,000 other jobs
- $14 billion added annual tax revenue

Texas Project Estimates: ExxonMobil
Industry Project Estimates: American Chemistry Council

The Talent Opportunity

- 1.2 million more STEM jobs by 2020
- Potential gap of 400,000 STEM graduates
- Need to build the skills that lead to job readiness
  - Student training
  - Mentoring

31

ExxonMobil
Taking on the world's toughest energy challenges.

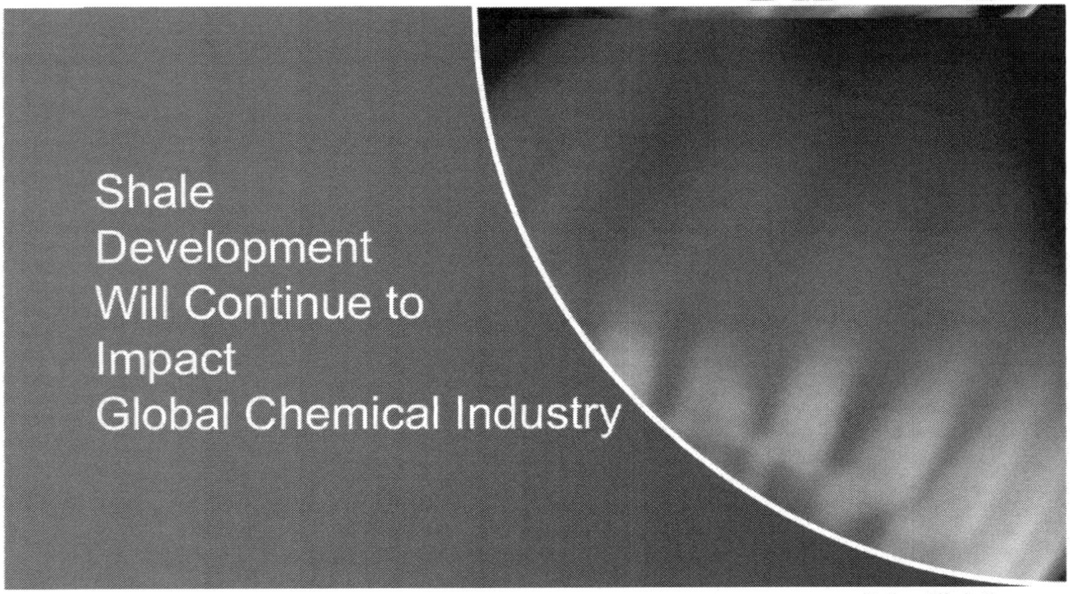

32

ExxonMobil
Taking on the world's toughest energy challenges.

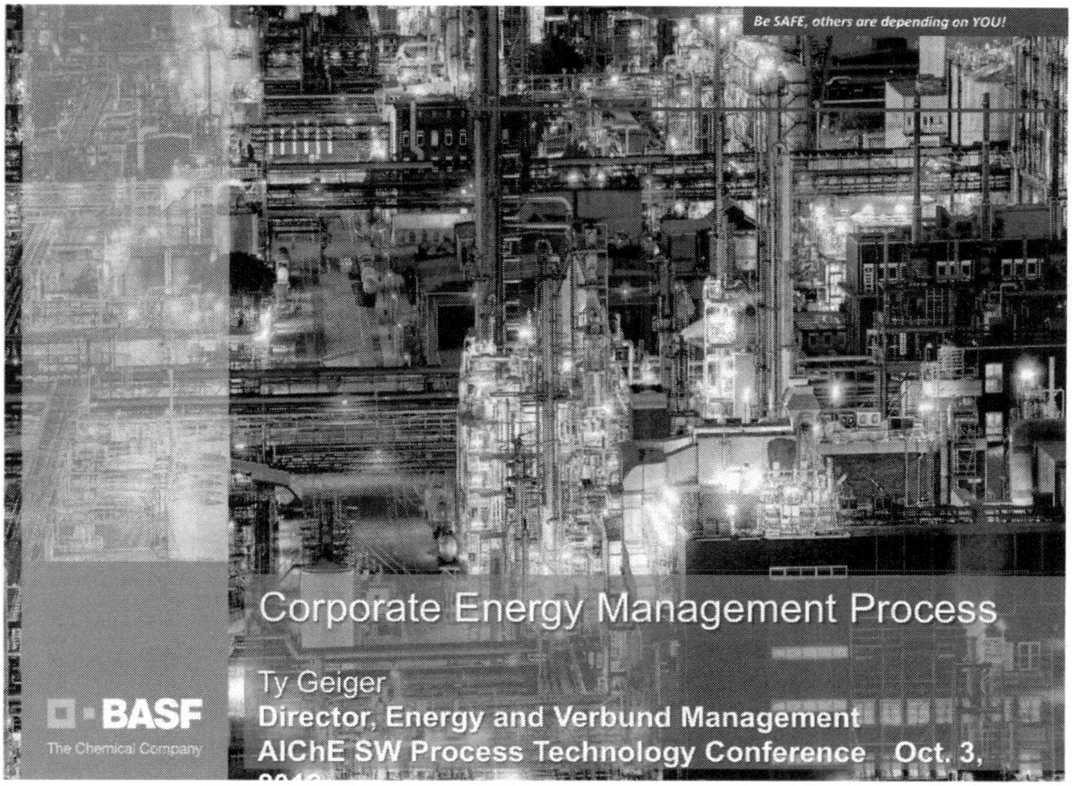

Be SAFE, others are depending on YOU!
Corporate Energy Management Process
Ty Geiger
Director, Energy and Verbund Management
AIChE SW Process Technology Conference Oct. 3,
BASF
The Chemical Company

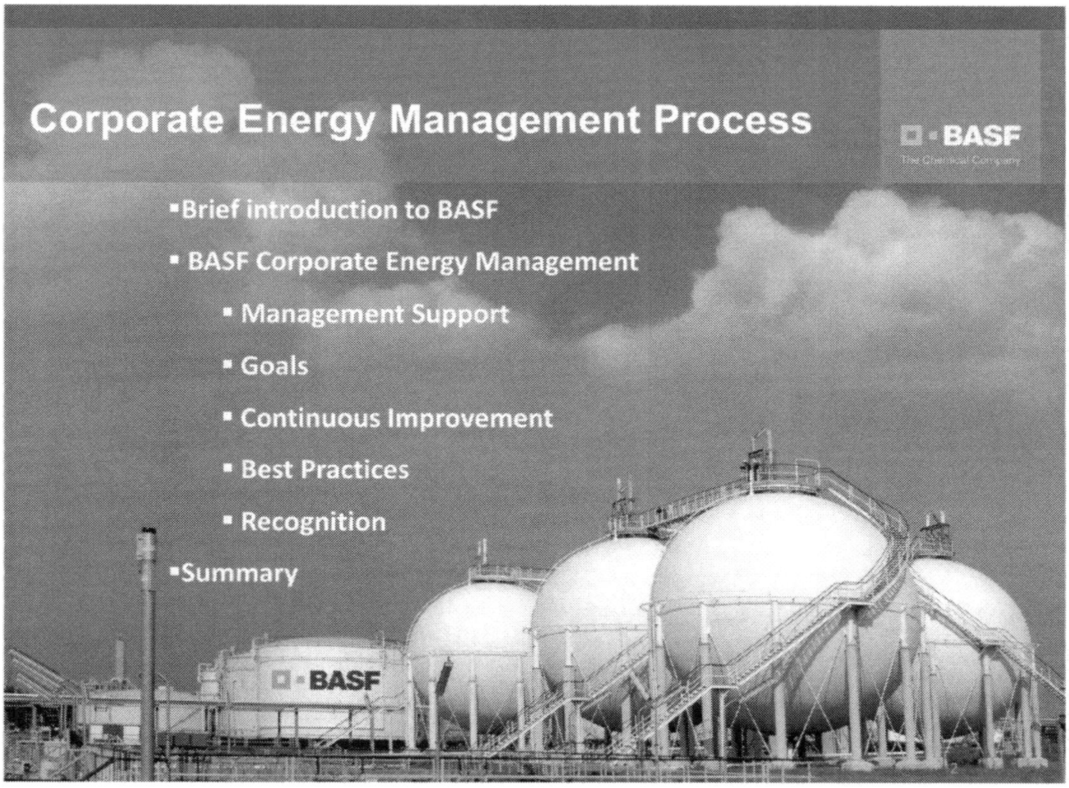

Corporate Energy Management Process
BASF
The Chemical Company
▪Brief introduction to BASF
▪ BASF Corporate Energy Management
▪ Management Support
▪ Goals
▪ Continuous Improvement
▪ Best Practices
▪ Recognition
▪Summary
BASF

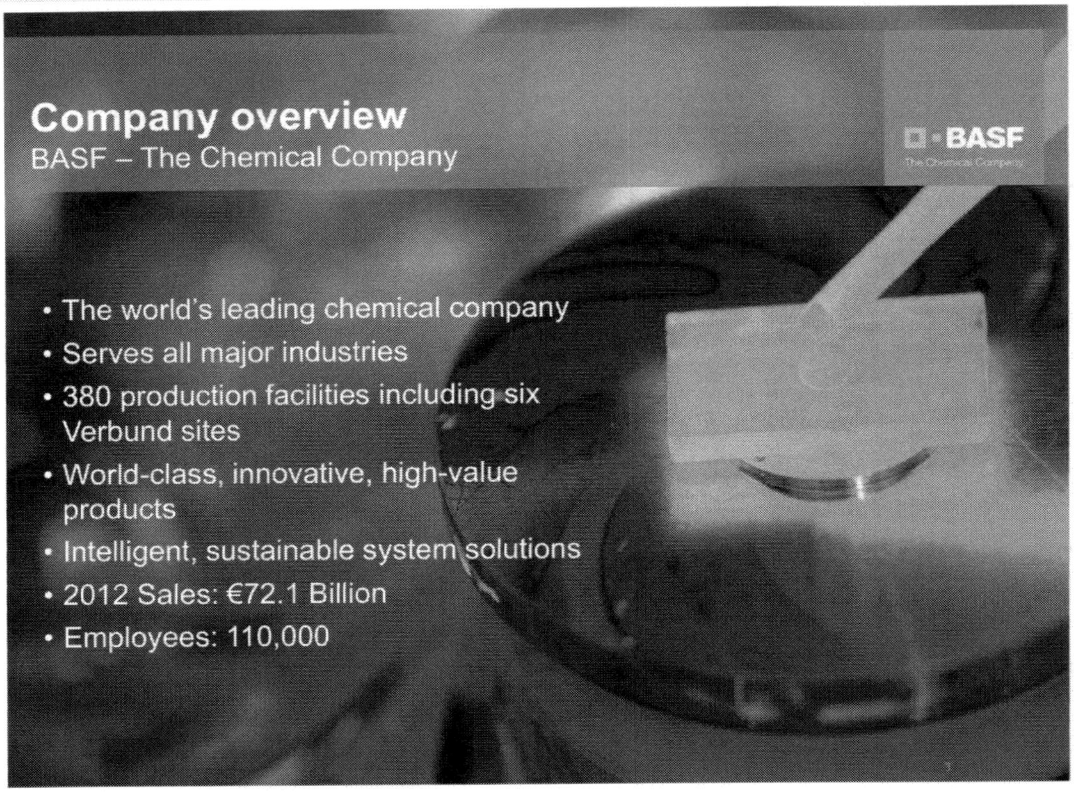
Company overview
BASF – The Chemical Company
BASF
The Chemical Company
• The world's leading chemical company
• Serves all major industries
• 380 production facilities including six Verbund sites
• World-class, innovative, high-value products
• Intelligent, sustainable system solutions
• 2012 Sales: €72.1 Billion
• Employees: 110,000

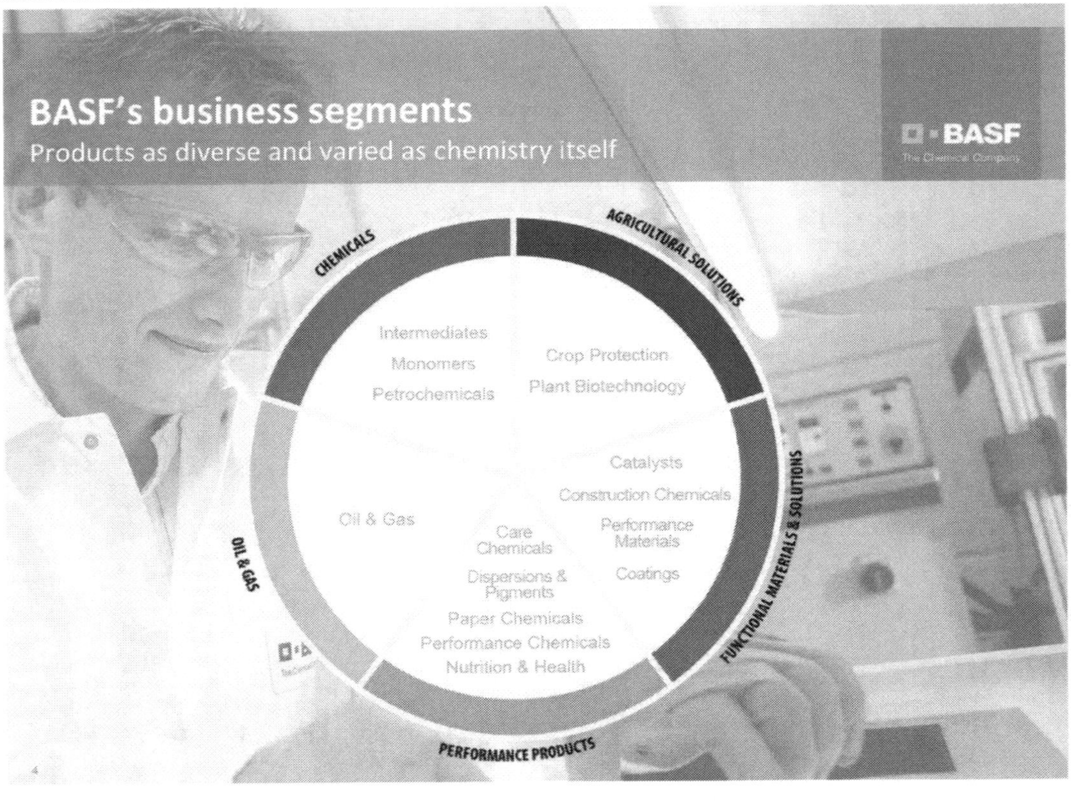
BASF's business segments
Products as diverse and varied as chemistry itself
BASF
CHEMICALS
AGRICULTURAL SOLUTIONS
FUNCTIONAL MATERIALS & SOLUTIONS
PERFORMANCE PRODUCTS
OIL & GAS
Intermediates
Monomers
Petrochemicals
Crop Protection
Plant Biotechnology
Catalysts
Construction Chemicals
Performance Materials
Coatings
Oil & Gas
Care Chemicals
Dispersions & Pigments
Paper Chemicals
Performance Chemicals
Nutrition & Health

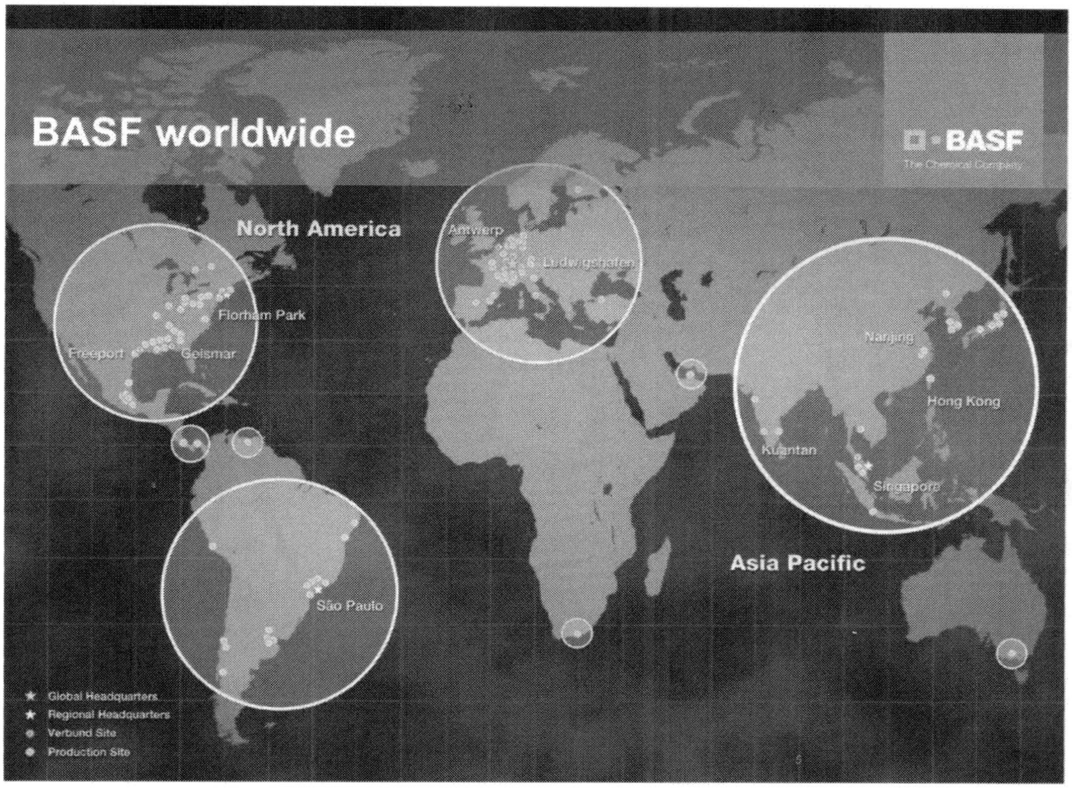
BASF worldwide
BASF
The Chemical Company
North America
Florham Park
Freeport
Geismar
Antwerp
Ludwigshafen
Nanjing
Hong Kong
Kuantan
Singapore
Asia Pacific
São Paulo
Global Headquarters
Regional Headquarters
Verbund Site
Production Site

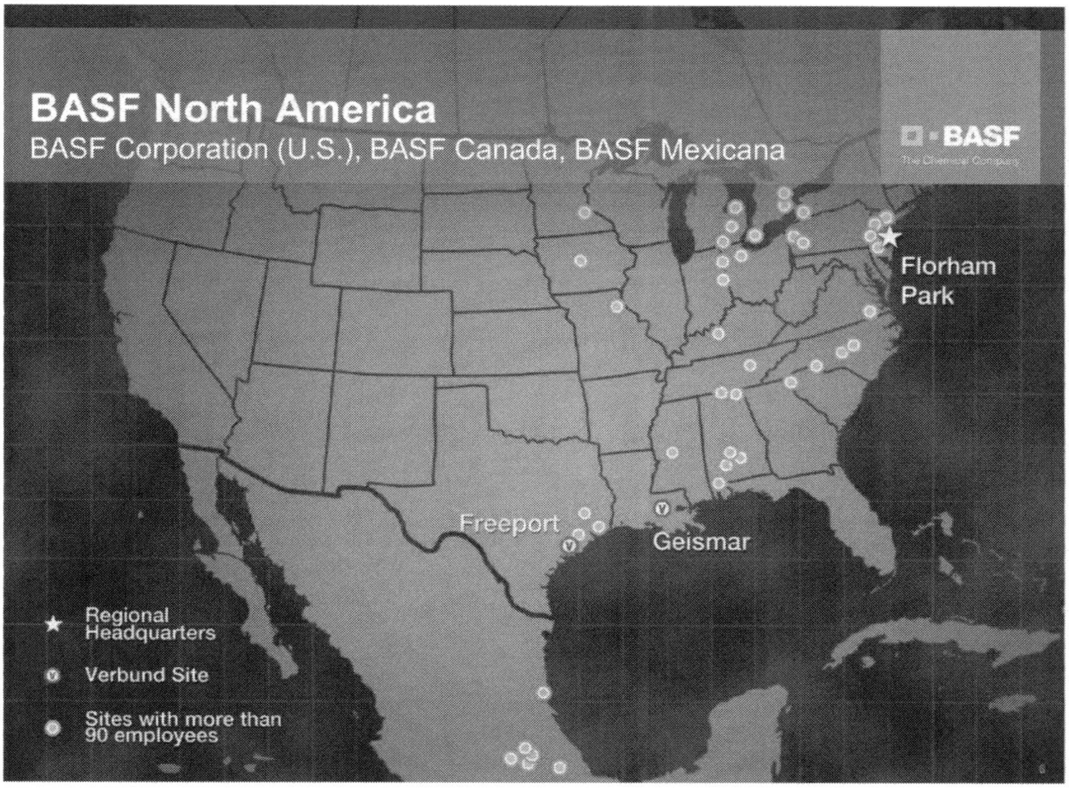
BASF North America
BASF Corporation (U.S.), BASF Canada, BASF Mexicana
BASF
The Chemical Company
Florham Park
Freeport
Geismar
Regional Headquarters
Verbund Site
Sites with more than 90 employees

BASF North America
BASF
The Chemical Company
2012 sales $18.5 billion
Employees 16,665
Production sites 99
R&D sites 27
Key customer industries
• Agriculture
• Automotive
• Chemicals
• Coatings
• Construction
• Health & Beauty
• Packaging

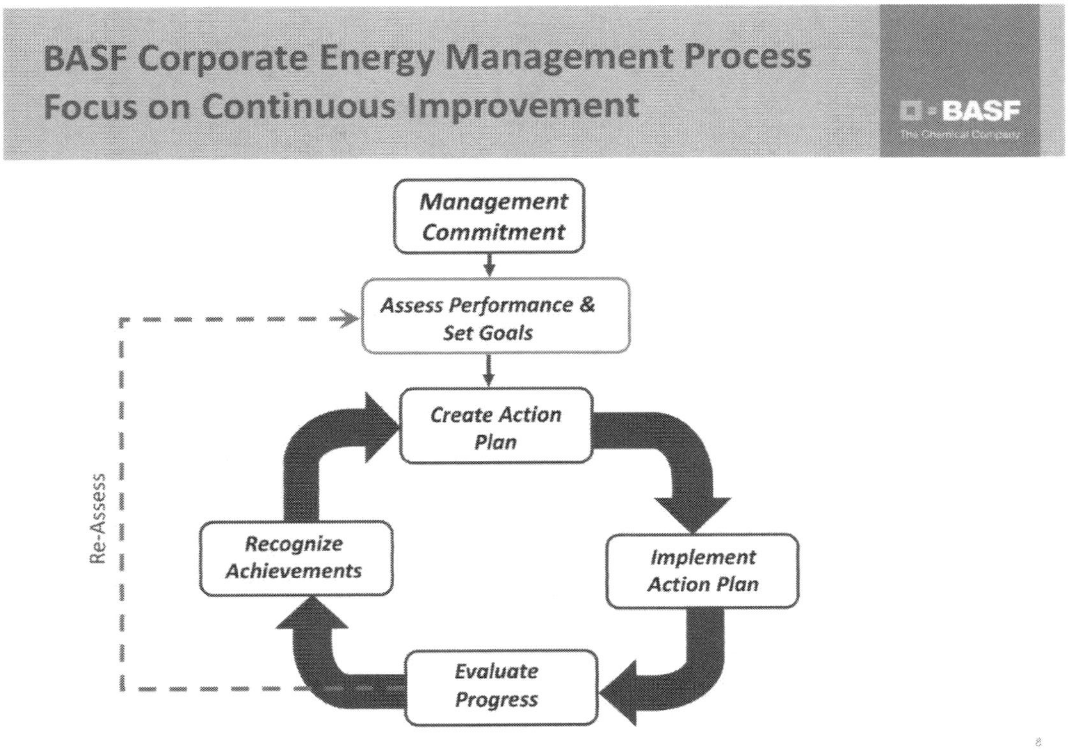
BASF Corporate Energy Management Process
Focus on Continuous Improvement
BASF
The Chemical Company
Management Commitment
Assess Performance & Set Goals
Create Action Plan
Implement Action Plan
Evaluate Progress
Recognize Achievements
Re-Assess
8

Management Commitment
Energy and climate protection goals

BASF The Chemical Company

Greenhouse gas emissions
per metric ton of sales product
by 2020 (baseline 2002)

-40% Status 2012: -31.7 %

Energy efficiency
in production processes
by 2020 (baseline 2002)

+35% Status 2012: +19.3 %

Near Term Performance

- Annual targets set for NA and by individual chemical production plants and sites
- Performance tracking and verification
- Management focus

9

Management Commitment
BASF Corporate Energy Management Department

BASF The Chemical Company

CHARTER

To assure reliable and economically competitive energy supplies to satisfy BASF Corporation's present and future needs and to promote the most efficient and economical use of energy throughout North America.

- **Lead BASF's corporate energy management process**
- **Promote/Facilitate development of energy optimization projects**
- **Stakeholder/Owner of energy and utility capital projects**
- **Energy technology consulting to manufacturing**
  e.g. Boilers, turbines, waste combustion, compressed air, wastewater treatment, demineralized water
- **Review BASF Corporation's capital and expense projects which consume or generate energy**
- **Regional platform for global energy expert group**

10

Effective Energy Management

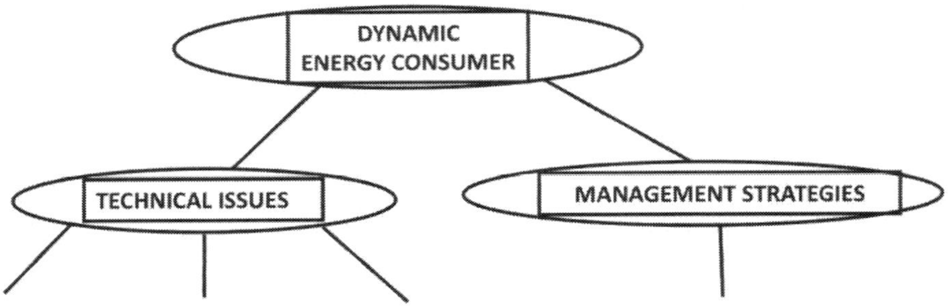

THERMAL
Boilers
Steam Distribution
  Leaks, Insulation, Traps
Heaters
Dryers
Ovens/Kilns
Waste Heat Recov.
Pinch Technology

ELECTRICAL
Motors
Air Compressors
Refrigeration/HVAC
Process Equipment
Fans/Blowers
Lighting
Load Management
Generation/Supply
VFD's

OTHER
Nitrogen
Fuels
  Natural Gas
  Fuel Oil, Wastes
WWTP
Water
  Potable
  Demineralized
Process Energy

STRUCTURE
Commitment to continual improvement
Conduct regular management reviews
Identify conversion cost factors
Defined requirements/Facility goals
Documentation
Project planning
Monitoring/measuring
Training
Stable maintenance/Proper purchasing

11

Team Work leads to Success

Energy and Process Optimization Workshops

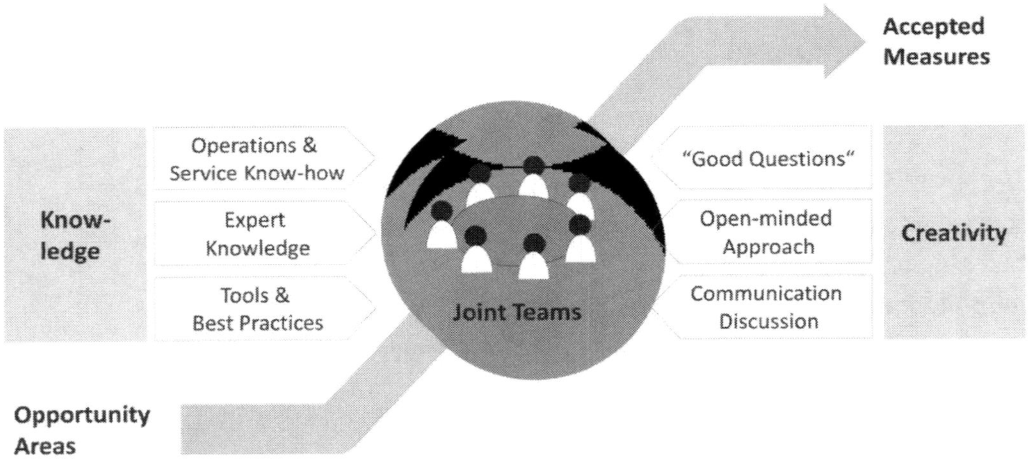

12

Best Practice and Technology sharing

BASF The Chemical Company

- Site Energy Teams for development and implementation
- Energy Workshops to identify new projects and review performance
- Implementation of Advanced Process Control technologies, KPI visualization and other automation technologies
- Frequent virtual meetings to share Best Practices and successes between locations
- Maintenance and Operations, e.g. steam systems, insulation, cooling towers, motor management, compressed air, lighting, etc.
- Waste heat recovery / heat integration
- Natural gas minimization for flares and incinerators
- Screen for new Combined Heat and Power Concepts
- Utility System reliability – internal and external

13

BASF Energy Projects
Combined Heat and Power Concept

BASF The Chemical Company

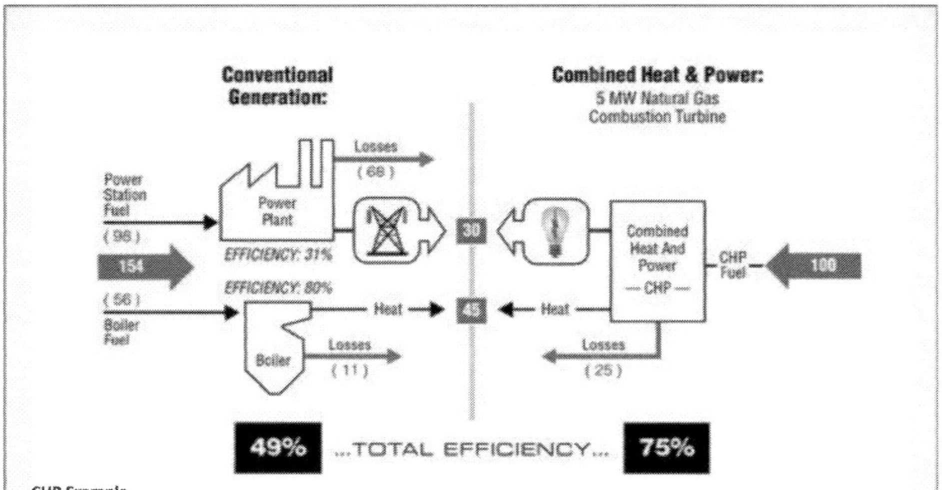

CHP Example

14

Recognition: American Chemistry Council Responsible Care® Energy Efficiency Awards

BASF The Chemical Company

1996	Energy Survey Program - Division and Site
1997	Energy Cost Reduction Program - HQ/Research
1998	Utilities Expansion Project - Cogen, Waste Fuels
1999	Corporate Energy Management Program
1999	Corporate Motor Management Guideline
2000	Geismar Cogeneration – AOG Fuel Utilization
2001	Freeport Cogeneration Project
2002	Portsmouth Site Energy Team
2003	Freeport ISBL Energy Audits
2004	Energy Management Group (EM)
2004	Geismar Flare Best Practice
2005	Site Energy Audits
2006	Multi-Site Energy Reduction Through Teamwork
2007	Energy Efficiency Improvements for Acquisitions
2008	Energy Efficiency Improvement Goal (EM)
2008	BFLP: Improved Hydrogen Utilization
2009	Freeport Steam Optimization (EM)
2009	Compressed Air Optimization (EM)
2009	Energy Efficiency Program (EM)
2010	Corporate Energy Efficiency Improvement Goal
2010	Geismar Ethylene Oxide Plant Multivariable Predictive Control Improvements
2010	Geismar Polyol Plant Steam Reduction
2010	Greenville Energy Optimization Initiative
2011	Corporate Energy Mgmt. Process (EM)
2011	Geismar EO Vent Gas Scrubber
2011	Geismar MDI Steam Pressure Optimization
2011	Jackson Flash Dryer
2011	McIntosh Production Schedule Optimization
2011	BFLP: LP Fuel System Optimization
2012	Corporate Energy Mgmt. Process (EM)
2012	Kankakee Site Wide Energy Conservation
2012	Geismar TDI Plant Steam Optimization
2012	Peekskill Instrument and Control Upgrade

EM: Exceptional Merit

15

Summary

BASF The Chemical Company

- **Strategy for success:**
  - **Management support and commitment**
  - **Corporate and global organization**
  - **Clear goals and objectives**
  - **Focus on Continuous Improvement**
  - **Communication of best practices and technologies**
  - **Recognition**

16

BASF
The Chemical Company

QUESTIONS?

17

BASF
The Chemical Company

18

Texas Petrochemicals

Using Statistical Analysis to Optimize Production within Utility Constraints

Oct. 3, 2013

Quality Products. Dependable Services.

Texas Petrochemicals

Forward-Looking Statements & Non-GAAP Financial Measures

This presentation may contain "forward-looking statements" within the meaning of the securities laws. These statements include assumptions, expectations, predictions, intentions or beliefs about future events, particularly statements that may relate to future operating results, existing and expected competition, market factors that may impact financial performance, and plans related to strategic alternatives or future expansion activities and capital expenditures. Although TPC Group believes that such statements are based on reasonable assumptions, no assurance can be given that such statements will prove to have been correct. A number of factors could cause actual results to vary materially from those expressed or implied in any forward-looking statements, including risks and uncertainties such as volatility in the petrochemicals industry, limitations on the Company's access to capital, the effects of competition, leverage and debt service, general economic conditions, litigation and governmental investigations, and extensive environmental, health and safety laws and regulations. More information about the risks and uncertainties relating to TPC Group and the forward-looking statements are found in the Company's SEC filings, including the Transition Report on Form 10-K, which are available free of charge on the SEC's website at http://www.sec.gov . TPC Group expressly disclaims any obligation to update any forward-looking statements contained herein to reflect events or circumstances that may arise after the date of this presentation.

This presentation may also include non-GAAP financial information. A reconciliation of non-GAAP financial measures to the most directly comparable GAAP financial measures, as well as additional detail regarding the utility of such non-GAAP financial information, is included in the Appendix.

Texas Petrochemicals

INTRODUCTION

- 257 acres along the Houston Ship Channel built in mid-40s
- Market leader in C3 and C4 Hydrocarbons ; largest producer of finished Butadiene, Butene -1 and Polyisobutylene in North America.
- Steam generation capacity of 1.4 MMPPH
- Fuel Gas requirements of about 40,000 MMBTU/Day
- Power production capacity of 33 MW

Butadiene

Butene-1

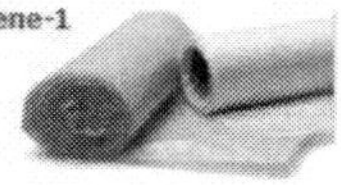

Fuel Products

Isobutylene Derivatives

Polyisobutylenes

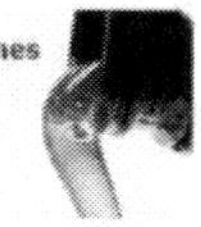

Nonene, Propylene Tetramer & Pentamer

Texas Petrochemicals

Opportunity Assessment

TPC Group Houston Operations Near Future	• Potential expansion of several existing production units as well as building new production units • Foresee major investments in infrastructure especially utilities • Plant was originally built in 1943
Objective	• Maintain supply for existing steam demand onsite • Plan for increase in demand • Maintain Reliability • Minimize Environmental impact
Steam Scenario	• AS IS : Existing steam generation is adequate to meet demand • TO BE : Assess alternatives to meet increase in steam demand
Opportunity	• Early in 2013 when steam production was lower than normal, there was an opportunity to analyze onsite steam demand under atypical circumstances

Using Statistical Analysis to Optimize Production within Energy Constraints 4

Texas Petrochemicals

Scenario Details

Lower than normal steam production onsite

- Oldest of three boilers down for maintenance
- 30% less steam produced than normal
- Leverage this opportunity to understand various "handles" on steam

Site Steam Demand

- 70 – 80 % of total steam demand set by two product lines
- Maintain steam export to customer
- Remaining steam demand fairly constant

Evaluate Impact

- Production
- Reliability
- Environmental
- Capital Investment for future

Texas Petrochemicals

Overall Impact

RELIABILITY

- Continue reliable operations with reduced steam capacity
- Insight into how near term strategic project will impact site steam demand
  - With all Boilers operating
  - With the Old Boiler down

ENVIRONMENTAL

- Impact on VOC Emissions/Credits
- Off gas

CAPITAL

- Opportunities for cost avoidance in utilities infrastructure

Texas Petrochemicals

Statistical Analysis:

Key Objectives:	Develop analytical relationship for production vs. steam consumption for two principal product lines	Maximize production while minimizing utility consumption	Do not violate internal or external constraints
Steps in Statistical Analysis:	Plot Data	Fit two(2) way ANOVA w / Interactions	Compare response groups
Advantages	Do not need to specify process details	Can handle interaction between independent variables	Easy to update and/or modify using software
Disadvantages	Can't extrapolate beyond range of data	Can't explain process details (i.e. "Black Box")	Can sometimes be hard to get a representative sample

Texas Petrochemicals

ANOVA : ANALYSIS OF VARIANCE

What is ANOVA?

- Two-way ANOVA – two independent variable.
- Interaction – synergistic effect between independent variables

General Form of Equation:

- **Y = C1*X1 + C2*X2 + C3*X1*X2**
- Y = Steam Cons. (MPPH)
- X1 = Product 1 Production (MMPPD / MMPPH)
- X2 = Product 1 Production (MMPPD / MMPPH)
- C1, C2, & C3 constants

Texas Petrochemicals

RESULTS

Actual by Predicted Plot

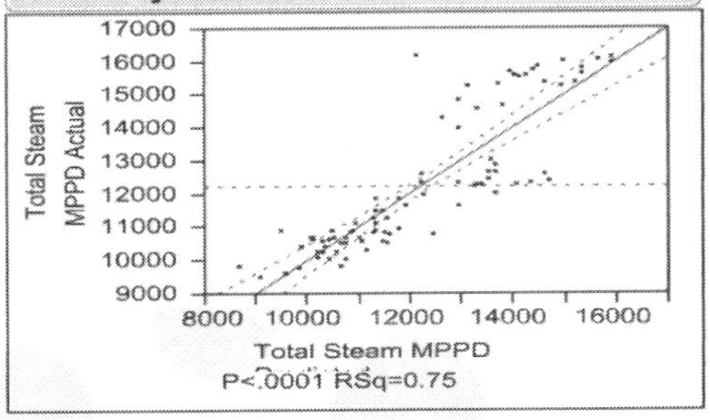

Summary of Fit

RSquare	0.749663
RSquare Adj	0.740828
Root Mean Square Error	1049.309
Mean of Response	12222.45
Observations (or Sum Wgts)	89

Fit of model is within normal range for a model based on historical data

75 % of the variability in total steam demand is explained by the model.

The standard deviation of the unexplained variability is the Root Mean Square Error (RMSE).

Texas Petrochemicals

RESULTS:

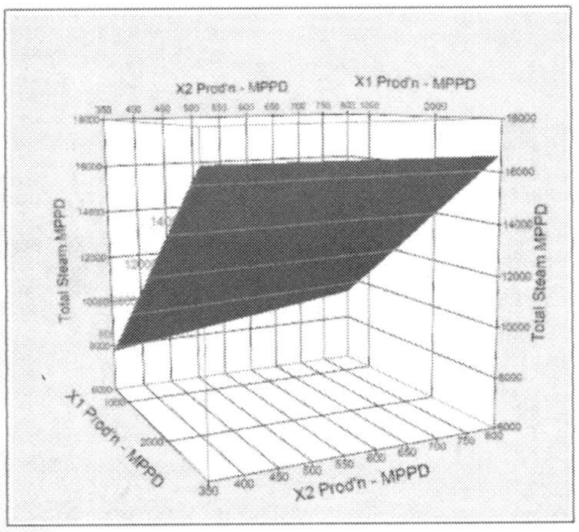

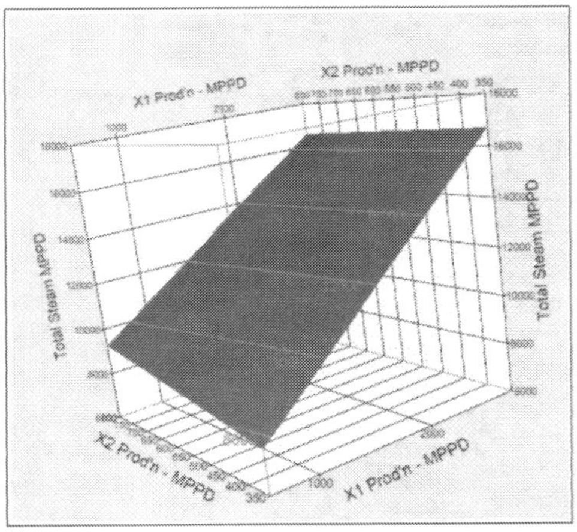

Picture of predicted response surface.

Twisted plane represents interaction.

Texas Petrochemicals

RESULTS:

Texas Petrochemicals

RESULTS

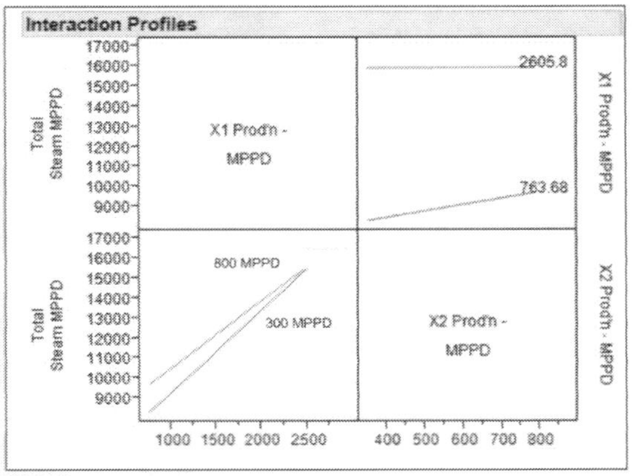

- Interaction between X1 and X2 is statistically significant although relatively minor.
- X2 impact on total steam demand greatest at low values of X1 production.
- X1 production has greatest impact on total steam production.
- Interaction profiles aid in optimizing production at given steam production to maximize profit

Texas Petrochemicals

CONCLUSIONS & RECOMMENDATIONS:

- ANOVA can be a powerful tool in modeling inputs & outputs in complex processes and energy forecasting.
- ANOVA is powerful because the details of the process (i.e. heat transfer coefficients, kinetics etc.) do not need to be known.
- Good process knowledge/familiarity is required to determine if a sample is truly representative.
- Validate, validate, validate.
- Continue to refine existing ANOVA models and explore potential new applications.

Using Statistical Analysis to Optimize Production within Energy Constraints 13

SOTEICA VISUAL MESA

Driving Sustainable Savings in Utility Operations through a Site-wide Energy Model

Case Study: CEPSA San Roque, Spain

Tyler Reitmeier, Solution Development Lead
tyler.reitmeier@soteicavisualmesa.com

VISUALMESA™

Agenda

- Introduction
- CEPSA Case Details
- Conclusions
- Q&A

VISUALMESA™

Capturing Energy Savings Opportunities…

VISUALMESA™

Soteica Visual Mesa LLC

- **25+ year focus on Energy Management Solutions**
- **Customer-Focused** Decision Support Solutions
  - Product Development
  - Sustainability/Support
- Specific Focus on **Loss Reduction**/Prevention
- Global Major **Industrial** Customer Base
- 2012-3: **Yokogawa** Largest Shareholder

VISUALMESA™

2013: Partnership with Yokogawa

- **Yokogawa Electric Corporation** now largest shareholder in Soteica Visual Mesa LLC (44%)
  - $5B company
  - Industrial Automation, Test and Measurement, Related Services
  - **VigilantPlant™** – Operational Excellence in Manufacturing over Entire Lifecycle
- **Visual MESA Energy Management Solution** enhances VigilantPlant™
  - Combined opportunities for integrated solutions
  - Extending Soteica Visual Mesa's industry-leading solutions

VISUALMESA™

Decision Support Solutions

Validating data for accuracy of calculated indicators driving performance improvement

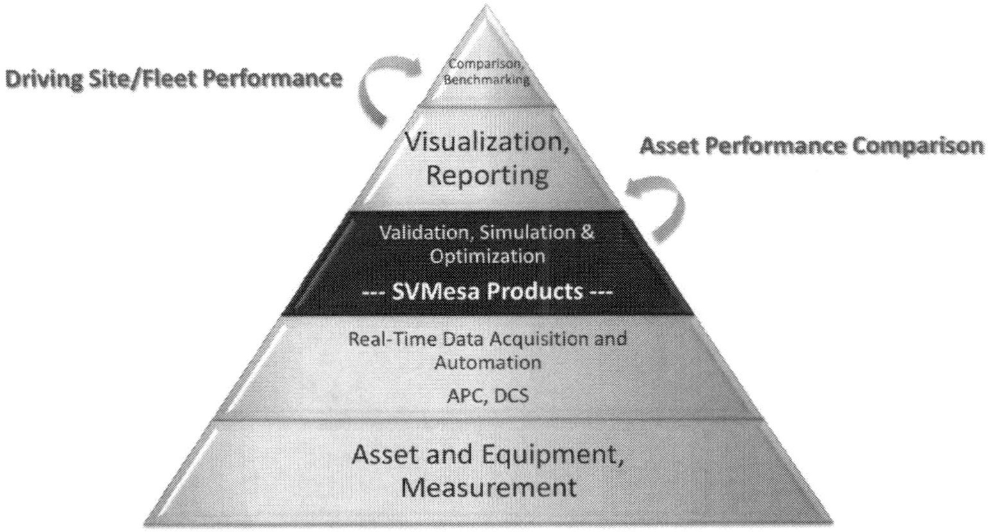

VISUALMESA™

Why is Energy Management needed?

Complexity of Utility Operations Decision Making

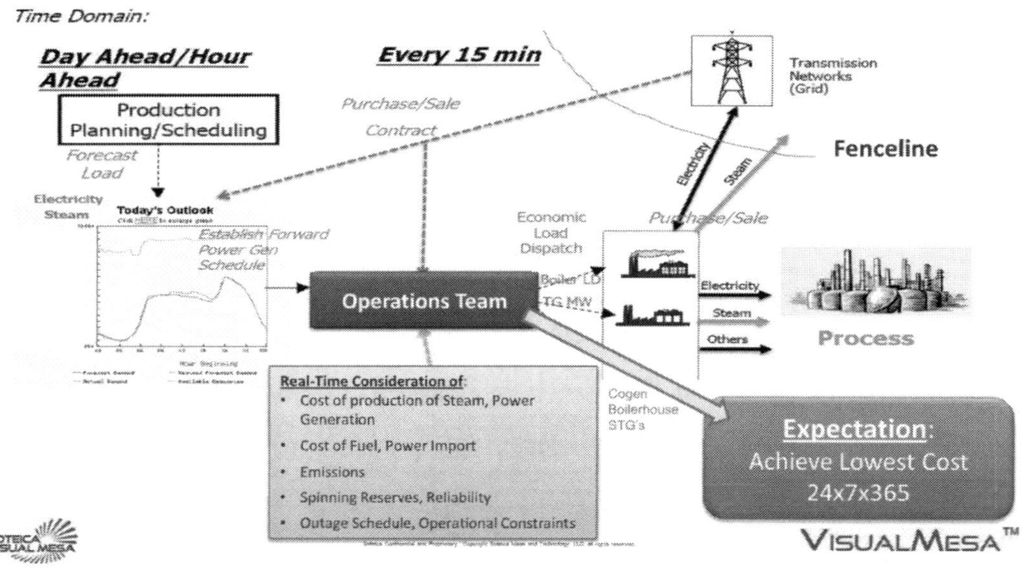

VISUALMESA™

Defining Energy Management

- ISO 50001 – International **Standard**
- Superior Energy Performance - **Certification** for compliance with 50001
- **Framework** taking Boardroom planning to Field implementation

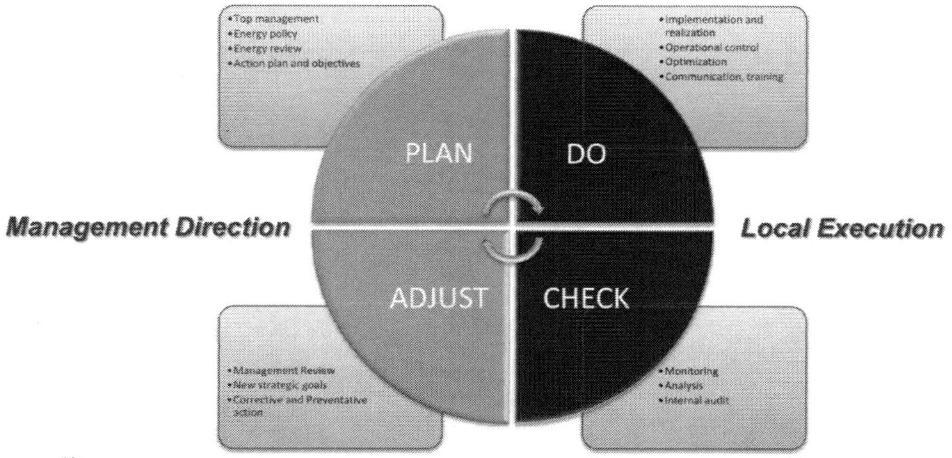

VISUALMESA™

Visual MESA-specific Basis

- "Process Gets What the Process Needs" - **Utility** Focus
- Objective: **Lowest Cost** operation within Constraints:
  - Safety, reliability
  - Contractual
  - Emissions
- Framework for **Real-Time** Performance Improvement
  - Direct, Actionable Operator Targets
    - Based on Current Operation
    - Automatically Updated
  - Monitoring and alerting through **Validated KPI's**
  - Driving **Best Practices**

VISUALMESA™

Enabling Energy Management

Visual MESA's Unique Value

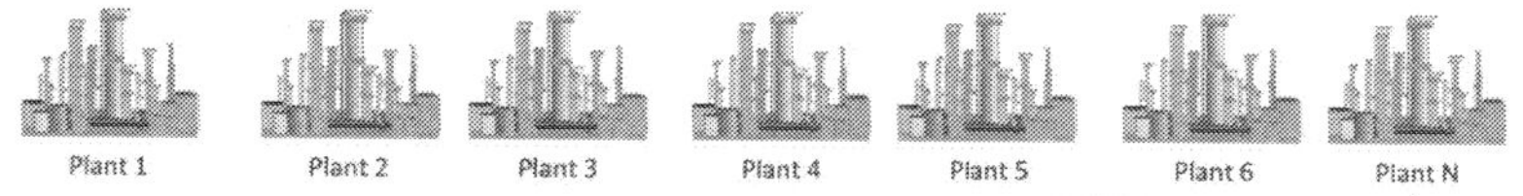

- Site-wide Energy Real-Time Optimization (**RTO**)
- **Validating** and using only Consistent Measurements
- **Accurate Calculation** of Energy Measures & KPIs
- **Driving Best Practice,** both site-wide and fleet-wide, through Visualization

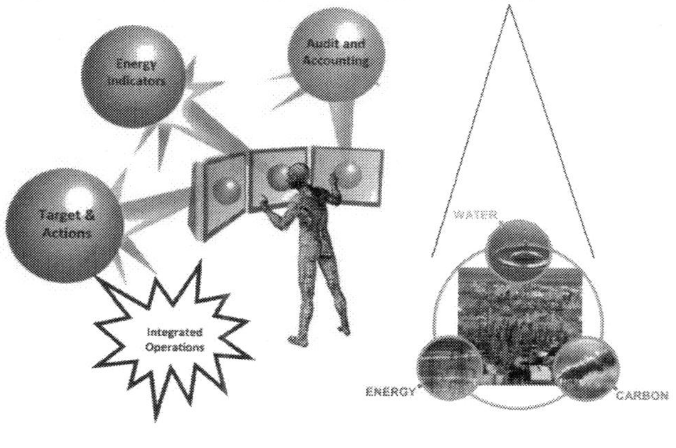

VISUALMESA™

Energy Real-Time Optimization (RTO)

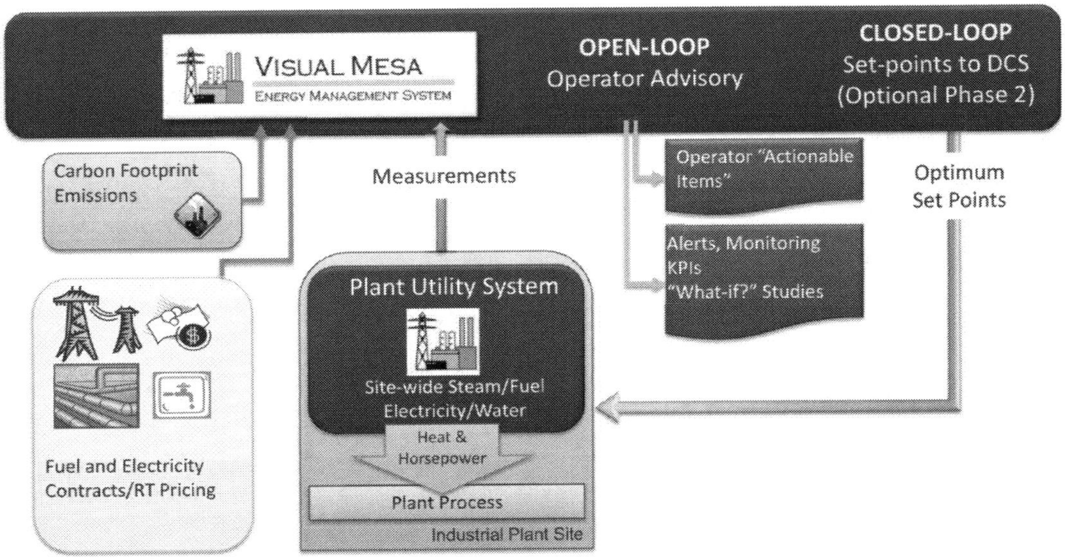

VISUALMESA™

Energy RTO Value

- Identifies and **Enables Capture of Savings Opportunities** in Current Operation
- **Energy Cost Gap:** Measure of Current Operation versus Best Achievable
- Typical Savings Realized: **2% to 5% of Total Energy Spend**

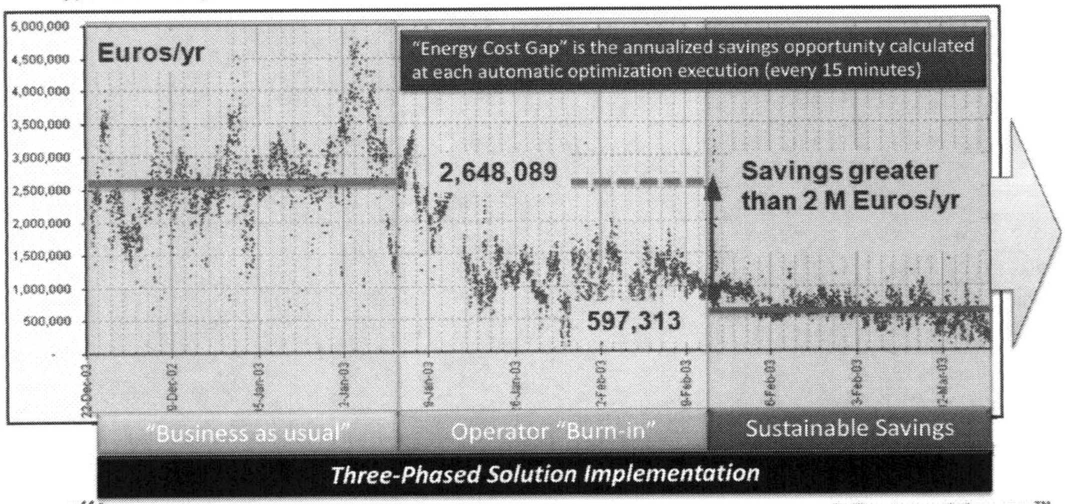

VISUALMESA™

CEPSA-Visual MESA Partnership

- Private Spanish Oil Company est. 1929
- 1987 - Purchased INTERQUISA
- 2008 - CEPSA Quimica formed, including INTERQUISA
- Only producer of PTA, PIPA (Polyester precursors) in Spain
- **2 Visual MESA implementations** at CEPSA in Spain in 2007 and 2011.
- **San Roque** – Chemical plant + integrated, joint venture Cogeneration Plant

SOTEICA VISUAL MESA

VISUALMESA™

San Roque Utilities

- **Steam system complexity**
  - 5 steam headers
  - 3 fired boilers
  - 2 hot oil furnaces
- Process heat exchangers: **hot oil and steam**
- **Fuel Gas** system complexity
  - Fuel gas purchased from 3rd party
  - Process off gas consumption
  - Natural gas make up
- **Cogen** (JV)

SOTEICA VISUAL MESA

VISUALMESA™

San Roque's Pain Points

- Optimized **heating of slurry**
- Boiler **load balancing**
- **3rd Party Purchase** versus production of Fuel Gas and Steam
- **Emissions** difficult to predict when considering changes to operations
- **Cogen** Power and Steam Generation
  - Joint Venture considerations
  - Turbo-generator operation

Unique Aspect for San Roque

Required Heat Supplied by Multiple Sources

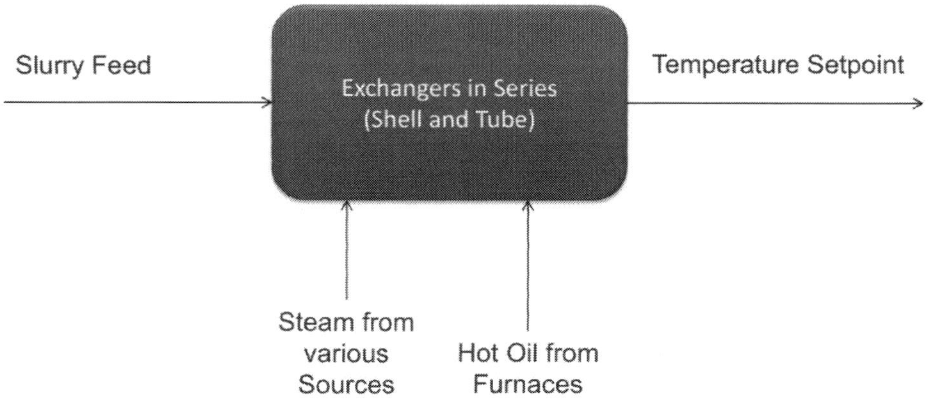

Site-wide Perspective Required

- **Interconnected Decisions** regarding Fuel, Electric and Steam
  - Fuel: Gas Turbine & Duct Firing Use; Boilers
  - Electric Power: Purchase from Grid, Cogen, Steam Turbine Generator
  - Steam: Waste heat from GT, Duct fired HRSG, Fired Boilers with Hot Oil tradeoff
- **Site-wide model** needed
  - To determine incremental cost of steam and other utilities
  - To drive effective optimization for lowest cost

Structured Model Implementation

1. Utility system Information Collection
2. Software Installation, Connection to Plant data
3. Utility Control System and Operation Review
4. Model Building: Steam, Fuel, Power, Water
   - Includes Optimization Definition and Configuration
   - Objective function: **Total Operating Cost** = Total Fuel Cost + Total Electric Cost + Σ Other Costs
5. Engineer Training and Model Review
6. Model "Burn-in" Testing and Acceptance
7. Operator Training
8. Commissioning – Hourly/Daily Use in Operations

Model Building Summary

- **Detailed** Site-wide model
- **Connected** to plant data from the start
- Built on Customer's system
  - Consistent **Collaboration** throughout building phase
- Optimization **Limits and Constraints** reflecting actual operation
- **Reports** to drive Solution Implementation for **Realized Savings** in Operations

SOTECA VISUAL MESA VISUALMESA™

CEPSA's Optimization Handles

- **Manipulated** Variables (Directly Controlled)
  - Boiler steam rate for boiler(s) on manual control
  - Start/Stop Steam Condensers (Fin Fan)
  - Steam and Fuel Gas Purchase from 3rd Parties
- **Controlled** Variables (Resulting from MV's)
  - Other boilers on pressure control
  - Hot Oil Use
  - Natural Gas import
  - Letdowns, vents

SOTECA VISUAL MESA VISUALMESA™

User Interface – Main Screen

- **Configuration directly from "template" of modeling blocks**
- **Customized to customer's needs**

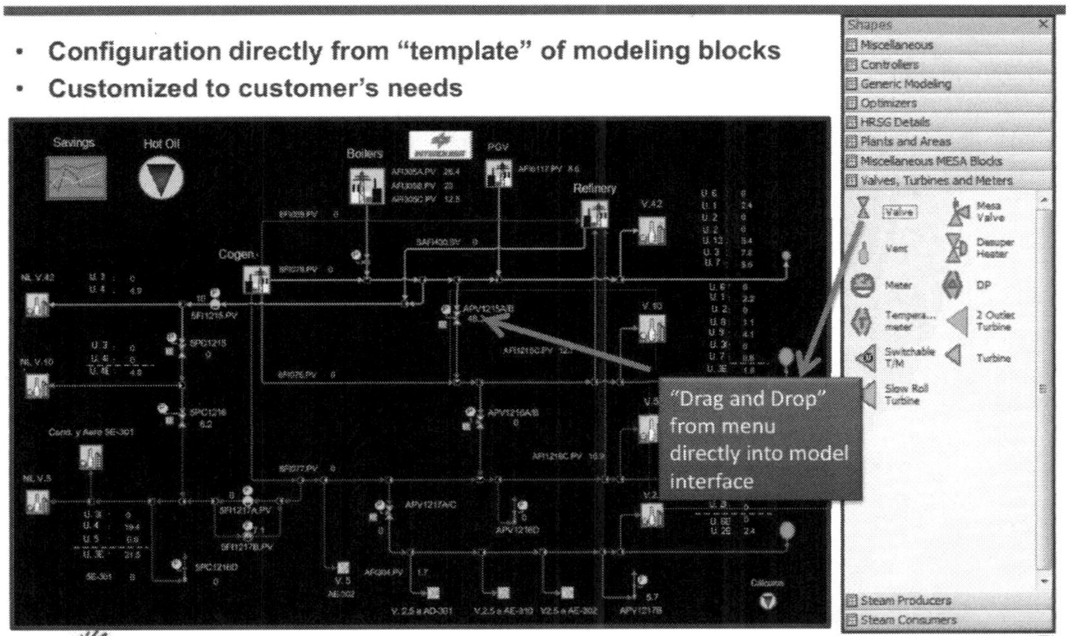

VISUALMESA™

CEPSA's Cogen and STG

Cogen block detail shows specific equipment in Cogen plant

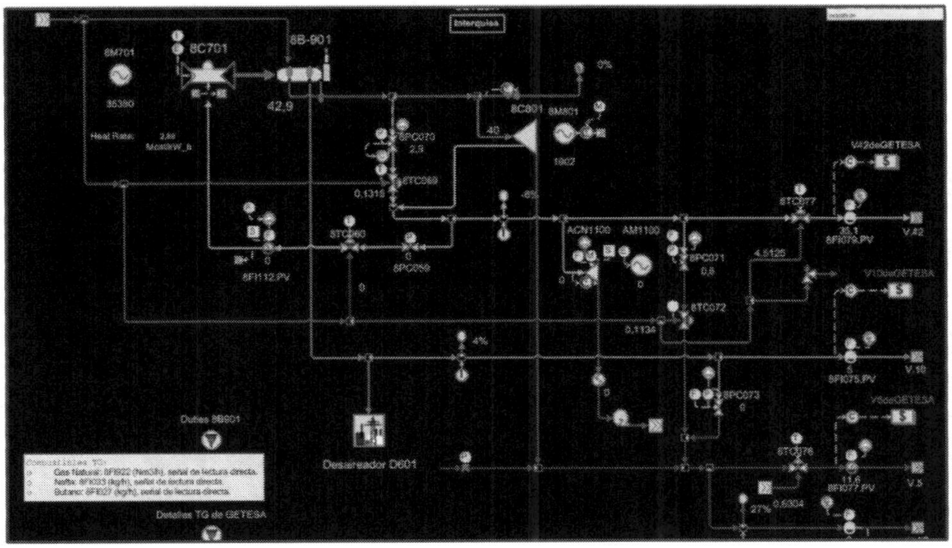

VISUALMESA™

Cogen in Visual MESA

Detailed Gas Turbine/HRSG modeling expands optimization opportunity

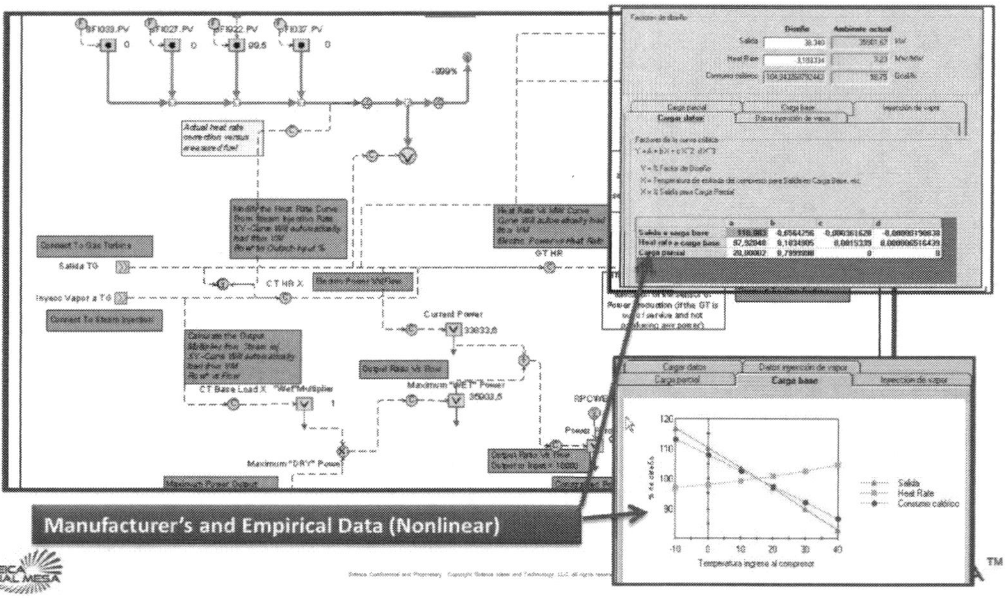

TM

Fired Boilers

Rigorous boiler efficiency increases savings in optimization through "biasing"

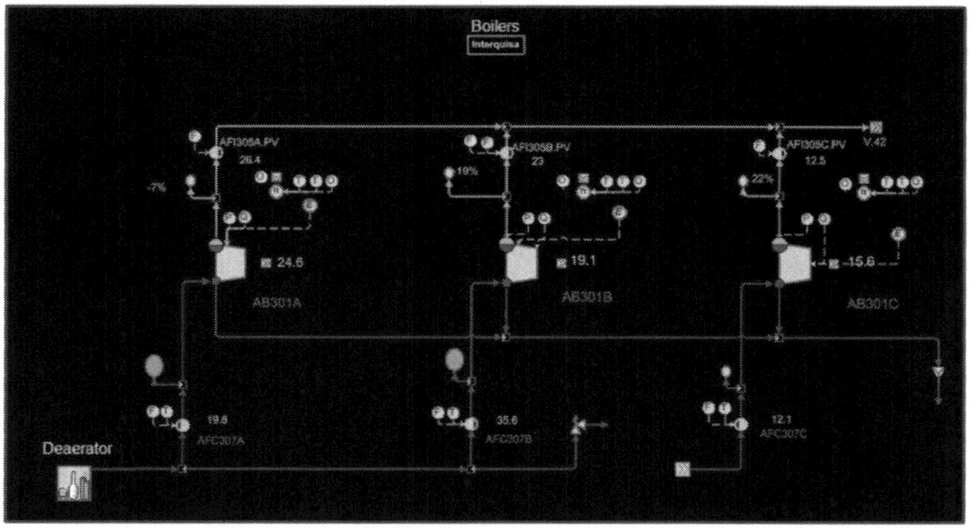

VISUALMESA™

Hot Oil Network

CEPSA: Alternative to steam creates unique optimization savings opportunity

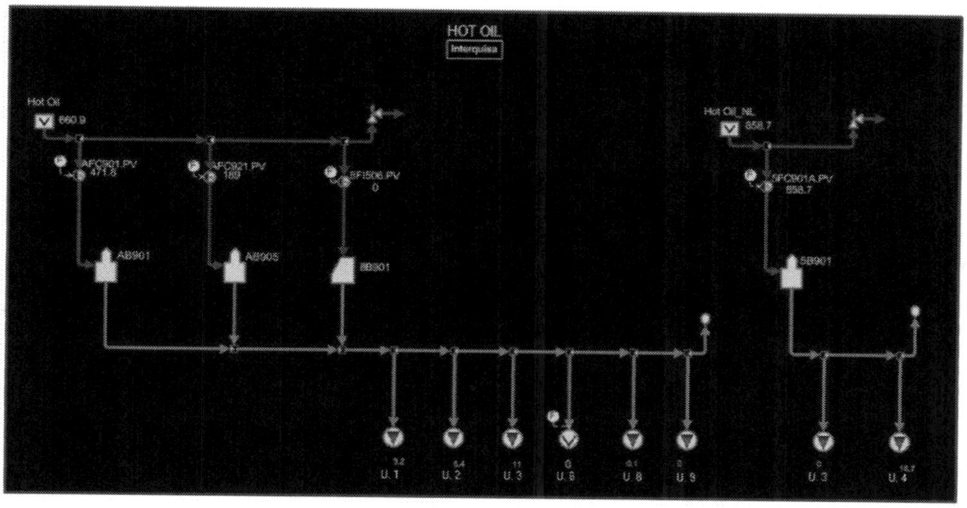

SOTEICA VISUAL MESA

VISUALMESA™

Emissions by Constituent

Tracking of Fuel by component to accurately constrain Emissions

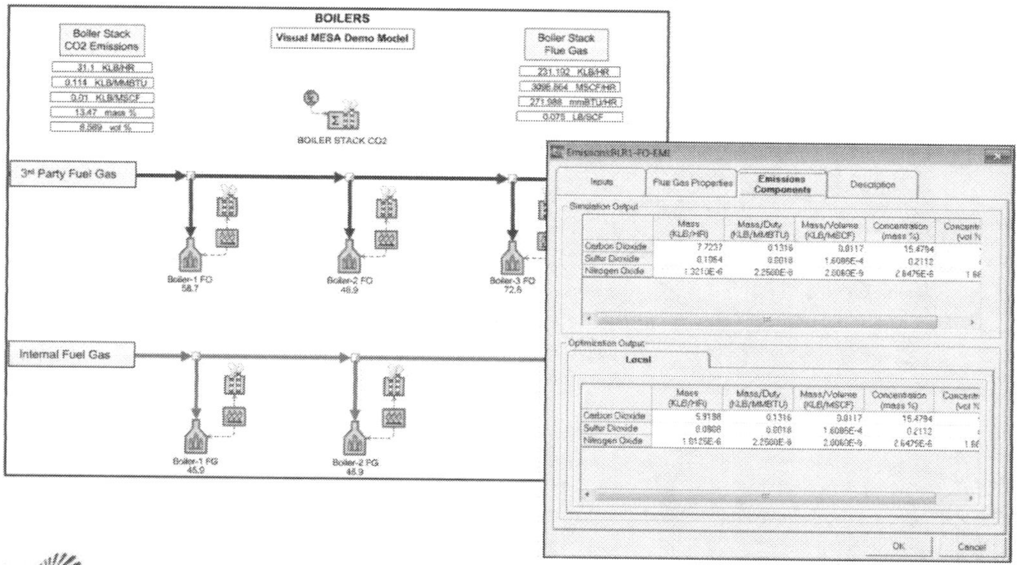

SOTEICA VISUAL MESA

VISUALMESA™

Flexible Operator Notification

Customized to Site-Specific Requirements

- Flexible, Excel-based Reports
  - Automatically updated recommendations
- DCS and/or Historian UI Integration
  - Energy RTO and Process Variables in single interface
- Dashboard
  - Validated Energy-related KPI's
- Model User Interface
  - Web Browser (Silverlight) or MS Visio
  - Engineering detail

VISUALMESA™

Operator Reports

- **Operators** focus on reliable "process" operation
  - Energy cost necessarily a lower priority
  - Minimize time requirement
- Provide **directly actionable** advice
  - Easily accessed
  - In expected context
  - Automatically updated
- **Show Benefit** of their effort ($/year)

VISUALMESA™

San Roque Operator Report Example

Show Benefit Captured by Specific Changes

Optimization Results for CEPSA — 3 de julio de 2012 — 20:45

GOOD SOLUTION

RESULTS	Current Operational Cost €/hr	Optimized Cost €/hr	SAVINGS €/hr	SAVINGS €/Year
COST	1,917	1,752	165	1,367,687

STEAM GENERATION	Current Operation t/h	Optimized t/h	Delta t/h	Eficiencia (PCI)
Vapor de 42 kg/cm2 en AB301A	0.7	0.0	-0.6	85.4
Vapor de 42 kg/cm2 en AB301B	1.0	0.0	-1.[illegible]	85.0
Vapor de 42 kg/cm2 en AB301C	0.0	0.0	0.0	88.3
Total Steam Generation	1.7	0.0	-1.6	

STEAM IMPORT FROM GETESA	t/h	t/h	t/h
Vapor de 42 kg/cm2	50.9	42.1	-8.8
Vapor de 10 kg/cm2	6.6	7.3	0.7
Vapor de 10 kg/cm2 dest. Artenius	0.5	0.5	0.0
Vapor de 5 kg/cm2	0.0	2.4	2.4
Total Import from GETESA	56.0	52.3	-5.7

STEAM EXPORT TO REFINERY	t/h	t/h	t/h
Vapor de 10 kg/cm2	0.1	0.0	-0.1
Vapor de 5 kg/cm2	1.9	0.0	-1.9
Vapor de 5 kg/cm2 Recalentado	6.8	11.9	5.1
Total Export to Refinery	8.8	11.9	3.1

SOTEICA VISUAL MESA

VISUALMESA™

"Actionable Items" Custom Report Example

Example from a Visual MESA Solution in Texas

Customer XYZ Report Version: 07h

Actionable Items

GOOD SOLUTION - SQP = 3

Last execution: 23-Sep-2013 at 13:30

Economic Summary	Actual Cost $/h	Optimum Cost $/h	Potential Savings $/h	Potential Savings $/year
Cost	88,098	87,808	289	2,535,599

Main Steam Production (t/h)	Actual	Optimized	Delta	Actual bias (%)	Optimized bias (%)
Boiler 1	119.8	106.9	-12.9	6.0	**10.0**
Boiler 2	117.2	121.3	4.1	10.0	**-2.1**
Boiler 3	119.1	116.2	-2.9	7.4	**2.1**
Boiler 4	117.9	106.9	-11.0	7.0	**10.0**
Boiler 5	121.1	130.7	9.6	-7.0	**-10.0**
Boiler 6	119.2	130.7	11.4	-7.0	**-10.0**
Boiler 7	0.0	0.0	0.0		
Total Steam Production	714.3	712.7	-1.7		

SOTEICA VISUAL MESA

VISUALMESA™

Conclusion

- CEPSA documents San Roque Operators deliver **3% to 6% Savings** with Visual MESA
  - Return on Investment in **less than 12 months**
- **5+ years** of **Sustained Savings** through Proactive Support Service
- **Visual MESA provides San Roque Operations a <u>Rigorous Basis</u> driving consistent <u>Best Practices</u> in their Energy Management**

VISUALMESA™

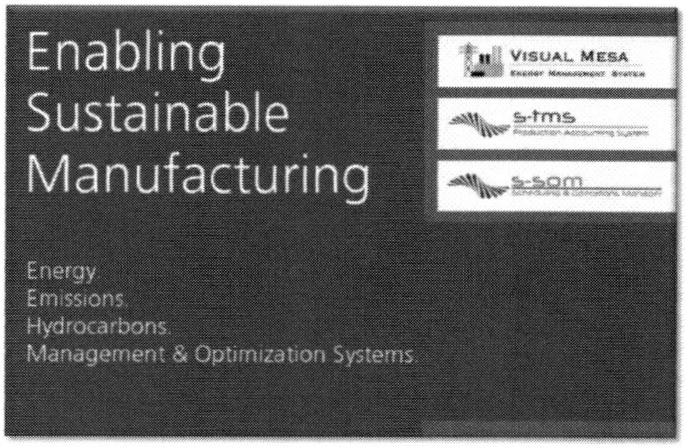

VISUALMESA™

IMPLEMENTATION EXPERIENCES IN ENERGY RETROFIT OF CRUDE UNITS USING AN ALTERNATIVE TECHNOLOGY

OK-Solutions

www.ok-solutions.org

Miguel Bagajewicz (Speaker), Gary Valtinson and DuyQuang Nguyen

Email: Miguel.bagajewicz@ok-solutions.org, **Phone:** (405) 360-5202

OK-Solutions

A Spin-Off Company from the University of Oklahoma

Dedicated to Transfer of New Technology (mostly developed at OU)

- Energy Retrofit of Plants (especially crude Units)
- Optimization of Crude Units
- Heat Integration across plants
- Cleaning Cycles for Heat Exchanger Networks
- Data Reconciliation
- Leak Detection in Plants and Pipelines
- Water Management
- Instrumentation Networks
- Investment Planning Models with Financial Risk Analysis

OUTLINE

- Energy retrofit highlights
- Pinch Technology Basics
- Alternative Method (HIT) Basics
- Example for one crude using HIT
- Comparison with Pinch Technology
- Retrofit Design for Multiple Crudes
- Conclusions

Energy Retrofit Basics

- An energy retrofit system must:
  - Reduce Energy Consumption.
  - Adjust desalter temperature if necessary.
  - Add and/or relocate actual exchangers appropriately.
  - Resolve hydraulic and potential fouling problems.
  - Calculate the costs of new proposed equipment.
  - Recalculate the new re-piping costs.
  - Optimize the profitability
  - CONSIDER THE USE OF DIFFERENT CRUDES **FOR THE SAME** HEAT EXCHANGER NETWORK → Sensitivity, hydraulics, and fouling ought to be analyzed for all options

Energy Retrofit

➢ Available Techniques:

- By inspection. (Inefficient even if based on pinch technology)

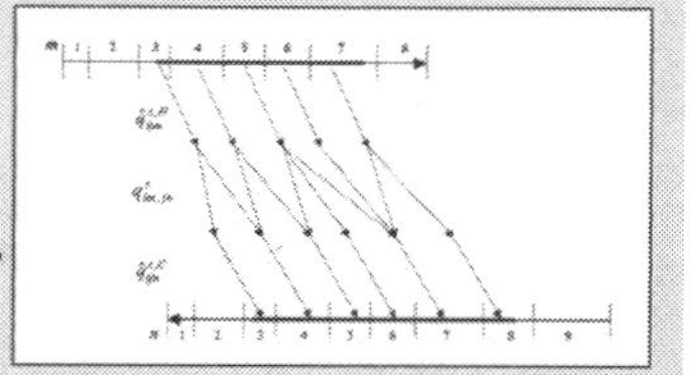

- HIT (Heat Integration Transportation Model).
- Pinch Technology.

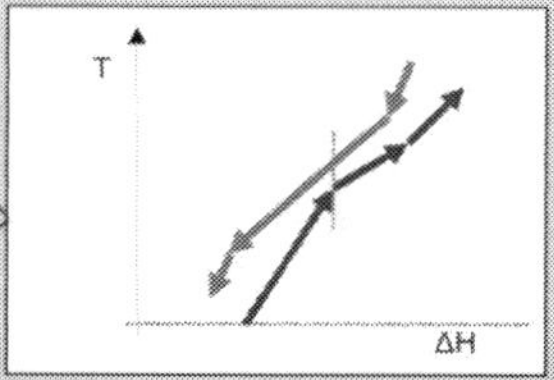

- Others (Mostly academic works not available for operation).

Special Conditions of Crude Units

- Crude units require **ONLY ONE** preheating train which must be used for crudes with different densities and with different minimum energy consumptions.

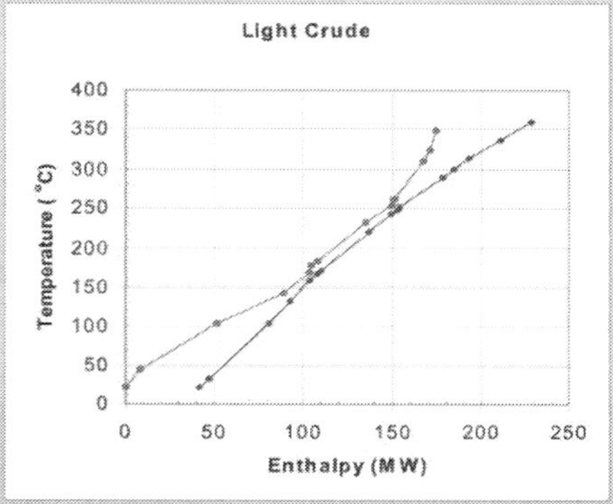

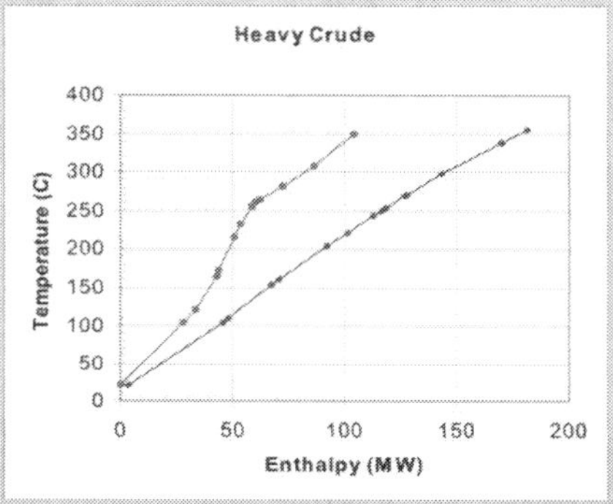

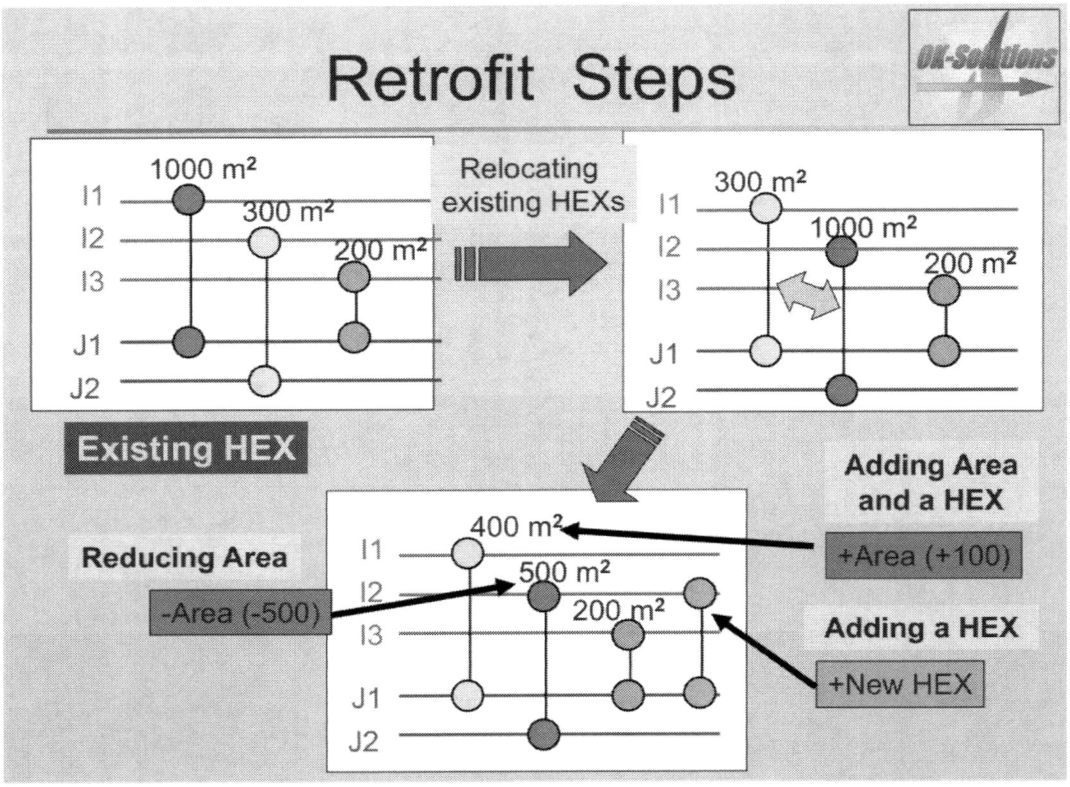
Retrofit Steps
1000 m²
300 m²
200 m²
I1
I2
I3
J1
J2
Existing HEX
Relocating existing HEXs
300 m²
1000 m²
200 m²
I1
I2
I3
J1
J2
Adding Area and a HEX
Reducing Area
400 m²
500 m²
200 m²
I1
I2
I3
J1
J2
-Area (-500)
+Area (+100)
Adding a HEX
+New HEX

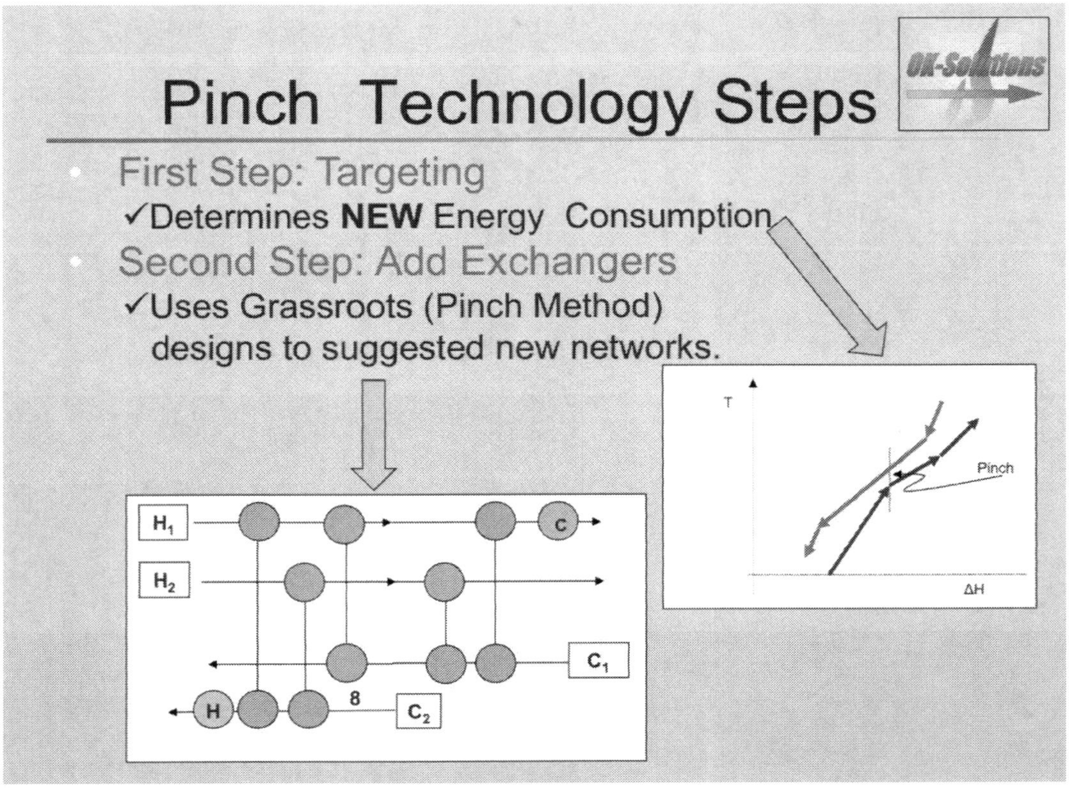
Pinch Technology Steps
First Step: Targeting
✓Determines NEW Energy Consumption
Second Step: Add Exchangers
✓Uses Grassroots (Pinch Method) designs to suggested new networks.
H1
H2
C
C1
H
8
C2
T
Pinch
ΔH

Targeting for Pinch

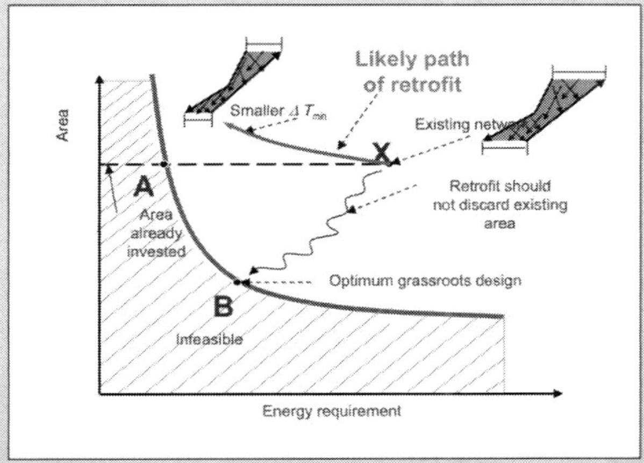

- ✓ It targets lower energy consumption because of its inability to assess capital investment properly
- ✓ It requires the use of especial techniques (loops and paths) to limit area reductions.
- ✓ Cannot guarantee good answers and becomes a game of trial and error; **thus, it may miss the real optimum.**

HIT
(Heat Integration Transportation Model)

✓It's based on old Transportation/Transshipment models

✓Developed from 2005 to 2008.

✓Automatic Retrofit (only one step, no targeting)

- Adds exchangers.(Shell & Tube, Alfalaval Compablocs, Fintube twisted tubes, etc.)
- Adds shells to pre-existing exchangers.
- Reduces area within a certain limit.
- Adds area to existing shells up to a limit (Using twisted tubes).
- Relocates streams (exchanges tubes with shells) in pre-existing exchangers.
- Considers limits on splitting of streams as well as clamp down certain matches.
- Takes into account the balance between real savings and costs.
- Generates automatically the second best, the third best solution, etc..
- Considers changes into pump-around loads and/or desalter temperature.
- Optimizes the column performance.

Several of the constraints of area reduction (due to pressure drop), as well as relocation and limits on splitting among others, are easily resolved by the **HIT** method.

HIT model

It is a Transportation model with about 150 equations and with the following structure:

- Divides each stream into small packs of a few degrees intervals of temperature.
- Creates connections between hot and cold packs "transporting" heat.
- Computes additional new area. Computes new shells.
- Computes cost of new equipment, splitting and new re-piping.
- Calculates the pay-out of the investment according to the chosen parameters.

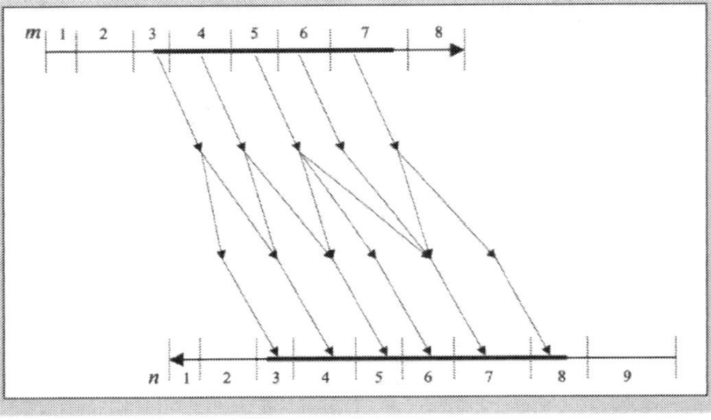

HIT model

Our experience with HIT:

We started to do these basic engineering projects in 2011, mostly in Asia. All have gone to the next detailed engineering phase, one is about to be implemented.

Statistics:

- Savings of about 25% (12 out of 40, 14 out of 35, 20 out of 100)
- Large IRR 20-50%
- Number of Exchangers added: 5-8 (this depends heavily of layout)
- Shells added to existing exchangers: 2-4
- Compablocs used (3-5)
- Desalter temperature: Sometimes changed

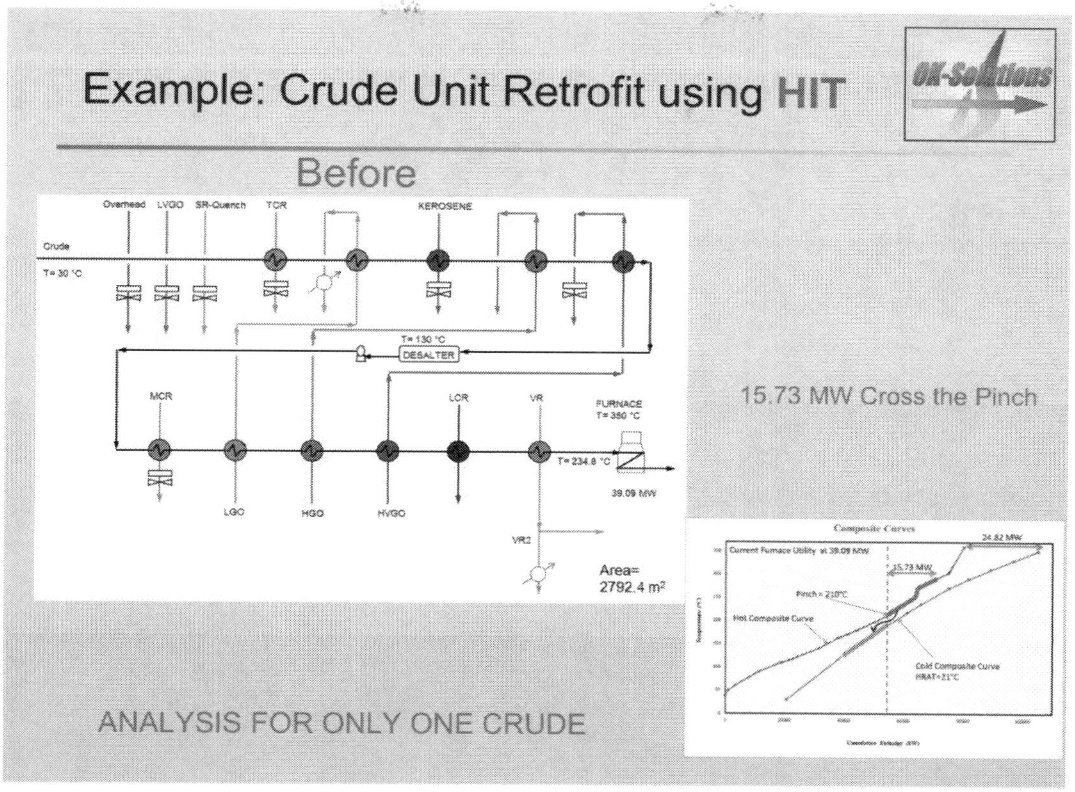
Example: Crude Unit Retrofit using HIT
OK-Solutions
Before
Overhead
LVGO
SR-Quench
TCR
KEROSENE
Crude
T= 30 °C
T= 130 °C
DESALTER
MCR
LCR
VR
FURNACE
T= 360 °C
T= 234.8 °C
39.09 MW
LGO
HGO
HVGO
VR2
Area=
2792.4 m²
15.73 MW Cross the Pinch
Composite Curves
Current Furnace Utility at 39.09 MW
24.82 MW
15.73 MW
Pinch = 210°C
Hot Composite Curve
Cold Composite Curve
HRAT=21°C
ANALYSIS FOR ONLY ONE CRUDE

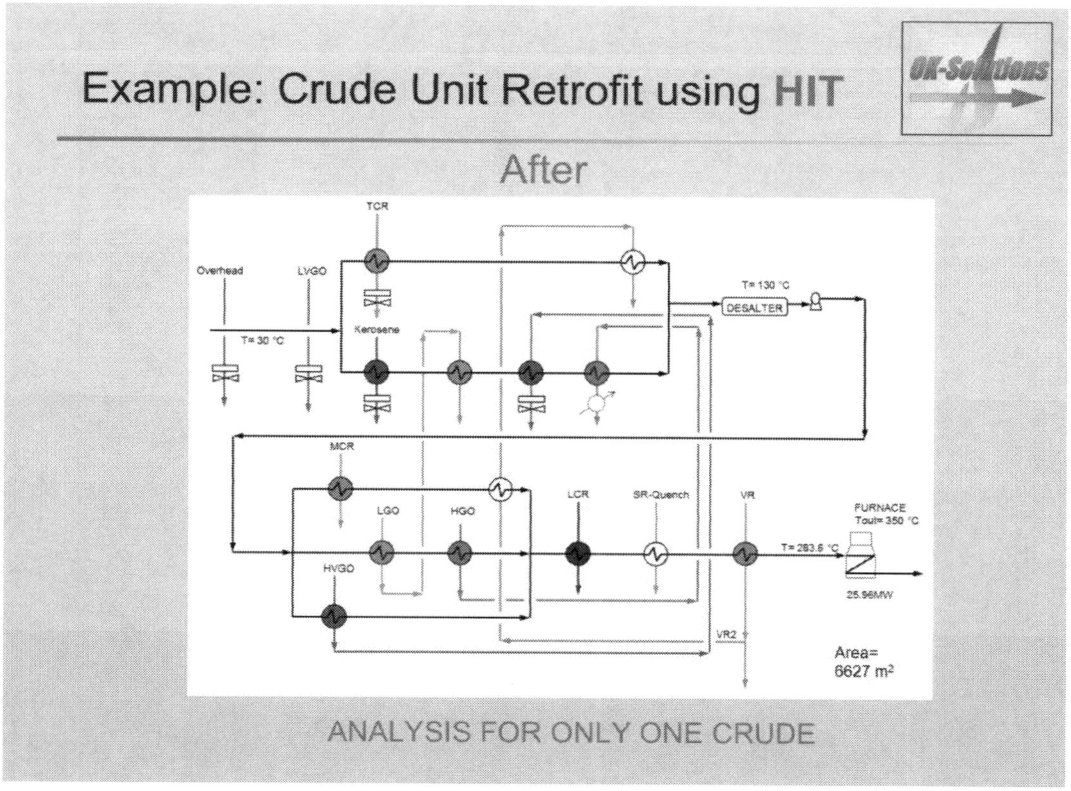
Example: Crude Unit Retrofit using HIT
OK-Solutions
After
TCR
Overhead
LVGO
T= 30 °C
Kerosene
T= 130 °C
DESALTER
MCR
LGO
HGO
LCR
SR-Quench
VR
FURNACE
Tout= 350 °C
T= 283.6 °C
25.96MW
HVGO
VR2
Area=
6627 m²
ANALYSIS FOR ONLY ONE CRUDE

Example: Crude Unit Retrofit using HIT

Before

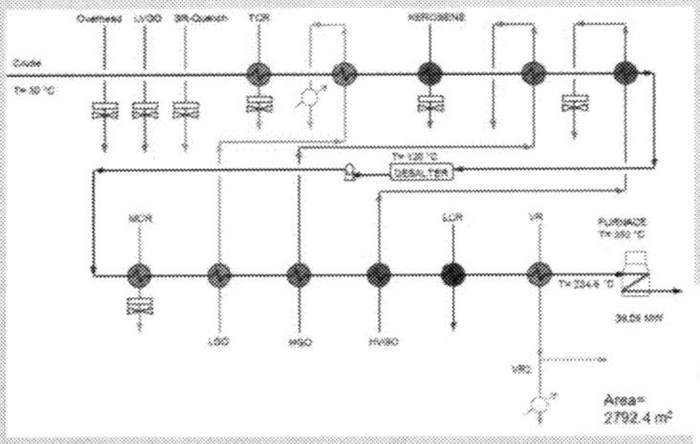

After

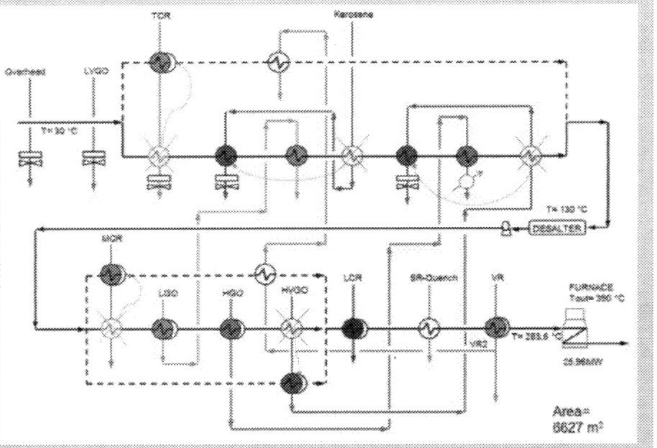

Can we avoid the re-piping suggested by HIT?

ONLY ONE CRUDE

Example: Crude Unit Retrofit using HIT

Before

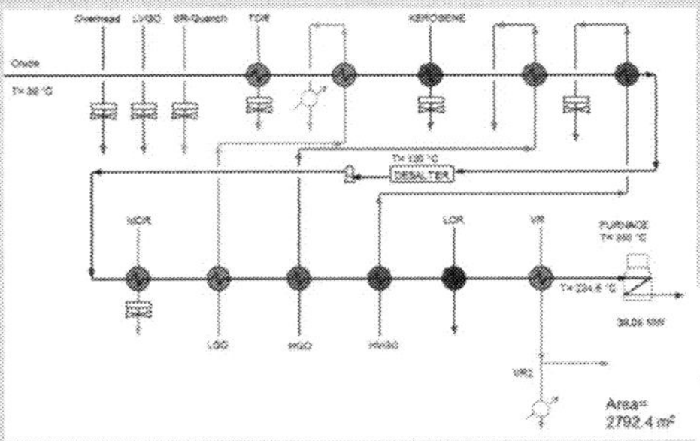

YES: Post HIT analysis confirms it

After

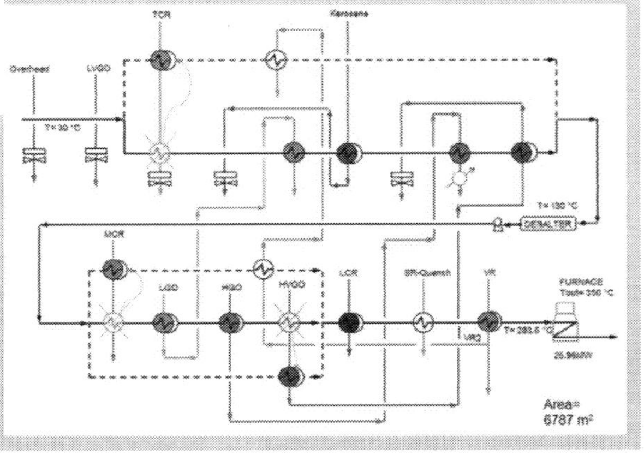

ONLY ONE CRUDE

Example: Crude Unit Retrofit using HIT

Savings:16.41 MW (56 MMBTU/hr) ($1.64M) FCI=$4.06M NPV(10yrs)=6.0M IRR(10yrs)=39%

HE	Original Load		Retrofit Load		Original Area		Retrofit Area		Area Addition		New Shells	Cost
Crude Match	MW	MMBTU/hr	MW	MMBTU/hr	m²	ft²	m²	Ft²	m²	ft²		USD$
TCR	7.52	25.769	6.58	22.45	707	7,612	760	8,184	53	572	1	**212,367**
LGO	2.0	6.82	1.23	4.28	63	676	53	574	(10)	(102)		
KERO	2.4	8.19	3.46	11.79	182	1,964	173	1,866	(9)	(98)		
HGO	2.0	6.82	1.93	6.57	96	1,031	96	1,031	0	0		
HVGO	6.07	20.73	4.26	22.45	411	4,421	382	4,117	(29)	(304)		
MCR	3.36	11.45	5.54	18.9	240	2,586	1206	12,985	966	10,399	2	**774,907**
LGO	1.99	6.8	3.19	10.89	74	801	305	3,280	231	2,479	1	**284,515**
HGO	1.88	6.43	1.37	4.68	68	728	269	2,891	201	2,163	1	**272,576**
HVGO	5.47	18.66	8	38.83	473	5,092	1,762	18,964	1,289	13,872	3	**1,097,023**
LCR	7.8	26.62	7.8	26.62	324	3,484	664	7,146	340	3,662	1	**329,267**
VR	4.64	15.83	4.64	15.83	154	1,662	336	3,617	182	1,955	1	**264,686**
VR2			2.53	8.64			216	2,321	216	2,321	1	**278,505**
VR2			2.41	8.22			273	2,942	273	2,942	1	**302,041**
SRQ			1.93	6.59			132	1,416	132	1,416	1	**244,268**
					2,792	30,057	6,627	71,338	3,883	41,781	13	**4,060,155**

Pinch Targeting: Inaccurate

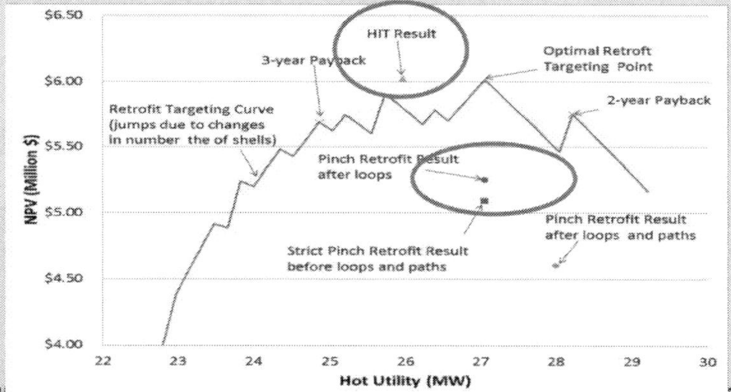

- ✓ NPV ta[illegible] results.
- ✓ Energy recovery levels are different for the optimum Pinch case and HIT
- ✓ HIT simulation is very close to predicted optimum with greater energy savings.

…and the network is too complicated

Why we don't use Pinch Technology?

- For Pinch Technology
  - A "continuous pinch" makes the pinch location fuzzy. Worse for light crudes.
  - Sometimes a second pseudo pinch is shown.
  - One cold stream vs. many hot streams requires splitting of the cold stream at the pinch point with the subsequent merging. This design is not quite practical and is inflexible to operational changes.
  - HIT does not exhibit these outcomes

Original Network

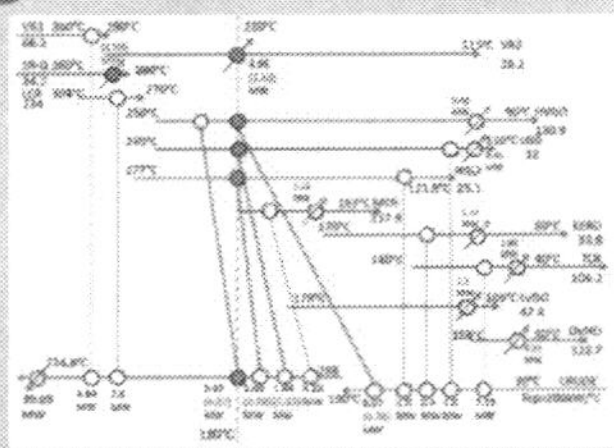

Pinch Retrofit

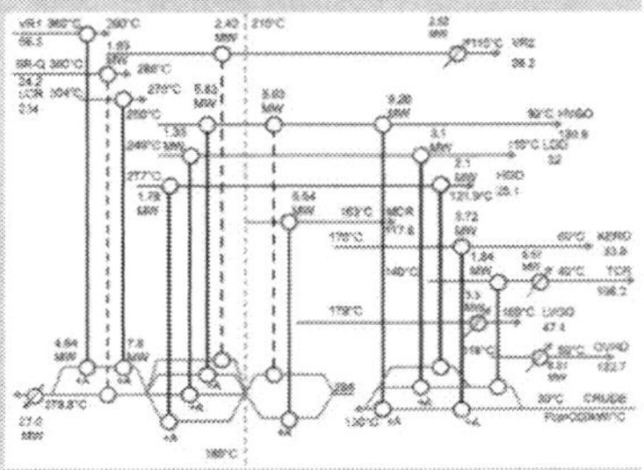

Why we don't use Pinch Technology?

- For Pinch Technology
  - It requires a few steps with no easy way to eliminate possible options (which there can be many: more than 200 sometimes).
  - Only an experienced engineer can create a reasonable, near optimal design using Pinch Technology.

Pinch Retrofit

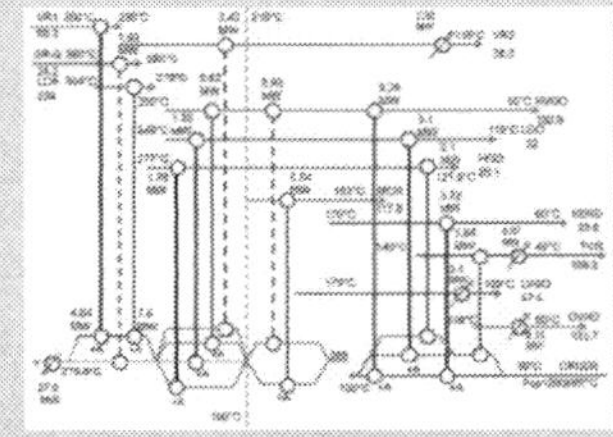

Pinch Network after loops

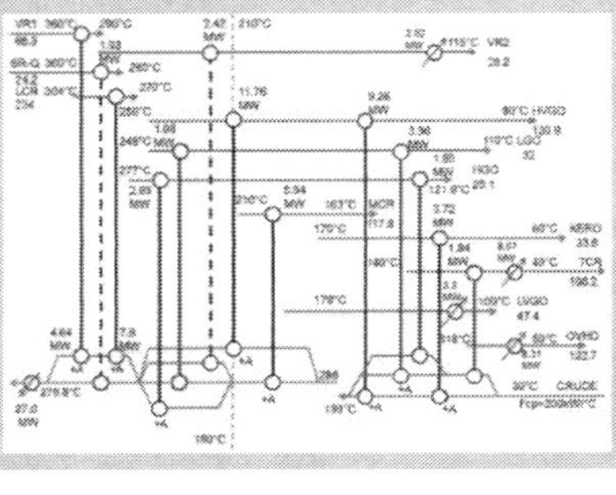

Simplified Pinch Network

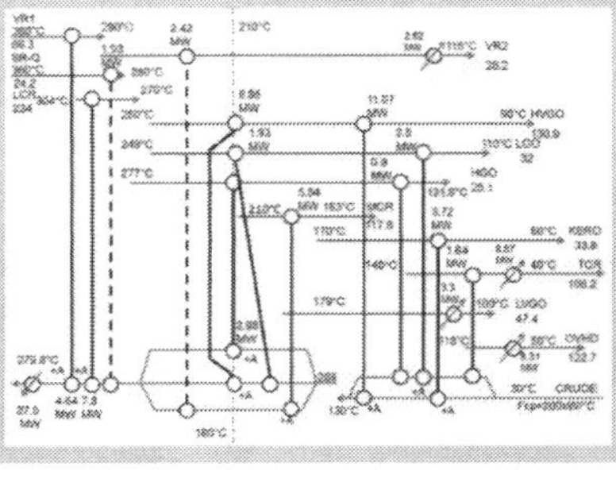

Pinch Technology

Original

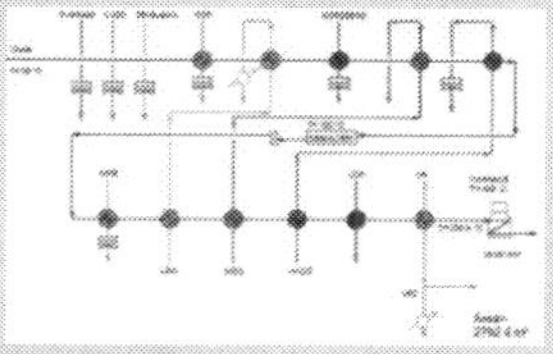

Pinch Design (excessive splitting/merging)

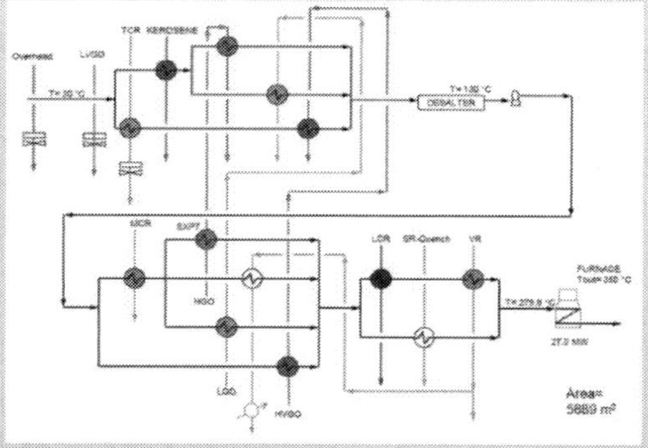

HIT Design (less complicated)

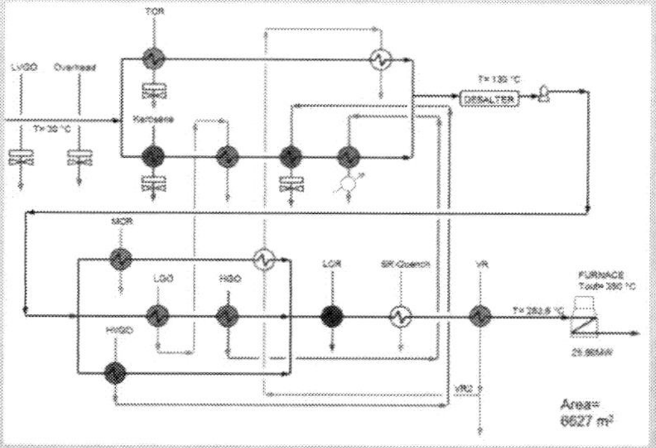

Retrofit Projects

Required to consider several designs, not just one.

In a typical CDU retrofit study:

- Several different crudes/blends have different optimal designs.
- These different optimal results should be compatible for all other crudes.
- Different NPV, ROI and FCI have to be considered.
- Hydraulics is usually favored by more branches, but branches benefit from balanced pressure drop
- Consider limiting the number of branches.
- Consider different throughput values.
- Consider specified exchanger locations.
- Different equipment may be considered (Fintube, Compablocs, S&T).
- Fouling needs to be tested.
- Lay-out and space issues need to be considered.

One project took us about **36** designs for **4** different blends and two different throughput values, narrowed down to **12** that were studied further.

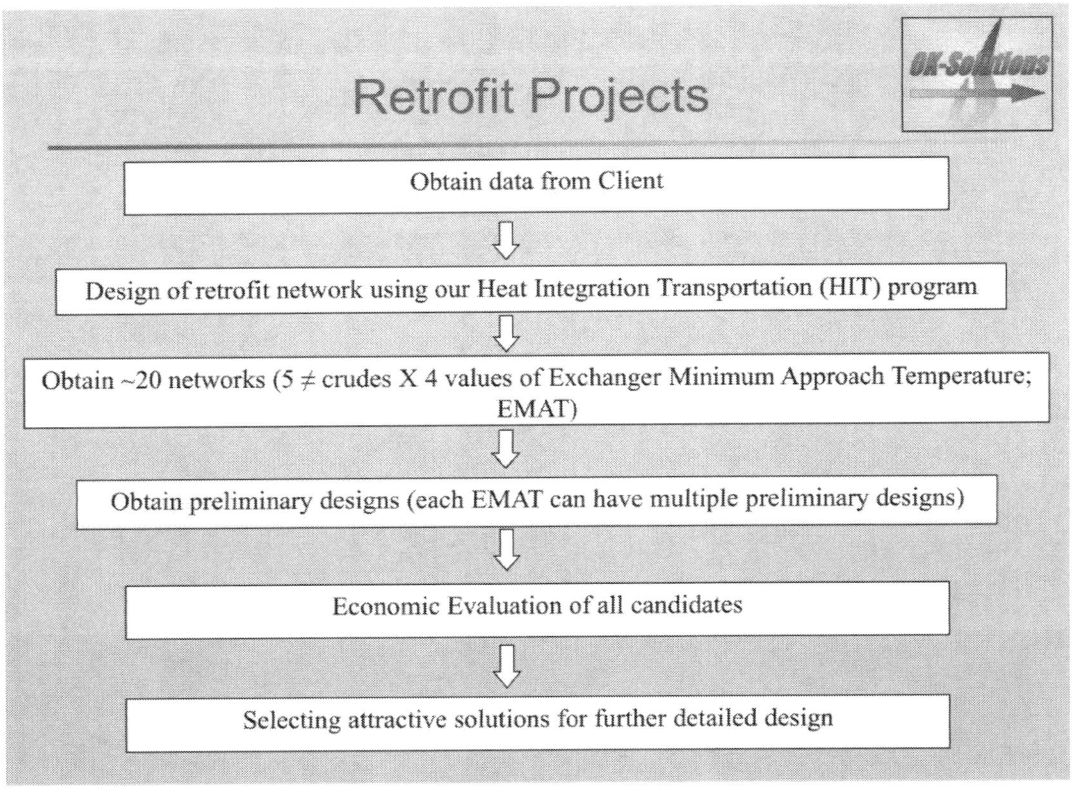
Retrofit Projects
OK-Solutions
Obtain data from Client
Design of retrofit network using our Heat Integration Transportation (HIT) program
Obtain ~20 networks (5 ≠ crudes X 4 values of Exchanger Minimum Approach Temperature; EMAT)
Obtain preliminary designs (each EMAT can have multiple preliminary designs)
Economic Evaluation of all candidates
Selecting attractive solutions for further detailed design

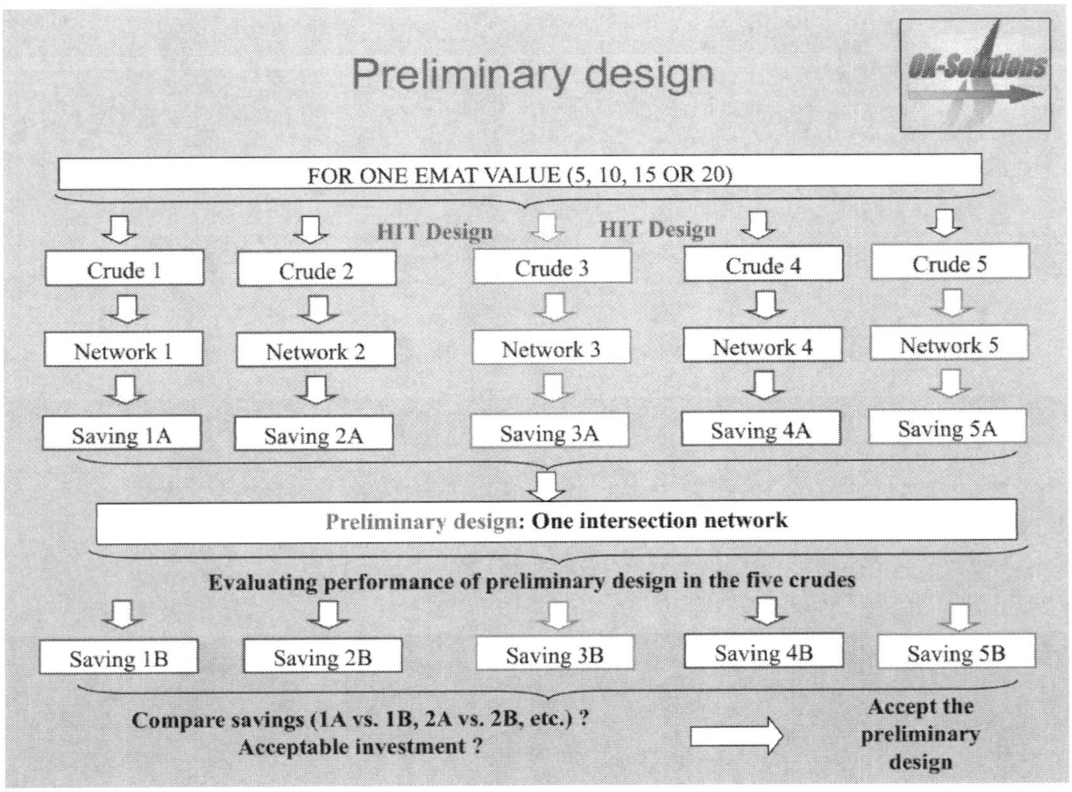
Preliminary design
OK-Solutions
FOR ONE EMAT VALUE (5, 10, 15 OR 20)
HIT Design
HIT Design
Crude 1
Crude 2
Crude 3
Crude 4
Crude 5
Network 1
Network 2
Network 3
Network 4
Network 5
Saving 1A
Saving 2A
Saving 3A
Saving 4A
Saving 5A
Preliminary design: One intersection network
Evaluating performance of preliminary design in the five crudes
Saving 1B
Saving 2B
Saving 3B
Saving 4B
Saving 5B
Compare savings (1A vs. 1B, 2A vs. 2B, etc.) ?
Acceptable investment ?
Accept the preliminary design

Example: Crude Unit Retrofit using HIT

Two Crudes!!

Crude 1: 31.8 API

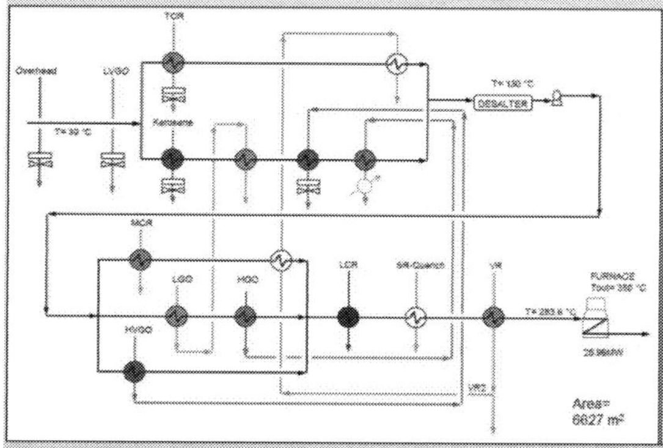

Crude 2: 37.8 API

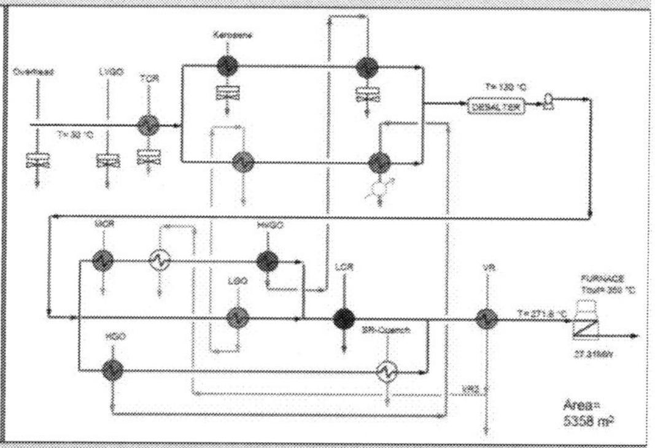

Two Designs, but … we need one

Example: Crude Unit Retrofit using HIT

Common Network

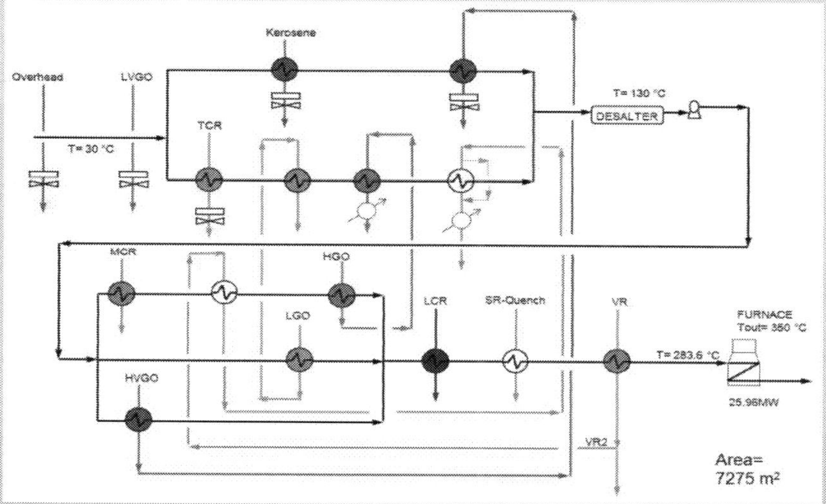

Checked using Simulation (Pro II; Invensys)

Example: Crude Unit Retrofit using HIT

Economics for two crudes

Network	API	Energy Savings		Area Addition		Total Area		CAPEX	Savings	NPV (10 years)	IRR
		MW	MMBTU/hr	m^2	ft^2	m^2	ft^2	USD$	USD$/yr	USD$	%
HIT Result	31.8	13.13	44.80	3,882	41,783	6,627	71,332	4,060,154	1,641,250	6,024,616	39%
HIT Result	37.8	11.78	40.20	2,678	28,827	5,358	57,673	2,807,090	1,472,500	6,240,785	52%
Common Network	31.8	13.13	44.80	4,482	48,246	7,275	78,303	5,258,207	1,641,250	4,826,564	29%
Common Network	37.8	11.78	40.20	4,482	48,246	7,275	78,303	5,258,207	1,472,500	3,789,668	25%

- We kept the predicted savings (we always try that first).
- Area of common network is higher
- As a result, CAPEX is higher and IRR and NPV lower (as expected)

Example: Crude Unit Retrofit using HIT

Typical Economics of CDU Retrofit with multiple crudes

Economic Details	Typical Range
Change in Total Area (Only S&T)	3-5 times
Change in Total Area (Only Compabloc)	Less than 2 times
Savings (Depends on existing furnace inlet temperature)	25%-35%
IRR	20%-50%
Payout	2-5 years

What to expect for a typical case

- For Light crudes.
  - If the furnace inlet temperature is 230 °C (446 °F) or lower, we believe we should be able to obtain savings larger than 20%.
  - Cases of larger furnace inlet temperature (>260 °C) can still achieve 10-15% return
- **We seek and use real HEX parameters and costs supplied by Alfa Laval, Koch Fintube and others.**
- **Different types of crude blends are always considered.**

Conclusions about HIT

- Based on our example, and our experience.
  - HIT is capable of producing a reasonable network without excessive splitting or merging.
  - HIT obtained a network 1 MW better in savings than the pinch targeting.
  - HIT economics are ~15% greater than the pinch results.
  - Provided multiple crudes, the HIT results can be merged into a unified network that works in either case.
  - HIT outperforms Pinch economically and structurally (less splits)

Summary

- A competitive technology (HIT), based on a new and powerful optimization method tested on the field was presented

- Our experience indicates that:
  - Complexity of multiple crudes requires multiple designs.
  - Several alternative multipurpose designs should be discussed.
  - Projects are interactive over 4-6 months
  - Column Optimization (fix pumparounds, steam injection, flash zone temperature, etc.) can also be performed to improve the energy recovery.

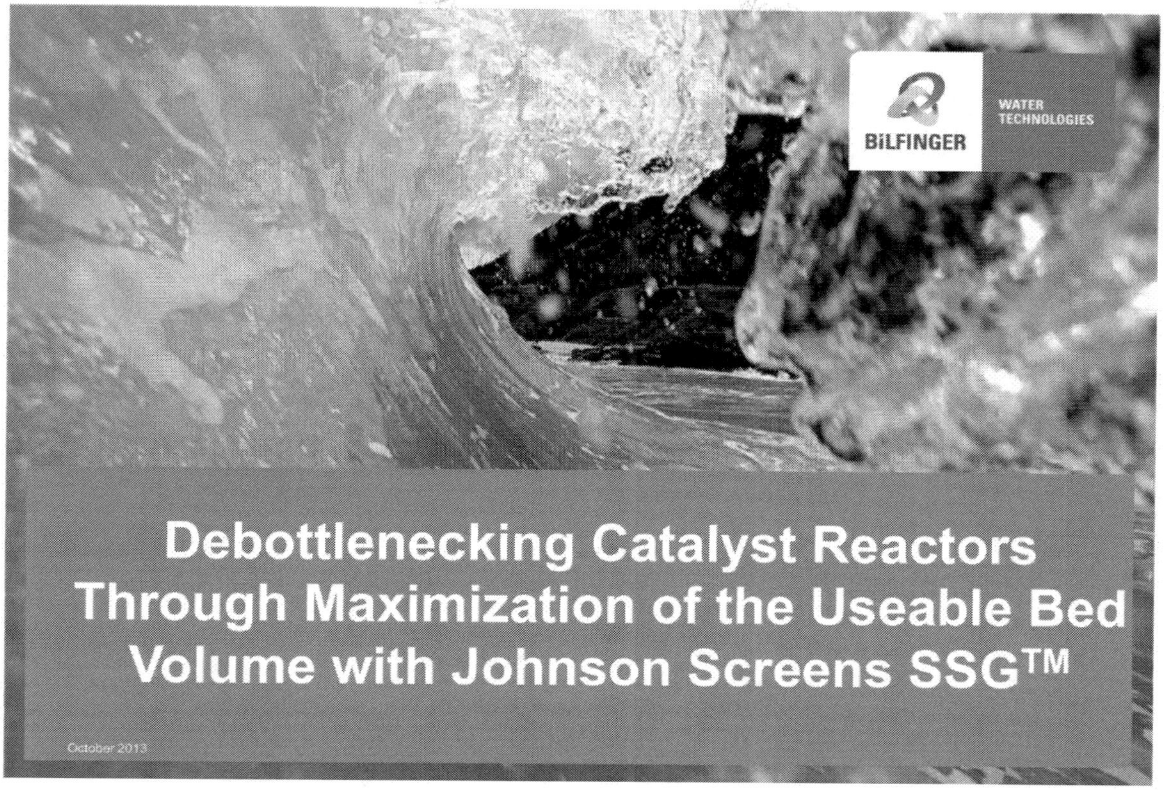

CONTACT INFORMATION

BILFINGER WATER TECHNOLOGIES

Benjamin Schmitt, PE
Senior Applications Engineer
Phone +1 651 638 3274
Mobile +1 651 253 1258
Benjamin.schmitt@johnsonscreens.com

Gene Eberhardt
Sales Manager
North America and Canada
Phone +1 651 638 3247
Mobile +1 612 562 5059
gene.eberhardt@johnsonscreens.com

Johnson screens®

OIL & GAS GLOBAL BUSINESS UNIT

PRESENTATION OVERVIEW

- Review of Common Design Configurations Using Support Grids and Outlet Collectors
- Introduction of the Shaped Support Grid SSG™
- Key Features of the Shaped Support Grid SSG™
- Closing Remarks
- Q & A Session

SUPPORT GRID GENERAL CONFIGURATION:

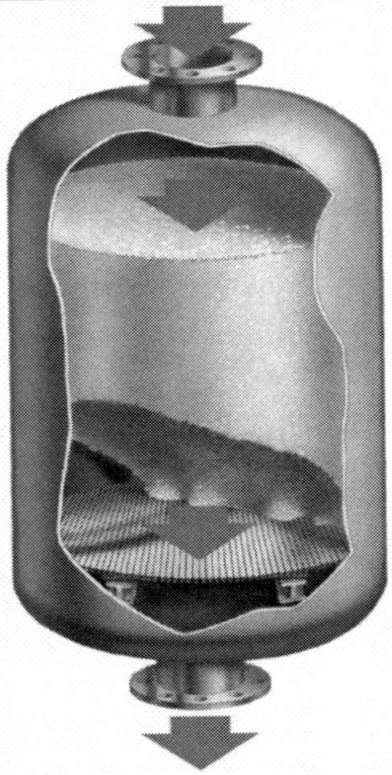

- Applicable in Catalytic Reactors
- Full Cross Sectional Catalyst Support
- Versatility of Design
- Ability To Support Catalyst In Direct Contact

OUTLET COLLECTOR GENERAL CONFIGURATION:

BILFINGER WATER TECHNOLOGIES

- Applicable in Catalytic Reactors
- Utilized Lower Reactor Head Volume for Catalyst / Inert Media
- Versatility of Design
- Ability To Support Catalyst In Direct Contact

GENERAL CONFIGURATION LOADING DIAGRAM

BILFINGER WATER TECHNOLOGIES

Diameter: 10' Tan-Tan: 20'	Support Grid	Outlet Collector
Catalyst Volume	1374 ft^3	1571 ft^3
Inert Material Volume	79 ft^3	262 ft^3
Un-Used Volume	459 ft^3	262 ft^3

CFD REVIEW: SUPPORT GRID

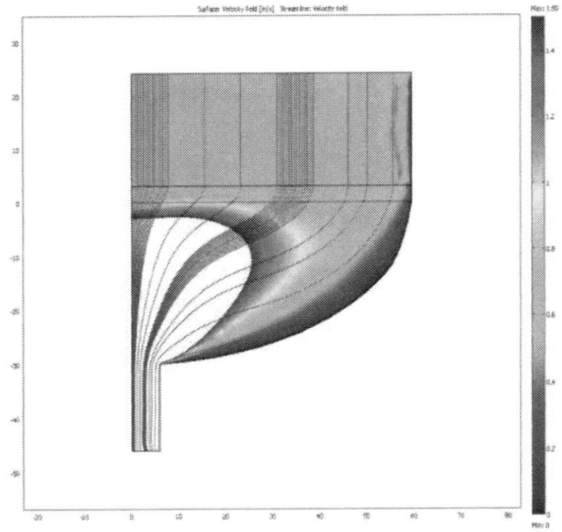

- **Modeling Assumptions:**
  - Down Flow
  - Homogeneous Catalyst Bed
  - Uniform Inlet Flow
  - 2:1 Elliptical Head Configuration
  - Support Grid Located At Tangent Line Of The Vessel
  - Shadowing Effect Of Support Beams Not Included In Analysis

CFD REVIEW: OUTLET COLLECTOR

Modeling Assumptions:

- Down Flow
- Homogeneous Catalyst Bed
- Uniform Inlet Flow
- 2:1 Elliptical Head Configuration
- Top Screen Type Outlet Collector

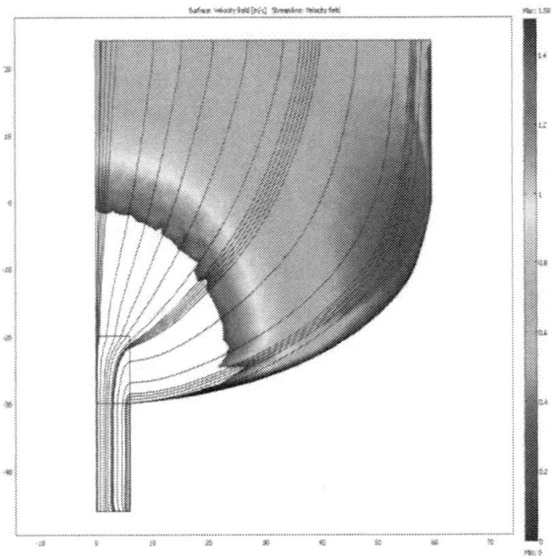

BENEFIT ANALYSIS SUPPORT GRID / OUTLET COLLECTOR

	Support Grid Configuration	Outlet Collector Configuration
Catalyst Volume		✓
Catalyst Usage	✓	
Fluid Distribution/ Hydraulic Efficiency	✓	
Mechanical Versatility		✓
Support Media Volume	✓	
Bed Containment Mechanical Reliability	✓	✓

JOHNSON SCREENS SHAPED SUPPORT GRID (SSG™)

GENERAL CONFIGURATION LOADING DIAGRAM

BILFINGER WATER TECHNOLOGIES

Diameter: 10' Tan-Tan: 20'	Support Grid	Outlet Collector	SSG™
Catalyst Volume	1374 ft^3	1571 ft^3	1751 ft^3
Inert Material Volume	79 ft^3	262 ft^3	82 ft^3
Un-Used Volume	459 ft^3	262 ft^3	82 ft^3
Catalyst Volume Increases		14.3%	27.4%

CFD REVIEW: SHAPED SUPPORT GRID (SSG™)

BILFINGER WATER TECHNOLOGIES

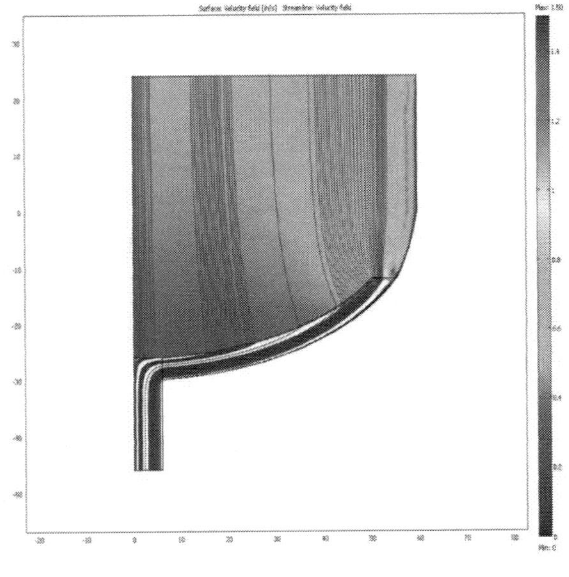

Modeling Assumptions:

- Down Flow
- Homogeneous Catalyst Bed
- Uniform Inlet Flow
- 2:1 Elliptical Head Configuration

CFD REVIEW: SHAPED SUPPORT GRID (SSG™)

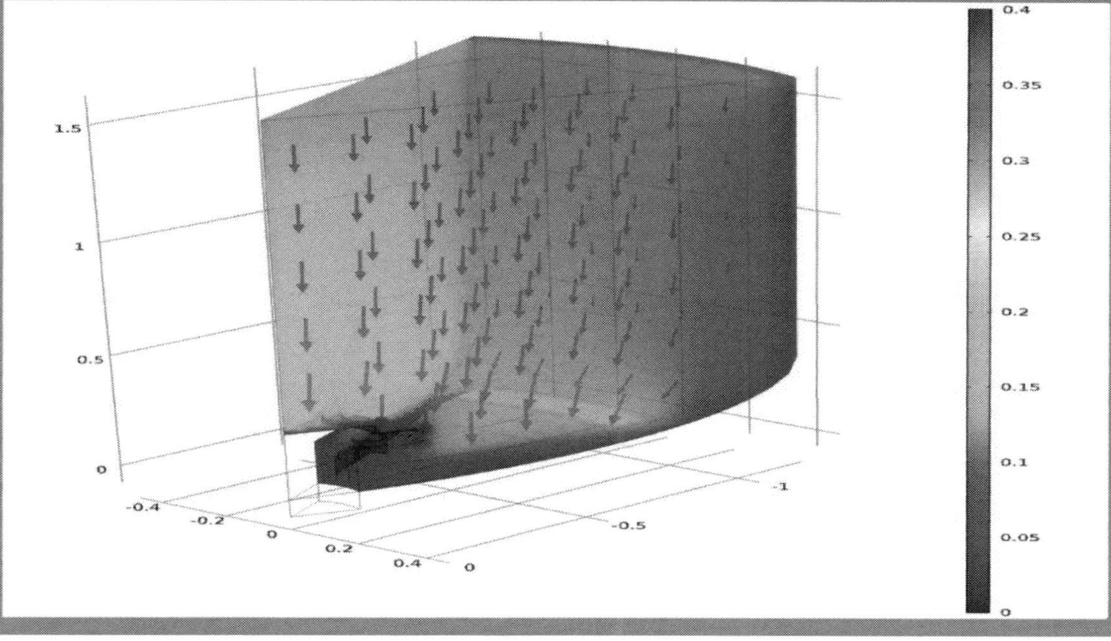

CONFIGURATION ANALYSIS:

	SUPPORT GRID	OUTLET COLLECTOR	SSG™
Catalyst Volume		✓	✓
Catalyst Usage	✓		✓
Fluid Distribution/ Hydraulic Efficiency	✓		✓
Mechanical Versatility		✓	✓
Support Media Volume	✓		✓
Bed Containment Mechanical Reliability	✓	✓	✓

JOHNSON SCREENS SHAPED SUPPORT GRID (SSG™)

JOHNSON SCREENS SHAPED SUPPORT GRID (SSG™)

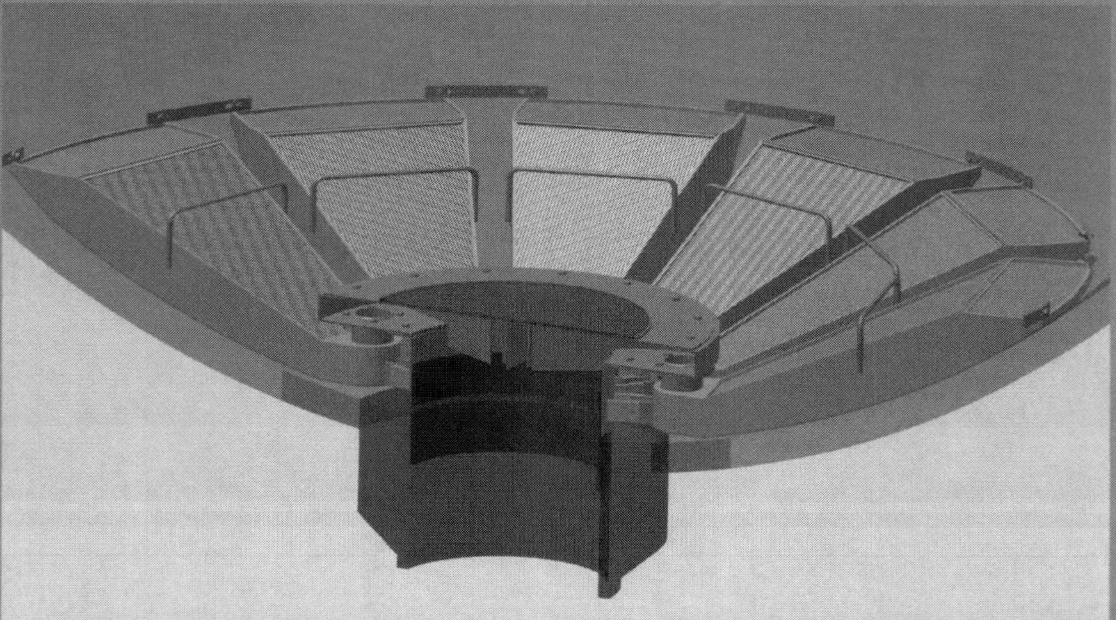

JOHNSON SCREENS
SHAPED SUPPORT GRID (SSG™)
BiLFINGER

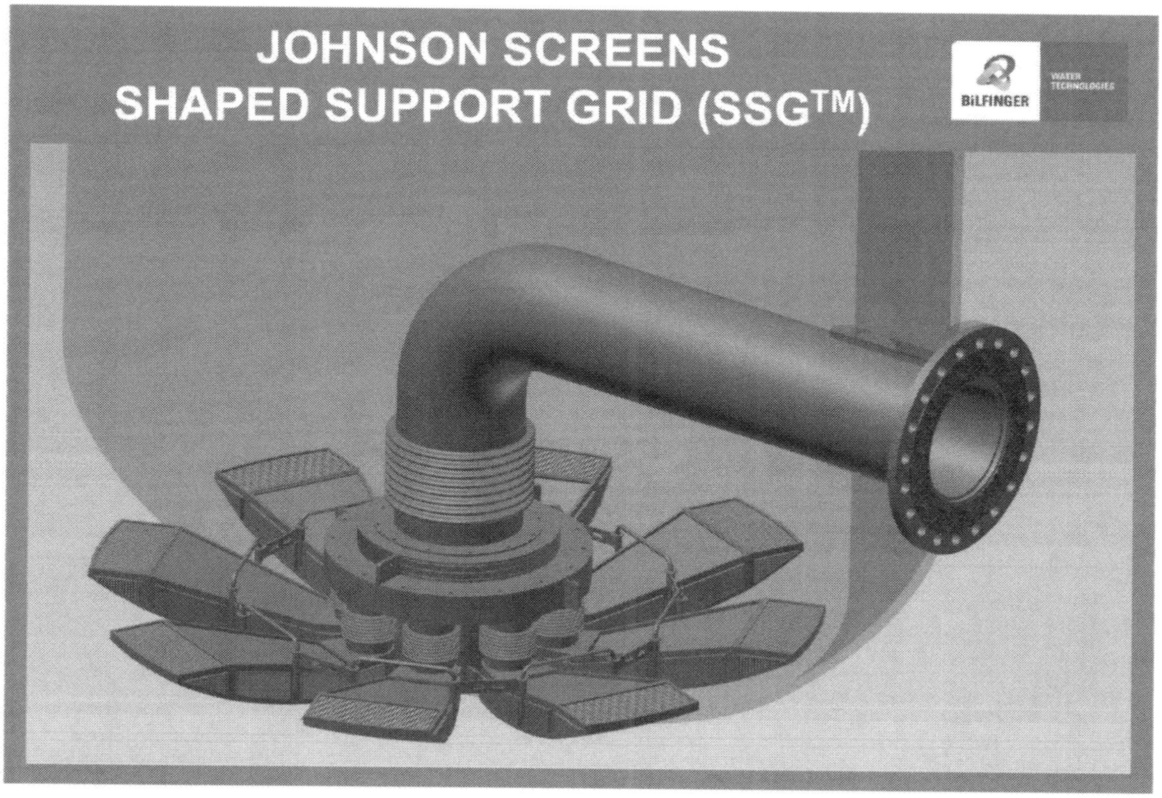
JOHNSON SCREENS
SHAPED SUPPORT GRID (SSG™)
BiLFINGER

JOHNSON SCREENS SHAPED SUPPORT GRID (SSG™)

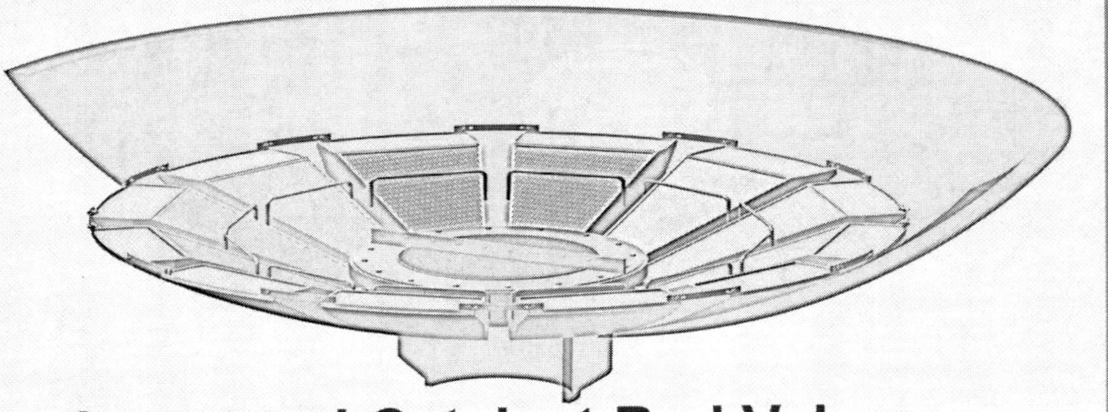

- Increased Catalyst Bed Volume
- Improved Hydraulic Design
- Design Efficiency

THANK YOU FOR YOUR ATTENTION!

QUESTIONS AND ANSWERS

Applying the Ergun Equation to the Degradation of a Reactor Catalyst

Bikram Sharda
James Sturnfield
The Dow Chemical Company

5th AIChE Southwest Process Technology Conference
Galveston, TX
October 3-4, 2013

Overview

- Ergun Equation
- Applying Ergun Type Equation to Process Data
- Changes in Pressure Drop
- Flow Adjusted Pressure Drop & Reference PD
- Degradation at Different Reactor Level
- CFD Modeling of Fines
- Conclusions

2

Overview of Ergun Equation

$$\frac{\Delta P}{L}=\frac{150\cdot\mu\cdot(1-\varepsilon)^2}{\varepsilon^3\cdot d_p^2}\cdot V+\frac{1.75\cdot(1-\varepsilon)\cdot\rho}{\varepsilon^3\cdot d_p}\cdot V^2$$

ε =void fraction
d_p =effective diameter of media components
ΔP =pressure drop across the media
L =length of the catalyst bed
μ =absolute (dynamic) viscosity of fluid
V =superficial velocity
ρ=density of the fluid

- Ergun Equation
  - Relates pressure drop to the velocity and the square of velocity
    - Superficial velocity is actual volumetric flow divided by cross sectional area
  - Developed for uniform packing of spheres
  - Expanded to other uniform shapes by replacing sphere diameter with an effective diameter
  - Usually assume an average density and viscosity
- Variety of alternative constants suggested
  - Varies depending on assumptions
  - Linear Constant 150 – 200
  - Quadratic Constant 1.75 -4.0

Actually an even wider range can be found

3

Ergun Type Equations

$$\frac{\Delta P}{V}=R_v\mu L+\frac{R_I}{2}\rho\, L\, V$$

R_V =viscous resistance
R_I =inertial resistance

ΔP =pressure drop across the media
L =length of the Catalyst Bed
μ =absolute (dynamic) viscosity of fluid
V =superficial velocity
ρ=density of the fluid

- Inertial and Viscous Resistivity
  - Slightly more general quadratic equation for pressure drop
  - Can replace velocity with mass flow by adjust terms
    - If density is assumed constant, then group with resistance parameters
    - If ideal gas assumed with fixed temp, MW, then replace ΔP with $(P_{in}^2-P_{out}^2)/2$

$$\frac{\Delta P}{F}=\frac{R_v}{\rho A}\mu L+\frac{R_I}{2\rho A^2}L\, F$$

F=mass flow
A= cross section area

Inertial resistance is related to the constriction of the flow
Viscous resistance is related to the inverse of permeance

4

Changes in Pressure Drop

- **Reactor Pressure Drop Grows with time**
  - Operation eventually reduces flow to limit PD
  - In order to separate PD from look at PD/Flow
  - Group into periods of common behavior
    - Actually based on plots of flow and PD/flow

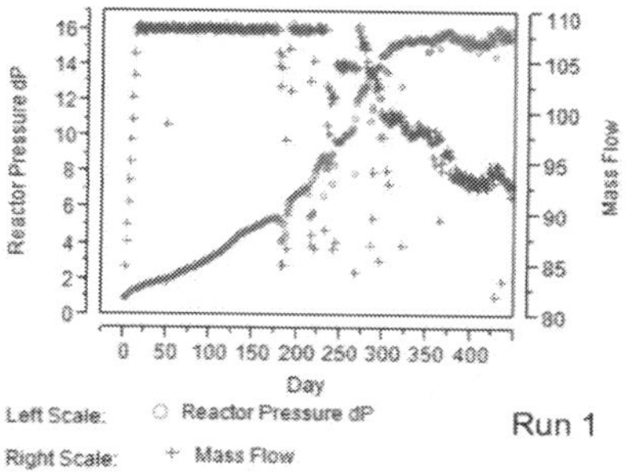

Run 1

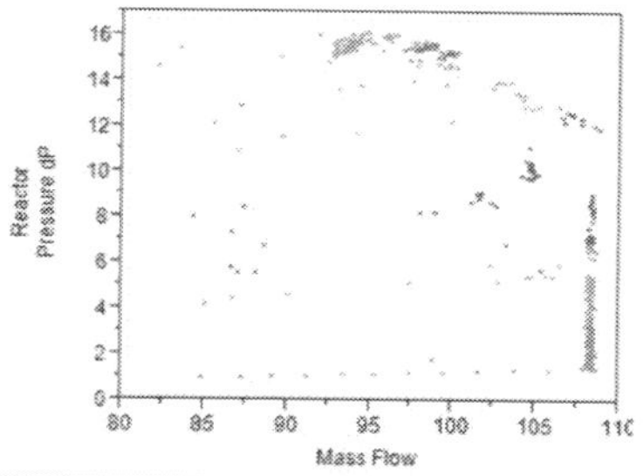

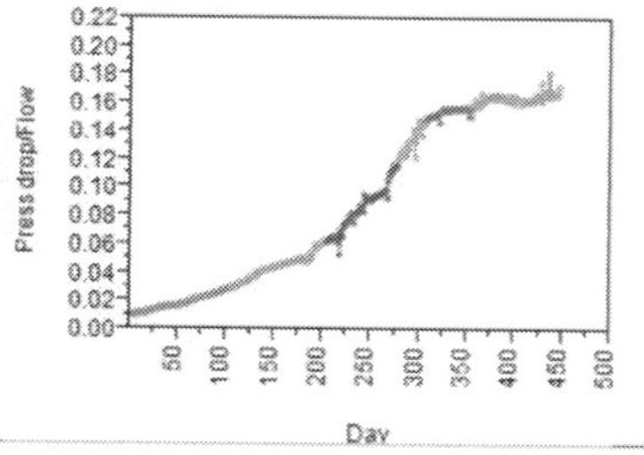

5

Changes in Flow-Adjusted Pressure Drop

- PD/Flow Shifting Higher
  - Using the Ergun Equation, the quadratic term is not apparently changing much
  - Change in viscous resistance term
  - Expect minimum channeling to occur

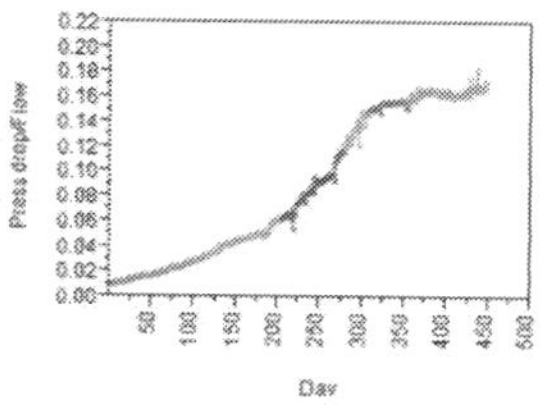

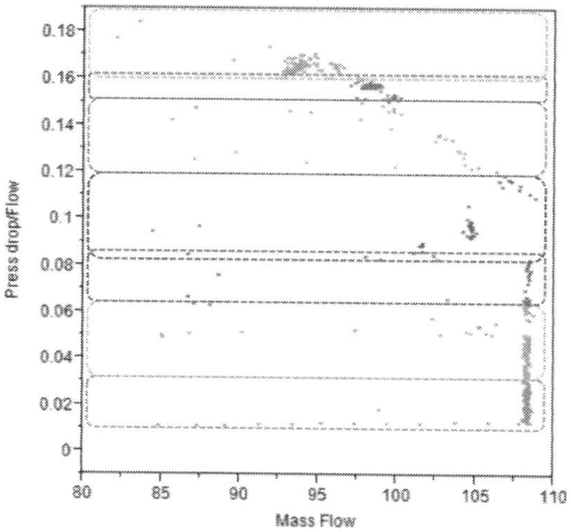

Changes within time period (color regions) are flat with changes in flow

6

Reference Pressure Drop

- Using Ergun equation with assumptions can adjust pressure drop to reference flow rate
  - Make assumptions on Inertial Resistance

$$\Delta P_{ref} = \left(\frac{\Delta P}{F} - R_I^*\right) F_{ref} + R_I^* F_{ref}^2 \qquad \text{where} \qquad R_I^* = \frac{R_I}{2\rho A^2} L$$

  - Even small variation in R_I^* can cause big changes in ΔP
  - When data has too much noise, can be as good to adjust with only by flow rate

$$\Delta P_{ref} = \left(\frac{\Delta P}{F}\right) F_{ref}$$

  - This will provides some insights to the reactor

7

Degradation at Different Level

- Fixed Bed Catalytic Reactor
  - Flow is from the top
  - Pressure reading spaced evenly apart
  - Pressure reading were taken daily at different level of Reactor
- Ergun Analysis
  - Analysis showed a linear flow correction was appropriate for this data
  - Flow adjusted pressure drop provided a clearer view of where the catalyst degradation was occurring

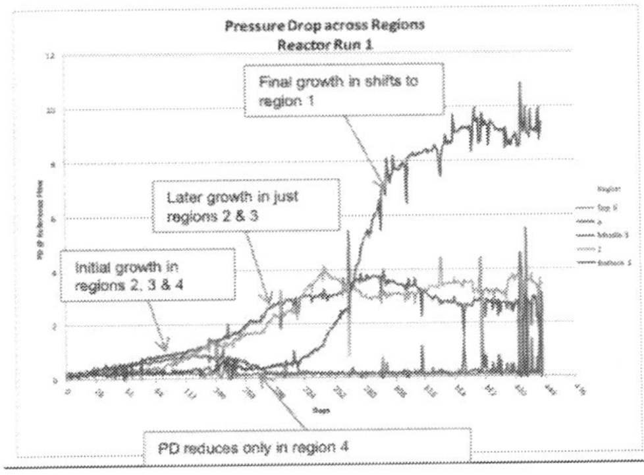

- For reactor run 1
  - Initial PD growth included the upper layer
  - Upper layer PD actual reduce in later run
    - Possibly fines shifted lower in reactor
  - Final PD growth only in lower region

8

Different Degradation from Different Run

- Adjusted PD by Flow Rate
- Run 2 had slower PD growth
  - Final pressure drop is less than run 1
  - Had big growth of PD after 2 years

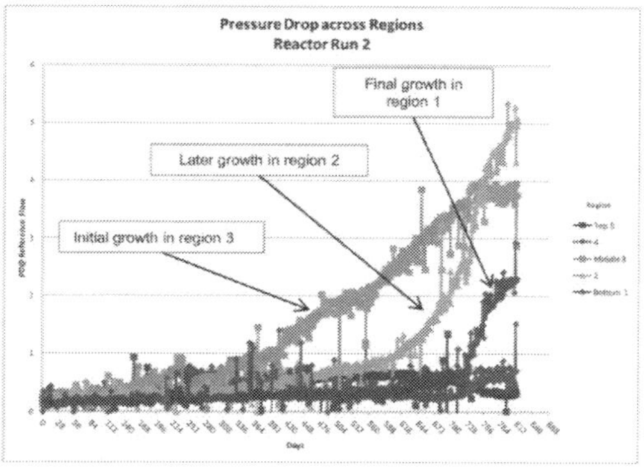

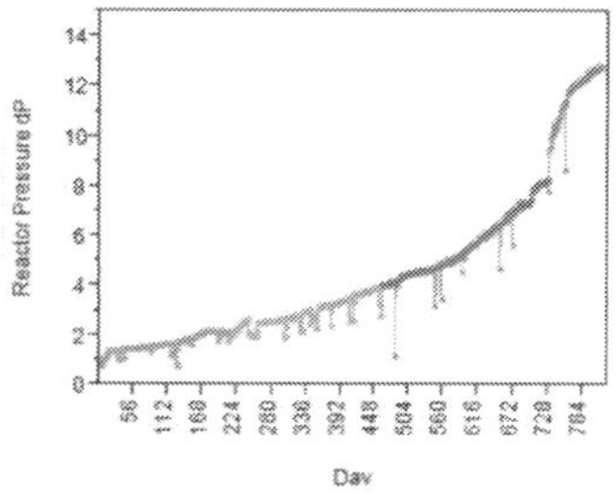

- For reactor run 2
  - Initial PD growth in middle region
    - Most active reaction region
    - Growth initial seen after a shutdown
  - No growth in upper level
  - Growth shifts to lower levels with time
  - Rate of PD growth increases with each growth shift
  - Final pressure drop contribution from 3 lower level

9

CFD modeling of Reactor with Fines

- Used GeoDict to Model Reactor Bed
  - Filled void with uniform fines loading
    - Catalyst void 38%
    - Loading measured as percent of volume
    - Fines size distribution matched spent catalyst sampling

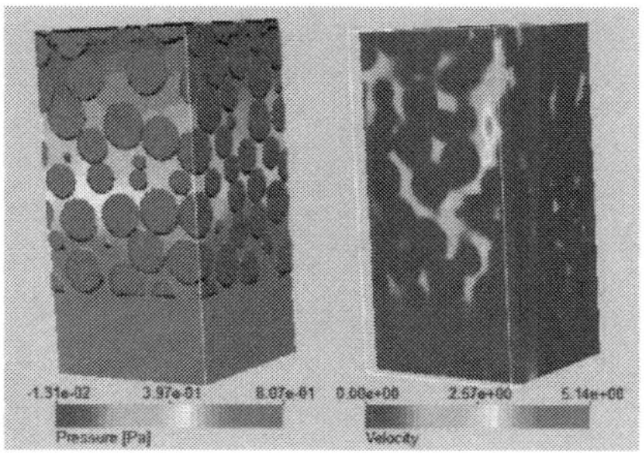

Velocity and Press Drop Calculated

10

Impact of Fines on Catalyst Bed

- Pressure drop growth as function of fines added
  - Doubling of pressure drop with 1% fines added
    - Would be even worse because of non-uniform distribution

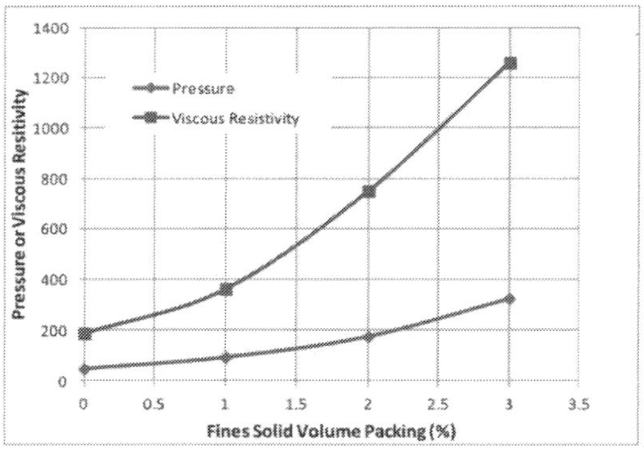

11

Conclusions

- Process data have many challenges for degradation analysis
- Pressure adjusted for flow rates can reveal more information about the changes
- The viscous resistance term of the Extended Ergun Equation can provide a good view of degradation
- Multiple pressure readings in the reactor can indicate were degradation is occurring
- Modeling can provide insights into the impact of degradation on pressure drop

12

Thank You

® Trademark of The Dow Chemical Company

13

Maximizing Capabilities in Flexible Operation with a New Hydrocracking Catalyst

Kulach, J.,
Wroblewski, R., Honeywell-UOP
Bricker, M. L., UOP - A Honeywell Company
Kumar, N., Honeywell-UOP
Roney, S., Honeywell-UOP
Originally presented on: 10/3/2013 11:00:00 - 11:30:00

A new maximum saturation hydrocracking catalyst, HC-470LT, has been developed to fill a gap in Honeywell-UOP's Unicracking™ catalyst portfolio in the flexible region between the naphtha and jet/distillate catalysts in order to better meet customer needs. This catalyst was tailored for customers who are (1) looking to increase distillate production from existing naphtha hydrocracking units, but also would like to retain the option to revert back to max naphtha operation and (2) converting distillate hydrotreaters to diesel upgrading seeking increased product volume swell.

During the initial phase of the project, four different zeolites were screened and evaluated against two commercial reference catalysts. The best-performing formulation was then chosen for scale-up activities. Scaled-up prototype catalyst performance was evaluated in a larger pilot plant and activity/yield curves were generated for the new catalyst along with the two references. Product samples were fractionated off-line and each cut was analyzed in order to compare product quality.

This talk will discuss the project test methodology, from the initial catalyst screening phase through the scale-up phase. Select catalyst performance results and product quality comparisons will be shown for the new flexible hydrocracking catalyst.

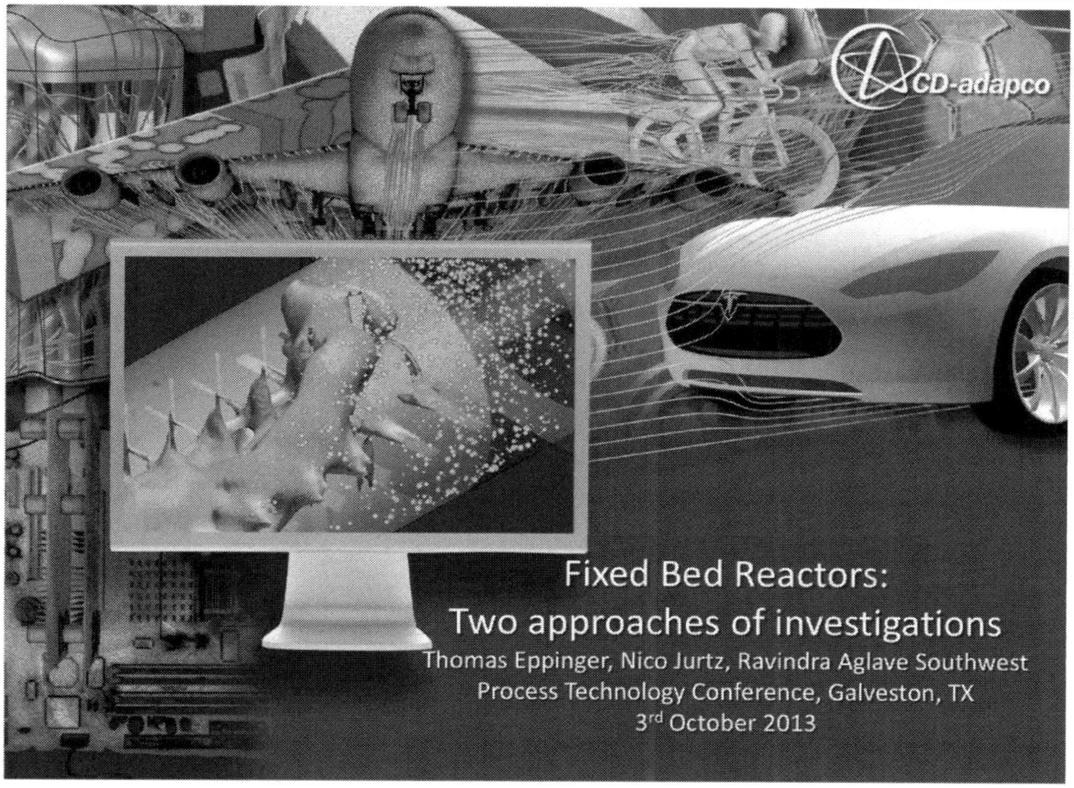

Fixed Bed Reactors: Two approaches of investigations

Thomas Eppinger, Nico Jurtz, Ravindra Aglave Southwest
Process Technology Conference, Galveston, TX
3rd October 2013

Outline

CD-adapco

- Introduction
- Motivation
- Approach 1: Porous Media (Black Box)
- Approach 2: Fully resolved
- Current Status
- Future: Open door for collaborative work

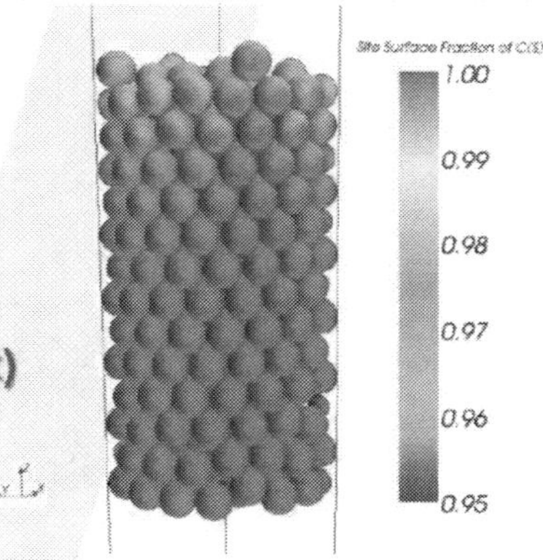

Deactivation of the catalyst:
Almost all active sites on the surface are covered with carbon

Introduction

CD-adapco

- **Packed Bed Reactors**
  - Higher conversion per weight of catalyst
  - Reaction rate is based on the amount of the solid catalyst
- **Design is challenging**
  - Reaction kinetics: measured, 1D models
  - Heat transfer: Hot spots can appear:
    - Mostly estimated using correlations
  - Ageing: Hot spots move downward
  - Flow & Mass transfer: Recirculation, Stagnation
    - Experiments, Dispersion correlations

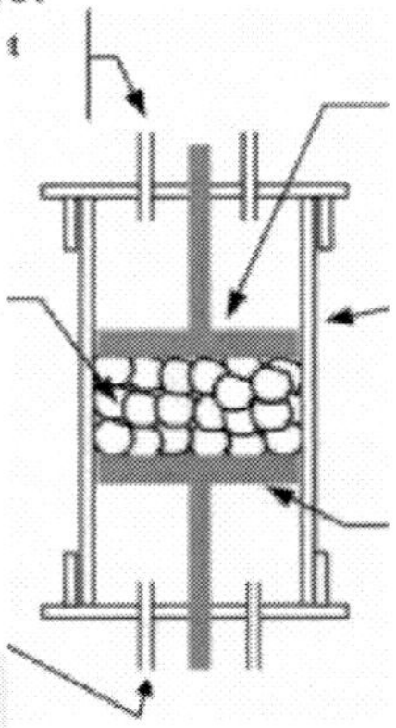

Motivation

CD-adapco

- **Can simulation supplement the information derived during experiments?**
- **Would resolving the geometry in 3D add any value?**
  - Prof. Dixon: WPI
  - Dr. Marshal Richmond, PNNL
- **Can it be an engineering tool or a university level research tool?**
  - Hurdles and solutions
- **Is a quicker process of screening catalysts shapes/size to evaluate the variables involved possible?**
  - Automation and speed
- **Can industry facilitate a more collaborative experimental-computation approach?**
  - Accuracy, Value

Example: Oxidative Coupling of Methane

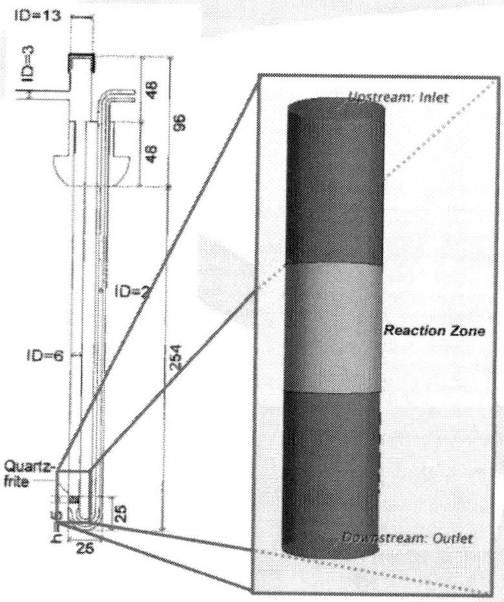

- **Promising route for the use of methane**

$$3CH_4 + 3O_2 \xrightarrow{\text{Cat}} C_2H_4 + 4H_2O + CO_2$$

- **Green feedstock for CPI**
  (Ethylene demand in W.-Europe ~22M t/a)

Temperature [°C]	700 – 955
p_{CH4} [kPa]	10 – 95
p_{O2} [kPa]	1 – 20
p_{CO2} [kPa]	1 – 8
$\dot{V}$ [ml / s]	4 – 13
$M_{cat} / \dot{V}$ [kg s / m³]	0.76 – 250

Z. Stansch, L. Mleczko, and M. Baerns. Comprehensive kinetics of oxidative coupling of methane over the Al2O3=CaO catalyst. Ind. Eng. Chem. Res., 36:2568 – 2579, 1997.

Reaction mechanism derived from experiments

Stansch et. al. developed a comprehensive reaction mechanism which is widely used for simulating OCM.

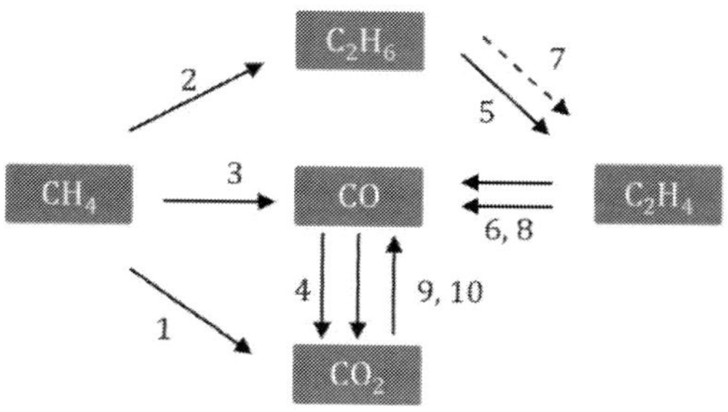

step 1: $CH_4 + 2O_2 \rightarrow CO_2 + 2H_2O$

step 2: $2CH_4 + 0.5O_2 \rightarrow C_2H_6 + H_2O$

step 3: $CH_4 + O_2 \rightarrow CO + H_2O + H_2$

step 4: $CO + 0.5O_2 \rightarrow CO_2$

step 5: $C_2H_6 + 0.5O_2 \rightarrow C_2H_4 + H_2O$

step 6: $C_2H_4 + 2O_2 \rightarrow 2CO + 2H_2O$

step 7: $C_2H_6 \rightarrow C_2H_4 + H_2$

step 8: $C_2H_4 + 2H_2O \rightarrow 2CO + 4H_2$

step 9: $CO + H_2O \rightarrow CO_2 + H_2$

step 10: $CO_2 + H_2 \rightarrow CO + H_2O$

Step 1, 3-6: Hougen-Watson rate equation
Step 2: Langmuir-Hinshelwood rate equation
Step 7-10: Arrhenius-type rate equation

Implementation in STAR-CCM+

CD-adapco

Chemical reaction modeling in STAR-CCM+:

- **Built-in functionality (e.g. EBU break-up, kinetics only)**
  - Arrhenius-type rate equation
  - Only Gas-phase reactions
- **Using DARS (detailed reaction mechanism)**
  - More species can be handled
  - More rate equation types are available
  - Gas- and surface reactions
- **User Coding / Field Functions**
  - Reactions are implemented by source terms
  - Reaction rates per mass catalyst – as used by Stansch et. al. – can be used.

Verification I

CD-adapco

- **Assumption:**
  - Fixed temperature profile similar to experiment
  - Flat flow profile
- Simulation Results
  - X = Conversion
  - S = Selectivity
  - Y = Yield

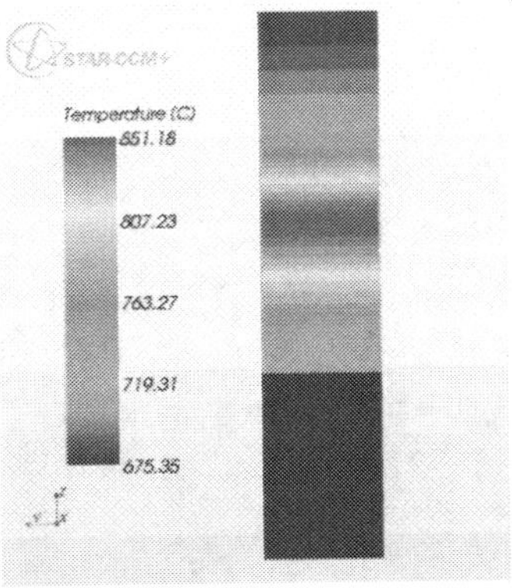

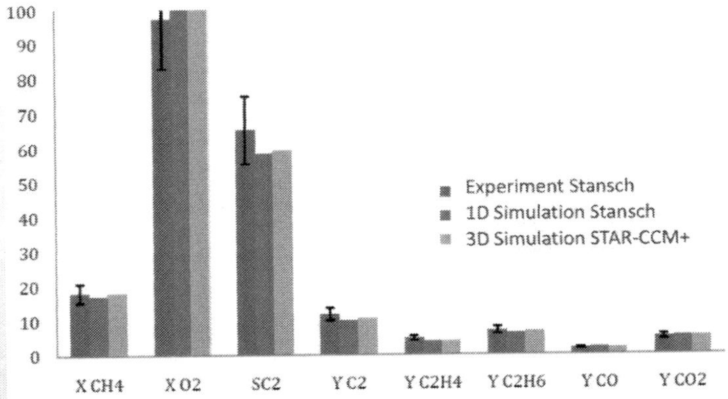

Validation II: Fully resolved flow profile

CD-adapco

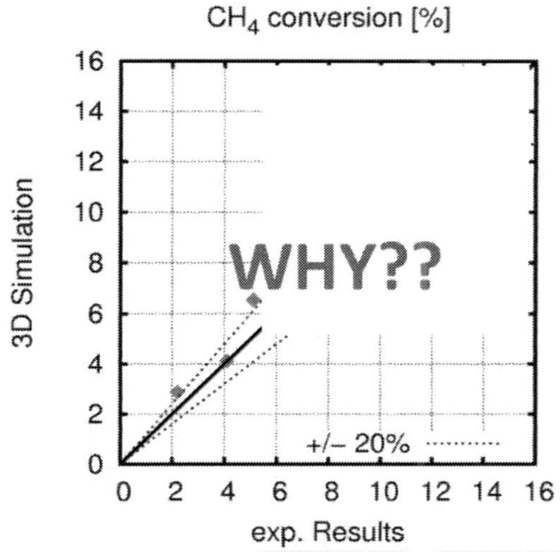

With increasing residence time (=higher temperature) the deviation between simulation and experiment increases.

Discrepancy

CD-adapco

- Stansch considers only an axial temperature profile
- A radial temperature profiles is neglected

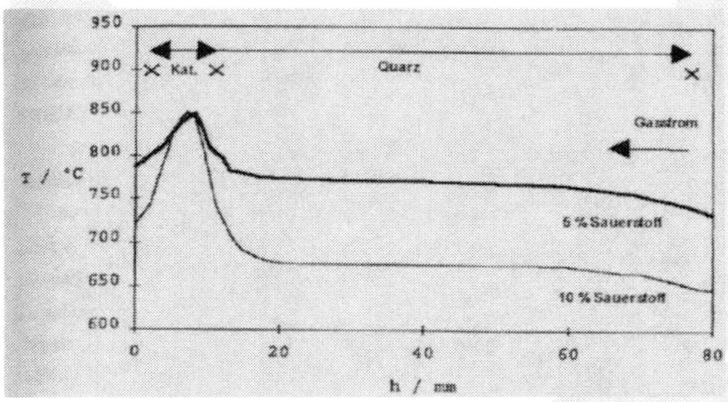

Measured axial temperature profile

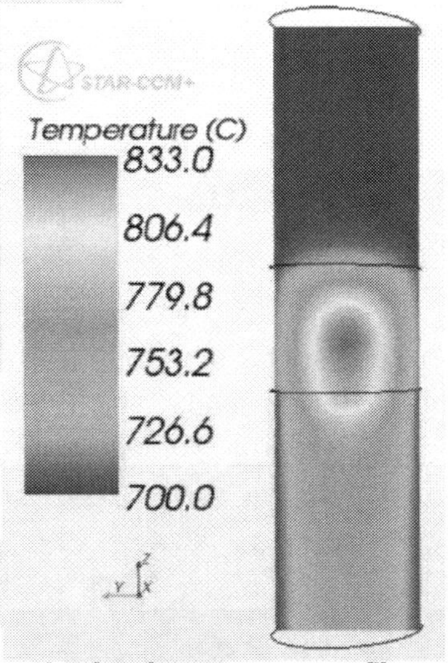

Simulated temperature profile

Part A: Conclusion

- ✓CFD uncovers a loose assumption in radial profile
- ✓When deriving kinetics, neglecting 3D effects can lead to erroneous kinetic parameters
- ✓CFD Simulation results can be correlated with kinetics to make a correction
- ✓Reliability of the reaction kinetics is a crucial point

Resolving the packed beds in 3D

- **Small d/D ratios**
  - Local wall effects are significant
  - Large variations from correlations
  - Pseudo-homogenous assumption ≠ Physical reality
- **Accurate Heat transfer description**
- **Use of CFD since mid 90's**
  - Mostly in Universities (Prof. Dixon, WPI)
- **Challenges:**
  - Generating representative random packing
  - Meshing (contact resolution)
- **A quicker simulation process can increase its value in combination with experimental work**

Scheme

DEM
(discrete element method)
Generation of a randomly packed bed

DARS
Chemistry solver.
For stiff systems

Spatially Resolved Simulation of Fixed Bed Reactors

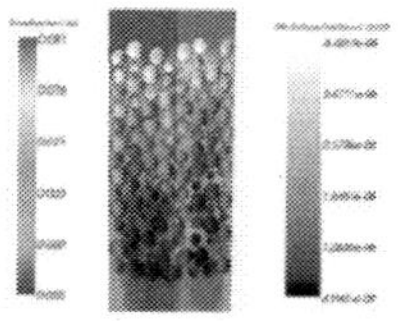

CFD
Calculation of fluid flow, temperature, species distribution
...

Optimate / Optimate+
Design of Experiment
Optimisation / Surface Resonse Methodology

Packing / Bed Generation Methods

- **Structured Packings**
- **Monte-Carlo-Method**
- **Experimental Data (CT/MRT)**
- **DEM**
  - Fast
  - Physics-based
  - Cost-efficient

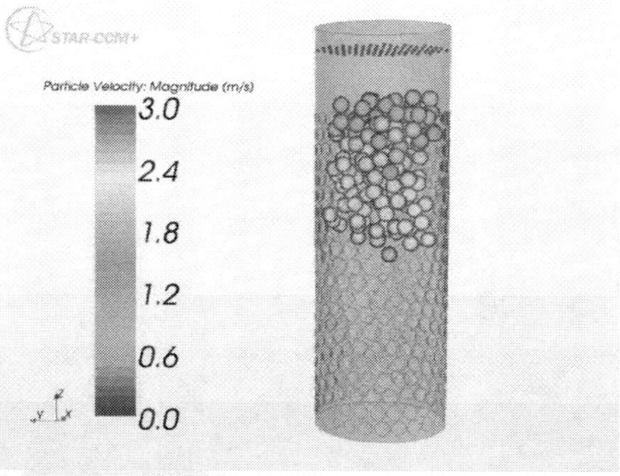

Contact Flattening

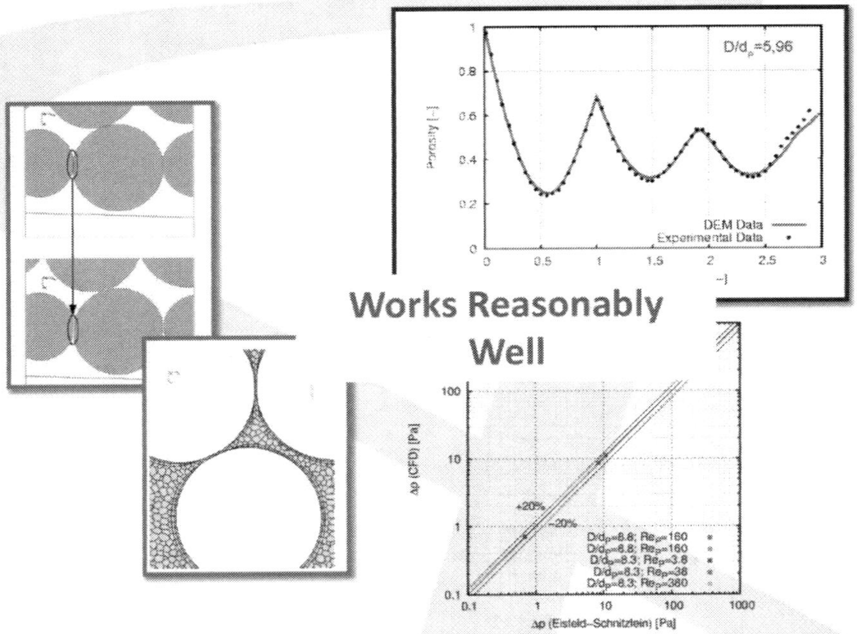

Eisfeld,B.; Schnitzlein, K., (2001), Chem. Eng. Sci. 56, 4321-4329

Validation of Convection & Heat Transfer

- D/d_P=5.45; Re=6804; T_w=100°C

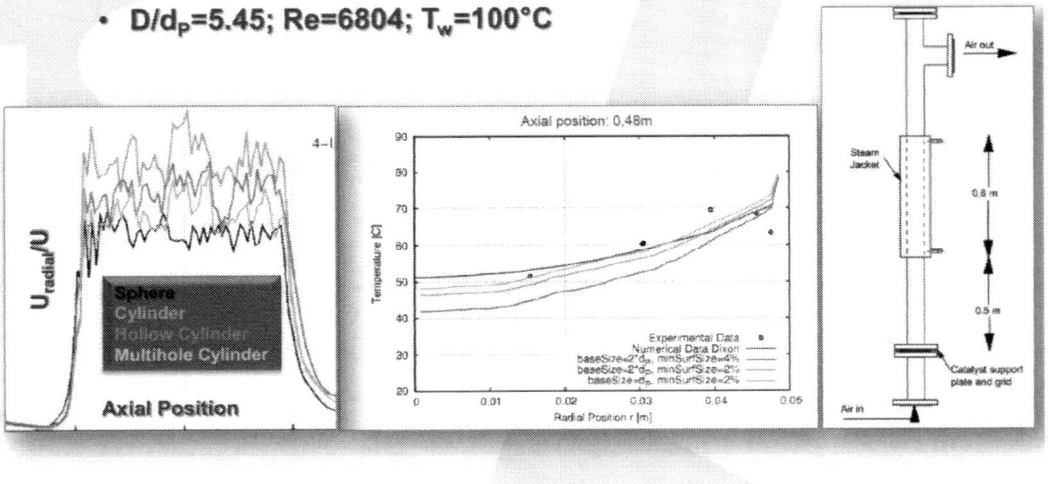

Dixon, A. G. (2012). *Experimental validation of high Reynolds number CFD simulations of heat transfer in a pilot-scale fixed bed tube*. Chemical Engineering Journal. 200-202:344-356

Stagnation and Recirculation Zones

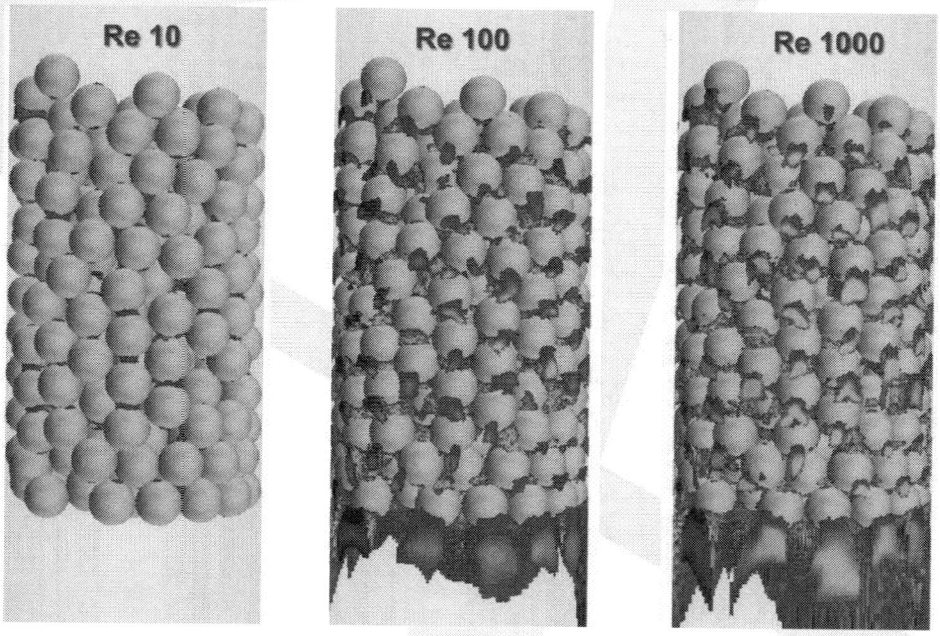

Axial Dispersion

- **Method of calculation:**
  - P=particle tracks;
  - T = tracer;
- **Tube-to particle diameter ratio:**
  - 3,4,5,6,8
- **length of the bed (l/d):**
  - 5,7,10,50

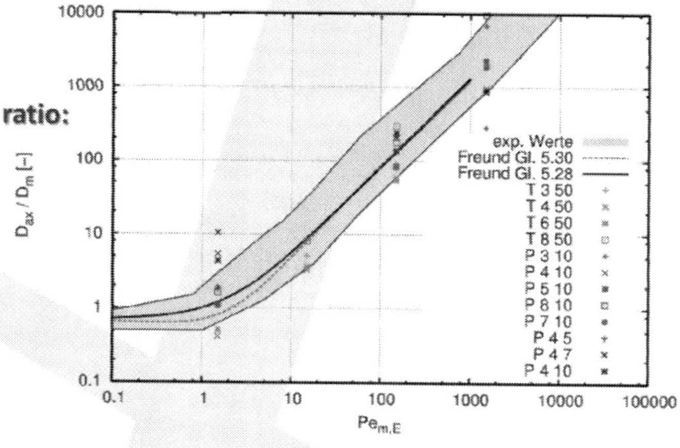

Current Status

- **Resolved CFD simulation giving reasonable results for important variables for spherical packing**
  - Porosity variation
  - Pressure drop
  - Heat transfer
  - Recirculation and stagnation
  - Axial dispersion
- **What about non-spherical packing?**
- **What about liquid flow?**
- **What about gas-liquid flows?**
- **How long does it take?**

Non-Spherical Packings

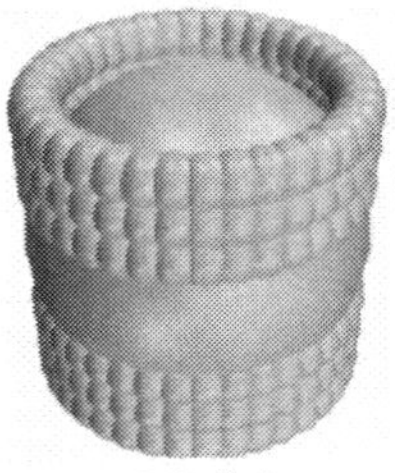

240 Spheres

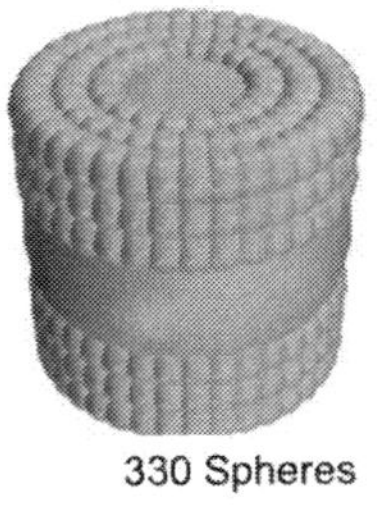

330 Spheres

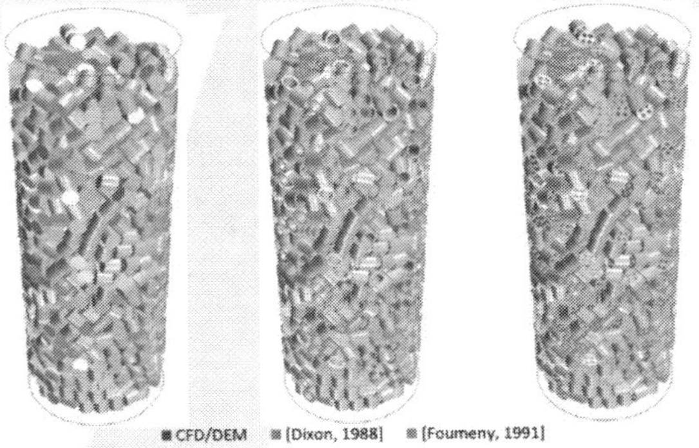

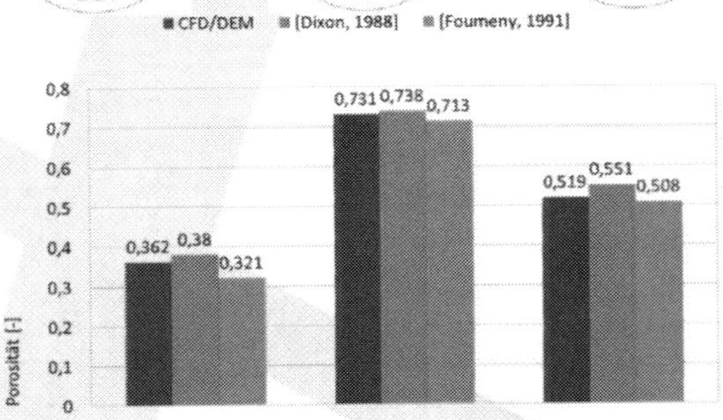

Dixon, A. (1988), Can. J. Chem. Eng. 66, 705-708
Fourneny, E.A et al. (1991), Can. J. Chem. Eng.69, 1010 -1015

Liquid Flow

CD-adapco

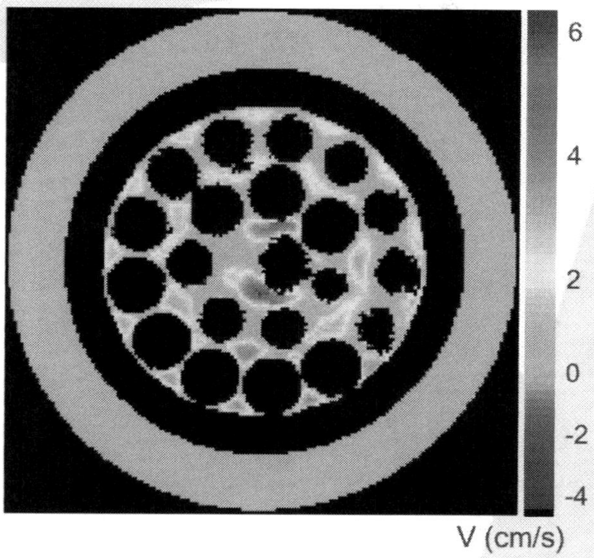

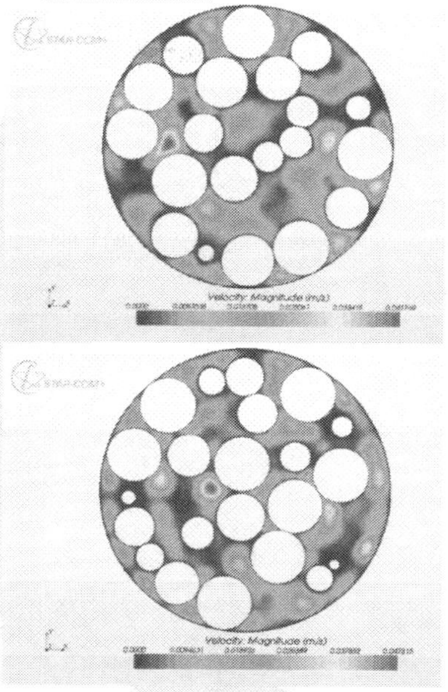

NMR Imaging of Water Flow in Packed Beds, *Wassim Salameh, Sébastien Leclerc, Didier Stemmelen, Jean-Marie Escanyé,* diffusion-fundamentals.org 14 (2010) 5, pp 1-5

Setup & Compute Times

CD-adapco

- Geometry creation for few hundred particles: 0.5 hrs
- Mesh generation: 1-3 hr
- Simulation of liquid flow: Overnight on a12 core workstation
- Are we getting to a point where a process engineer could run a few simulation at rapid pace?
  - We are definitely in the "engineering solution" phase rather than a "university research" phase

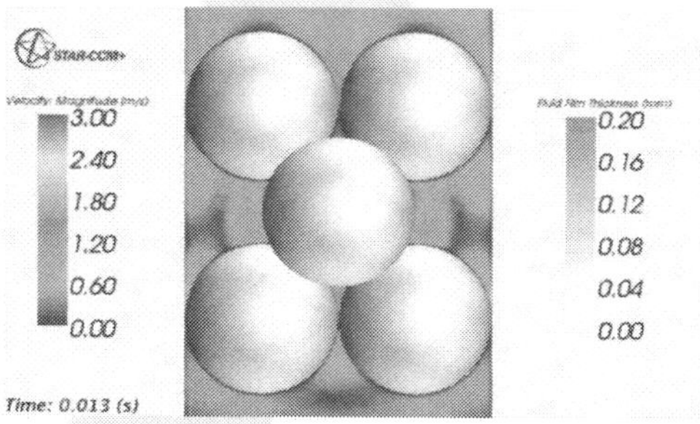

Future Direction

CD-adapco

- Further work with multiphase flows
- Film formation
- Dispersed gas-liquid
- Industry co-operation /collaboration welcome

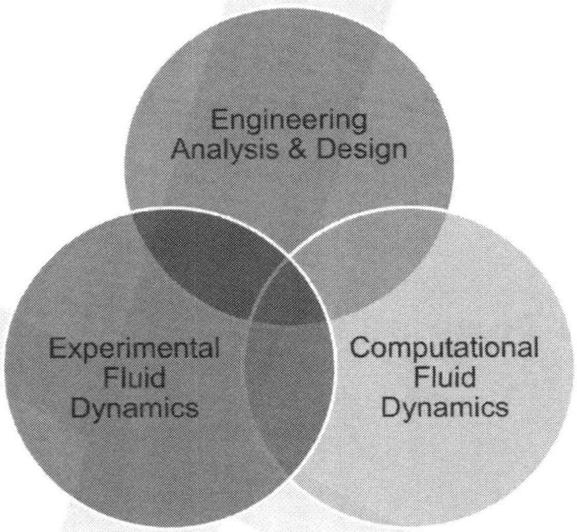

Dynamic Data Reconciliation – What You Can Know And When You Can Know It

Beyleveld, M., Virtual Materials Group USA, Inc.

Originally presented on: 10/3/2013 9:30:00 - 10:00:00

Dynamic data reconciliation allows for improved performance compared to steady state reconciliation because the dynamic model has accumulation terms built into into the model. Dynamic reconciliation models also allow for rejection of high frequency noise in an intuitive way. There are limitations on observing unmeasured variables. Two principal causes are correlation between unmeasured variables, and process lag between the unmeasured variables and the measured variables being used to infer values. The paper explores these issues, and to what extent one can mitigate them using variable transformations.

amoeba

Optimization of Reactor Flows using Discrete Tangent Based Analysis

Vibha Srinivasan
Prabhu Sathyamurthy
Sanjay Mathur

Amoeba Technologies, Inc.
Austin, TX

AIChE Southwest Process Technology Conference, 2013
Galveston, TX

amoeba

Reacting Flows

- Reactors are governed by many operating parameters: reactor geometry, flow rates, stream compositions, reaction mechanisms, pressures, and temperatures.
- CFD methods provide detailed modeling of reactor geometry and flow conditions.
- Ultimate interest is improved reactor design to achieve a target performance (example, a target mixture composition or temperature).

Sensitivity Analysis

- What inputs (operating conditions, geometry, material properties etc.) will yield desired behavior of the system?
  - How does a particular input influence all outputs? (forward mode or tangent problem)
  - How do all inputs influence a particular output? (reverse or adjoint problem)
- What are the effects of uncertainties and tolerances?
- How can a single point simulation/solution be improved?
  - Where should mesh/discretization accuracy be refined if we are interested in specific output quantities?
  - What are the effects of modeling parameters and discretization? (error estimation)
  - How can iterative solution speed be improved?
- Current CFD tools require a large number of single-point simulations to map the influence of the entire parameter space on the reactor performance.

Amoeba Solver: Discrete Tangent Methodology

- Derivatives of solution variables are key to obtain sensitivity information
- Specify sensitivity parameter
- operating condition, e.g., inlet stream temperature
- solver setting, e.g., underrelaxation factor
- material property, e.g., heat of formation
- Solver computes gradients of all solution fields with respect to the sensitivity parameter
  - Algorithmic differentiation to compute gradients

Sensitivity Analysis

Richer Data with Every Simulation
Fewer simulations to characterize system behavior

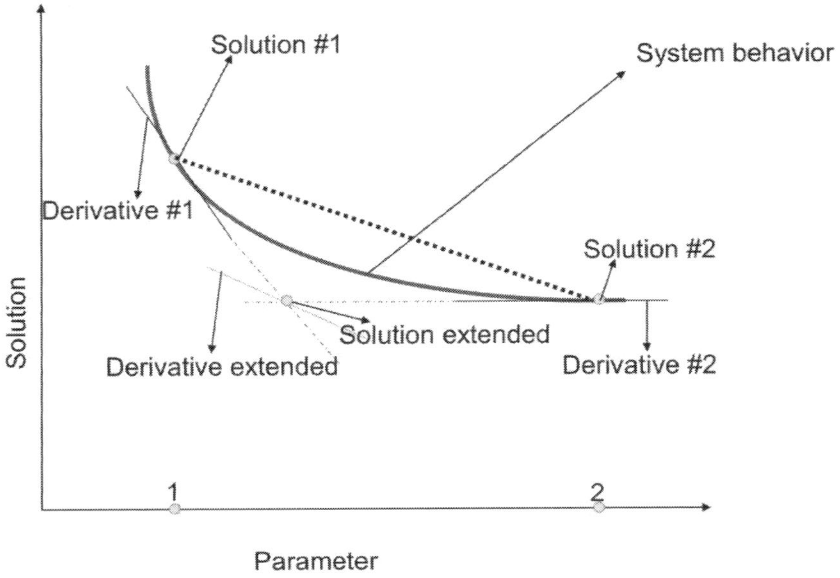

Reaction Rate Models: Eddy Dissipation Model

- Fast Chemistry Assumption

- Reaction rate limited by turbulent mixing rate, which is related to the *eddy lifetime*, $\tau=\kappa/\varepsilon$, where κ =turbulent kinetic energy and ε=turbulent dissipation rate.
- *Kinetic rates not taken into account*

- Suitable for simple single or two-step mechanisms.

- Net rate is minimum of reactant mass fraction-based rate and product mass fraction-based rates.

Case: Methane Combustion in PFR

Geometry: 1 m X .1 m X .1 m
Fuel and Air inlet streams separate

Single Reaction: $CH_4 + 2O_2 = CO_2 + 2H_2O$
Reaction Rate Model: Eddy Dissipation Model

Parameters Studied for Sensitivity:

- **CH4 Inlet Stream Temperature (ch4Temp)**
- **CH4 Inlet Concentration (ch4In)**
- **Air Inlet Velocity (inVel)**

Methane combustion in PFR: Computed Composition Profile

H_2O Composition in the PFR

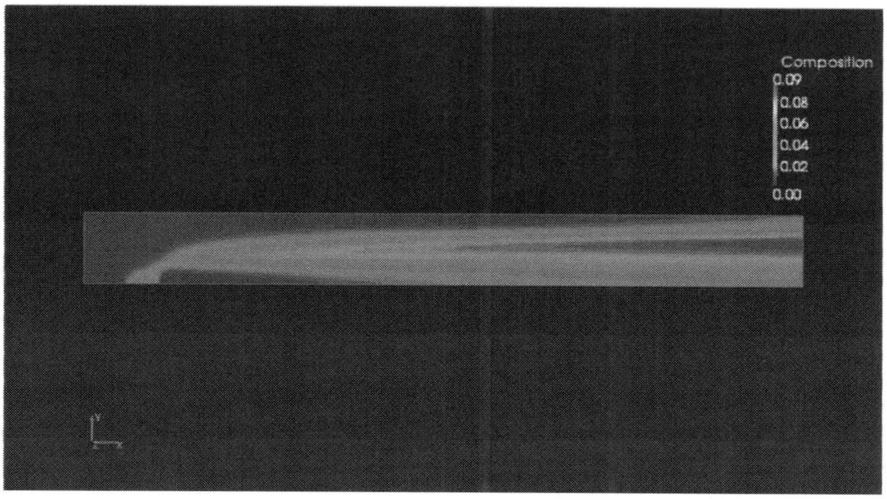

Parameters Values:
- **inVel = 1 m/s, Ch4Temp = 300 K, Ch4In = 0.8**

Methane combustion in PFR: Computed Composition Profile

CH_4 Composition in the PFR

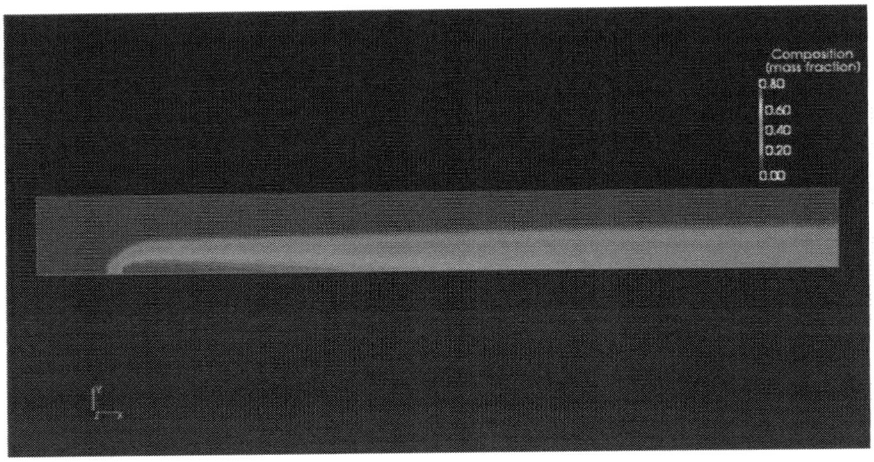

Parameters Values:

- **inVel = 1 m/s, CH_4Temp = 300 K, CH_4In = 0.8**

Methane Combustion in PFR: Computed Temperature Profile

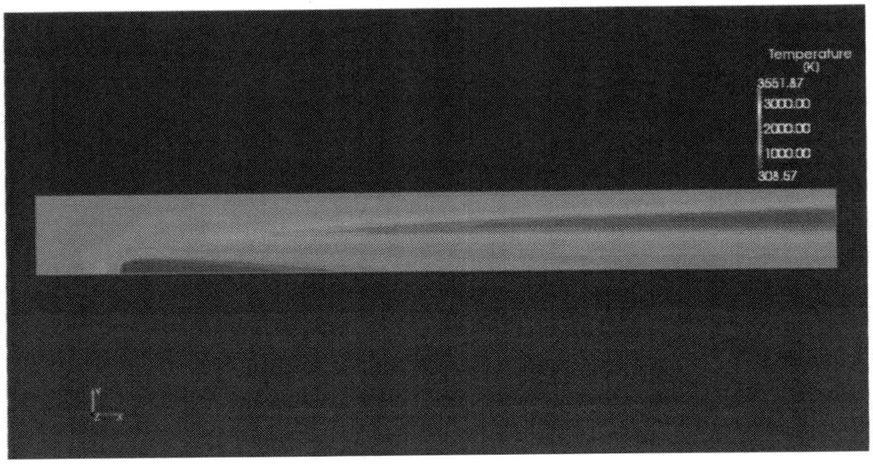

- **Parameters Values: inVel = 1 m/s, CH_4Temp = 300K, CH_4In = 0.8**
- **Computed T_{max} = 3552 K**

Methane Combustion in PFR: Computed Temperature Profile

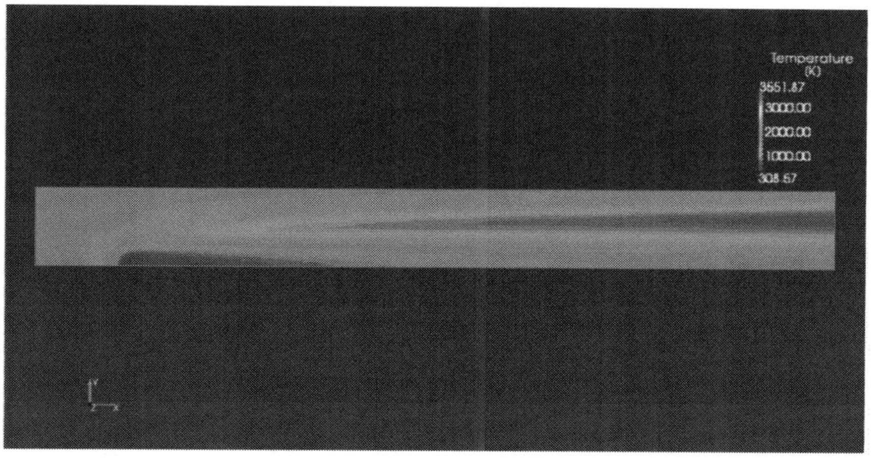

- **Parameters Values: inVel = 1 m/s, CH_4Temp = 300K, CH_4In = 0.8**
- **Computed T_{max} = 3552 K**

Sensitivity to CH4 Inlet Temperature

Extrapolated Temperature Profiles

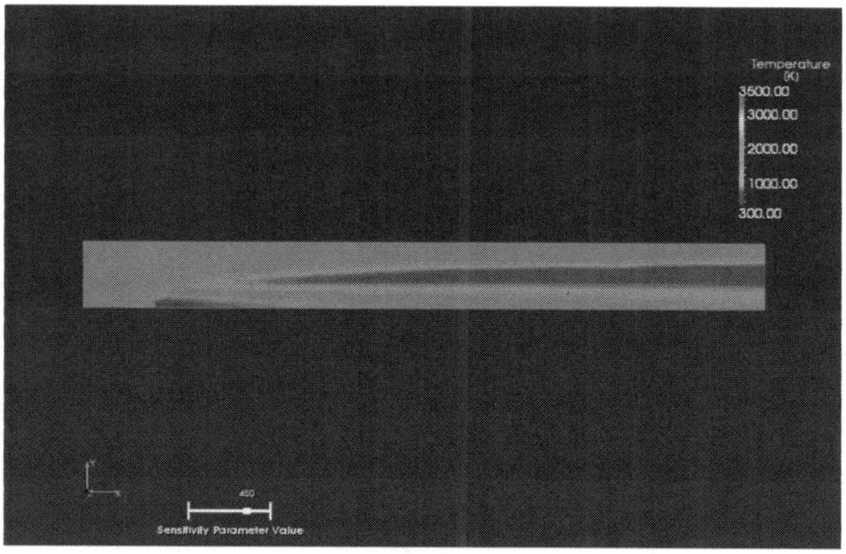

- Gradients computed at CH_4Temp = 300K
- Extrapolation: CH_4Temp varying from 300 K to 500 K

Sensitivity to CH4 Inlet Temperature

Extrapolated H2O Compositions

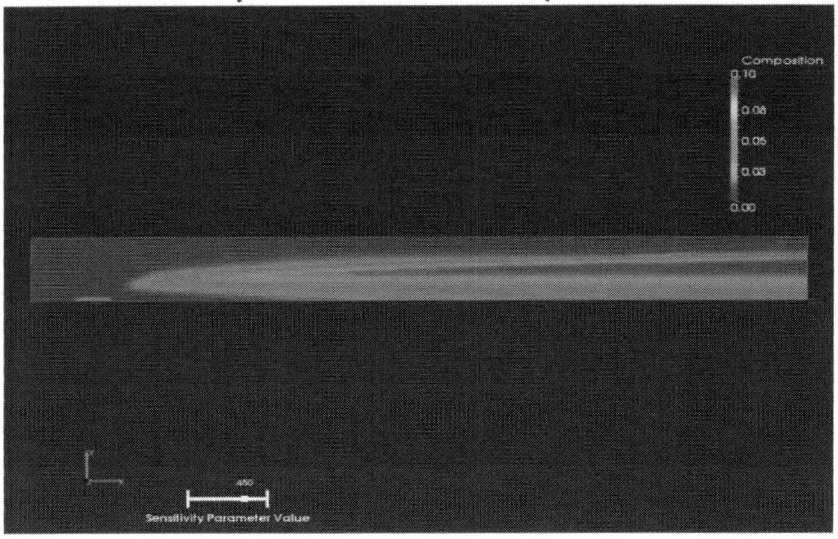

- Gradients computed at CH_4Temp = 300K
- Extrapolation: CH_4Temp varying from 300 K to 500 K

Sensitivity to CH_4 Inlet Composition

Extrapolated Temperature Profiles

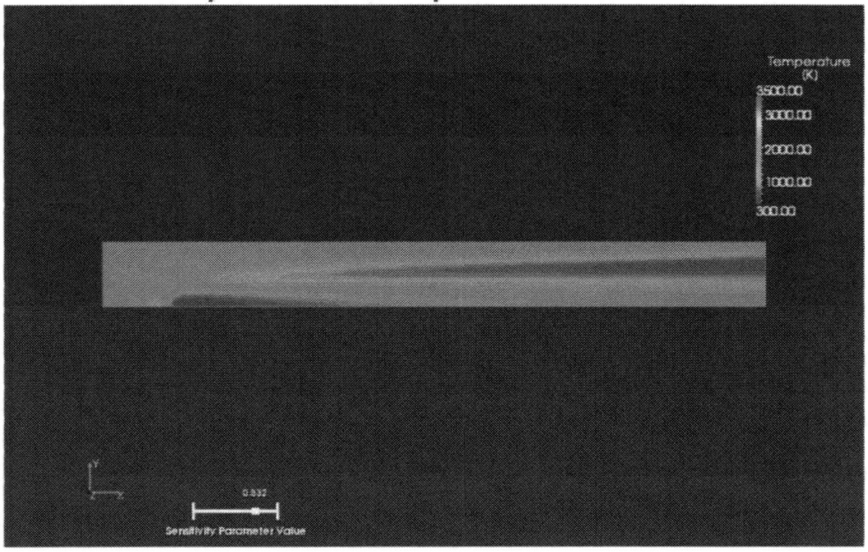

- Gradients computed at CH_4In=0.8
- Extrapolation: CH_4In varying from 0.6 to 0.9

Sensitivity to CH_4 Inlet Composition

H2O Composition Profile: Extrapolated Values

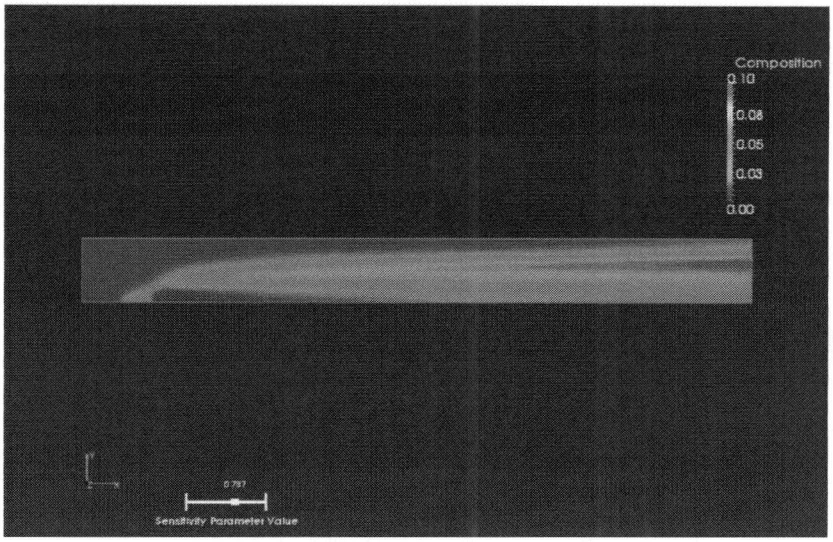

- Gradients computed at CH_4In=0.8
- Extrapolation: CH_4In varying from 0.6 to 0.9

Sensitivity to Air Inlet Velocity

Temperature Profile: Extrapolated Values

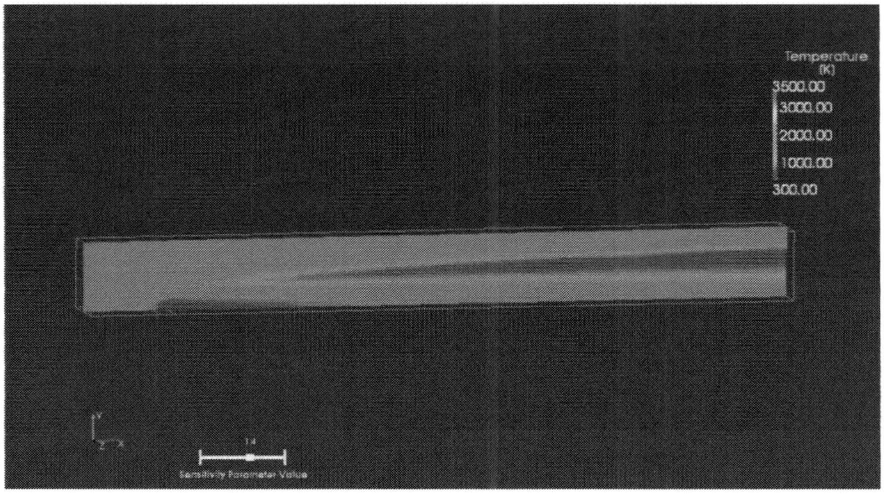

- Gradients computed at inVel=1m/s
- Extrapolation: inVel from 0.5 to 2.0 m/s

Sensitivity to Air Inlet Velocity

Extrapolated CH_4 Composition Profile

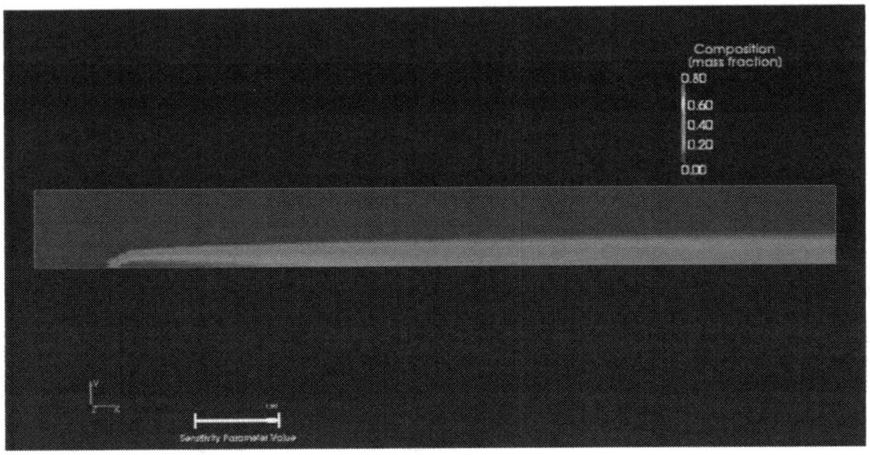

- Gradients computed at inVel=1m/s
- Extrapolation: inVel from 0.5 to 2.0 m/s

Case: Premixed CH_4+air in PFR

Geometry: 1 m X .1 m X .1 m

Sensitivity Parameters:
CH_4 Composition in the stream: ch4In
Inlet Velocity: *inVel*

Premixed Case: Sensitivity Results

Extrapolated H_2O Composition Profile

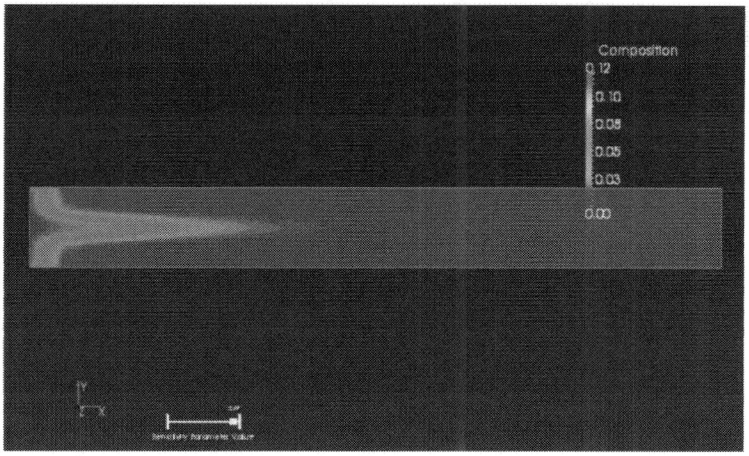

Gradients computed at CH4In=0.5
Extrapolation for CH4In from 0.3 to 0.7

Premixed Case: Sensitivity Results

Temperature Profile

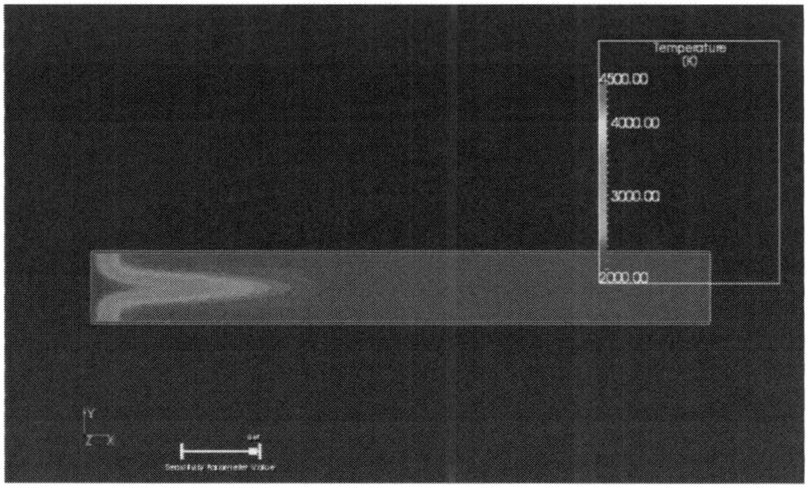

Gradients computed for CH4In=0.5
Extrapolation for CH4In from 0.3 to 0.7

Premixed Case: Sensitivity Results

H_2O Composition Profile: Extrapolated Values

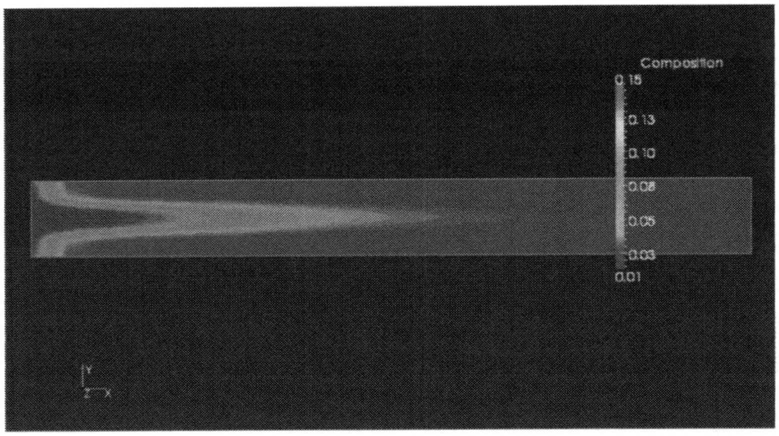

Gradients computed for *inVel*=1m/s
Extrapolation for *inVel* varying from 0.5 to 2 m/s

Premixed Case: Sensitivity Results

Temperature Profile: Extrapolated Values

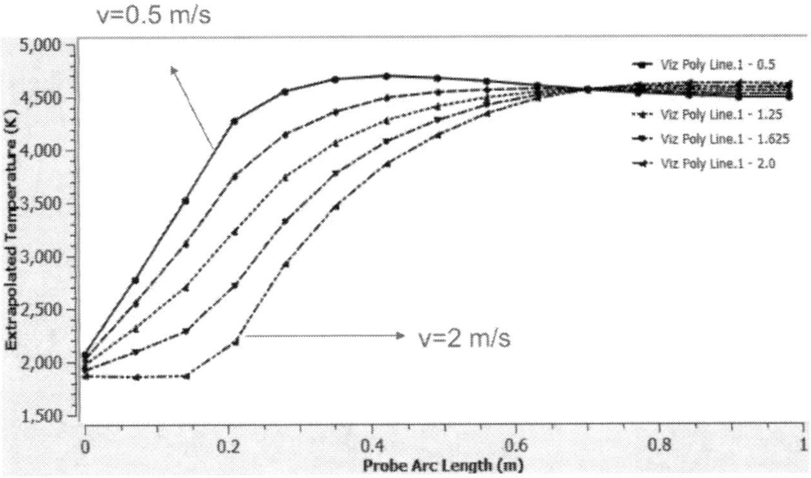

Extrapolation for *inVel* varying from 0.5 to 2 m/s

Amoeba Solver: Multi-Step Reactions

- Reaction source terms computed using a third-party tool, **Cantera**
  - Provides reaction mechanisms and integrates the ODE's for the network of reactions to get the final compositions.
- Species i source term computed as:
  - $Source_i = (X_{final,\, i} - X_{initial,i})/timescale$
- Source term calculation can be sped up using ISAT (*in situ* adaptive tabulation: Pope et al)
  - During the source term calculation, build up a table of the accessed composition-temperature space.
  - As calculation proceeds, look up this table to search for the closest match in the composition-temperature space (within a specified tolerance, ε)
  - If no match is found, add the current composition-temperature point as a new entry in the table.

Case: H_2O_2 species set in PFR

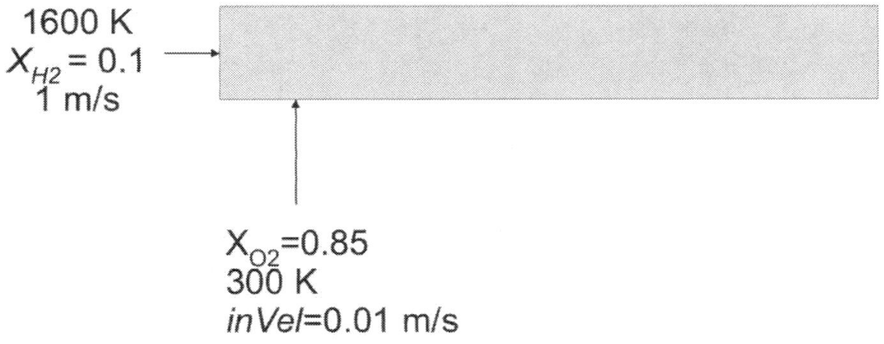

- 9 species set: H_2, O_2, H_2O, OH, H, O, HO_2, H_2O_2, Ar
- 27 reactions
- Separate fuel and air streams

Sensitivity to Fuel Inlet Velocity

Extrapolated H2O Composition Profile

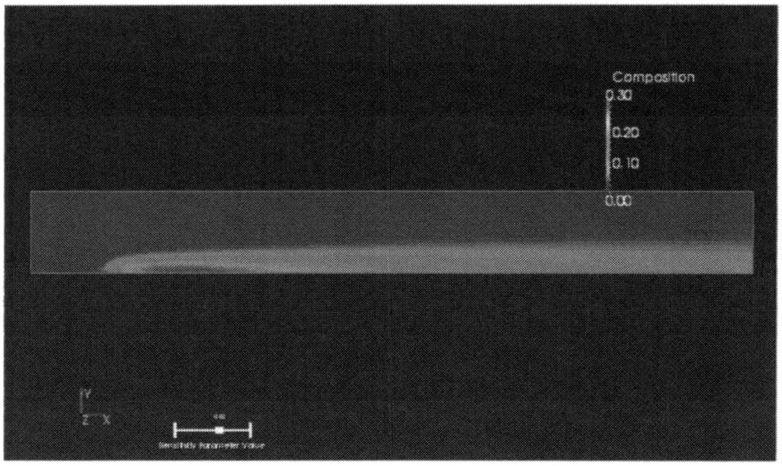

Gradients computed for *inVel*=0.01 m/s
Extrapolation for *inVel*=0.005 to 0.03 m/s

Sensitivity to Fuel Inlet Velocity

Extrapolated Temperature Profile

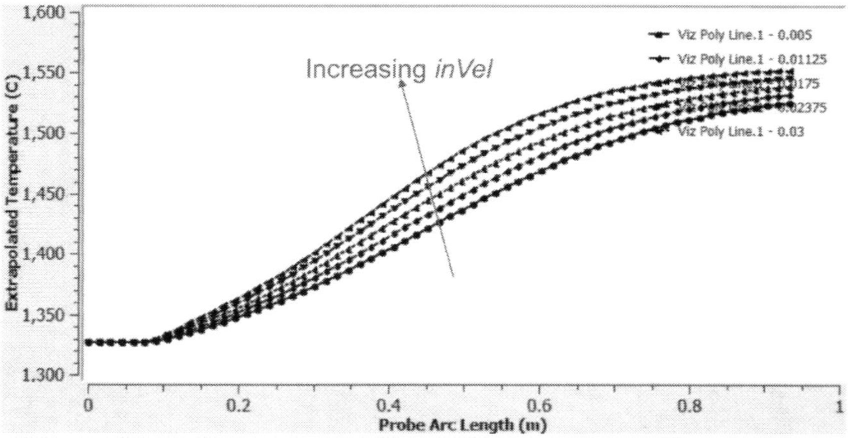

Gradients computed for *inVel*=0.01 m/s
Extrapolation for *inVel*=0.005 to 0.03 m/s

Case: H_2O_2 species set in PFR

H_2O_2 set
1600 K
$h_{2in} = 0.01$
$X_{O2} = 0.03$
inVel = 1m/s

- 9 species set: H_2, O_2, H_2O, OH, H, O, HO_2, H_2O_2, Ar
- 27 reactions

Computed Temperature Profile

H_2O_2 species set in PFR: Computed H_2O Profile

Computed: H_2O Max = 0.0337, T_{max} = 2260 K
Fluent gives: H_2O Max = 0.0336, T_{max} = 2180 K

amoeba

Temperatures Along the PFR

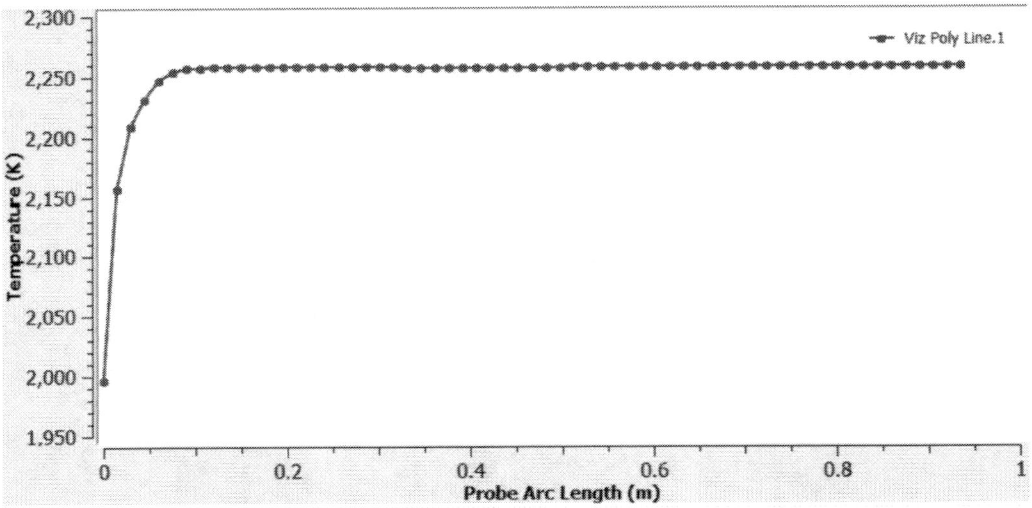

amoeba

Extrapolated H_2O Profiles

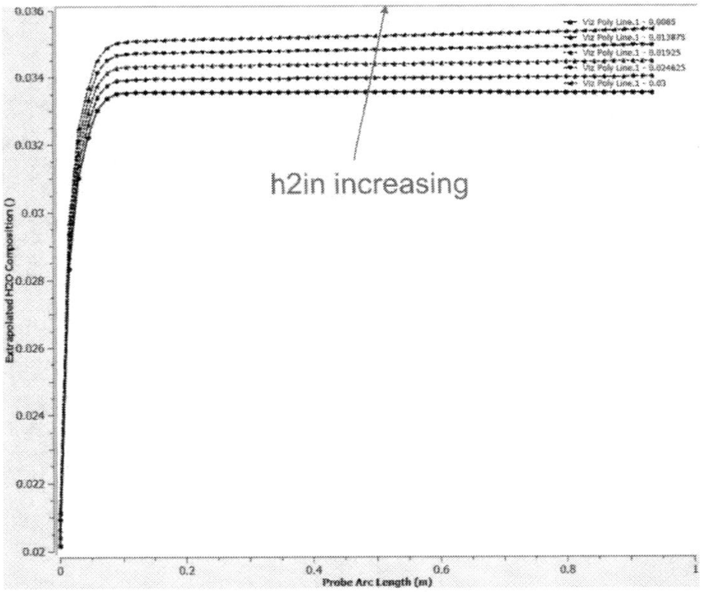

Gradients computed for h2in = .01
Extrapolation for *h2in from .0085 to .03*

Extrapolated Temperature

Temperature along the reactor for different values of *h2in from .0085 to .03*

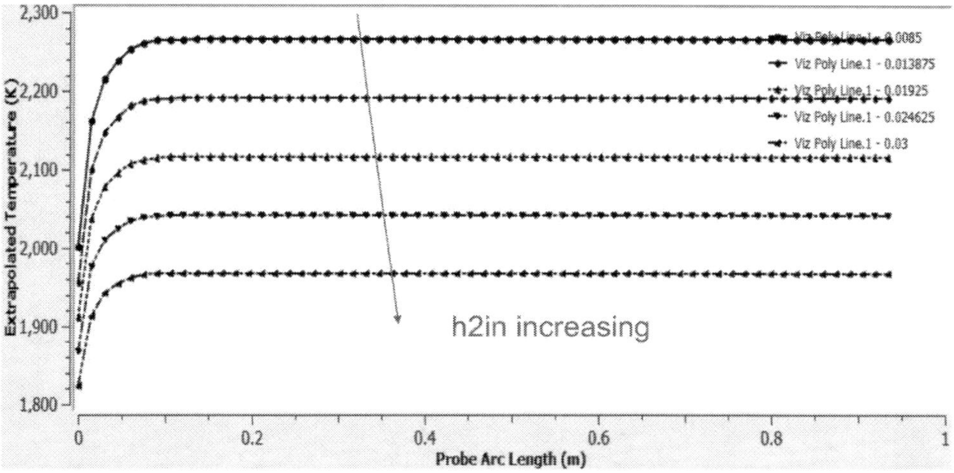

Computed Temperature Profile for h_{2in}=0.03

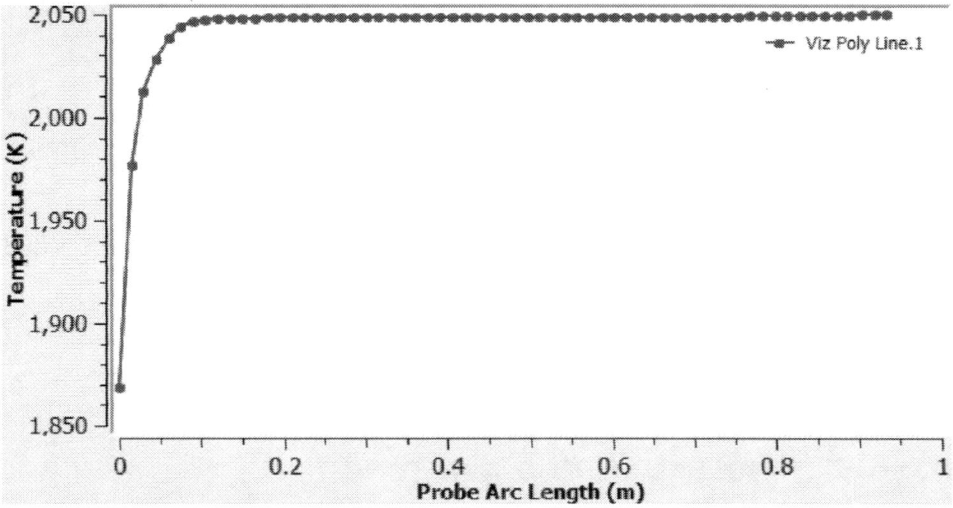

Computed Tmax = 2050 K
Extrapolated Tmax (using gradients computed at h2in=0.01)=1960 K

amoeba

Sensitivity Analysis

Richer Data with Every Simulation

Fewer simulations to characterize system behavior

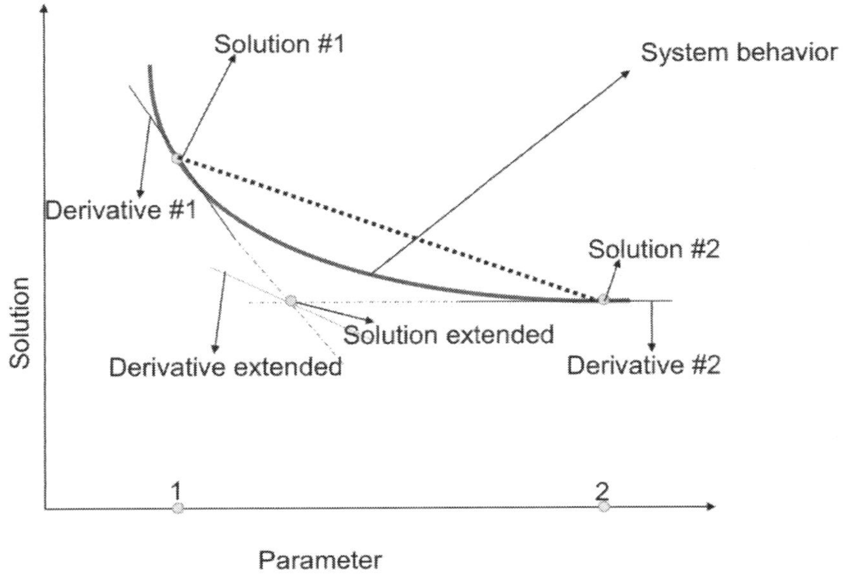

amoeba

Summary

- Performed sensitivity analysis for a PFR reactor with Eddy Dissipation Model and detailed chemistry model.

- Sensitivity Analysis using Discrete tangent approach gives a far richer information set.
  - How do species compositions and temperature vary with changing operating parameters?
  - Sensitivity analysis can be done for multiple parameters quickly by freezing the main solution while computing gradients.
  - Useful for design and optimization.

- Computational cost of computing gradients is minimal compared to the main single state simulation.

Adiabatic Mixing of Two Hydrocarbons: CFD and Thermodynamics

Dakshinamoorthy, D., Wayne State University
Wu, C.,
Shah, R.,
Originally presented on: 10/3/2013 11:00:00 - 11:30:00

Hydrocarbons are regularly mixed in gas pipelines to meet certain homogenous conditions for various purposes. Mixing two components in single-phase (gas or liquid) offers many challenges and lot of researchers have studied and reported many useful findings. However very little information is available in literature about mixing a gas phase component with a liquid phase component while keeping in mind the VLE behaviors of these two components. Depending on the state (T, P and mole-fraction of component), the gas component may condense when mixed with liquid or the liquid component may vaporize when mixed with gas. Condensation and Vaporization may happen at different sections of the pipe depending on the flow characteristics. In order to understand the entire process a simultaneous calculation of CFD and thermodynamics is required. This talk will showcase the details of this approach and also present CFD results that will highlight the value of doing such simulations.

Practical Thermodynamics

AIChE 5th PTC
Moody Gardens Oct 2013

Chemstations™ www.chemstations.com

Process Simulator Basics...missing

- There is a gap between college and professional use of a simulator
- Many engineers are not learning / forgetting important details for simulation
- Most dangerous gap:
  - not understanding thermodynamics selection

Chemstations™

Outline

- What's a thermodynamics?
- Basics of thermodynamics
- Selecting a [decent] VLE model
- Validating a model selection
- Learning more

Chemstations™

Thermodynamics in Process Simulation

A simulator calculates mass and energy balances around equipment.

- Pure component properties...
  - what's the boiling point? Viscosity? Heat capacity? Critical points?
- Phase equilibria
  - how much alcohol stays in the whiskey liquid?
- Enthalpy balance
  - how much energy to heat, to burn, or to mix it?

Chemstations™

Physical Properties

- Which heat capacity curve is correct?
- Do you even care?

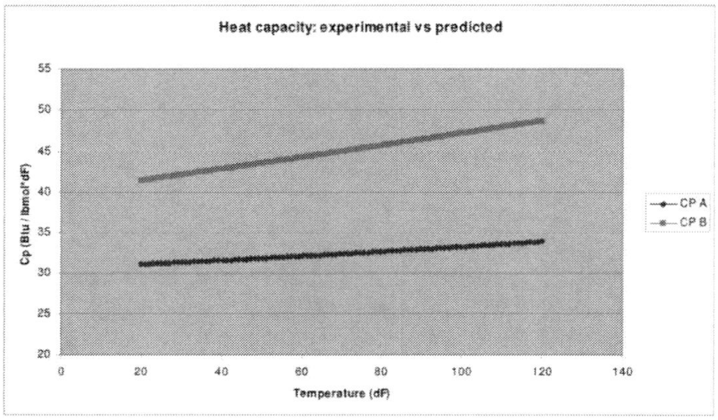

Chemstations

Phase Equilibria Models

K Values: Ki=yi/xi

- Mathematical models which describe the composition of the vapor and/or liquid phases of a mixture
- Models can be based on physical effects, thermodynamic constraints, or fit to data.
- There are many models!

Chemstations

Equilibrium curves from different models

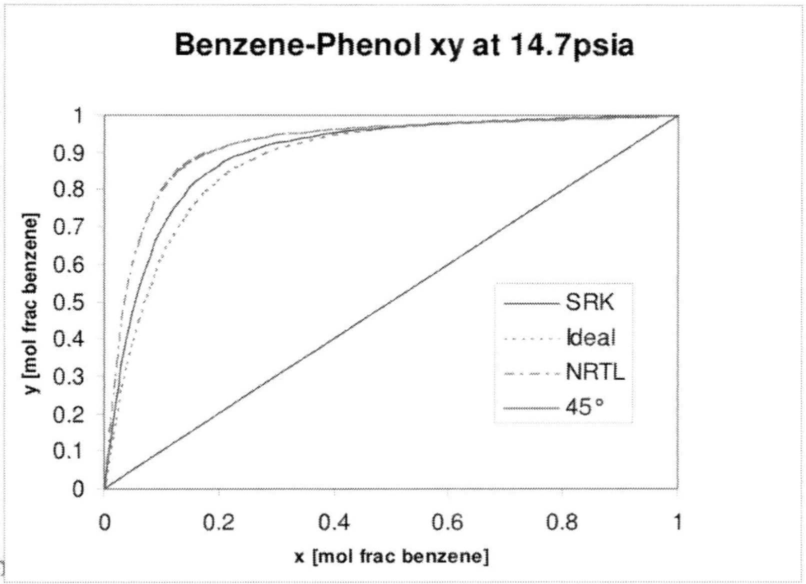

Ch

Why there are different thermodynamic models

- No [precise and accurate] universal model exists.
- Different systems have different dominating effects.
- While some systems are well understood, many are not.

Chemstations™

Which model - why do you care?

- Simulation programs include many different thermodynamics models
- Computers are poor at engineering judgment
- **Bad thermodynamic choices means your model won't fit reality**
  - **Which means you might not be able to build it**
  - **Or it might catch fire**
  - **Or it might be to big**

Chemstations

Equilibrium curve from different models

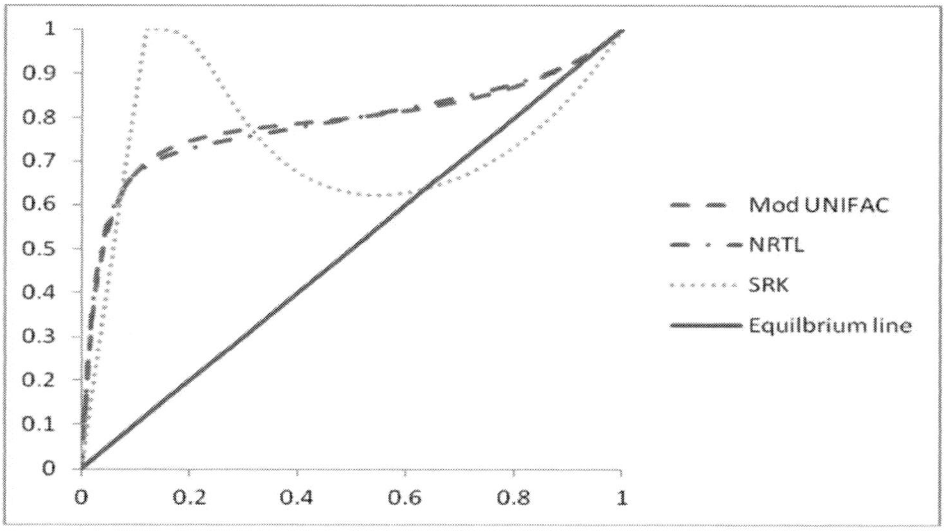

Chemstations

Outline

- Basic thermodynamics
- Selecting a [decent] VLE model
- Validating a model selection

Chemstations

Perfect (Ideal) Gas Behavior

- Ideal gases are point masses with no attractive or repulsive force
- No energy is exchanged during collisions

$$P_1V_1/T_1 = P_2V_2/T_2 \sim PV = nRT$$

Chemstations

Real Gas behavior

- Real gases have volume
- Real gases have attractive / repulsive forces between particles
- Ideal gas law must be corrected to account for the interactions and volume

$$\left(P+\frac{n^2a}{V^2}\right)(V-nb)=nRT$$

Chemstations

Liquid Ideal Solution behavior

- Particles are distributed randomly
- All interactions are (on average) the same
- Described by Raoult's law
  - $y_iP=x_ip_i^{sat}$

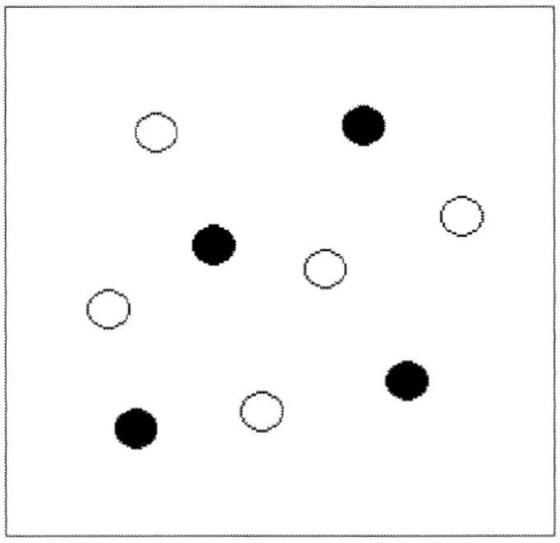

Chemstations

Regular Liquid Solution

- Particles are distributed randomly
- Interactions are not the same
  - Physical collisions are affected by size differences
  - (weak) Intermolecular forces exist
- Sizes are only moderately different

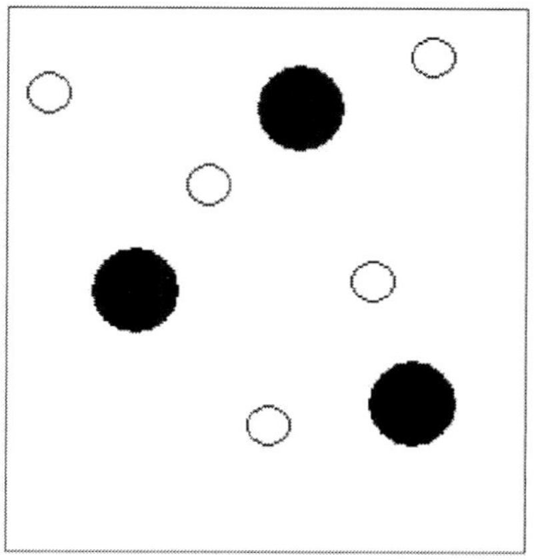

Chemstations™

Polar Non-ideal Liquid Solution

- Particles are not distributed randomly
- Intermolecular forces of attraction or repulsion dominate
- Hydrogen bonding can be significant
- Azeotropes may occur

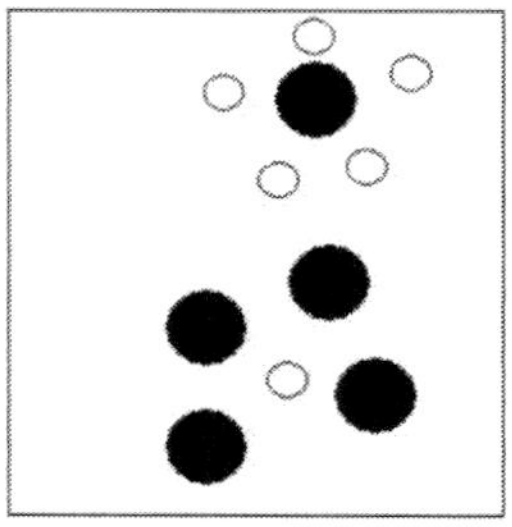

Chemstations™

Other Special Systems

- Electrolytes -- dissociation is a reaction
- Dissolved gases in a solute
- Reactive systems
- Polymers
- Solids

Chemstations™

Categories of Thermo Models

- Equations of State
- Activity coefficient methods
  - Correlated
  - Predictive
- Combined models
- Special purpose / empirical
- Advanced models (SAFT, ESD, SPEAD, COSMO, etc)

Chemstations™

Major Thermodynamic Methods

- Activity Coefficient Models
  - UNIFAC (LLE, VLE)
  - Modified UNIFAC, PSRK
  - UNIQUAC,
  - NRTL, Van Laar
  - Wilson, Margules
  - Regular Solution
- Equations of State
  - Peng-Robinson
  - Grayson Streed, Twu
  - SRK, API SRK, TSRK,
  - MSRK, PSRK, BWRS
- Other
  - Sour Water
  - TEG Dehydration
  - Amines
  - ESSO

Chemstations™

Outline

- Basic thermodynamics
- Selecting a [decent] VLE model
- Validating a model selection

Chemstations™

Equation of State: use scenarios

Equations of state are based on PVT relations

- Real gases and regular solutions are well modeled by equations of state
- Molecules should be moderately similar size
- Calculates compressibility effect; pressure effect on enthalpy
- Well suited for hydrocarbon mixtures
- These models need pure component properties: critical properties and acentric factor
- Mixing terms allow fine tuning for regular solutions

Chemstations™

Activity Coefficient Models: use scenarios

Activity models are based on thermodynamic constraints on the liquid ($\Delta G = RT\ln\gamma$)

- **Assume ideal gas** (corrections can be made)
- **Use for non-ideal (liquid) solutions**
- Molecules may have drastically different size
- Binary parameter models are fit to data
- Predictive models based on structure
- Can predict LLE
- Extended versions can handle electrolytes, polymers
- Interaction parameters are important

Chemstations™

Effects of phase equilibria model selection

- How easy is it to separate two components? Different models give you different results (TPxy plots...)
- What comes out the top?
- How much heat from the reactor?
- The wrong model will not describe the physical process
  - Qualitatively
  - Quantitively

Chemstations

Workhorse thermodynamic models

- Equations of state – SRK and Peng Robinson
- Activity Coefficient models—NRTL and Wilson
- Predictive Activity – PRSK, VTPR, Modified UNIFAC, UNIQUAC
- Associating – SAFT

Chemstations

Special Systems

- electrolytes methods
- amines and sour water treatment empirical models

Chemstations™

How conditions affect selection

- Single phase?
- Moderate to high pressure over nonideal liquid
- Dissolved supercritical gases
- Immiscibles?

Chemstations™

Environmental Problems

- At very low concentrations, precision becomes important
- Equations of state have inherent error in the model
- Activity coefficient models can't extrapolate to infinite dilution without data
  - Gamma data
- Predictive models (UNIFAC, PSRK) can predict inf. dilution

Chemstations™

Outline

- Basic thermodynamics
- Selecting a [decent] VLE model
- Validating a model selection

Chemstations™

Validating your model selection

- Verify property data is appropriate
- Inspect the results from a model
- Compare to external data
- Compare to similar models and predictive models
- Compare to similar systems

Chemstations

Equilibrium curves from different models

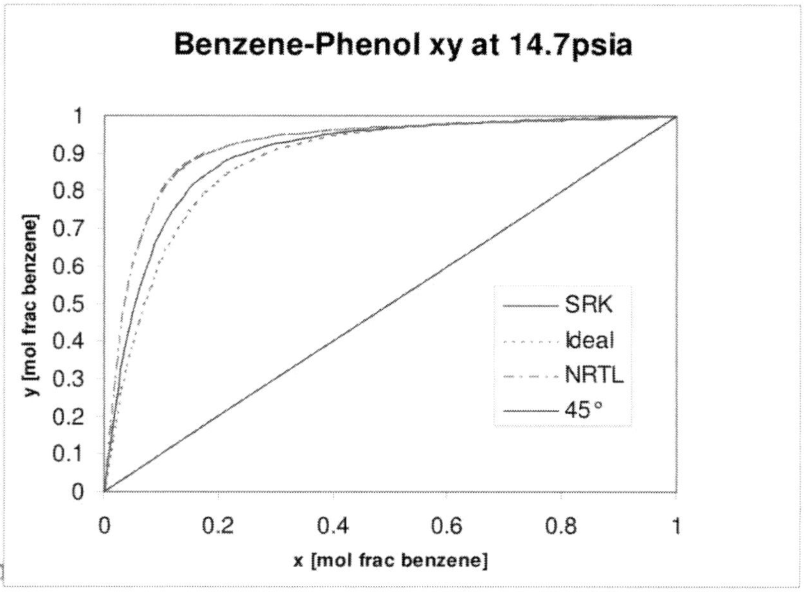

Ch

Validating models - your friend,Txy diagrams

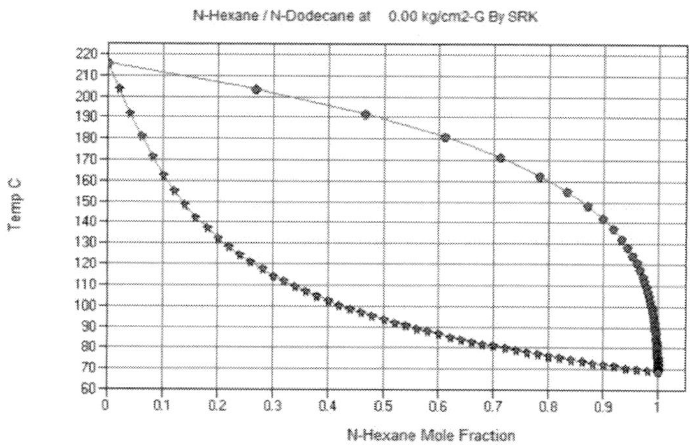

Chemstations

Verify property data available

- For EOS, need good Tc, Pc, Acentric Factor
- For activity methods, need binary interaction parameters and saturated pressures
- For subgroup methods...need subgroups

Chemstations

Inspecting your results

- Check TPxy plots for key component pairs
- Check residue curve maps for azeotropes
- Check binodal plots for LLE
- Perform simple flash separations to confirm expectations

Chemstations™

TPxy plots show the binary separation

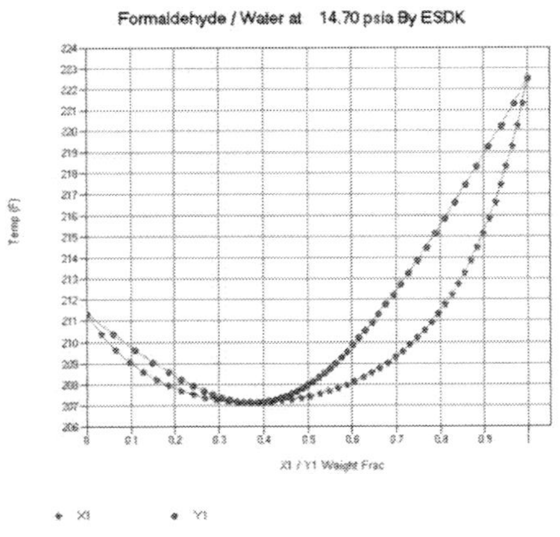

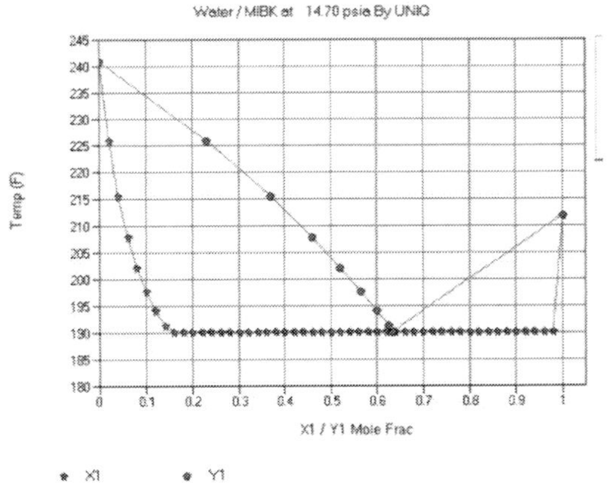

Chemstations™

Compare to Predictive model

- Predictive methods often 'come close' to the right answer

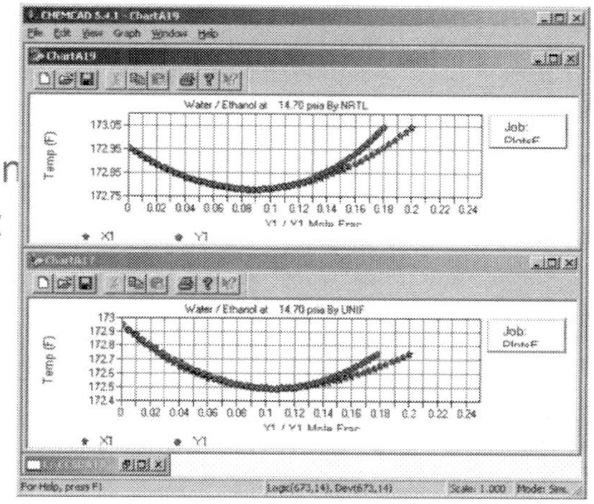

Chemstations

Comparisons of similar models

- Models with similar basis should have similar results
- If models disagree.. investigate why

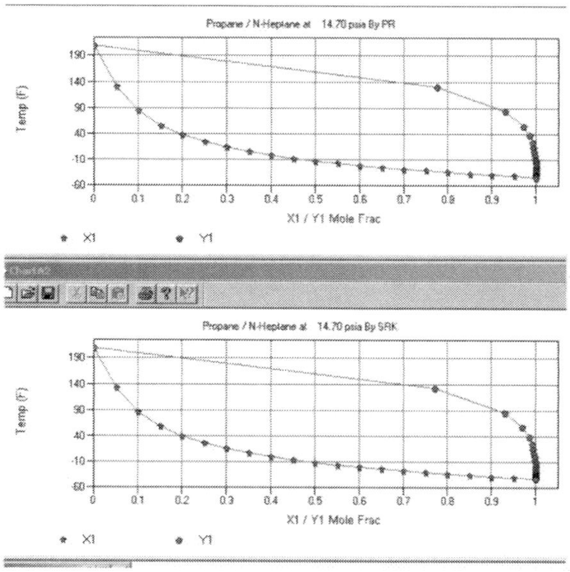

Chemstations

Four Azeotropes

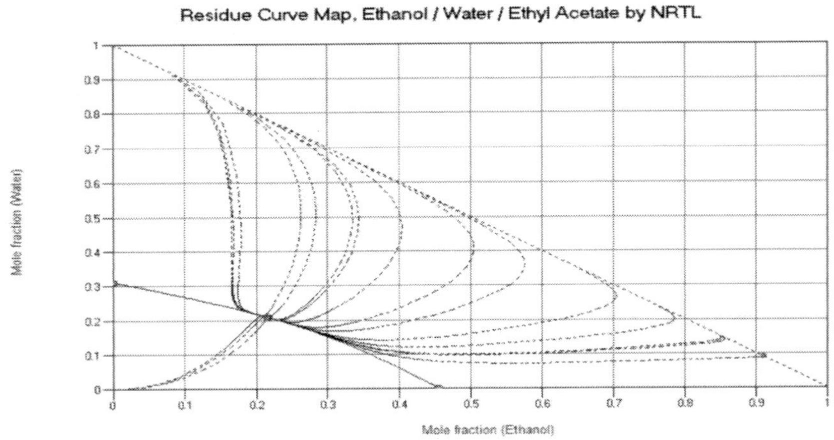

Bubble point at 14.70 psia
Ethanol = 172.922 F
Water = 212.000 F
Ethyl Acetate = 170.708 F

Binary Azeotrope
(Water,Ethyl Acetate) = (0.311,0.689) at 160.696 F
(Ethanol,Ethyl Acetate) = (0.453,0.547) at 161.267 F
(Ethanol,Water) = (0.909,0.091) at 172.763 F
Ternary Azeotrope
(0.215,0.210,0.576) at 159.735 F

Confirming Thermo

For the system Ethanol, Ethyl Acetate, Water three Binary Azeotropes exist as well as a ternary azeotropes. NRTL predicts the azeotropes below:

Bubble point at 14.70 psia
Ethanol = 172.922 F
Water = 212.000 F
Ethyl Acetate = 170.708 F

Binary Azeotrope
(Water,Ethyl Acetate) = (0.311,0.689) at 160.696 F
(Ethanol,Ethyl Acetate) = (0.453,0.547) at 161.267 F
(Ethanol,Water) = (0.909,0.091) at 172.763 F
Ternary Azeotrope
(0.215,0.210,0.576) at 159.735 F

UNIFAC predicts the azeotropes tabulated below

Bubble point at 14.70 psia
Ethanol = 172.922 F
Water = 212.000 F
Ethyl Acetate = 170.708 F

Binary Azeotrope
(Water,Ethyl Acetate) = (0.314,0.686) at 159.297 F
(Ethanol,Ethyl Acetate) = (0.466,0.534) at 159.816 F
(Ethanol,Water) = (0.893,0.107) at 172.467 F
Ternary Azeotrope
(0.099,0.301,0.600) at 158.510 F

The significant difference between NRTL and UNIFAC with respect to Ethanol and Ethyl Acetate ternary azeotropes composition needs to be confirmed by experimental data.

Chemstations

Get Data to Validate Choice

- Even a few data points valuable
  - Regress Physical Properties
  - Get best match between Reality and model
- "Good" Data
  - Plant scale trumps Pilot Plant
  - Pilot Plant trumps bench work
  - Bench measurements better that Literature
  - Literature values better than estimates

Chemstations™

Sources of data

- Books and journals
  - DECHEMA, Fluid Phase Eq., J.Che.Eng.D., JANAF tables, Int'l Critical. Tables,
- Data companies
  - AIChE\DIPPR, DECHEMA/DETHERM, Knovel.com, PPDS
- Predictive packages
  - don't use them blindly! Read about the model before applying it
  - Vacuum Processing Caveat

Chemstations™

Sources of new experimental data

- Lab data
  - internal chemist
  - corporate chemist
- A handful of companies like Wiltec do custom labwork
- Labs of large corporations becoming available for contract
- Few universities run analytical labs

Chemstations™

Absent any Valid Data...

- The known unknowns...things we don't know that we know we don't know
- Homologue plots
- Qualitative behavior and expectations

Chemstations™

Some Important Caveats

- Can't do ppm with models that have error
- H for EOS is not from the Cp
- EOS for a nonregular pure component can give funny results
- EOS is dependent on Tb,Tc, Pc, acentric
- Activity methods depend on Psat, bips
- Bips are generally regressed at 1atm
- For vacuum or high pressure, subgroup models lose reliability
- Pure components don't have a phase envelope

Chemstations

Learn More

- Hill, Justice, "Practical Thermodynamics for Process Simulation," CEP, Dec 2011.
- Carslon, "Don't Gamble with Physical Properties for Simulations," CEP, Oct 1996.
- Gmehling, Chemical Thermodynamics for Process Simulation
- Poling, Properties of Gases and Liquids
- Walas, Phase Equilibria in Chemical Engineering

Chemstations

Thank You

Chemstations™

Important properties for distillations

- Vapor Liquid equilibria (separation)
- Heat capacities (calculating heat duties and temperatures)
- Heat of vaporization

Chemstations™

Can't a simulator do this for me?

- Thermodynamics hasn't solved everything
- Simulators and prediction packages are CALCULATORS, not the Great Oz
- The benefits of computers are speed, not intelligence

Chemstations™

Enthalpy for an equation of state

- Equations of state calculate the liquid enthalpy as a deviation from the vapor...not from liquid heat capacity.
- Latent heat models calculate liquid enthalpy from liquid heat capacity.
- Latent heat models may not always be consistent, as Hvap+Hliq not eq Hgas

Chemstations™

Framework for Creating the Best-in-Class Safety Culture

Dr. M. Sam Mannan, PE, CSP, DHC
Regents Professor and Director
Holder of T. Michael O'Connor Chair I

Mary Kay O'Connor Process Safety Center
Chemical Engineering Department
Texas A&M University System
College Station, Texas 77843-3122, USA
(979) 862-3985, mannan@tamu.edu
http://process-safety.tamu.edu

Mary Kay O'Connor Process Safety Center

Overview

- Practical attributes of Best-in-Class safety culture
- Different safety philosophies or management school of thought
- Consensus principles and learning from broad spectrum of safety philosophies
- Is it possible to possess all of the attributes or employ all of the techniques
- Guidelines that an organization can use as part of its overall self-assessment program

Mary Kay O'Connor Process Safety Center

BIC Safety Culture ⇒ Automatically Results in Effective Ground-Level Programs, such as

- Maintenance and inspection practices
- Hazard identification procedures
- Risk assessment
- Lock-out tag-out protocols
- Mechanical safety and relief mechanisms
- Process safety management standards
- Emergency response plans
- Job safety analysis/Safe work plans
- ....

Mary Kay O'Connor Process Safety Center

Best-in-Class Attributes

- Leadership
- Culture and Values
- Goals and Policies
- Organization and Structure
- Employee Engagement and Behaviors
- Resource Allocation
- Systems, Standards and Processes
- Metrics and Reporting
- Performance Management
- Learning
- Verification and Audit

Mary Kay O'Connor Process Safety Center

Leadership

- Do leaders at all levels communicate, by word and deed, that safety is a *core value*?
- Are leaders at all levels actively engaged and visible on safety?
- Do leaders at all levels get the right information about safety policies and performance?

Mary Kay O'Connor Process Safety Center

Culture and Values

- What is the culture of the company and what are its core values?
- Do the culture and values support the behaviors needed to ensure safety?
- How are the company's values communicated to employees, whether explicitly or implicitly?
- Does the company set high expectations for safety and communicate them clearly?

Mary Kay O'Connor Process Safety Center

Goals and Policies

- How are safety goals and policies set -- with or without input from the field?
- How is implementation of goals and policies assured from the top to the ground level?
- How is compliance with goals and policies tracked and encouraged?

Mary Kay O'Connor Process Safety Center

Organization and Structure

- Are there well-defined roles and accountabilities related to safety?
- Is there an appropriate balance between business unit autonomy and central control?
- Is there an effective system for integrating newly acquired companies regarding safety?

Mary Kay O'Connor Process Safety Center

Employee Engagement and Behaviors

- Do front-line supervisors feel (a) empowered; and (b) accountable on safety?
- Do employees feel motivated with regard to safety and pride in their workplace?
- Do employees feel empowered to identify and address issues?
- Do employees exhibit good teamwork and collaborate across boundaries?

Mary Kay O'Connor Process Safety Center

Resource Allocation

- Does the organization effectively allocate resources to fulfill safety needs and values?
- How are staffing levels determined and monitored?
- Are technology and technical people deployed properly?

Mary Kay O'Connor Process Safety Center

Systems, Standards and Processes

- Are there clearly defined processes?
- Are the processes rigorously executed?
- Are the processes effective?

Mary Kay O'Connor Process Safety Center

Metrics and Reporting

- Are the right metrics (both leading and lagging) being used?
- Are metrics aggregated in an appropriate manner for each level of the organization?
- Is reporting encouraged?
- Is it accurate and effective?

Mary Kay O'Connor Process Safety Center

Performance Management

- How is performance measured and managed (using carrots and sticks)?
- Is safety tied to compensation and if so how?
- Is safety performance seen as relevant to advancement within company?
- Does the performance management process drive the right behaviors?

Mary Kay O'Connor Process Safety Center

Continuous Learning Organization

- Is there a strong, well-resourced training and development program?
- Is there a strong program for the investigation of incidents and "near-misses?"
- Is there an effective feedback loop for continuous learning from reporting, audit, incident investigation, learning from other plants, organizations, external sources?
- Are there mechanisms to stay abreast of current technology and emerging issues?

Mary Kay O'Connor Process Safety Center

Verification and Audit

- What is evaluated and how?
- Do auditors have the requisite skills?
- Is there sufficient follow-up to resolve identified issues?
- Who sees what results at each level of the company?

Mary Kay O'Connor Process Safety Center

Process Safety Management
21 Years Plus or Minus

"What I Wish I'd Known Then and
What We Can't Afford to Forget Now!"

Alan C. Brackey
Valero
One Valero Way, San Antonio, TX 78249-1616
alan.brackey@valero.com

1

Alan C. Brackey

1979 Iowa State University: B.S. Chemical Engineering, EIT
1979 Cargill: Project/Operations Supervisor
1982 Terra: Process Engineer
1984 Celanese: Process/Project/Production Engineer
1994 Celanese: PHA/PSM Coordinator/PSM Risk Advisor
2005 AcuTech: Principal Engineer
2006 TPC: PSM Manager
2008 Valero: Sr. Manager Process Safety Audit

2

The Most Important Takeaway

- Over the years, a number of significant process safety events have occurred in the industry.
- One good source of information is CCPS publication *"Incidents that Define Process Safety"* by John Atherton and Frederic Gil.
- For me, there have been a few other incidents a little closer to home...

3

The Most Important Takeaway

- A small dust explosion led to an equipment fire in a fiber handling system.
- An explosion in a reactor piping system led to a major hydrocarbon release, formation of a large vapor cloud that found an ignition source, ignited and exploded.
- A reactor explosion occurred while a unit was down for maintenance.
- An explosion and fire in a compressor train during a unit startup.
- A large hydrocarbon release from a failed piping system found an ignition source.

4

The Most Important Takeaway

- ***Bad things don't always happen to somebody else!***
- ***"If you think it can't happen to you, you may be setting yourself up to be next".***

  - Glenn Erwin *"Anatomy of a Disaster - Explosion at BP Texas City Refinery March 23, 2005"*, US Chemical Safety Board Safety Video

- **We must *never* lose a sense of our own vulnerability!**

5

Heart and Soul

- Fourteen "working elements" (parts) in 29 CFR 1910.119 (commonly referred to as the OSHA PSM standard). All are important and must work together.
- Prevention program requirements in the EPA Risk Management Plan (RMP) rule (40 CFR Part 68) are similar to those in the OSHA PSM standard.
- Outside the US, other PSM regulations are in place.
- Certain elements are sometimes viewed as being the ***"heart and soul"*** of PSM.

6

Heart and Soul
Process Safety Information (PSI)

- Relief system design and design basis information
  - ➢ Maintaining accuracy is a continuing challenge.
  - ➢ Flare header and recovery systems add complexity.
  - ➢ Impacts of process changes must be considered.
- Significant effort required to effectively manage relief system programs.
- Definition/development of ***"safe limits"*** for each process and associated equipment.

7

Heart and Soul
Process Hazards Analysis (PHA)

- Key challenge is to achieve *consistent* results
  - ➢ Study to study across the facility.
  - ➢ Facility to facility across the organization.
- Increased use of various semiquantitative risk analysis techniques and ***"risk matrices"*** will continue to drive the need for consistency.

8

Heart and Soul
Operating Procedures (OP)

- Requires significant employee involvement.
- Risk is often greatest during process startup.
- Development of ***"operating limits"*** for each process, including points of operational *transition* from "normal operations" to "emergency operations".
- Additional guidance - refer to Section V of the *"OSHA VPP Application Supplement for Sites Subject to PSM"*.

9

Heart and Soul
Management of Change (MOC) and Pre-Startup Safety Review (PSSR)

- Two of the most critical elements for maintaining strong PSM programs - protect against "drift".
- Variations on the "one size fits all" approach may enhance change management system effectiveness and efficiency, while continuing to meet all technical and regulatory requirements.

10

Heart and Soul
Mechanical Integrity (MI)

- Foundation built upon effective inspection, testing and preventative maintenance (ITPM) systems for all safety critical equipment and instrumentation.
- Availability of safety critical instrumentation plays a key role in providing required layers of protection.
- Must identify Recognized and Generally Accepted Good Engineering Practices (RAGAGEP) that may be applicable to various ITPM activities.
- Effective deficiency management systems are essential.

11

Heart and Soul
Management Systems/Performance Audits

- More than just the triennial PSM compliance audits.
- Monthly, quarterly, or annual management system reviews will drive continual program improvement.
- Can begin by focusing on the current status of a few key elements and systems, then adjusting scope of future reviews based on results of initial reviews.
- Recent CCPS publications provide excellent guidance
  - *"Guidelines for Auditing Process Safety Management Systems"*
  - *"Guidelines for Risk Based Process Safety"*

12

PSM practice <u>can</u> be frustrating!

- The OSHA and EPA PSM regulations are performance based, not prescriptive. They state ***<u>what</u>*** needs to be done, but mostly let the users decide ***<u>how</u>*** to do it.
- As with many technical endeavors, the PSM devils are in the details, and those details touch most every aspect of the entire operation!
- The desire on the part of some to want to try to "score" everything.
- We can really never prove a negative, in other words prove that something we did (or money we spent) actually caused something bad ***<u>not</u>*** to happen.

13

Looking Ahead

- Continue to emphasize appropriate risk-based approaches to key process safety activities
  - ➢CCPS publication *"<u>Guidelines for Risk Based Process Safety</u>"* provides practical insights and information.
  - ➢No such thing as unlimited resources, thus we must utilize sound assessment models that appropriately address all aspects of risk to personnel, the community, the environment and the business.

14

Looking Ahead

- Work to develop, implement and maintain effective, functional systems that are not any more complicated than they need to be.
  - ➢ Simple systems work better than complicated ones.
- Must never be satisfied with anything less than full and proper execution of these systems and activities.
  - ➢ No "perfect" systems ("Swiss Cheese" analogy).
  - ➢ Strive to identify/address any system weaknesses.

15

And Finally...

- Significant progress over the past 20+ years, but
  - ✓ ***We're not there yet!***
- Must never lose a sense of our own vulnerability.
  - ✓ ***Bad things don't always happen to somebody else.***
- The things we do <u>absolutely</u> make a difference.
  - ✓ ***Lives, livelihoods and our licenses to operate...***

16

Functional Safety

Implementation of a Safety Instrumented System (SIS) in a Project

aE Solutions

Presented by aeSolutions
Date: October 3, 2013

aesolns.com

Mike Scott, PE, CFSE

- VP, Process Safety with aeSolutions
- BS Degree in Mechanical Engineering
- Masters of Engineering
- Registered Professional Engineer in SC,GA,IL and AK
- Certified Functional Safety Expert (CFSE)
- Author of numerous technical papers related to process safety
- ISA Course Developer / Instructor
  - SI-BMS Webinar
  - SI-FGS Webinar
  - SI-BMS Class ES 17
- ISA SP84 Voting Committee Member
  - BMS Subcommittee core member
  - FGS Subcommittee chair
- ISA Safety Division FGS chair
- IEC 61511 committee member

aesolns.com 2 aE Solutions

Presentation Overview

- Define the Safety Lifecycle
- Discuss Project Execution
- Review Keys to Success
- Discuss Lessons Learned

aesolns.com 3 aE Solutions

Industry Standards for Safety Instrumented Systems (SIS)

- Instrumentation, Systems, and Automation Society (ISA), ANSI/ISA S84.01, *Application of Safety Instrumented Systems for the Process Industry*, 1996 (revised 2004).
- International Electrotechnical Commission (IEC), IEC 61511, Functional Safety: Safety Instrumented Systems for the Process Sector

Performance Oriented Standards

aesolns.com 4 aE Solutions

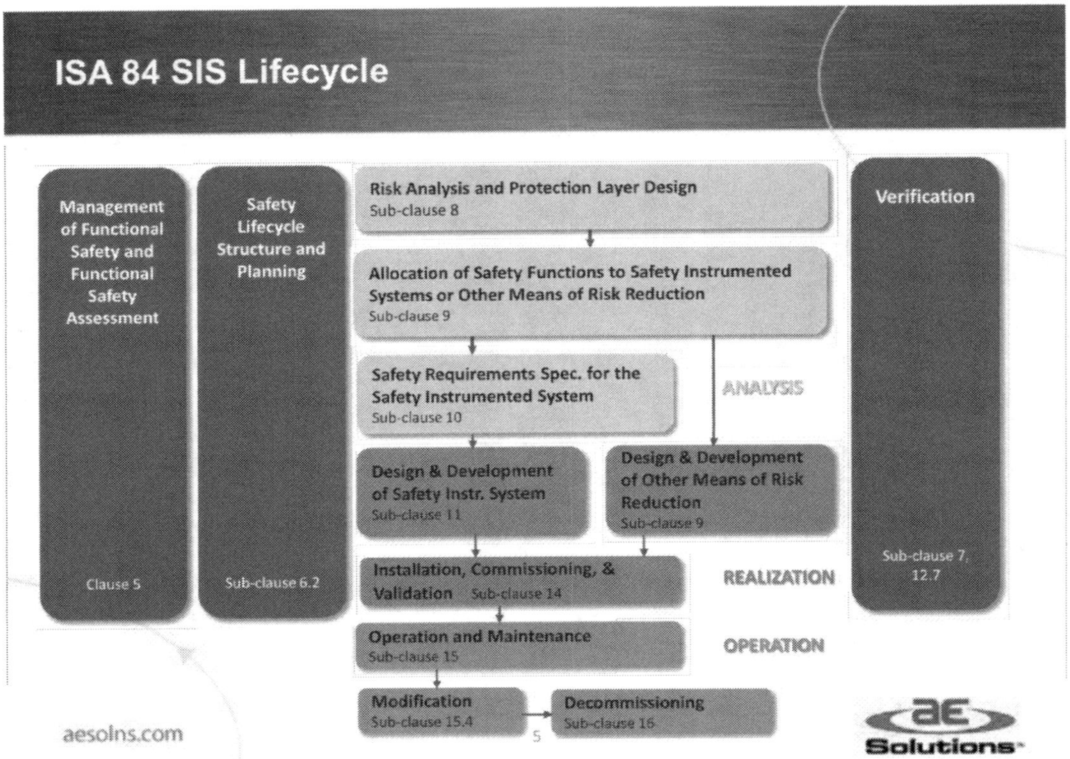
ISA 84 SIS Lifecycle
Management of Functional Safety and Functional Safety Assessment
Clause 5
Safety Lifecycle Structure and Planning
Sub-clause 6.2
Risk Analysis and Protection Layer Design
Sub-clause 8
Allocation of Safety Functions to Safety Instrumented Systems or Other Means of Risk Reduction
Sub-clause 9
Safety Requirements Spec. for the Safety Instrumented System
Sub-clause 10
ANALYSIS
Design & Development of Safety Instr. System
Sub-clause 11
Design & Development of Other Means of Risk Reduction
Sub-clause 9
Installation, Commissioning, & Validation Sub-clause 14
REALIZATION
Operation and Maintenance
Sub-clause 15
OPERATION
Modification
Sub-clause 15.4
Decommissioning
Sub-clause 16
Verification
Sub-clause 7, 12.7
aesolns.com
5
aE Solutions

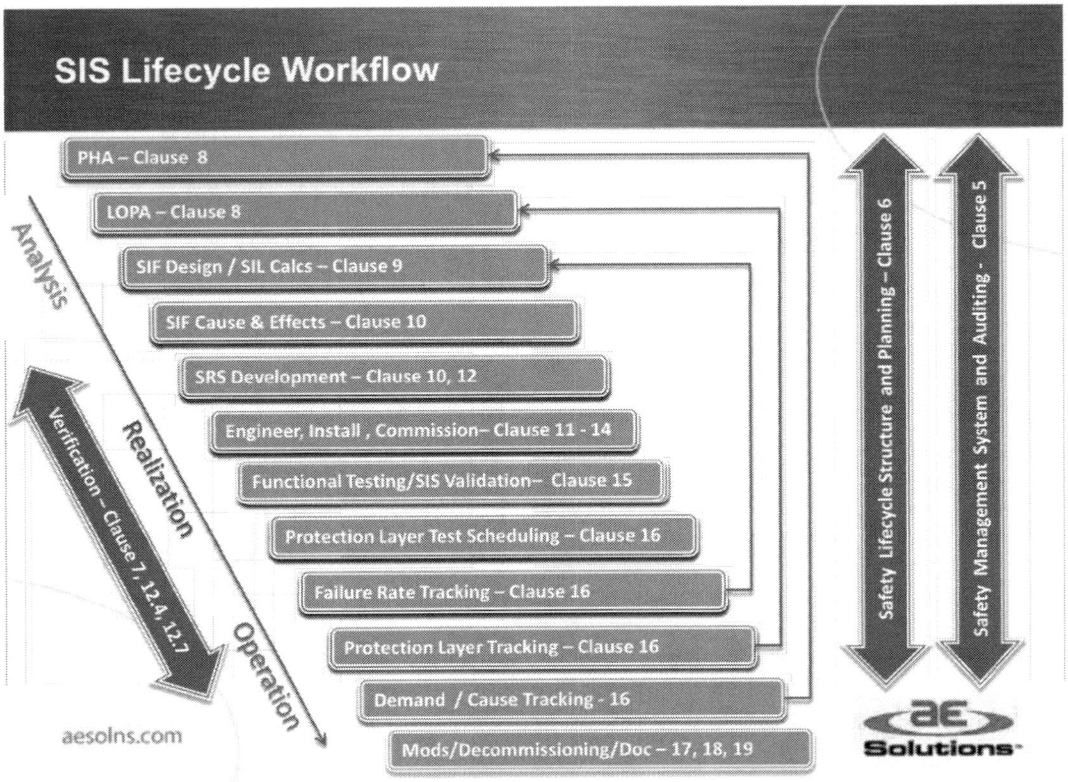
SIS Lifecycle Workflow
PHA – Clause 8
LOPA – Clause 8
SIF Design / SIL Calcs – Clause 9
SIF Cause & Effects – Clause 10
SRS Development – Clause 10, 12
Engineer, Install , Commission– Clause 11 - 14
Functional Testing/SIS Validation– Clause 15
Protection Layer Test Scheduling – Clause 16
Failure Rate Tracking – Clause 16
Protection Layer Tracking – Clause 16
Demand / Cause Tracking - 16
Mods/Decommissioning/Doc – 17, 18, 19
Analysis
Realization
Operation
Verification – Clause 7, 12.4, 12.7
Safety Lifecycle Structure and Planning – Clause 6
Safety Management System and Auditing - Clause 5
aesolns.com
aE Solutions

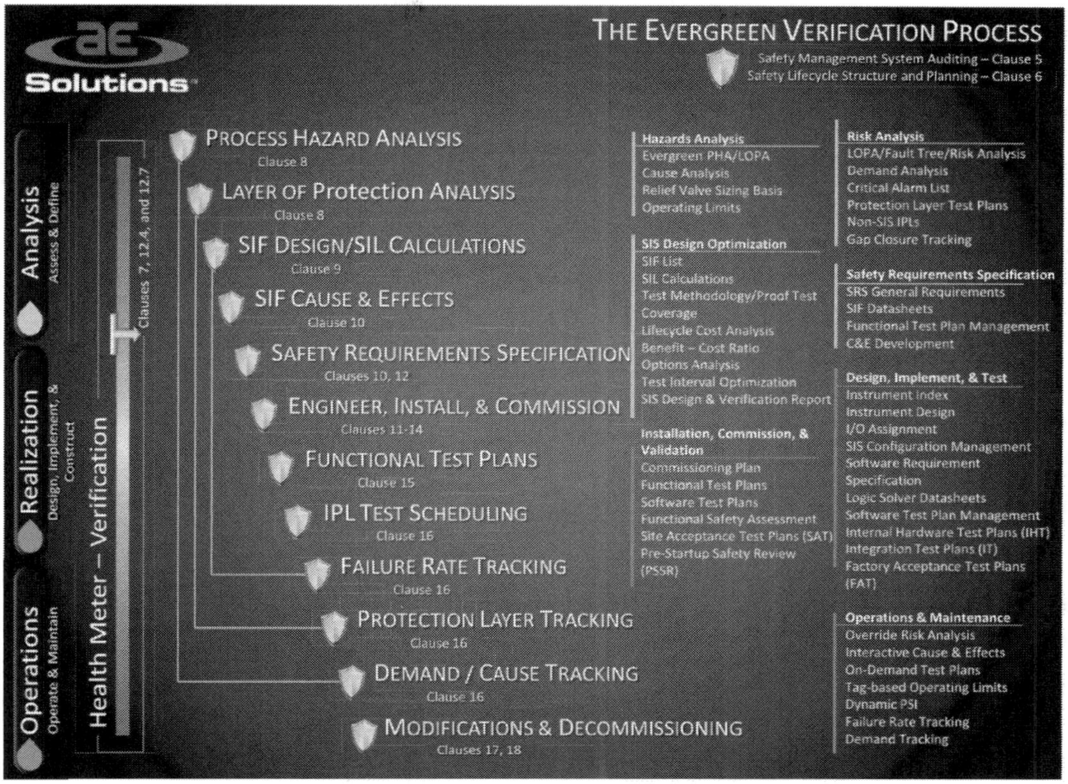

Increasing safety.... Increasing cost

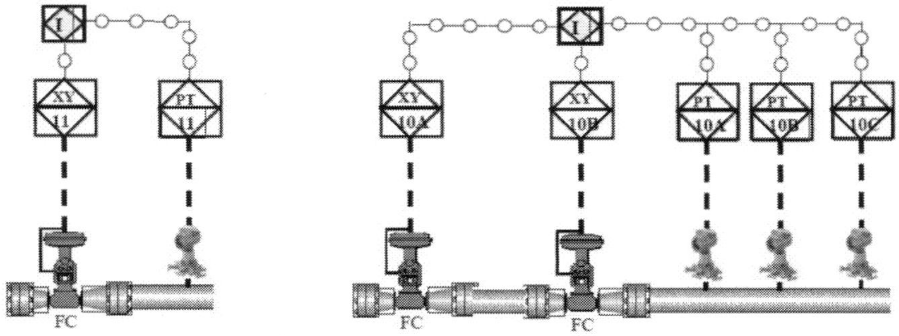

Typical SIL 1

- inexpensive
- high spurious trip rate
- not especially safe

Typical SIL 3

- more expensive
- low spurious trip rate
- safe

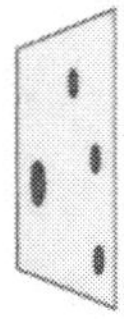

aesolns.com

8

Solutions

CAPITAL PROJECTS: Keys to Success

- Safety Lifecycle Plan is critical
- Preliminary PHA / LOPA early in the project
- Development of SIS Conceptual Design
- SIS Knowledgeable staff available to project team
- Use of SIS Tools in development deliverables
- Development of SIS Guidance Documentation
- Functional Safety Assessments are Critical Tools
- Management of Change within the project
- SIS Engineering as a consultant vs discipline

aesolns.com 9 aE Solutions

Potential Project Impacts

- If SIS Design Requirements are NOT addressed at the right stage of the project one could expect:
  - Schedule Delays
  - Cost Impacts
  - Failure to adequately mitigate risk

aesolns.com 10 aE Solutions

Thank you for your attention. Questions???

aesolns.com

aesolns.com

Mike Scott, PE, CFSE

VP of Process Safety

907-301-3111

Mike.Scott@aesolns.com

Risk Assessment Challenges 20:20

Concept by Angela E. Summers, PhD, PE, President

Presentation by Eric Randecker

SIS TECH

PSM Goal

- To prevent unwanted releases of hazardous chemicals especially into locations that could expose employees and others to serious hazards.
- An effective process safety management program requires a systematic approach to evaluating the whole chemical process.

OSHA Process Safety Management Guidelines for Compliance

SIS TECH

Process Industry Goal

Organizational Behavior Assigns Value

Zero Loss of Containment Events
and
Zero Injuries and Fatalities
and
Zero Environmental Impact

SIS-TECH

Recognize High Hidden Cost

- Human costs in trauma, loss work productivity
- Asset losses
- Loss production
- Loss of market share
- Loss of good will and public approval
- Loss of stock value
- Fines from regulatory agencies
- Environmental cost
- Insurance premiums

SIS-TECH

Standard of Care

- Level of care that a reasonable person would exercise in given circumstances to prevent harm.

Demonstrated level of care must pass
red face test

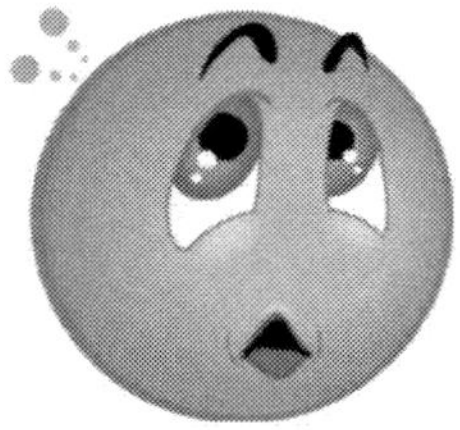

SIS TECH

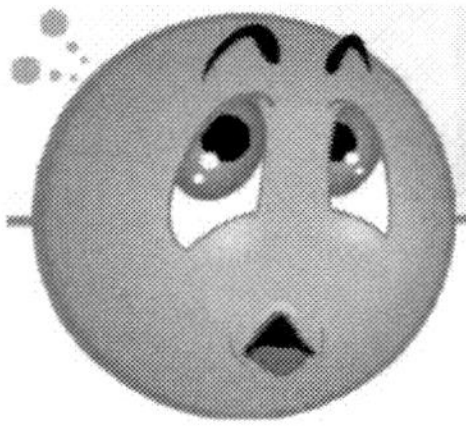

Red Face Test = ALARP

- As low as reasonably practicable (ALARP)
- Loss Prevention Plan
  - Identify safeguards
  - Design to address identified risks
  - Demonstrate risk reduction

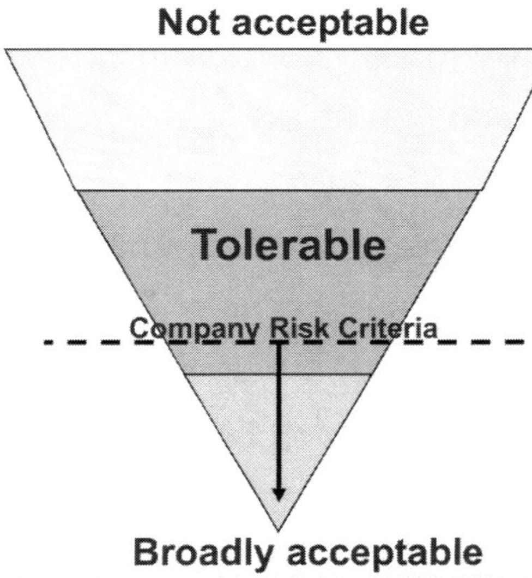

SIS TECH

Things to think about...

- Will you be able to justify your choices?
- Do you have a strong technical basis?
- Have you addressed what you know?
- Have you considered the <u>*limits of what you don't know*</u>?

Too focused on fatality events?

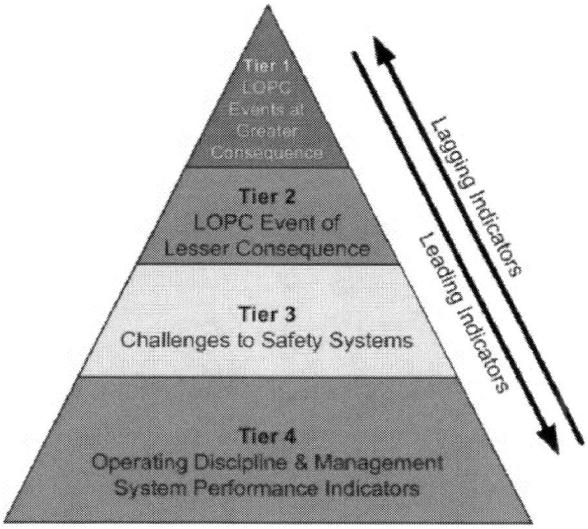

API 754 graphic

Tier 1 – **Monitor events of Greater Consequence**

- *Worst Lagging indicator*
- Most threatening to company value and existence

Tier 2 – **Monitor Events of Lesser Consequence**

- *Lagging indicator*
- May reveal underlying systematic problem

SIS TECH

Reduce Abnormal Operation Frequency and Demands on Safeguards

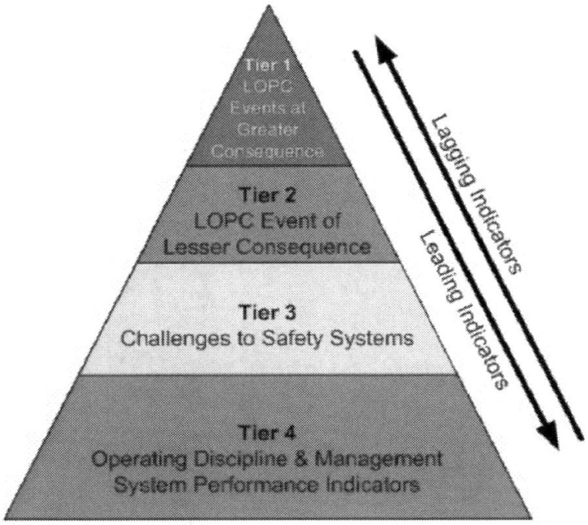

API 754 graphic

Tier 3 – **Monitor challenges**

- Safe Operating Limit excursions
- Demands on safeguards, e.g., alarms, shutdowns, PRV lifts
- Safeguard test failure

Tier 4 – **Monitor degradation**

- Abnormal operation
- Action item closure
- Safeguard inspection and diagnostic failure

SIS-TECH

Reduce Tier 1 and 2 by Monitoring Tier 3 and 4

- Accept that abnormal operation more than once per year is indication of a process reliability problem.
  - You are not meeting industry norms.
  - Get help from process control professionals.
- Implement defense in depth strategy to prevent abnormal operation from becoming a loss event
  - Functional diversity
  - Work group diversity
  - Technology diversity

SIS-TECH

Challenges 20:20

1. Avoid scenario tunnel vision
2. Beware of complexity of human and automation interaction
3. Avoid single jeopardy paradigm
4. Beware of hypnotic lure of calculations
5. Think outside of the box

SIS TECH

Avoid scenario tunnel vision

- Occurs when one incident pathway is identified and others are not fully examined
  - All events are foreseeable given sufficient analysis
  - Yet, HSE incident analysis (>10 years of data) shows that 25% of loss events are not identified in hazard studies.
- Occurs when you believe that if you mitigate the worst incident, the lesser ones are taken care of
  - Even a minor event can escalate into a large event under the wrong conditions at the wrong time

SIS TECH

Beware of complexity in human and automation interaction

- What does the operator see when an incident is occurring?

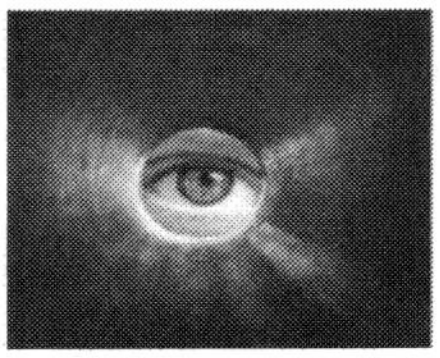

  - Do changes in the process conditions correlate obviously with a change in hazard state?
  - Are there multiple process conditions that can be relied upon to detect the change in hazard state?
- How does the process respond as process conditions change?
  - When an operator intervenes, what happens and how will the operator recognize a good response from a bad one?

SIS TECH

Avoid single jeopardy paradigm

- Reality is that many single and numerous multiple causes can lead to the loss event
  - For every loss event occurrence, there are many possible root causes or underlying systematic conditions.
- Foundation for loss events is laid with process design and operating discipline.
  - Systematic problems with process operation can be seem in operating records.

SIS TECH

Beware of hypnotic lure of calculations

- Takes 10 years and a billion $ to build a deepwater rig, but only a few minutes to sink it.
- Risk calculations are good for estimating things that can be easily measured.
- Risk calculations rarely account for the human element.

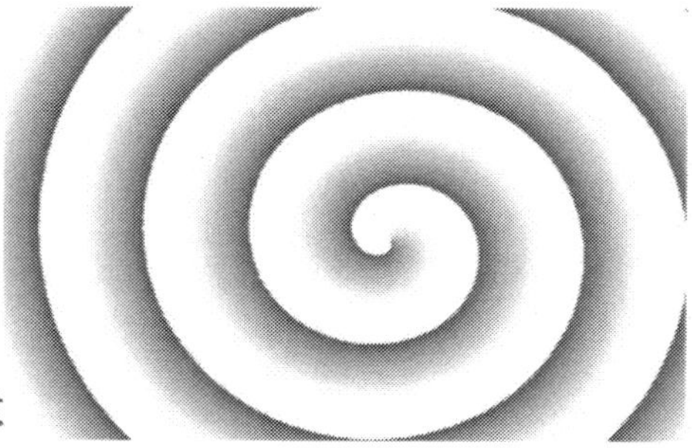

SIS TECH

Think outside of the box

- Don't limit protection layers to what is found in the risk analysis.
- Update process simulations to match actual process response during normal and abnormal operation.
  - Test process response to changes in process conditions and control strategy.
  - Use process simulators for training operators on how process behaves in response to changing process conditions.

SIS TECH

PSM Effectiveness

- Depends on rigor of the **systematic approach** applied to develop a loss event prevention plan and the **organizational discipline** that implements the plan.
- Measured by **actual performance** over the **life of a process**.

SIS-TECH

Actual Performance

- Assess the demand rate on safeguards during actual operation to verify the risk assessment assumptions.
  - Examine abnormal situation, alarm management system, safeguard initiation, and trip reports
- Monitor and assess whether reliability parameters agree with specifications.
  - Examine operating and maintenance records

SIS-TECH

Improve Performance

- Identify and prevent systematic failures which could jeopardize safety.
  - Use documented practices and procedures for installation, commissioning, maintenance and operation of process equipment and safeguards.
- Define the necessary corrective action if the failure rate of any equipment is greater than assumed in design basis.
  - Implement quality assurance processes for safeguards

SIS-TECH

Develop comprehensive plan

- Document the process operation
  - Design and technology
  - Non-routine activities and procedures
  - Routine operating and maintenance procedures
  - Safe operating limits
- Document a loss prevention plan
  - Abnormal operation response procedures
  - Safeguard design with required operation and maintenance procedures
  - Emergency preparedness plans and procedures

SIS-TECH

Questions

SIS-TECH

AIChE
Southwest Process Technology Conference

5th AIChE Southwest Process Technology Conference

October 3-4, 2013 • Galveston, TX
Moody Gardens Hotel and Convention Center

TANKS A LOT

A VIEW OF THESE VITAL VESSELS IN CASE HISTORIES

by: Roy E. Sanders

Free Lance Process Safety Consultant

Lake Charles, Louisiana

Classification of Tanks

- **Low Pressure Tanks**

(Shaped like a Soup Can or Tuna Fish Can)

  - API 650 Welded Steel Tanks – These are large, with typically near atmospheric ratings
  - API 620 Welded Low Pressure - Rated to 15 PSIG
  - API 12 D & 12 F Smaller Tanks for Production Liquids

- **Pressure Vessels**
  - ASME Code 15 to 3,000 PSIG

(Shaped like a Fat Sausage or Sphere)

- **Others**

(Transportation, Shop Built, Underground, Plastic, Cylinders, Non-Code etc.)

PG

According to Dr. Trevor A. Kletz

in all Five Editions of

What Went Wrong?

"No Piece of Equipment is Involved in More Accidents than the Storage Tank, Probably because Storage Tanks are Fragile and Easily Damaged by Slight Overpressure and Vacuum.

Tanks A Lot

will review some common problems that Tanks Experience including:

- Poor Vent Design
- Tampering with Well Designed Vents
- Coping with Vacuum
- Overfill Problems
- Poor Management of Change
- Lack of or Improper Maintenance
- Tank Siting

A Small Fiberglass Tank is Destroyed

A Small Fiberglass Tank Was Destroyed During a Routine Operations

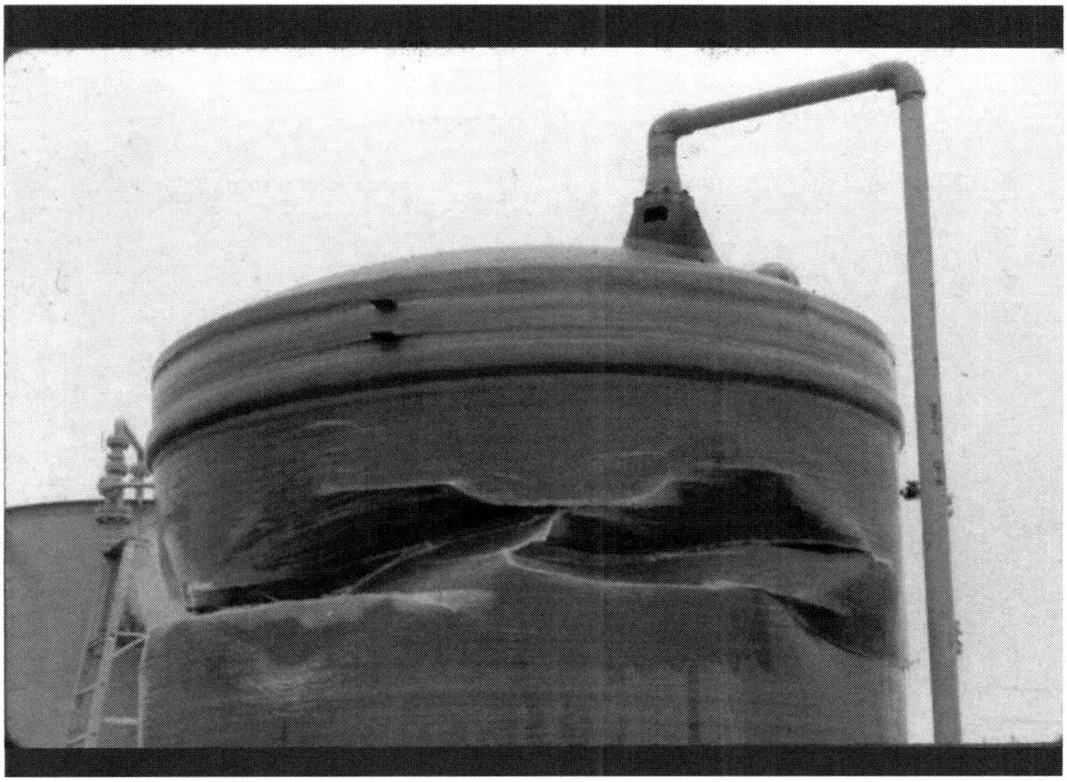

Coker Drum Collapses
at a Major Refinery

Before Initial Startup

Brand New Distillation Column Topples

82 Foot Tall Stainless Column Collapses During Initial Startup

Improvement Planned

Environmental "Improvement" Triggers Tank Collapse

- Solvents Tank Size
  - 20 Feet in Diameter
  - 30 Feet High
- **Tank Was Destroyed Beyond Repair**

Simple Overfilling of Tanks

- Designers must provide effective level indicators.
- Operating Personnel must have Process Safety Information & Understand Liquid Level Limits.
- If during a transfer, the level in the receiving tank fails to increase, stop the transfer.
- Strong consideration must be made for Overflow Protection including High-High Level Protection.
- Safety Critical Alarms and Interlocks must be regularly tested.

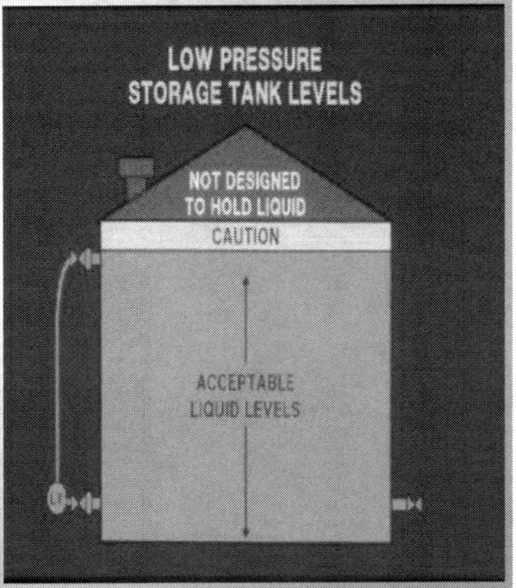

It is Hard to Believe that a
Can of Soup or Can of Beans
Will Withstand 4 to 8 times more Pressure than a
Low Pressure Tank

Quoted from Tank Farm an (Un)Loading Safety Booklet No. 12
BP's Sharing the Experience

Over / Under Pressurizing Tanks

1. Venting systems must be professionally designed to codes for all known normal operating and upset conditions scenarios.
2. Plant personnel must routinely check for plugging of vents in fouling services.
3. Plant personnel must routinely inspect and test over-pressure and vacuum relief devices for continued proper operations .
4. Well intended individuals can easily block some tank venting systems. Tamper-proof vent systems are ideal.
5. Any modification to a tank vent system must be authorized by some type of Management of Change System or other evaluation/approval system.

Understand Local Hazards !

Ensure Tank Reliability

An Internal Inspection often takes lots of preparation, but can be very revealing.

Other Tank Fundamentals

1. Tank spacing and layout are critical. Various property insurance publications offer help and the National Fire Protection Association offer guidance for flammable liquids.
2. Local conditions, such as the possibility of flooding, hurricanes, etc. which can affect low pressure tanks should be considered.
3. Tanks containing incompatible chemicals must not be in the same diking system.
4. Fire protection features including static electricity dissipation, vapor space inerting, protective foam generators, fire water spray, overpressure protection and dike design demand professional engineering input.
5. Tank bottoms should be sloped to facilitate drainage for easier inspection.
6. Trained professionals should periodically inspect all tanks for external and internal corrosion and other damages.
7. Any modification to a tank vent system must be authorized by some type of Management of Change System.

Reference Material for Increased Learning

- The "Beacon" AIChE's CCPS monthly publication. By my count about 20 – 25 % of the Beacons contain information on Tanks
- "What Went Wrong?" 5th ed. Trevor Kletz Elsevier - 2009

Reference Material for More Detail

- Chemical Process Safety: Learning from Case Histories - Third Edition – Elsevier 2005
- Treat Tanks with Care - Putman's Chemical Processing Nov 2010
- Tanks A Lot – Trevor: For Enduring Process Safety Contributions - Journal of Loss Prevention in the Process Industries – Sept 2012

Fractionation Research, Inc.
RESEARCH - KEY TO BETTER DESIGN

Column Revamps

Southwest Process Technology Conference (SPTC)

Galveston, Texas

Mr. Mike Resetarits
FRI(SM) Technical Director

3 October 2013

Acknowledgements

FRI

- Koch-Glitsch
- Sulzer Chemtech
- Shell
- UOP

3

Column Revamps

4

Column Revamps

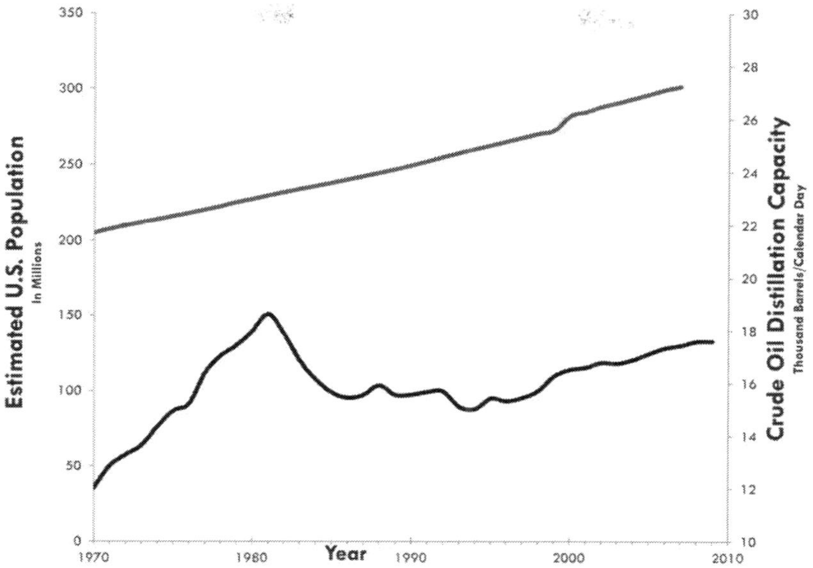

5 Column Revamps

The BANANA Rule

6 Column Revamps

BANANA Rule

You can not

Build Anything

Near Anybody or

Near Anything

Bottlenecked Column

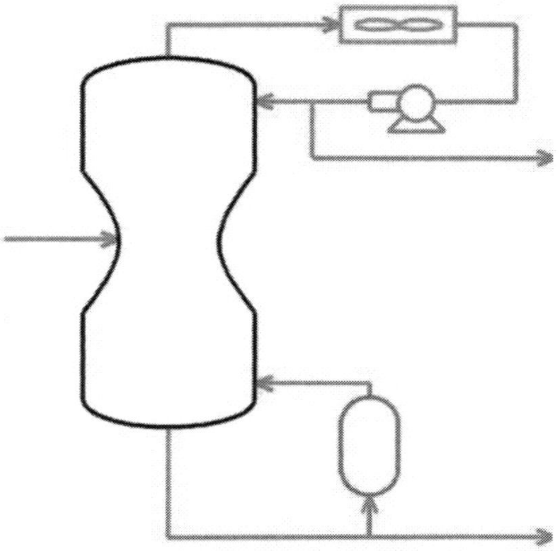

Bottlenecked Column and Ancillary Equipment

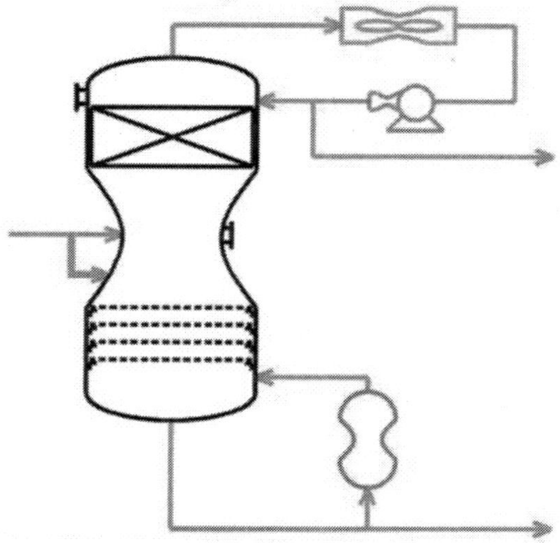

9 Column Revamps

Existing Trayed Column

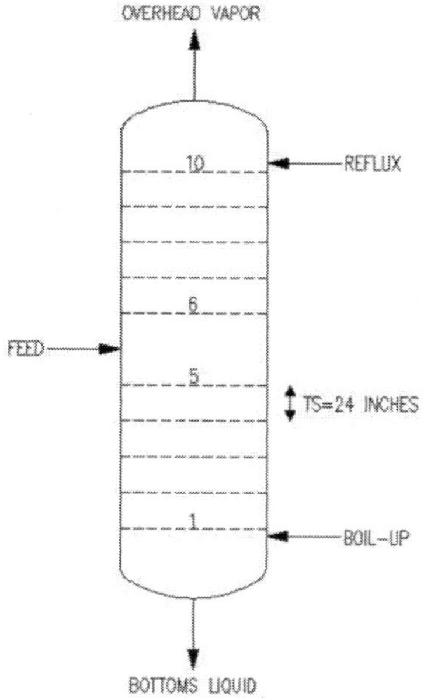

Situation:	Trays are limiting capacity
Goal:	Increase capacity by 20%
Options:	Larger and smaller spacings

10 Column Revamps

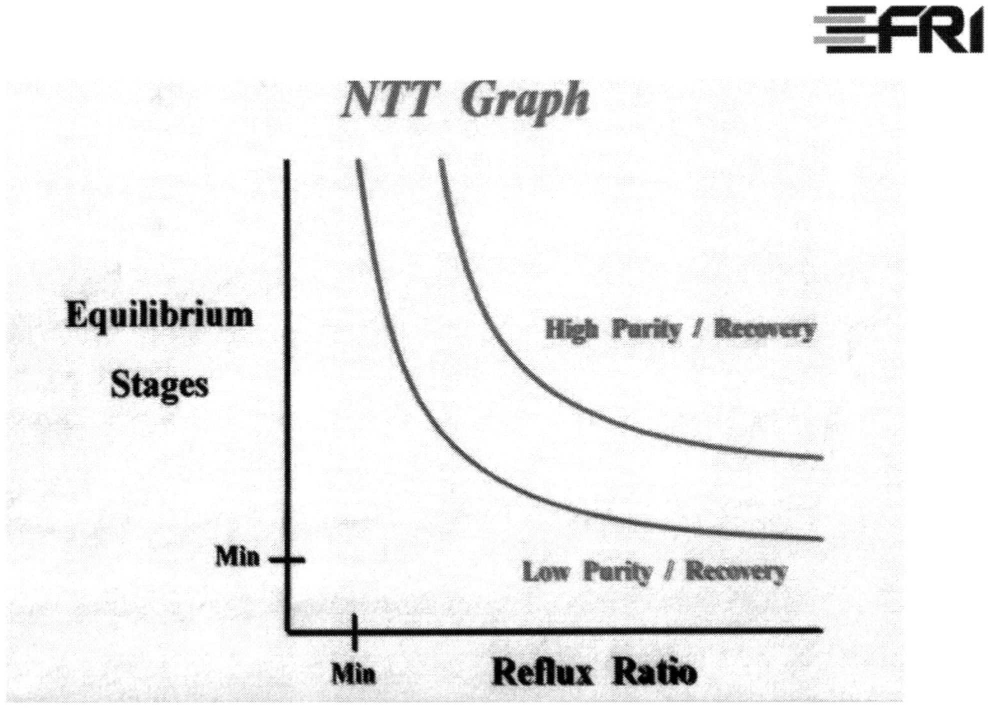

11 Column Revamps

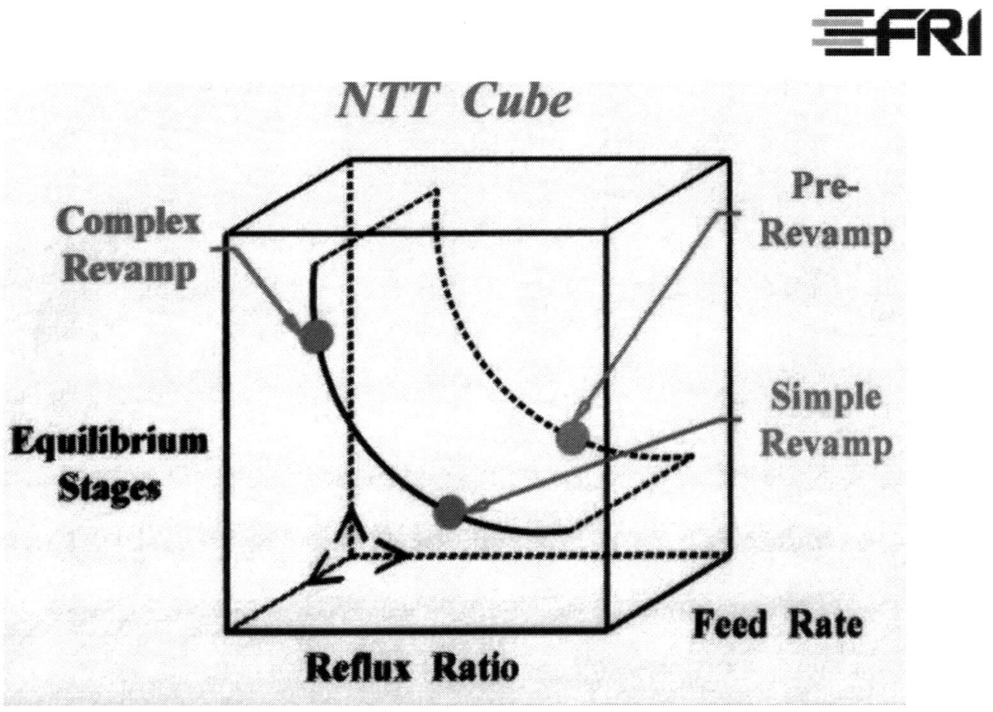

12 Column Revamps

Revamp Option

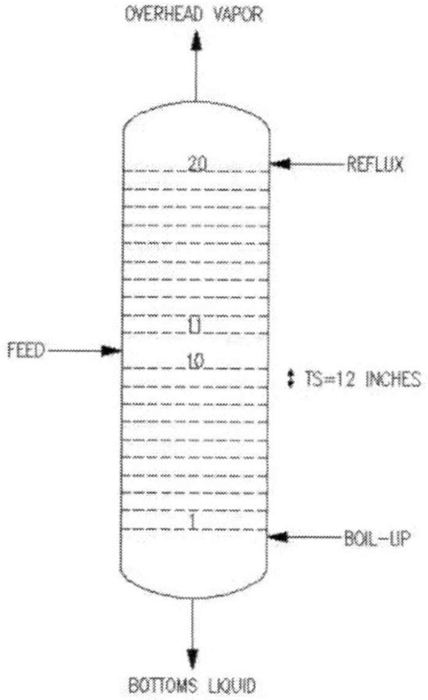

Option 2: Increase actual and theoretical trays
Reduce reflux requirement

Question: Will the reduced capacity of the closely-spaced trays offset the gains from the reduced reflux?

13 Column Revamps

Column Revamps

- Trays Replace Trays
- Packing Replaces Packing
- Packing Replaces Trays

14 Column Revamps

Generations of Trays

- Crossflow
- Augmented Crossflow
- Countercurrent Flow
- Cocurrent Flow

15 Column Revamps

Existing Packed Column

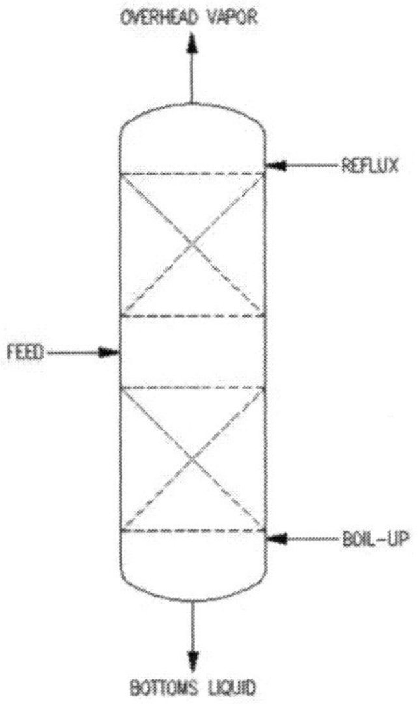

Situation: Packing is limiting capacity

Goal: Increase capacity by 20%

Option: Larger packing elements

16 Column Revamps

Random Packing Options

- 2 Inch Rings have the following compared to 1 inch Rings:
  - Higher capacity
  - Lesser efficiency
  - Lesser pressure drop
  - Less cost
- And superior packings achieve superior results

17 Column Revamps

Structured Packing Options

- 250 m^2/m^3 has the following when compared to 500 m^2/m^3:
  - Higher capacity
  - Lesser efficiency
  - Lesser pressure drop
  - Less cost
- And superior packings achieve superior results

18 Column Revamps

Packing Replaces Trays

- Especially structured packing
- Especially in vacuum service
  - Pressure drop critical
  - Packing pressure drop very low
  - Example: Ethylbenzene-styrene

THANK YOU FOR YOUR ATTENTION

Hebron: An Exercise in Separation Technology Deployment

Steinhauser, A., ExxonMobil Upstream Research Company
Grave, E., ExxonMobil Upstream Research Company
Originally presented on: 10/3/2013 14:00:00 - 14:30:00

Equipment module weight and footprint are two main cost drivers when designing and constructing offshore oil & gas installations. Offshore installations producing heavy crude are even more cost-challenged, as traditionally, to achieve product specifications (e.g. BS&W, WIO) when treating unconventional resources, long residence times in large pressure vessels were specified. However, in the past decade, industry has been researching, developing, and deploying new technologies aimed at producing on-spec fluids while minimizing vessel size.

ExxonMobil's Hebron project, located offshore of Eastern Canada, will produce heavy, emulsion-prone oil. In an effort to minimize the cost impacts associated with traditional heavy-oil treating methods, and thus improve project economics, new high-efficiency, compact separation equipment was specified for a number of oil separators and compressor suction scrubbers. This paper will explore the challenges of qualifying, specifying, and deploying this equipment in an offshore environment.

Pitfalls and Safeguards in Distillation Tower Designs

Xu, S. X.,
Mancillas, E.,
Originally presented on: 10/3/2013 14:30:00 - 15:00:00

With the aid of computers and internet applications many methods and equations for designing distillation towers, including absorbers and strippers, have been automated. Engineers have easy access to various modeling or rating software tools for tower designs. However, surveys have shown that the number of tower malfunctions is not declining, in spite of the huge progress in computational capability. Instead of focusing on specific technical aspects of the tower designs, this article will highlight some pitfalls and safeguards in tower design steps, from process definition, simulation, internals selection and sizing, to tower layout. Tower design cases in oil refining, petrochemical and gas processing services will be discussed. The intention of this article is to help process engineers in the design of distillation equipment step by step efficiently and correctly.

The New UFM Tray provides High Capacity AND Flexibility

Summers, D. R., Sulzer Chemtech USA, Inc.

Originally presented on: 10/3/2013 15:30:00 - 16:00:00

Sulzer has developed a new type distillation tray that provides both high capacity and operational turndown. In the past, for towers to gain additional capacity, the trade-off was to sacrifice turndown and also give up some tray efficiency. Now with the advent of Sulzer's latest tray innovation, trayed distillation towers can realize the extra capacity of High Capacity trays without giving up operational flexibility or performance. The UFM valve, which has been on the market for more than a year, has now demonstrated repeatedly in operating plants that it delivers on both claims. Laboratory data, independent testing data and field operating data will be provided in this presentation to substantiate Sulzer's claims.

AIChE Southwest Process Technology Conference - 2013

Choosing Tower Internals for LNG, Shale Gas, and Tail Gas Treating

Ralph H. Weiland
Optimized Gas Treating, Inc.
P.O. Box 125
Clarita, OK 74535

Nathan A. Hatcher
Optimized Gas Treating, Inc.
Technical Development Center
142 Cimarron Park Loop, Suite C
Buda, TX 78610

Jenny Seagraves
INEOS Oxide
A Division of INEOS Americas LLC
Plaquemine, LA 70765

Background

- LNG:
  - Focus on deep CO_2 removal
  - Often piperazine promoted MDEA (fast reaction)
  - Frequently uses packed absorbers
- Shale Gas & Tail Gas
  - Focus on either high or maximum selectivity
  - Selectivity affected by internals
    - Tray design
    - Packing

Background

- Performance of tower internals
  - Hydraulics (pressure drop, flooding, capacity) well understood
  - Mass transfer performance:
    - Largely neglected because efficiencies and HETPs have served so well in HC distillation
    - Efficiency approach hopeless in amine treating
    - Tower internals have huge effect on treating
  - Tray design details matter
  - Reliable packed tower simulation a necessity

Questions

- How does tray geometry affect selectivity?
- Using packing in gas treating?
  - When?
  - What applications?
  - What kind of packing?
  - What size?
  - Direct simulation

What Determines Selectivity

- Similar solubility of H_2S and CO_2 in amines
- Primary factors:
  - Reaction kinetics of CO_2 in solvent
  - Mass transfer characteristics of contactor
- Kinetics:

$$H_2S \rightleftharpoons H^+ + HS^-$$

  - Instantaneous
  - H_2S absorption is gas-phase controlled
  - Gas-phase mass transfer characteristics

What Determines Selectivity

- Kinetics:

$$CO_2 + H_2O \rightleftharpoons H^+ + HCO_3^-$$
$$R_1R_2R_3N + H^+ \rightleftharpoons R_1R_2R_3NH^+$$

  - Amine protonation instantaneous
  - CO_2 hydrolysis is slow
  - H_2S kinetic instantaneous, CO_2 kinetics slow
  - Favors H2S absorption
- Liquid phase controls CO2 absorption rate
  - Liquid-phase mass transfer characteristics dominate

Selectivity vs. Deep CO_2 Removal

- Kinetics
  - Highest selectivity when no CO_2 reaction with amine, and alkalinity (OH^-) lowest
  - Increasing CO_2 reaction rate lowers selectivity
  - A fast promoter (e.g., piperazine) speeds up CO_2 absorption – allows deep CO_2 removal
- Internals Characteristics
  - High liquid-side resistance (low k_L) and low gas-side resistance (high k_G) favour selectivity
  - Lower liquid side resistance (high k_L) favors deeper CO_2 removal.

What Internals for Which Application?

- Internals Characteristics
  - High liquid-side resistance (low k_L) and low gas-side resistance (high k_G) favour **selectivity**
    - Structured packing (also for FPSO & FLNG)
    - Random packing (higher capacity than trays)
    - Special tray designs with lower k_L & higher k_G
  - Low liquid side resistance (high k_L) favors **deeper CO_2 removal**.
    - Well-designed trays

Tray Design for Selectivity

- High acid gas (very sour) streams make many applications high Liquid Flow
- High liquid flows need big downcomers and long weirs, hence, multi-pass trays
- At normal liquid loads, long weirs cause sprays
- At low liquid loads, normal weirs cause sprays
- Sprays are small liquid droplets in gas.
  - Drops are stagnant (no internal movement)
  - Diffusion only → very high liquid resistance
  - Reject more CO_2 & improve selectivity by 10X
  - *"The Legend of Low Capacity"* and a bad ROT

Packing: Why & When

- Advantages:
  - Low pressure such as TGTU, AGE
  - Mechanically more stable in Quench
  - Tower capacity often higher than trays
  - Smaller column diameters advantageous
  - In FPSO and FLNG a tighter design can yield large cost advantages – minimum column diameter with right packed depth
  - Foaming usually not as big a problem
  - Hydraulics & capacity well understood

Packing: Why & When

- Downside:
  - Weaker experience base in amine service
  - Mass transfer performance, especially with chemical reaction, specialized – early days
  - Mass transfer performance poorly known
  - Discomfort factor because of inexperience
  - Design reliability? Discomfort makes it more challenging to guarantee design
  - Huge range of random & structured packings
    - Sizes, shapes, materials, surface treatments

Packing: Simulation Reliability

- Is simulating packed columns really harder?
  - Great database of hydraulic & mass transfer characteristics,
  - Database & models backed by 30-40 yrs of research at Delft, UT Austin, elsewhere, in commercial sized equipment
  - Mass transfer rate-based packed column simulation just as easy as trays
  - Backed by positive comparisons with much plant performance data

Comparison with Commercial Data

Example Case

- LNG plant revamp using GAS/SPEC 2020
- 2.24% CO_2 gas at 4 Mpa
- Conventional valve trays
- Capacity at 70% of nameplate
- Supposed cause: foaming
- Treated gas:
  - Measured: 21.2 ppmv CO_2
  - Simulated using ProTreat®: 18.2 ppmv CO_2

Comparison with Commercial Data

Revamp:

- Cause of supposed foaming unknown
- Regenerator at 85% flood even at 70% plant capacity ← a likely culprit
- Revamp both columns
  - Larger diameters (known)
  - Structured packing (not a known brand)
  - Here's all the information we have on the packing.....

Comparison with Commercial Data

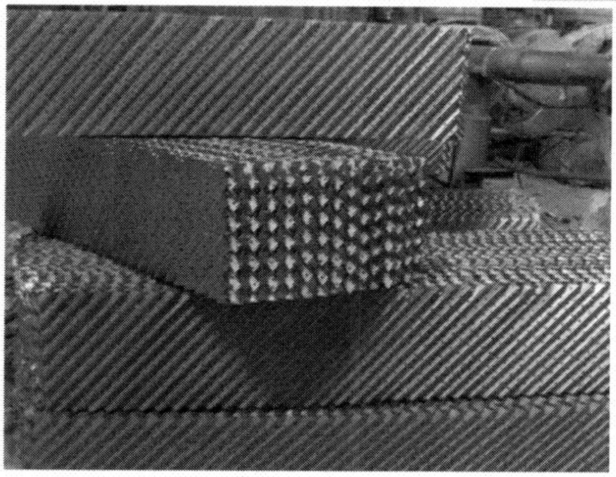

Comparison with Commercial Data

Comparison with Commercial Data

- Absorber & regenerator now running at 23% and 36% flood
- Packing crimp size estimated using photographs and ruler
- Treated Gas:
  - Measured: 7 ppmv CO_2
  - Simulated: 1-2 ppmv CO_2
- More discrepancy than we expected
- Possible causes: Mis-estimate of packing size, poor liquid distribution, no wall wipers

Effect of Packing Size on Performance

- FLEXIPAC 1X, 1.4X, 2X, 3X, 4X, 15250 mm bed
- Increasing crimp, decreasing surface area
- Different mass transfer coefficients
- Same reactions affect absorption differently

Conditions	
Temperature (°C)	37.2
Pressure (bara)	64.4
Flow (Nm³/d)	6,700,000

Composition (mol%)			
H_2S	0.0001	iC4	0.020
CO_2	2.000	nC5	0.006
C1	94.331	iC5	0.010
C2	1.900	nC6	0.030
C3	0.170	CH_3SH	0.0007
nC4	0.030	N_2	1.500

Effect of Packing Size on Performance

FLEXIPA® (15.25 m or 50 ft)	Dry Area	Treated Gas CO_2	Bulge Temperature	Bulge Position from Top	Wetted/Dry Area Ratio
Packing Size	(m²/m³)	(ppmv)	(°C)	(m or Tray No.)	(unitless)
1X	440	0.45	88.9	13.4	0.614
1.4X	340	0.46	96.9	13.1	0.741
2X	220	1.79	114.8	9.0	0.973
3X	110	108	110.3	7.0	1.332
4X	55	516	105.4	7.0	1.691
25 Trays		0.93	88.9	Tray 20	

Effect of Packing Size on Performance

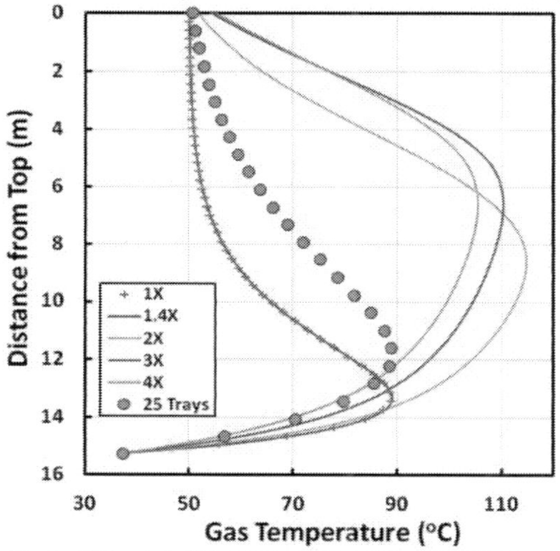

Effect of Packing Size on Performance

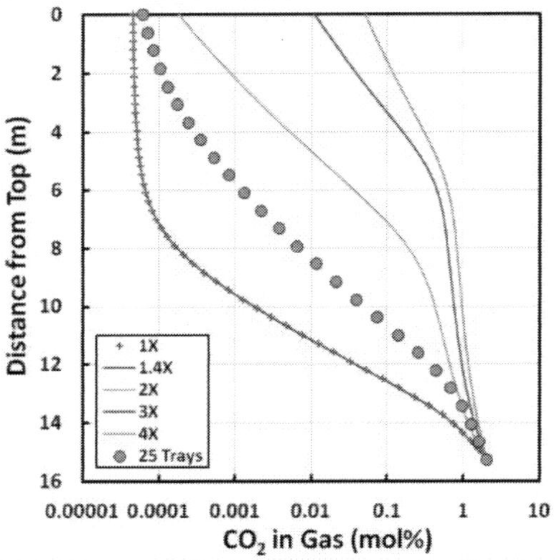

Effect of Packing Size on Performance

- Moderate size packings in this application gives bulge-pinched performance

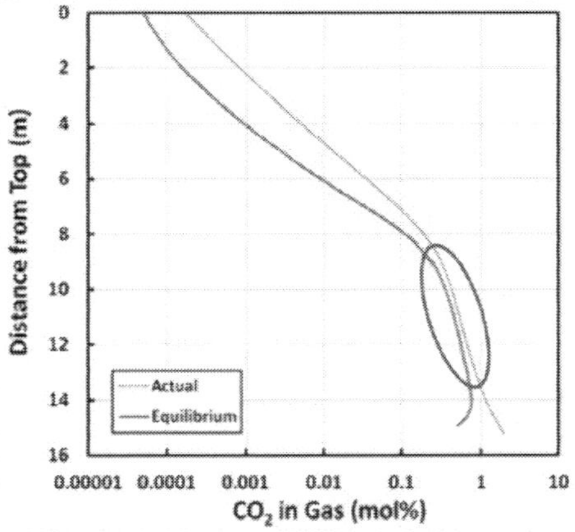

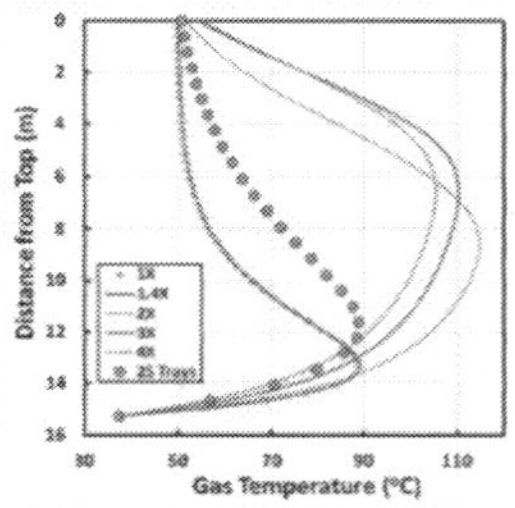

Consider Structured Packing

- Now being more widely applied
- Floating structures (FPSO, FLNG)
- Resists (but does not prevent) maldistribution from rocking
- Always consider for offshore applications
- Use the right distributor – no free surfaces, use only high-pressure-drop type for offshore
- No fouling service, needs good amine hygiene
- Naturally resists foaming
- High vapor capacity
- To lower pressure drop, use biggest crimp that provides adequate treating
- Reliable design with mass transfer rate based simulation

Summary

- Out-of-the-box ProTreat® simulation automatically matches good plant performance data "spot on"
- Generates a virtual plant
- No estimates, no guesses, no experience
- Highly reliable packed column simulation
- Select the right packing type, right size and specify the right amount to achieve treat.
- Benchmarked in wide range of operations using structured and random packing.

Enhanced Life of Field Recovery with Subsea Processing

Wiles, D.,

Originally presented on: 10/3/2013 13:30:00 - 14:00:00

Subsea boosting enables enhanced life of field recovery. Fields producing high vapor content fluids, which challenge efficient multi-phase pumps, can nonetheless be boosted subsea, if subsea separation is implemented. Production fluids from fields with low to medium reservoir pressures (5ksi – 10 ksi) can be treated by conventional gas/liquid separators, with enhanced internal features. Enhanced conventional separators deliver excellent separation efficiency, low complexity and varied supply chain options. At higher pressures multiple conventional separators can be operated in parallel or enhanced multi-pipe technology can be applied. The net result is that boosting technology can be implemented in fields that would otherwise be ineligible. In addition boosting power consumption can be reduced due to the reduction in pump fluid vapor content, lowering production operating costs.

Development of a Pipe Heat Transfer Correlation for use in Sea Bed Proximity

Dharmadhikari, C., Tridiagonal Solutions Inc.
Vedapuri, D., Tridiagonal Solutions Inc.
Golczynski, T.,
Originally presented on: 10/3/2013 14:00:00 - 14:30:00

Accurate prediction of Overall Heat Transfer Coefficient (U Value) is of vital importance to predict potential for flow assurance problems in subsea pipelines and design programs to mitigate such problems. In gas dominant wells it is important to predict the onset of condensation of water. In oil dominant wells producing waxy crudes the onset of wax deposition is of interest. An error in U value prediction would lead to incorrect prediction of these phenomena and hence incorrect design of flow assurance strategies. Several correlations based on shape functions have been developed to predict U Value for different burial depths. Accuracy of these correlations depends upon accurate input for convection heat transfer coefficient by sea water, hext.

External Heat transfer coefficient over pipe depends on the flow properties and velocity distribution around the pipe. Existing correlations assume symmetrical flow pattern across the pipe. However when the pipe is in close proximity to the sea bed, flow pattern above and beneath the pipe is not identical. Flow pattern on the bottom side of the pipe depends upon the gap between the pipe and the sea bed. During partial pipe burial, contribution to heat transfer from the bottom side of pipe is significantly less compared to top side. Several CFD simulations were performed for different burial depths and Reynolds number. Correlation for hext was modified by including burial depth as new parameter. When the pipe is far away from the sea bed, the correlation predicts hext as that of freely suspended pipe. On the other hand when the pipe is completely buried, the correlation predicts hext close to that of flow over a flat plate. The correlation was validated against data generated from CFD simulations.

Application of Heat/Mass Transfer and Thermodynamic Principles to Wax Deposition Modeling and Management

Huang, Z., Multiphase Solutions Kenny, Incorporated

Originally presented on: 10/3/2013 14:30:00 - 15:00:00

Wax deposition has been a significant flow assurance challenge in the past decades. Significant costs have been attributed to its remediation techniques. The applications of these techniques often require appropriate understandings of the fundamental knowledge from multiple disciplines. Currently different industrial users have developed their own methodologies to tackle wax deposition. However, often due to limited resources available, many of these guidelines are incomplete or are highly based on empirical evidence. This presentation is to present a general summary of industrial understandings on wax modeling management philosophies. Commonly used wax deposition models will be presented and the difference between their assumptions/results of these models will be analyzed. The last part of this presentation identifies the key issues that should be better understood for wax deposition modeling.

Multiphase Flows in Remote Gas-Liquid Processing Systems

Dakshinamoorthy, D., Wayne State University
Subramani, H. J., Chevron Energy Technology Company
Originally presented on: 10/3/2013 15:30:00 - 16:00:00

Subsea and/or remote separation of the oil, water, gas, and solid streams is an integral part of deepwater oil/gas production. Often times, improper characterization of inlet multiphase flow conditions to these separators lead to poor design and thereby lower efficiency and production. This talk will focus on such issues and provide a computational fluid dynamic (CFD) approach with experimental validation to characterize the flow in remote processing systems for reliable and efficient performance.

An Innovative Advanced Control Solution for Slug Mitigation

Nath, R., Honeywell Process Solutions
Sharma, S., Honeywell
Originally presented on: 10/3/2013 16:00:00 - 16:30:00

In Oil production processes, production from oil wells comprising a mixture of oil, water and gas flows from the wells to the surface Separators. Often, oil and water phases separate from the gas phase as production flows from the well bore to the Separators. This separation of phases can cause an undesirable operating condition when production flow becomes discontinuous with periods of large volume of liquid phase followed by periods of predominantly gas phase. This phenomenon is called slugging and causes the liquid flow into the Separators to swing significantly. Slugging is disruptive as it can lead to unstable operation of the downstream process and can sometimes cause flaring incidences or shutdown of the production process.

Traditionally, PID controllers are employed to maintain the oil level in the separators. Since PID algorithm is oblivious of the operating high and low limits, these controllers are tuned very tightly to ensure that the level remains within safe limits at all times. Doing so however propagates the fluctuations in oil flow to downstream process which is not desirable.

A solution that is sometimes employed to minimize slugging is to monitor the process operating conditions and manipulate the separator inlet choke as soon as slugging is detected. Although such strategy may mitigate slugging, it also has the undesirable side effect of choking production.

In this paper, an innovative advanced control solution that effectively mitigates slugging is presented. The proposed solution employs non-linear control technology to more fully utilize the capacity in the Separators to reduce fluctuations in the oil flow to downstream process. A case study is presented that reduces oil flow variance by 80% without curtailing production.

Active Galaxy NGC 1275
Exploring Space and Investing in the Future: Perceptions and Experiences of a Retired NASA Astronaut
Bonnie J. Dunbar, PhD NAE
University of Houston
Professor, Mechanical Engineering
Director, UH STEM Center
Director, Aerospace Grad. Pgm

The Milky Way: Inspiring Dreams

BRITISH TERRITORY
OREGON COUNTRY
SPANISH TERRITORY
UNITED STATES
Missouri
Rio Grande
President Thomas Jefferson
CORPS OF DISCOVERY 1804-06
Departed Camp Dubois on May 14, 1804, in 28 months, covered 8000 miles
William Clark
Meriwether Lewis

Robert Goddard 1882 - 1945

"It was one of the quiet, colorful afternoons of sheer beauty which we have in October in New England, and as I looked toward the fields at the east, I imagined how wonderful it would be to make some device which had even the ***possibility*** of ascending to Mars, and how it would look on a small scale, if sent up from the meadow at my feet."

"I was a different boy when I descended the tree from when I ascended, for existence at last seemed very purposive.“ 1899

October, 1957
Sputnik Launched by the Soviet Union:

Soviet Union--Yuri Gagarin:
First Human to Orbit the Earth:
April 12, 1961
CCCP
The Huntsville Times
Man Enters Space
'So Close, Yet So Far,' Sighs Cape
Soviet Officer Orbits Globe In 5-Ton Ship
Praise Is Heaped On Major Gagarin
Reds Win Running Lead In Race To Control Space

United States: May 5, 1961
Alan Shepherd launched Sub-orbital

Mercury 7

United States: John Glenn launched into orbit Feb 20, 1962

Flies again on Space Shuttle at age 77

President Kennedy Commits the Nation

Rice University, Houston, TX - 9/12/62

- *"...This generation does not intend to founder in the backwash of the coming age of space. We mean to be a part of it – we mean to lead it."*

President Kennedy Commits the Nation

Rice University, Houston, TX - 9/12/62

- *...This country was conquered by those who moved forward –and so will space...We choose to go to the moon in this decade and do the other things, not because they are easy, but because they are hard...The growth of our science and education will be enriched by new knowledge of our universe and environment....*

July 20, 1969: Neil Armstrong walked on the Moon

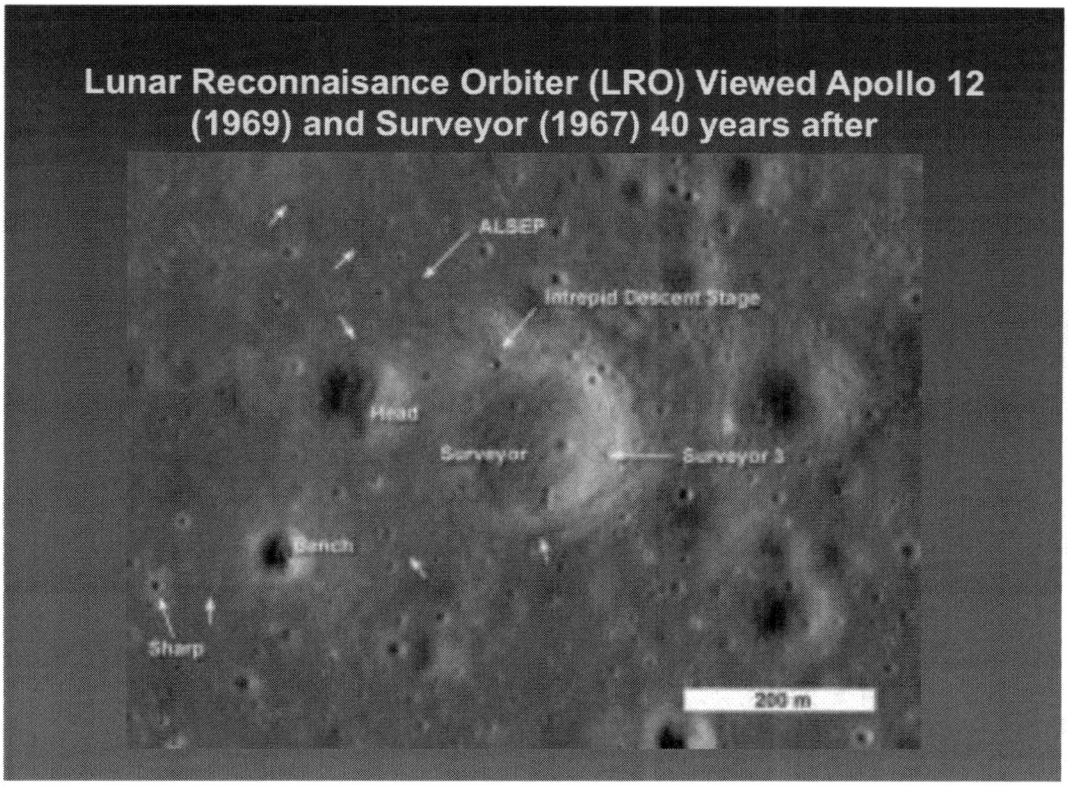
Lunar Reconnaisance Orbiter (LRO) Viewed Apollo 12
(1969) and Surveyor (1967) 40 years after
ALSEP
Intrepid Descent Stage
Head
Surveyor
Surveyor 3
Bench
Sharp
200 m

Skylab
UNITED STATES OF AMERICA

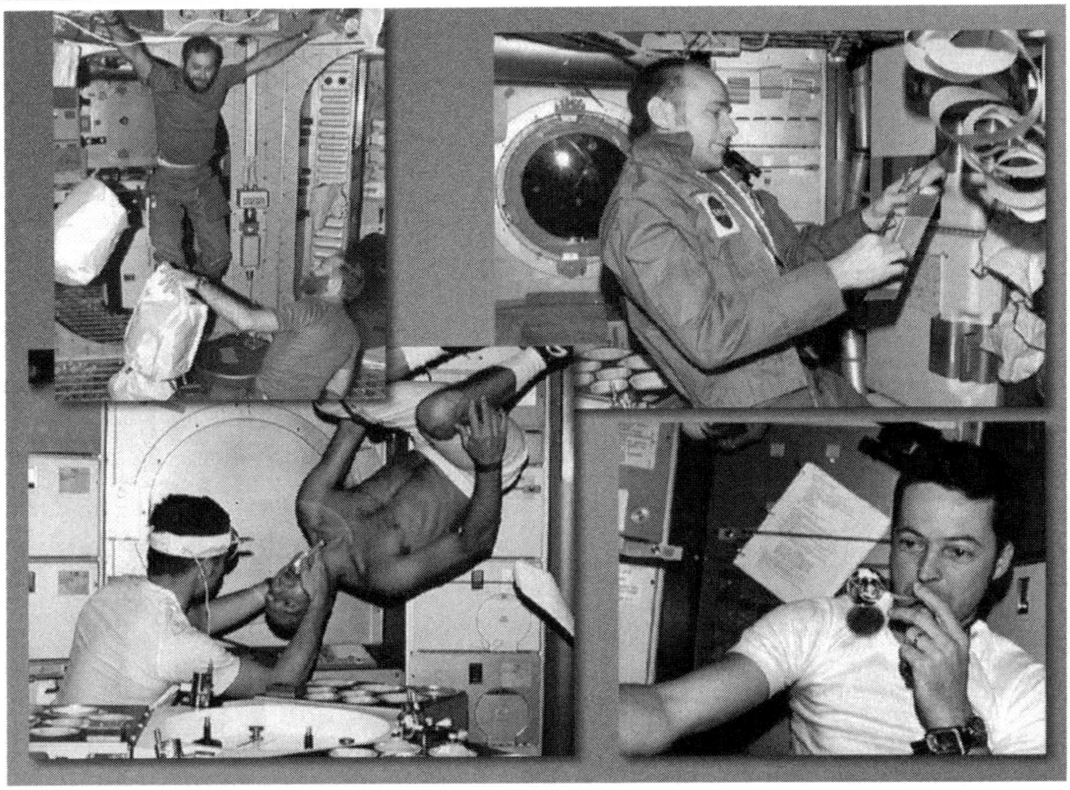

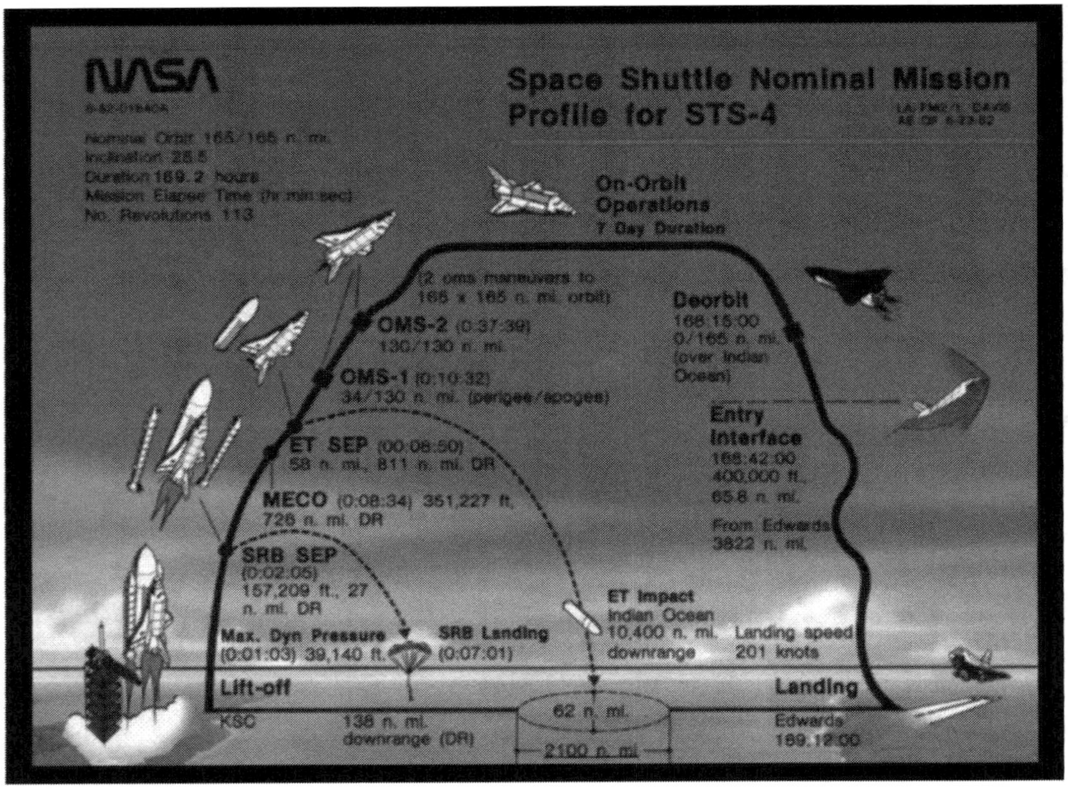
NASA
Space Shuttle Nominal Mission Profile for STS-4
Nominal Orbit 165/165 n. mi.
Inclination 28.5
Duration 169.2 hours
Mission Elapse Time (hr:min:sec)
No. Revolutions 113
On-Orbit Operations
7 Day Duration
(2 oms maneuvers to 165 x 165 n. mi. orbit)
OMS-2 (0:37:39)
130/130 n. mi.
OMS-1 (0:10:32)
34/130 n. mi. (perigee/apogee)
ET SEP (00:08:50)
58 n. mi., 811 n. mi. DR
MECO (0:08:34) 351,227 ft.
726 n. mi. DR
SRB SEP
(0:02:05)
157,209 ft., 27
n. mi. DR
Max. Dyn Pressure
(0:01:03) 39,140 ft.
SRB Landing
(0:07:01)
Lift-off
KSC
138 n. mi.
downrange (DR)
62 n. mi.
2100 n. mi.
ET Impact
Indian Ocean
10,400 n. mi.
downrange
Deorbit
168:15:00
0/165 n. mi.
(over Indian
Ocean)
Entry
Interface
168:42:00
400,000 ft.
65.8 n. mi.
From Edwards
3822 n. mi.
Landing speed
201 knots
Landing
Edwards
169.12:00

First Flight
April 12, 1981
•135 flights
•More than 2000 Experiments
•More than 200 Universities
•More than 200 K-12 Classes
•More than 3000 Graduate Students
•More than 500 Companies
Only Space Vehicle that Returned Cargo back to Earth
•Only Reusable Space Flight Vehicle (100X)
•Five Shuttles:
•Columbia
•Challenger
•Discovery
•Endeavor
•Atlantis
•Critical for human and robotic space Exploration Research

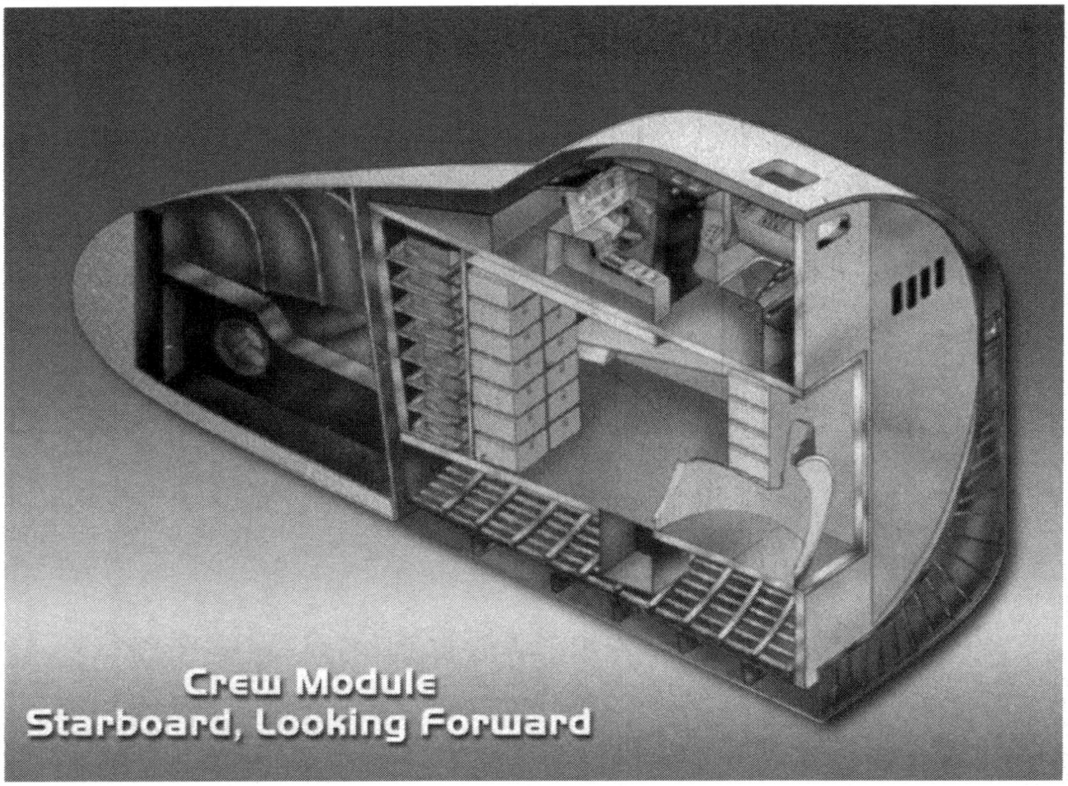
Crew Module
Starboard, Looking Forward

Reentry (Artist Concept c.1968)

Dr. James I. Mueller (Doc)

University of Washington

Mining, Metallurgical and Ceramic Engineering

Bonnie Dunbar
Has Her Sights On
the Stars
Rockwell International

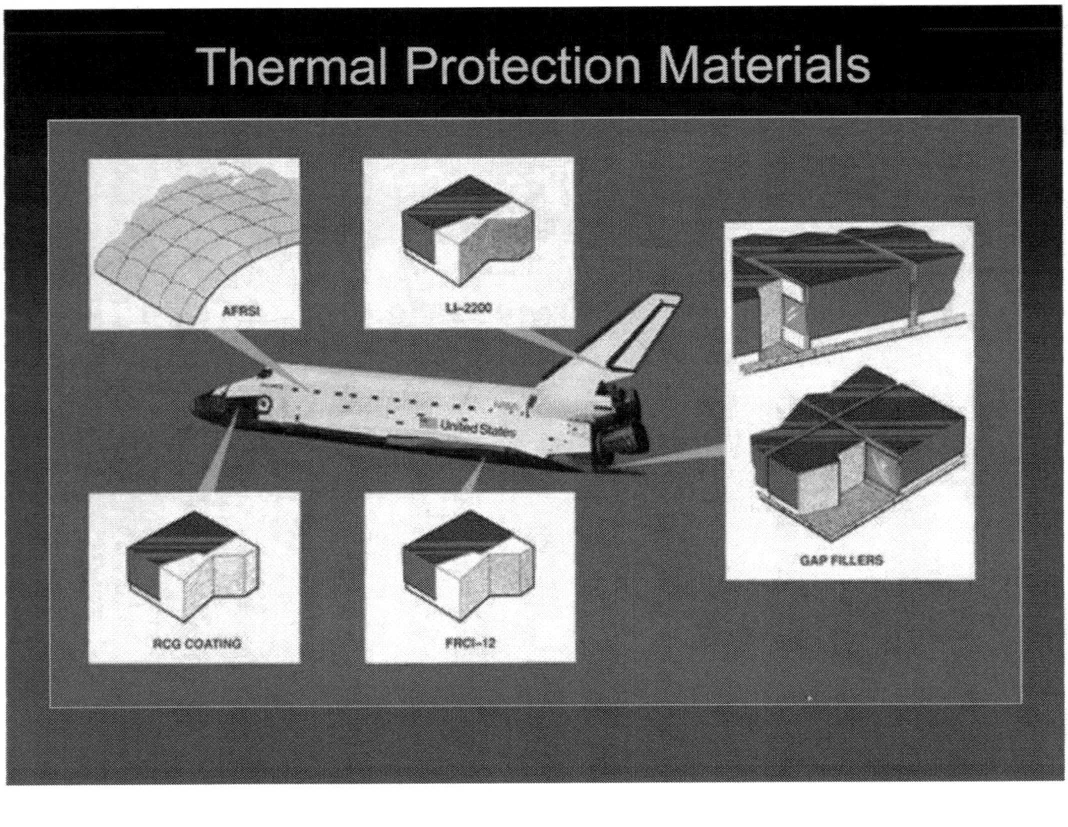
Thermal Protection Materials
AFRSI
LI-2200
RCG COATING
FRCI-12
GAP FILLERS
United States

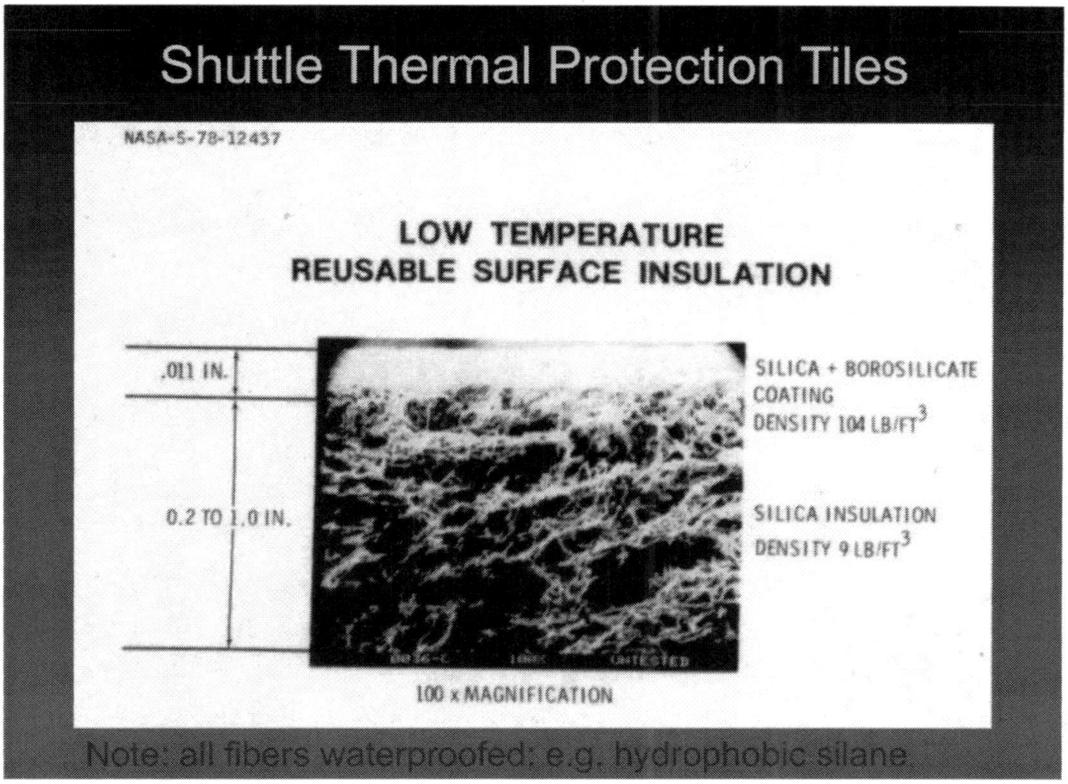
Shuttle Thermal Protection Tiles
NASA-5-78-12437
LOW TEMPERATURE
REUSABLE SURFACE INSULATION
.011 IN.
0.2 TO 1.0 IN.
SILICA + BOROSILICATE
COATING
DENSITY 104 LB/FT3
SILICA INSULATION
DENSITY 9 LB/FT3
100 x MAGNIFICATION
Note: all fibers waterproofed: e.g. hydrophobic silane

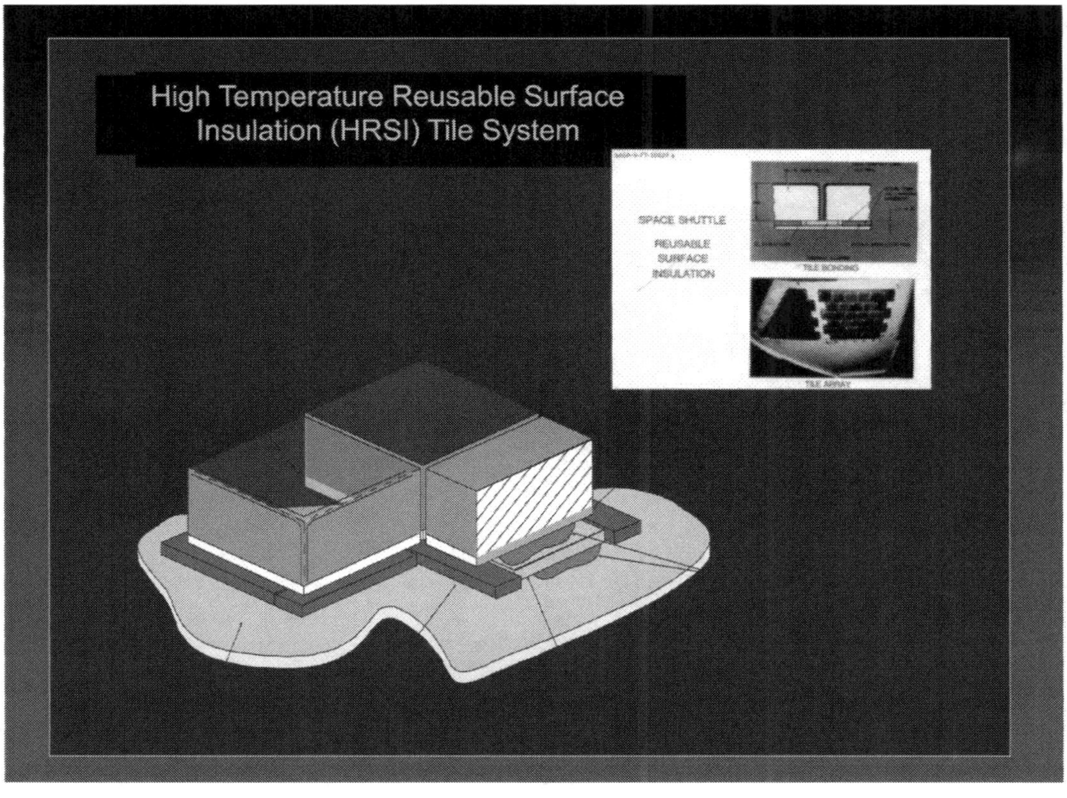
High Temperature Reusable Surface
Insulation (HRSI) Tile System
SPACE SHUTTLE
REUSABLE
SURFACE
INSULATION
TILE BONDING
TILE ARRAY

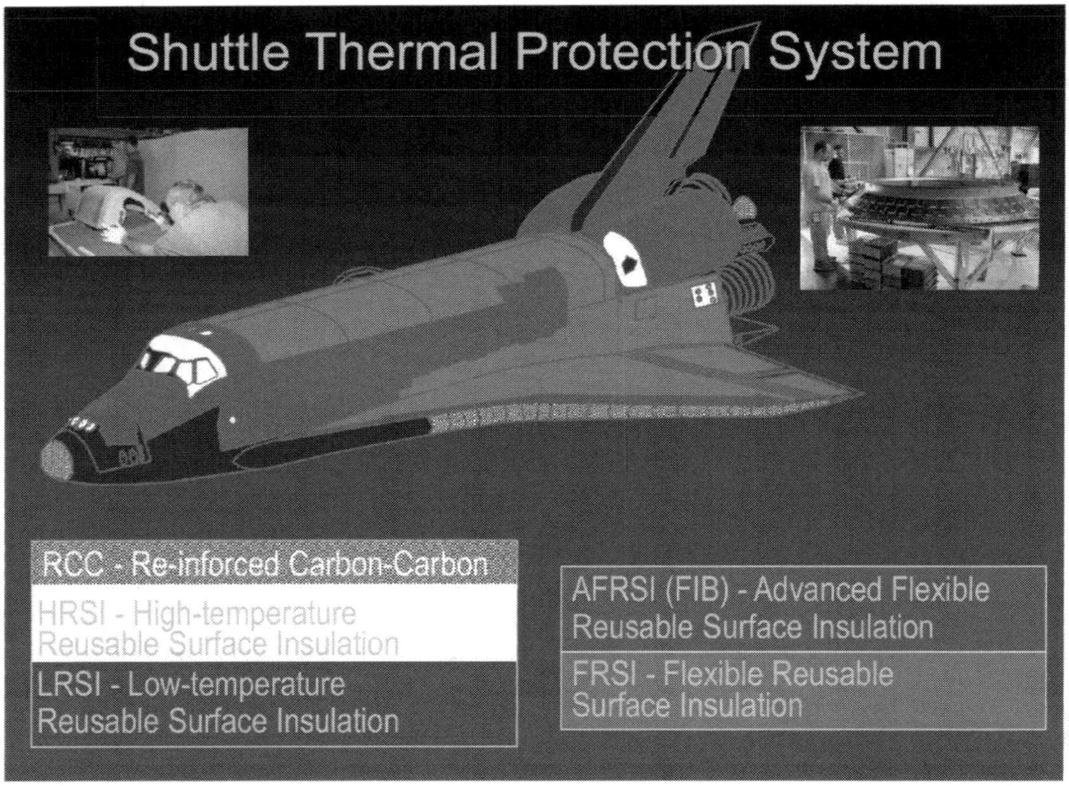
Shuttle Thermal Protection System
RCC - Re-inforced Carbon-Carbon
HRSI - High-temperature Reusable Surface Insulation
LRSI - Low-temperature Reusable Surface Insulation
AFRSI (FIB) - Advanced Flexible Reusable Surface Insulation
FRSI - Flexible Reusable Surface Insulation

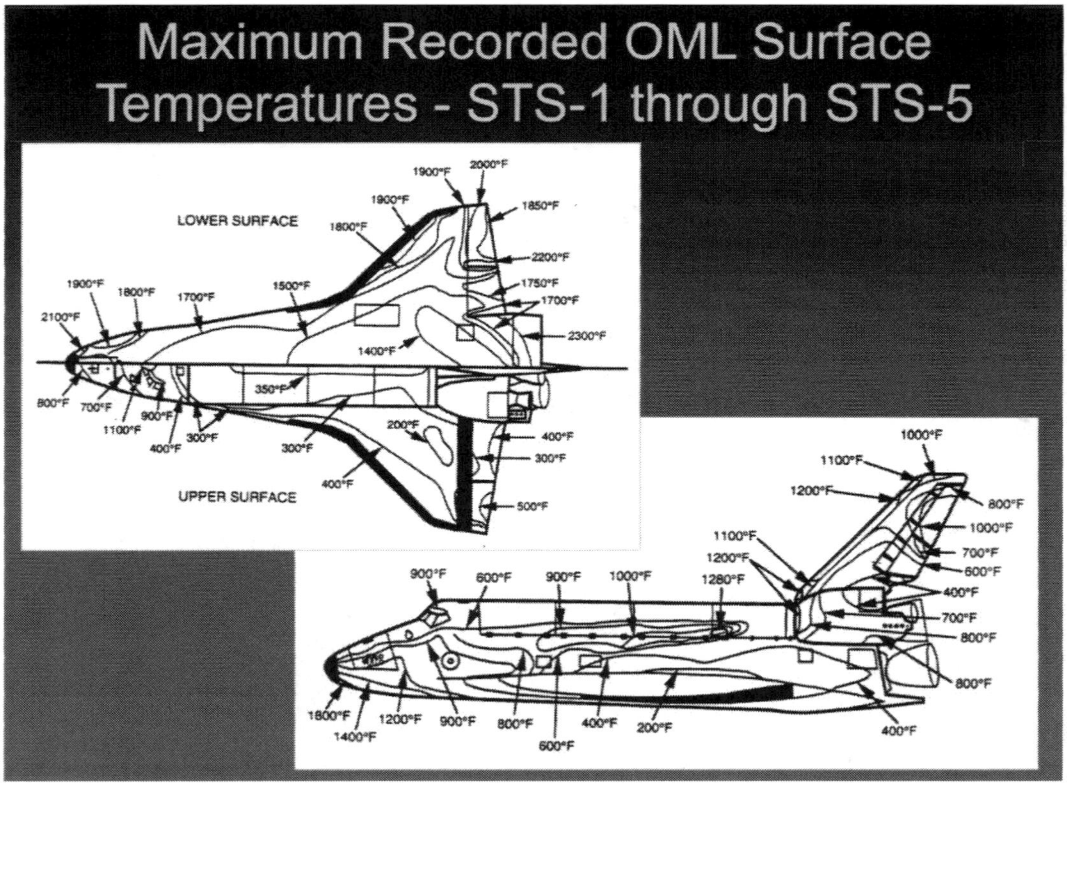
Maximum Recorded OML Surface Temperatures - STS-1 through STS-5
LOWER SURFACE
UPPER SURFACE
1900°F
2000°F
1900°F
1850°F
1800°F
2200°F
1750°F
1700°F
2300°F
1900°F
1800°F
1700°F
1500°F
2100°F
1400°F
350°F
800°F
700°F
1100°F
900°F
400°F
300°F
300°F
200°F
400°F
400°F
300°F
500°F
1000°F
1100°F
1200°F
800°F
1000°F
700°F
600°F
400°F
1100°F
1200°F
1280°F
900°F
600°F
900°F
1000°F
700°F
800°F
800°F
1800°F
1400°F
1200°F
900°F
800°F
600°F
400°F
200°F
400°F

Dr. James I Mueller ("Doc")
Property of the University of Washington Libraries, Special Collections Division
University Of Washington
Chair, Mining, Metallurgical and Ceramic Engineering (Materials Science and Engineering)

Astronaut Class 1980
NASA
19 80

Five Space Shuttle Flights: 50 Days
Columbia (x2), Challenger, Atlantis, Endeavor

STS-61A, D-1 10-30-1985
7 Days , Challenger
HARTSFIELD
SPACELAB
BUCHLI
NAGEL
D1
BLUFORD
DUNBAR
MESSERSCHMID
OCKELS
FURRER

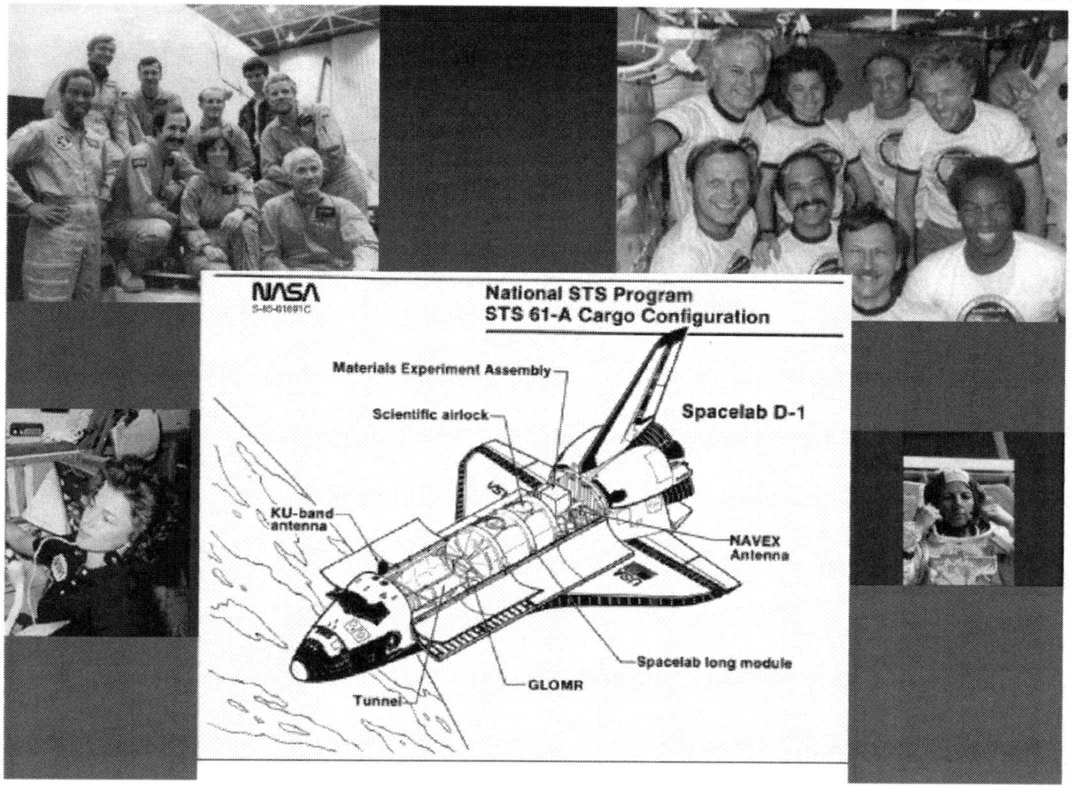
NASA
National STS Program
STS 61-A Cargo Configuration
Materials Experiment Assembly
Scientific airlock
Spacelab D-1
KU-band antenna
NAVEX Antenna
Spacelab long module
GLOMR
Tunnel

STS-32 January 1990
Columbia 11 days

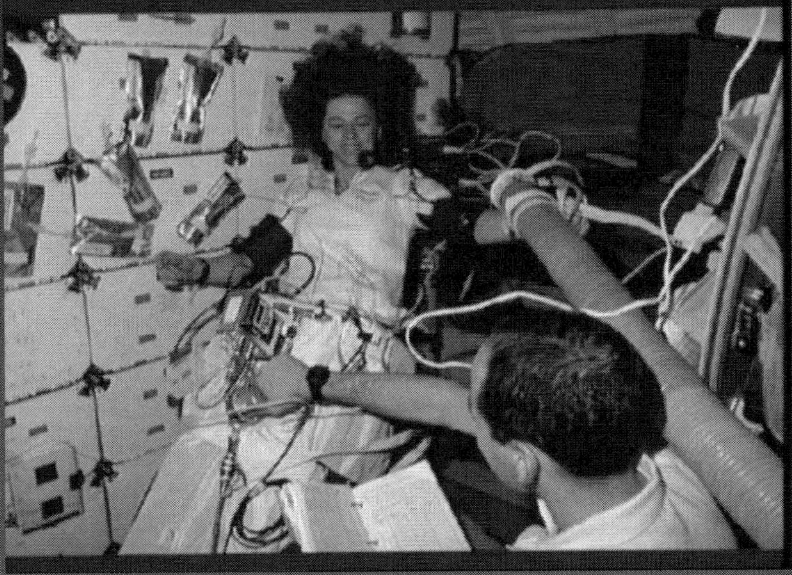

STS-50 USML-1
(United States Microgravity Lab)
June 1992 13 Days

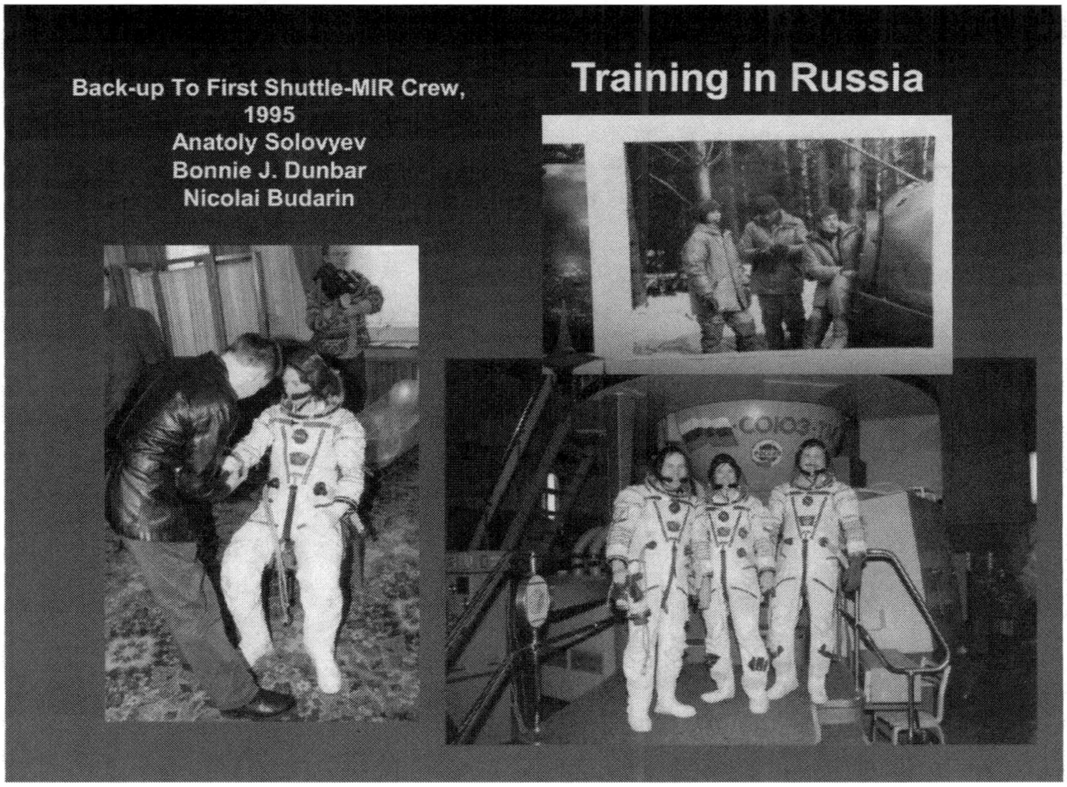
Back-up To First Shuttle-MIR Crew,
1995
Anatoly Solovyev
Bonnie J. Dunbar
Nicolai Budarin
Training in Russia

Valentina Tereshkova (USSR)
Roberta Bondar (Canada)
Mary Cleave (USA)
Svetlana Saviskaya (USSR)
Bonnie J. Dunbar (USA)
Helen Sharman (UK)
Meeting international pioneers in Russia

Winter Survival in Siberia

STS-71 First Shuttle-MIR Docking Mission June 1995
PRECOURT GIBSON
ДЕЖУРОВ СТРЕКАЛОВ
BAKER HARBAUGH DUNBAR СОЛОВЬЕВ БУДАРИН THAGARD

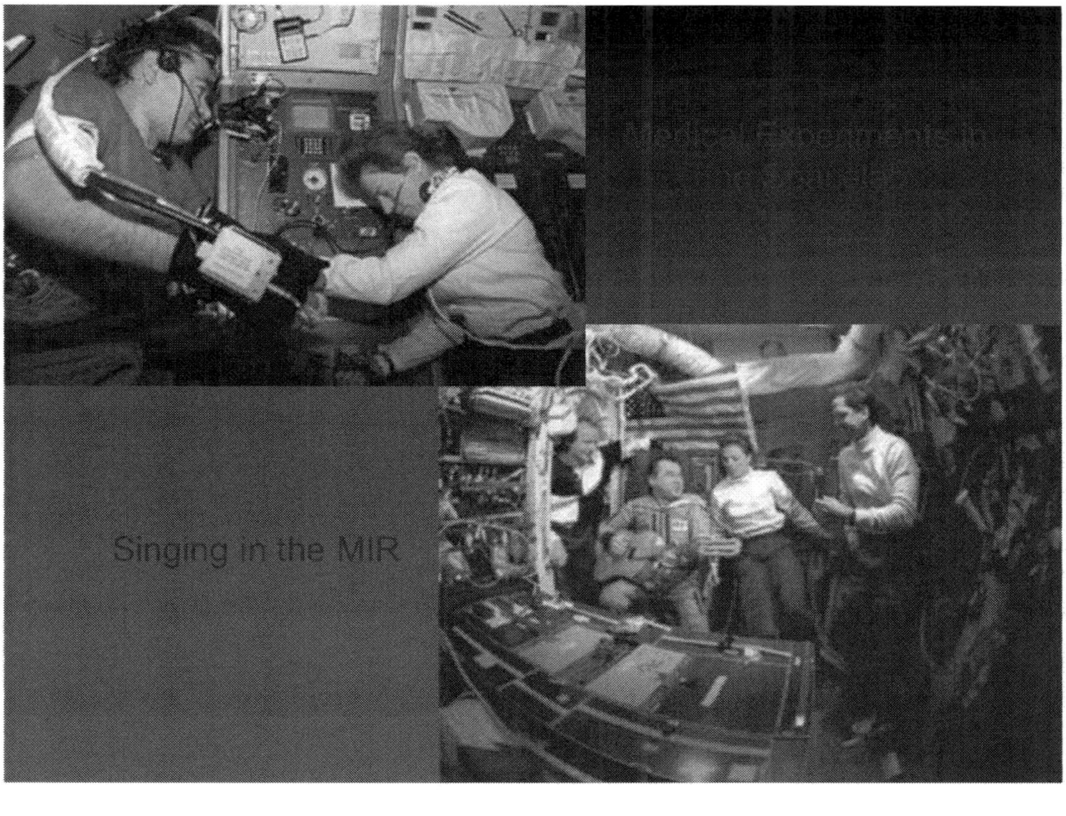
Medical Experiments in
Singing in the MIR

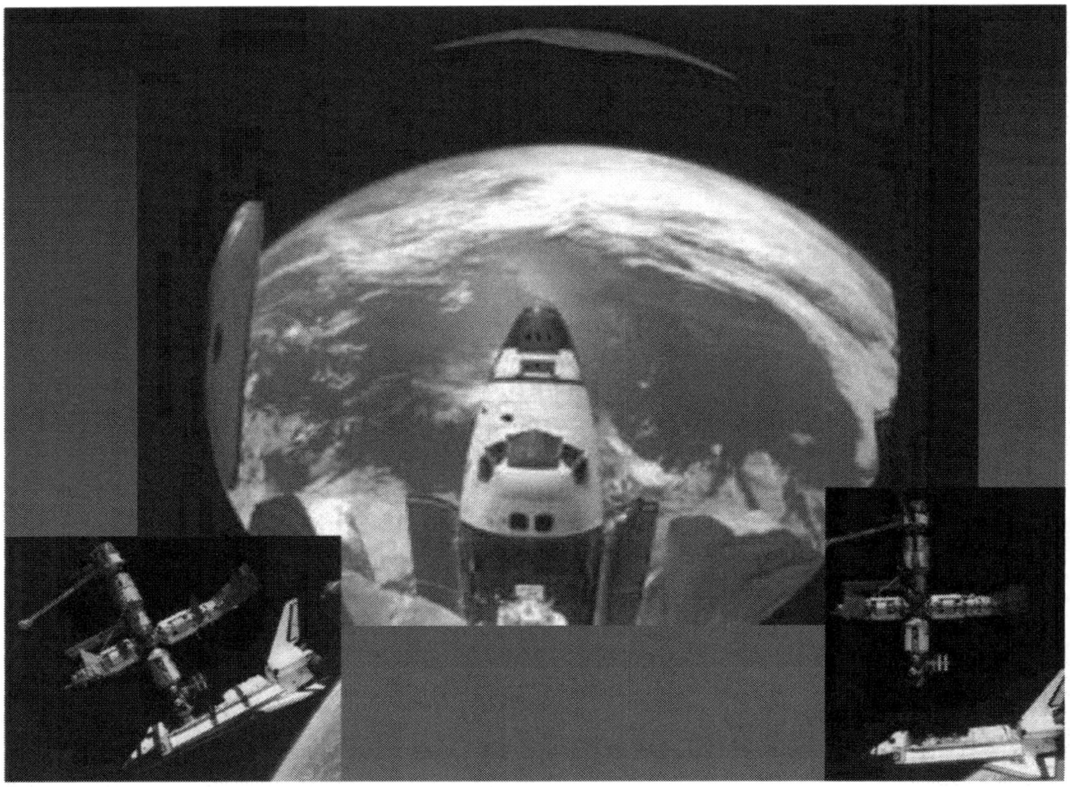

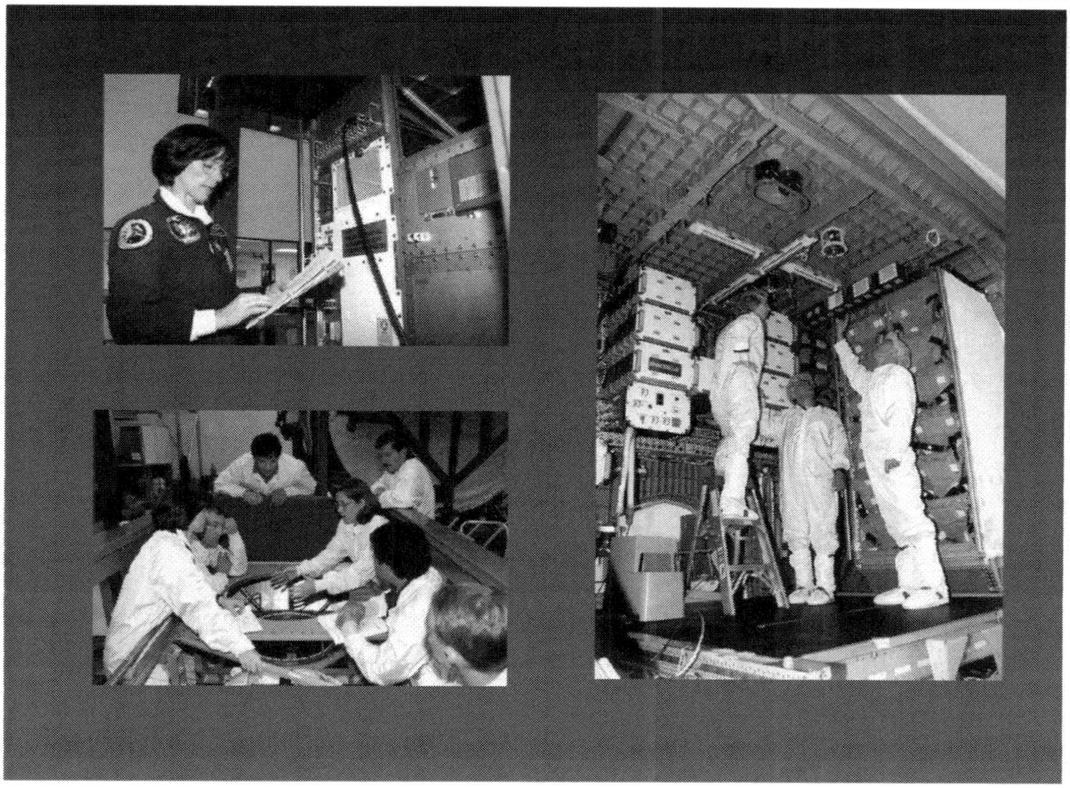

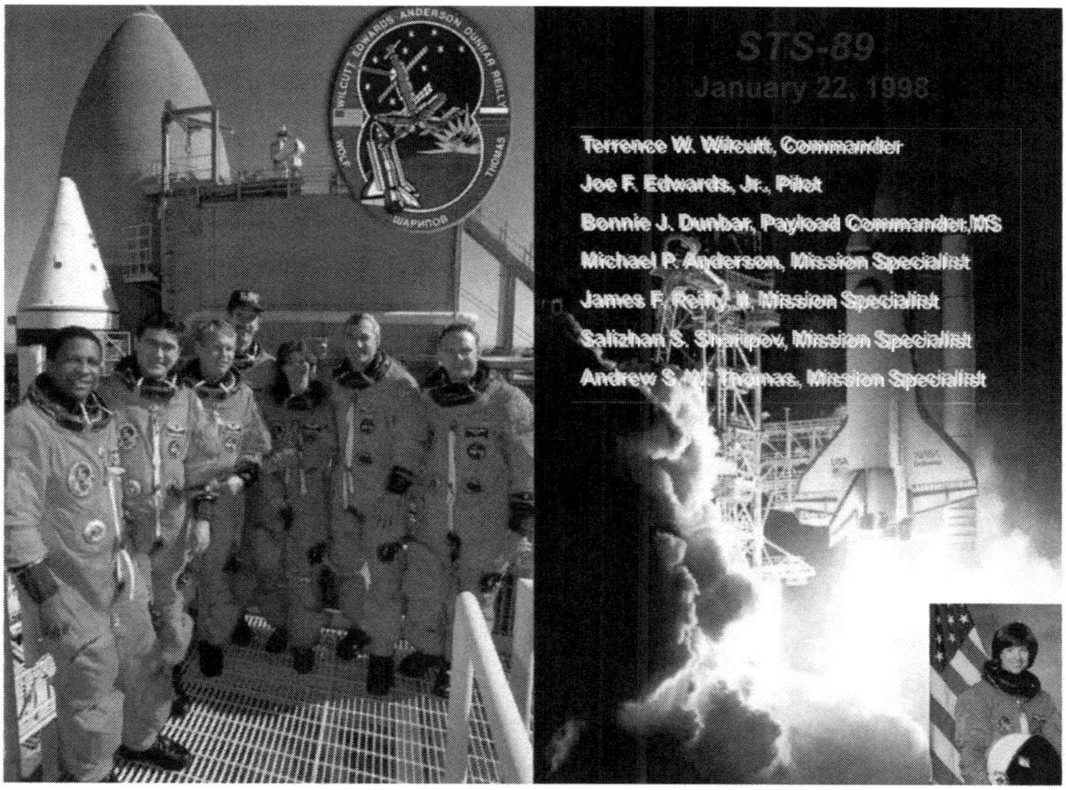
WILCUTT EDWARDS ANDERSON DUNBAR REILLY
THOMAS
ШАРИПОВ
STS-89
January 22, 1998
Terrence W. Wilcutt, Commander
Joe F. Edwards, Jr., Pilot
Bonnie J. Dunbar, Payload Commander,MS
Michael P. Anderson, Mission Specialist
James F. Reilly, II, Mission Specialist
Salizhan S. Sharipov, Mission Specialist
Andrew S Thomas, Mission Specialist

NASA

I ♥ EAA

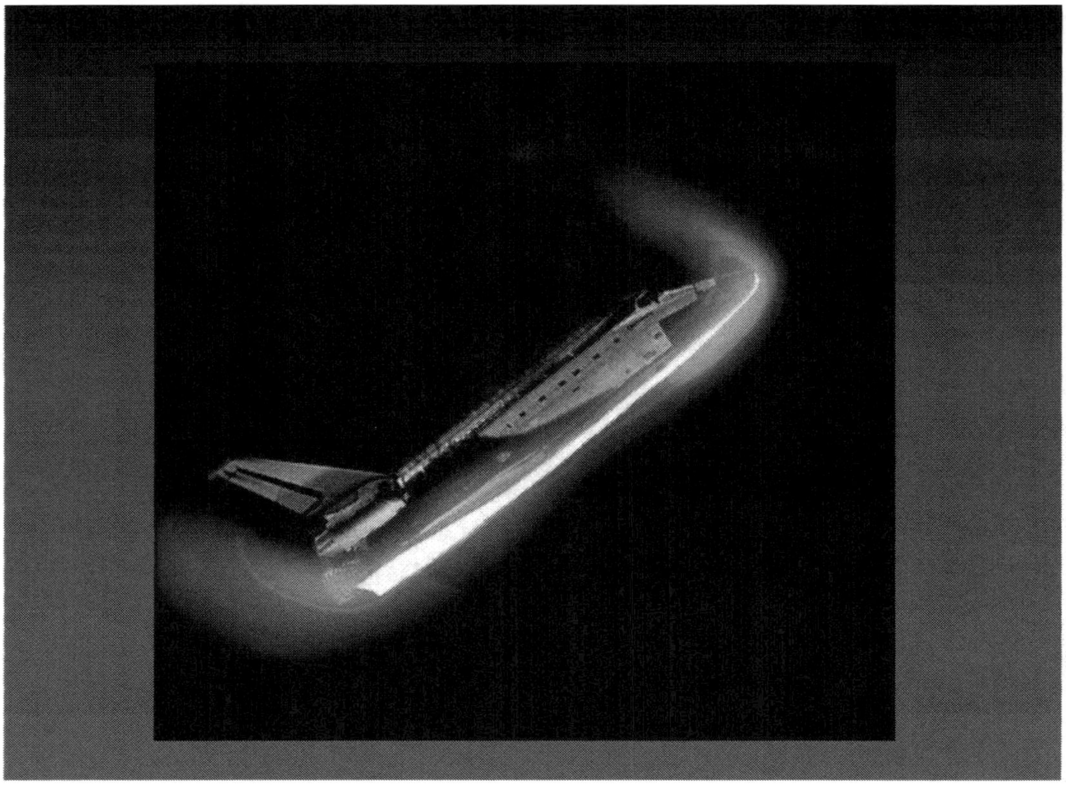

United States

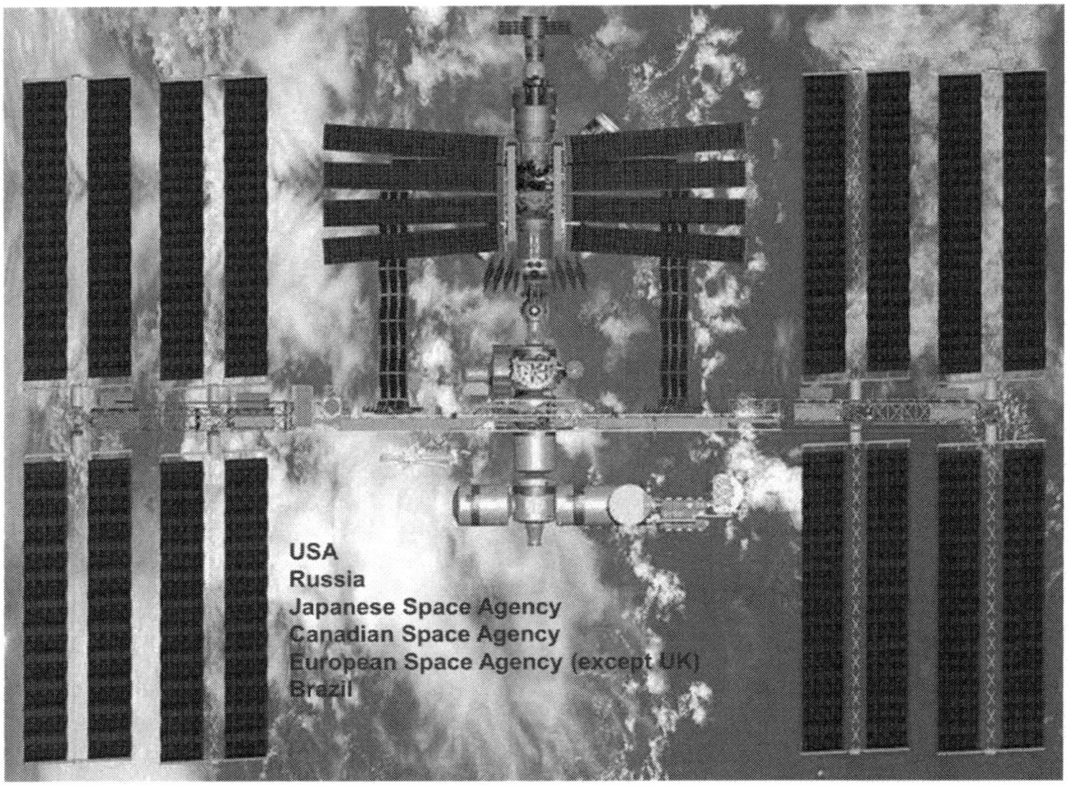
USA
Russia
Japanese Space Agency
Canadian Space Agency
European Space Agency (except UK)
Brazil

"I'm not afraid of storms...
for I am learning to sail my ship."
Louisa May Alcott

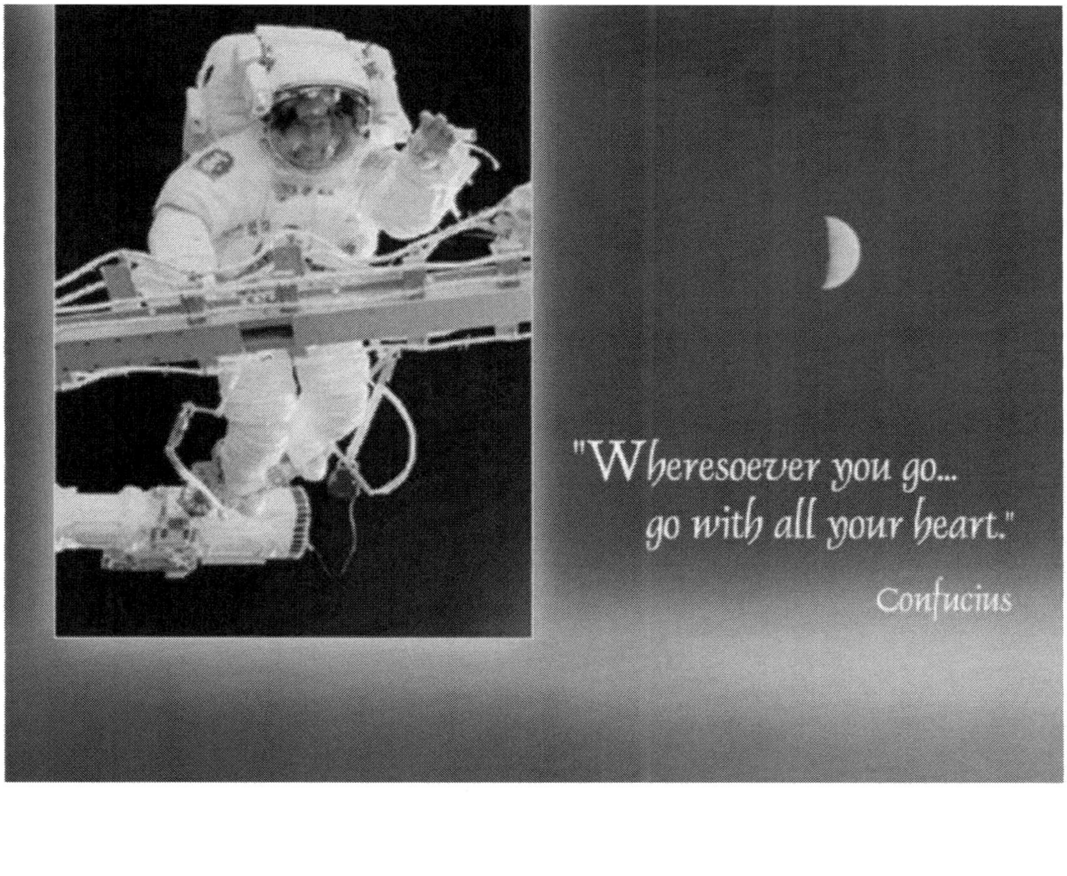
"Wheresoever you go...
go with all your heart."
Confucius

Canada

"I never see
what has been done...
I only see
what remains to be done."
Marie Curie

"Every great
advance in science
has issued from
a new audacity...
of the imagination."
John Dewey

Atlantis
UNITY

"What makes
the desert beautiful",
said the little prince,
"is that somewhere
it hides a well..."
Antoine De Saint Exupery

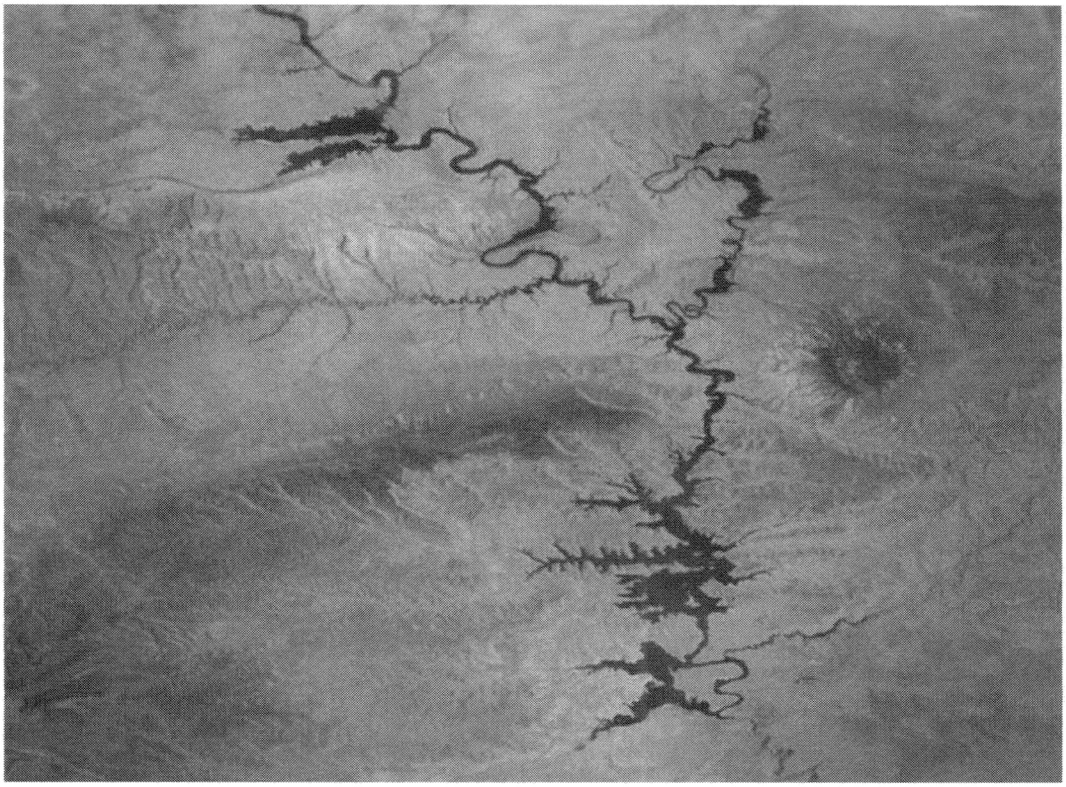

"Am I not destroying
my enemies
when I make friends
of them?"
Abraham Lincoln

"We must become the change we want to see."
Gandhi

International Space Station (ISS)...

http://www.spaceflight.nasa.gov

aurora
lightning
Chicago
Minneapolis/St. Paul
Appalachians
Des Moines
St. Louis
Omaha
N

THE INTERATIONAL SPACE STATION:
OVER TEN YEARS OF CONTINUOUS HUMAN OCCUPATON

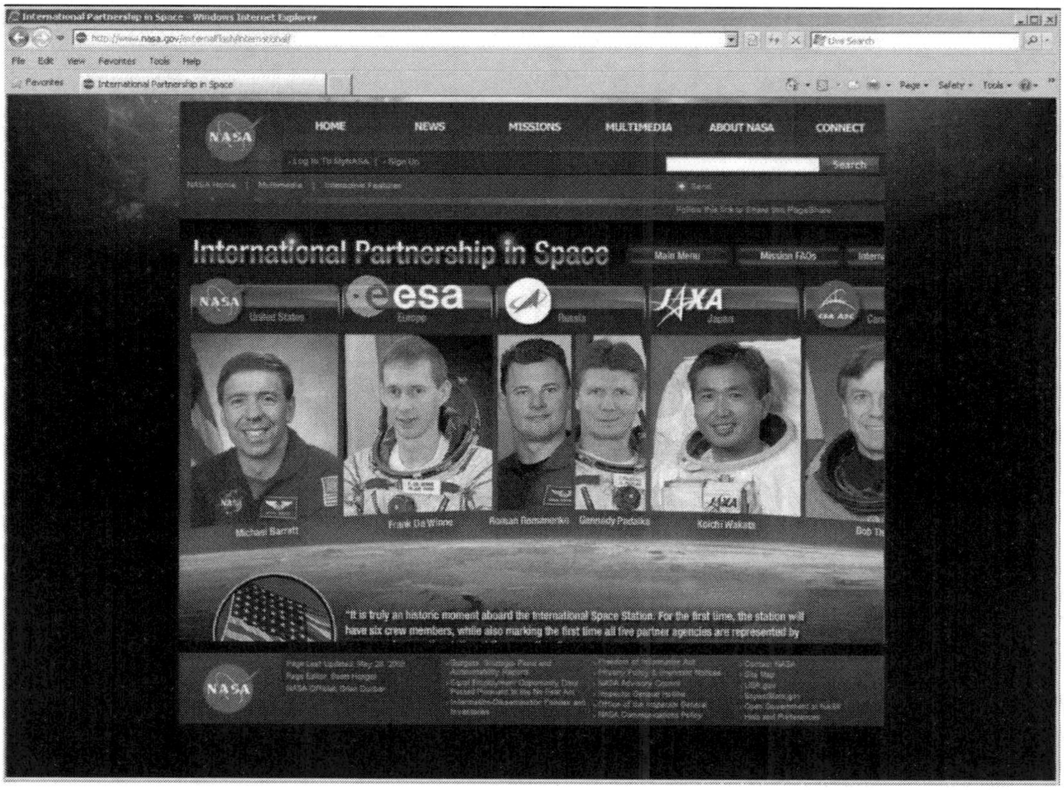
International Partnership in Space - Windows Internet Explorer
HOME
NEWS
MISSIONS
MULTIMEDIA
ABOUT NASA
CONNECT
Search
International Partnership in Space
Main Menu
Mission FAQs
esa
JAXA
United States
Europe
Russia
Japan
Michael Barratt
Frank De Winne
Roman Romanenko
Gennady Padalka
Koichi Wakata
"It is truly an historic moment aboard the International Space Station. For the first time, the station will have six crew members, while also marking the first time all five partner agencies are represented by

Discovery Rolls to Pad for Last flight:
February, 2011

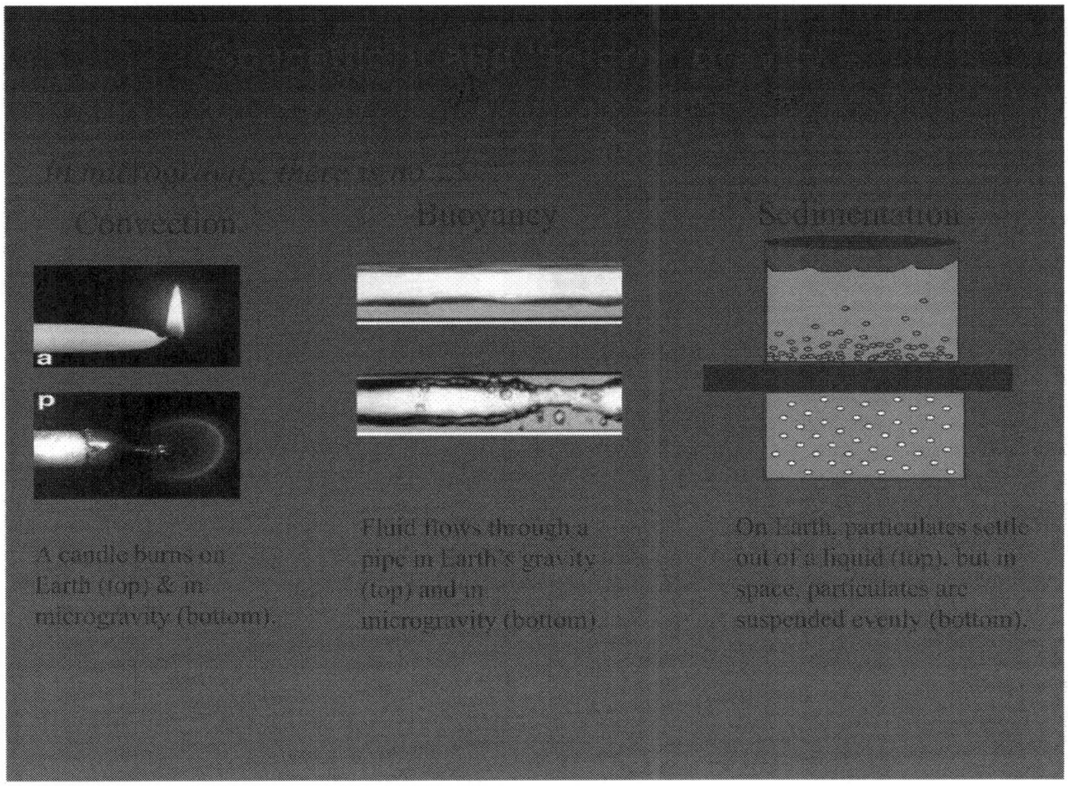
In microgravity, there is no…
Convection
Buoyancy
Sedimentation
a
b
A candle burns on Earth (top) & in microgravity (bottom).
Fluid flows through a pipe in Earth's gravity (top) and in microgravity (bottom).
On Earth, particulates settle out of a liquid (top), but in space, particulates are suspended evenly (bottom).

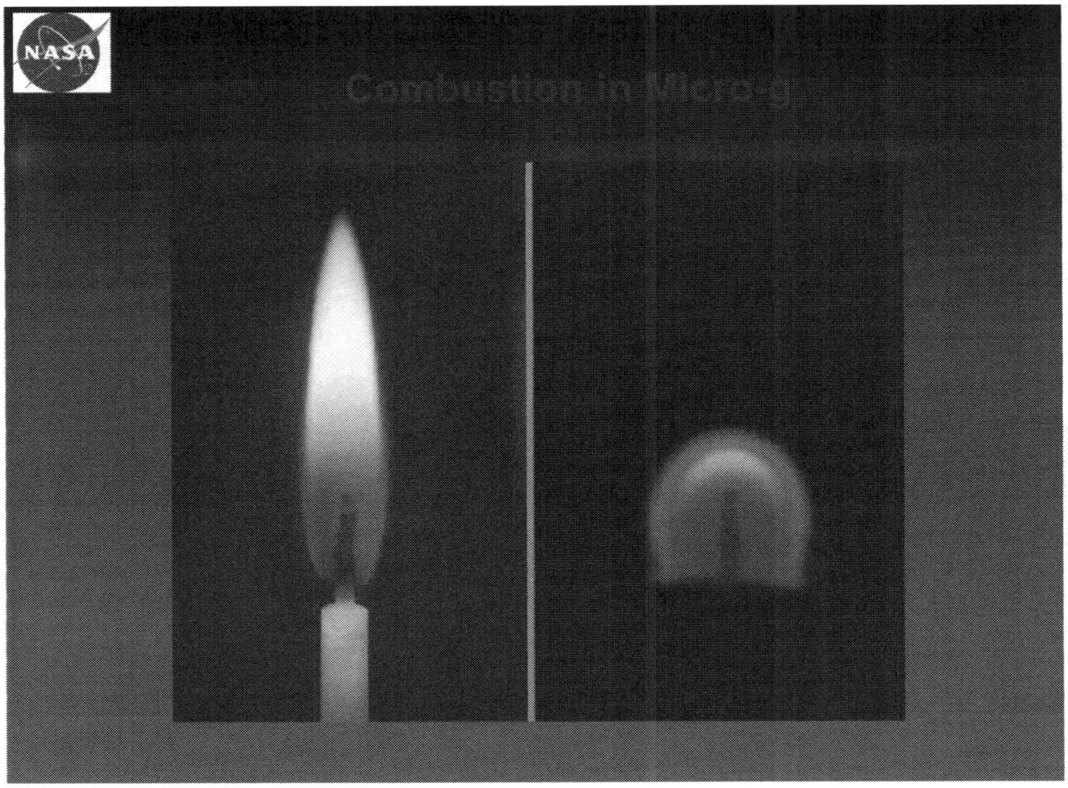
Combustion in Micro-g

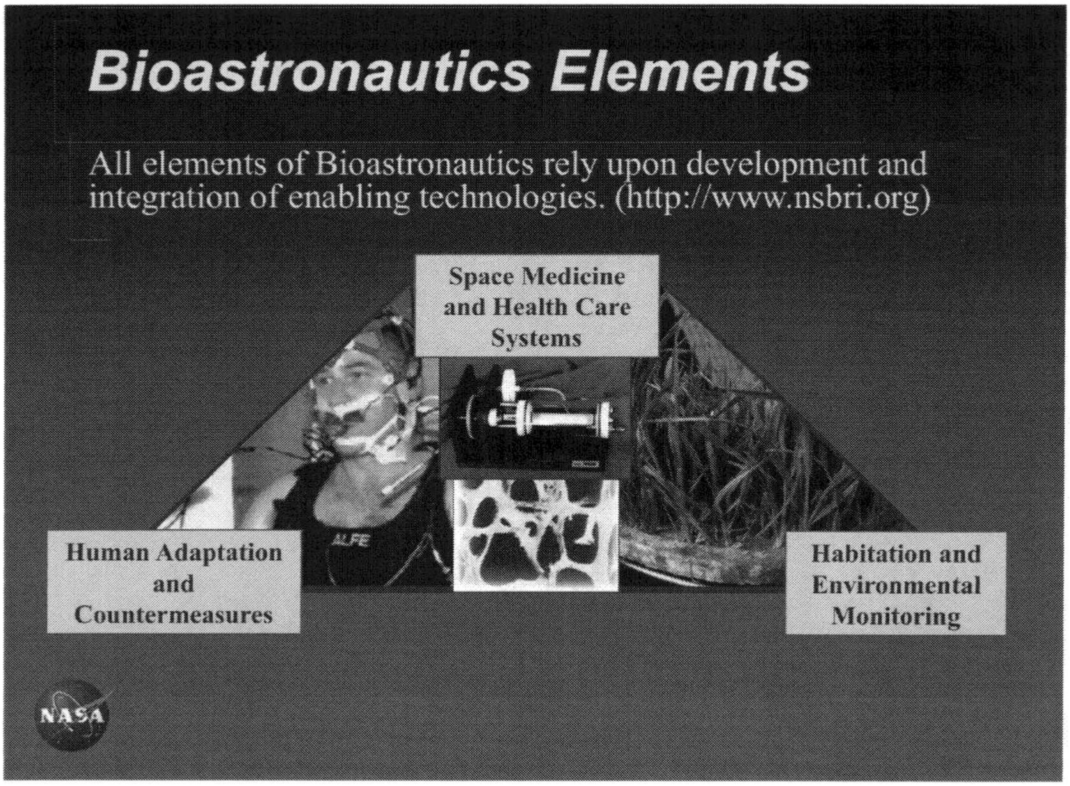
Bioastronautics Elements
All elements of Bioastronautics rely upon development and integration of enabling technologies. (http://www.nsbri.org)
Space Medicine and Health Care Systems
Human Adaptation and Countermeasures
Habitation and Environmental Monitoring
NASA

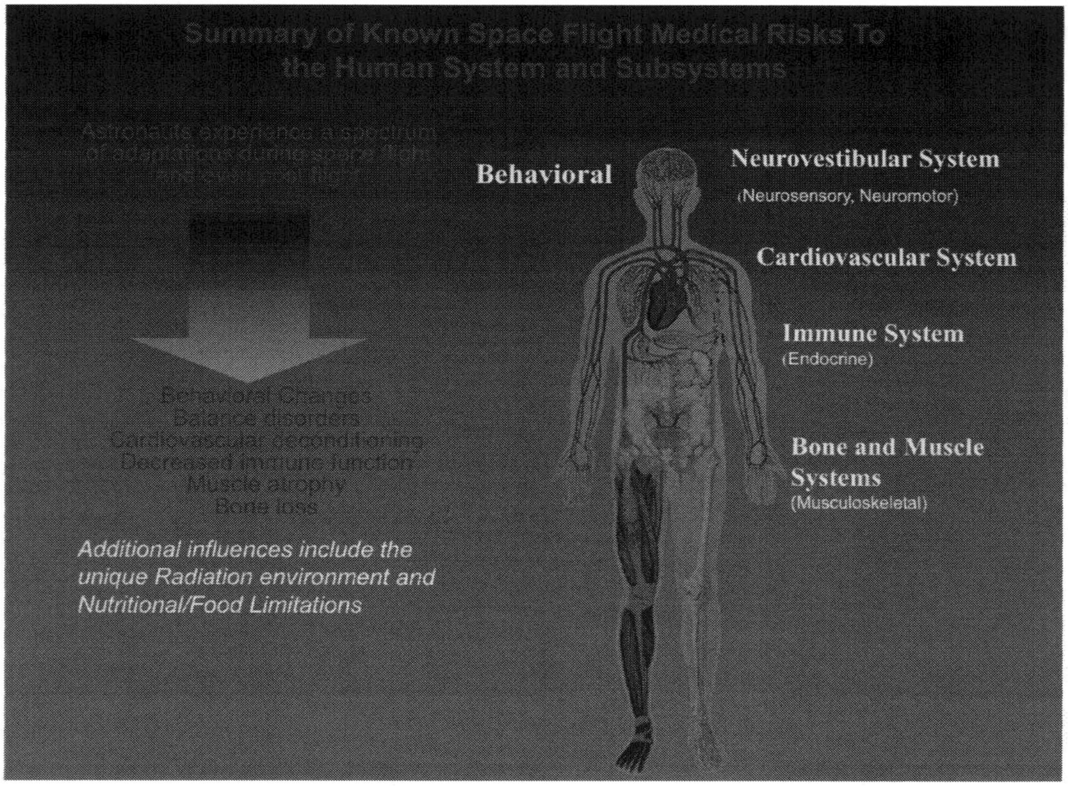
Summary of Known Space Flight Medical Risks To the Human System and Subsystems
Behavioral
Neurovestibular System
(Neurosensory, Neuromotor)
Cardiovascular System
Immune System
(Endocrine)
Bone and Muscle Systems
(Musculoskeletal)
Behavioral Changes
Balance disorders
Cardiovascular deconditioning
Decreased immune function
Muscle atrophy
Bone loss
Additional influences include the unique Radiation environment and Nutritional/Food Limitations

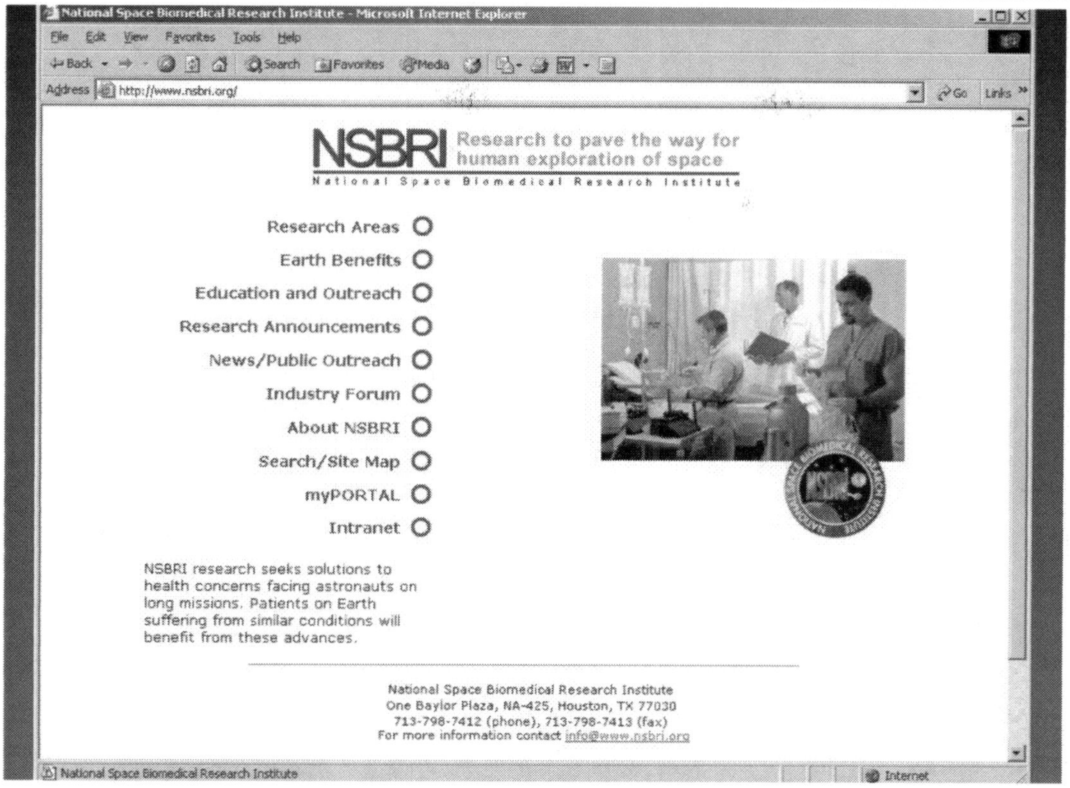
National Space Biomedical Research Institute - Microsoft Internet Explorer
http://www.nsbri.org/
NSBRI
Research to pave the way for human exploration of space
National Space Biomedical Research Institute
Research Areas
Earth Benefits
Education and Outreach
Research Announcements
News/Public Outreach
Industry Forum
About NSBRI
Search/Site Map
myPORTAL
Intranet
NSBRI research seeks solutions to health concerns facing astronauts on long missions. Patients on Earth suffering from similar conditions will benefit from these advances.
National Space Biomedical Research Institute
One Baylor Plaza, NA-425, Houston, TX 77030
713-798-7412 (phone), 713-798-7413 (fax)
For more information contact info@www.nsbri.org

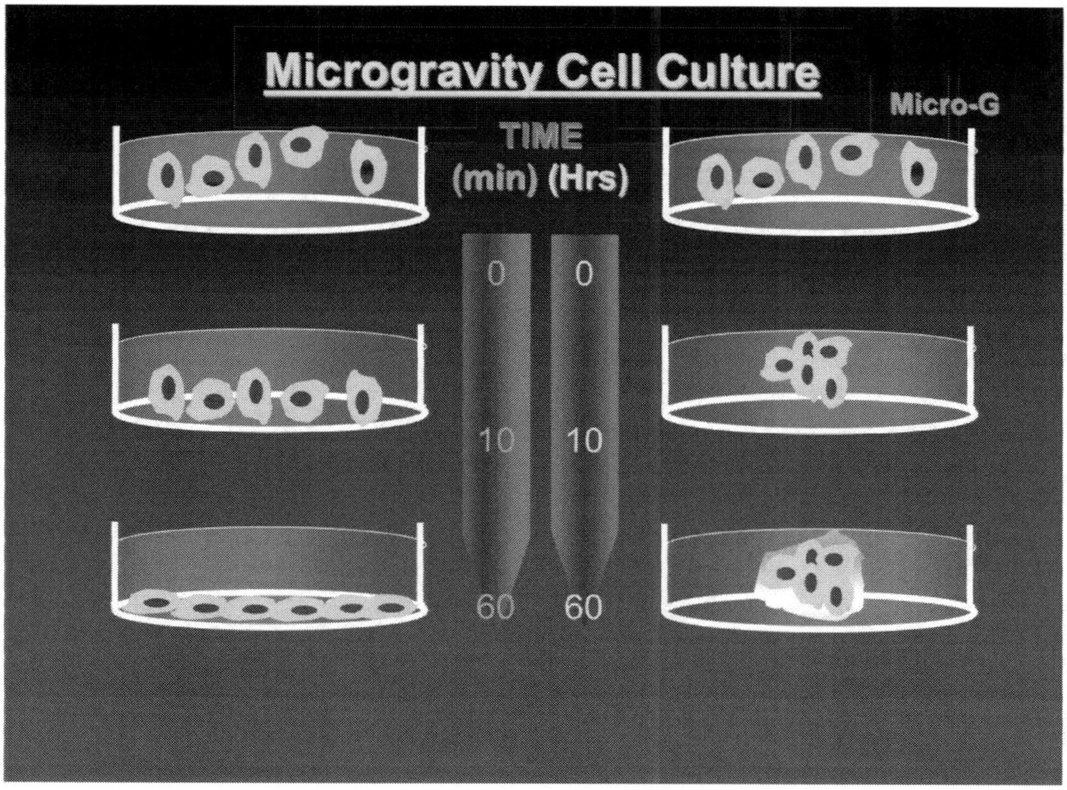
Microgravity Cell Culture
Micro-G
TIME
(min) (Hrs)
0
0
10
10
60
60

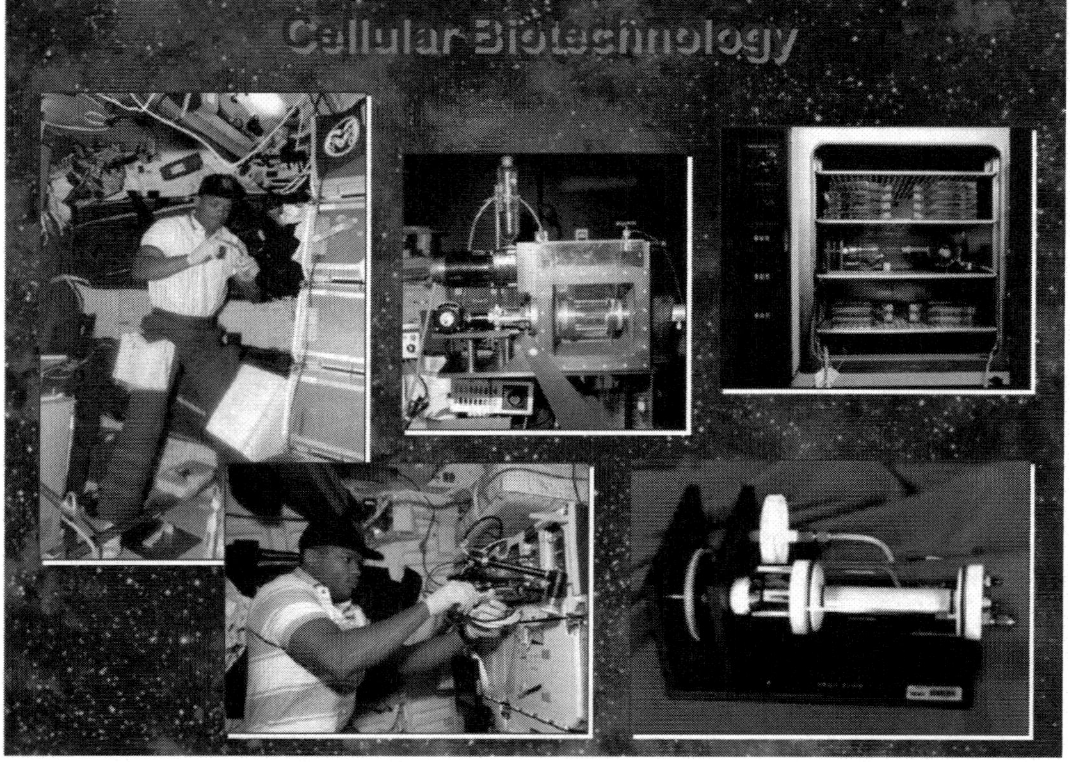
Cellular Biotechnology

What's Next?

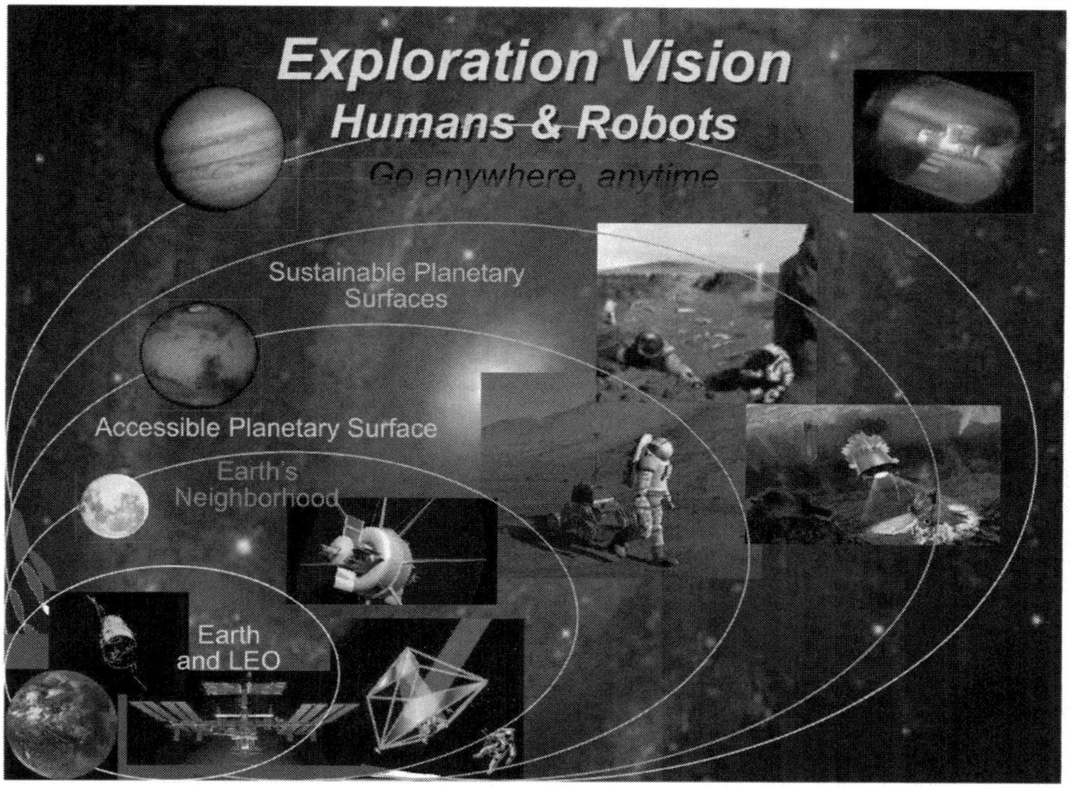
Exploration Vision
Humans & Robots
Go anywhere, anytime
Sustainable Planetary Surfaces
Accessible Planetary Surface
Earth's Neighborhood
Earth and LEO

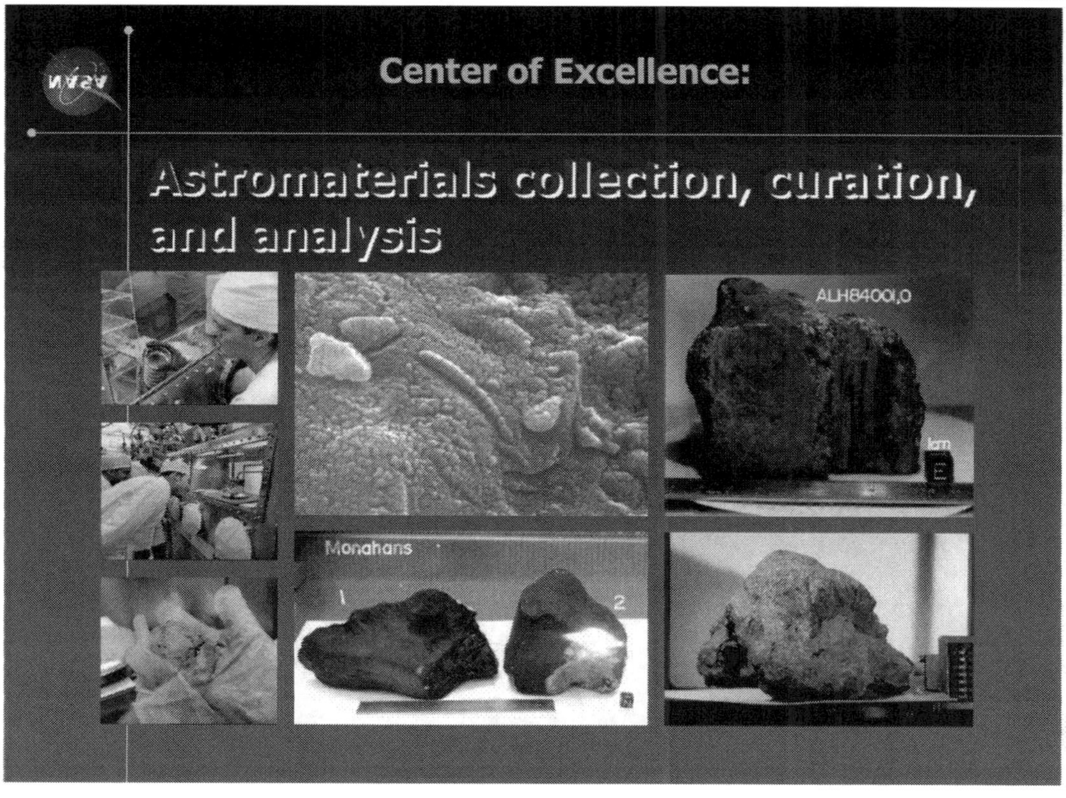
Center of Excellence:
Astromaterials collection, curation, and analysis
ALH84001,0
Monahans

Lunar Base

Martian Base

Original 2010 Budget: Constellation Program Components

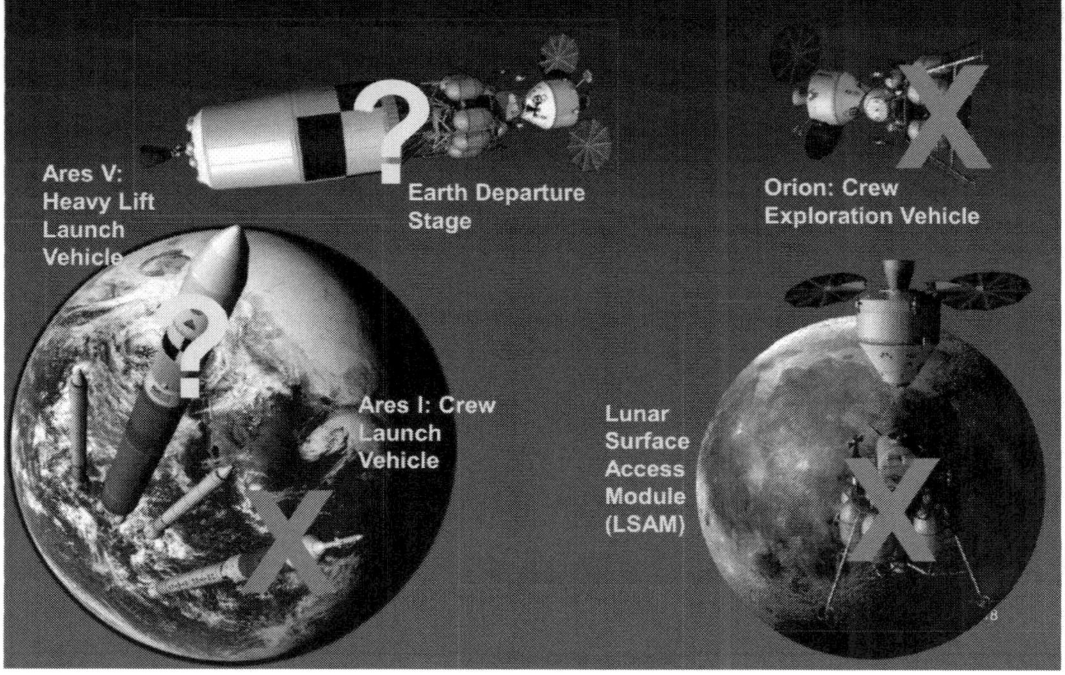

Original 2010 Budget: Constellation Program Components
Ares V: Heavy Lift Launch Vehicle
Earth Departure Stage
Orion: Crew Exploration Vehicle
Ares I: Crew Launch Vehicle
Lunar Surface Access Module (LSAM)

Russian Soyuz: Only method of reaching ISS after 9/2010
• Crew Transport Vehicle (3 crewmembers)
• Launches from Kazakstan

National Engineering Education Challenge

Engineering Degrees 1945-2001

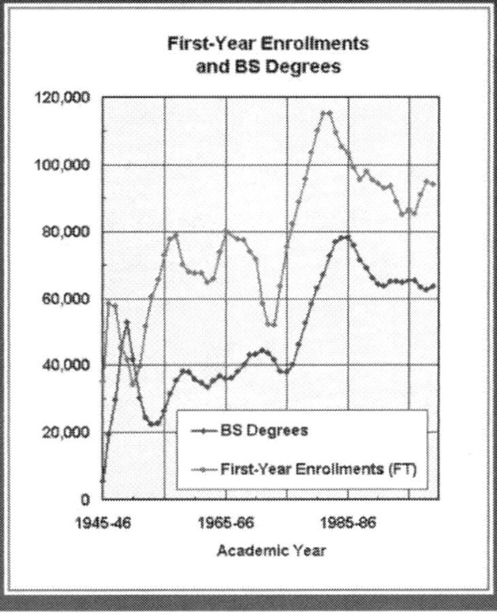

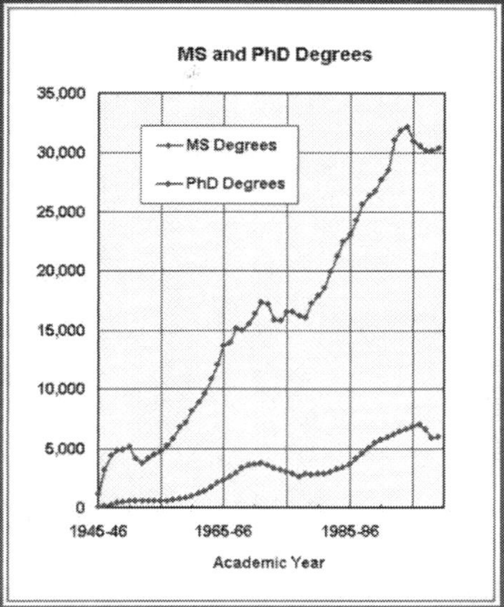

The U.S. Engineering and Physical Sciences *Workforce* "Crises" (*Not new News)*

- 1945 – 2012: National Science Foundation "Indicators":
- 1989 Space Policy Institute Report on Origins of Scientists and Engineers
- 2001 Hart Rudman: Commission on National Security/21st Century
- 2002 Walker Aerospace Commission Report:

2005: National Academy of Engineering

Norman R. Augustine
Craig Barrett
Gail Cassell
Stephen P. Chu
Robert M. Gates
Nancy Grasmick
Charles Holliday, Jr
Shirley Ann Jackson
Anita K. Jones
Joshua Lederberg

Richard Levin
C. D. (Dan) Mote
Cherry Murray
Peter O'Donnell, Jr.
Lee R. Raymond
Robert C. Richardson
P. Roy Vagelos
Charles M. Vest
George M. Whitesides
Richard N. Zare

NAE 2010

Rising Above the Gathering Storm, Revisited: Rapidly Approaching Category 5

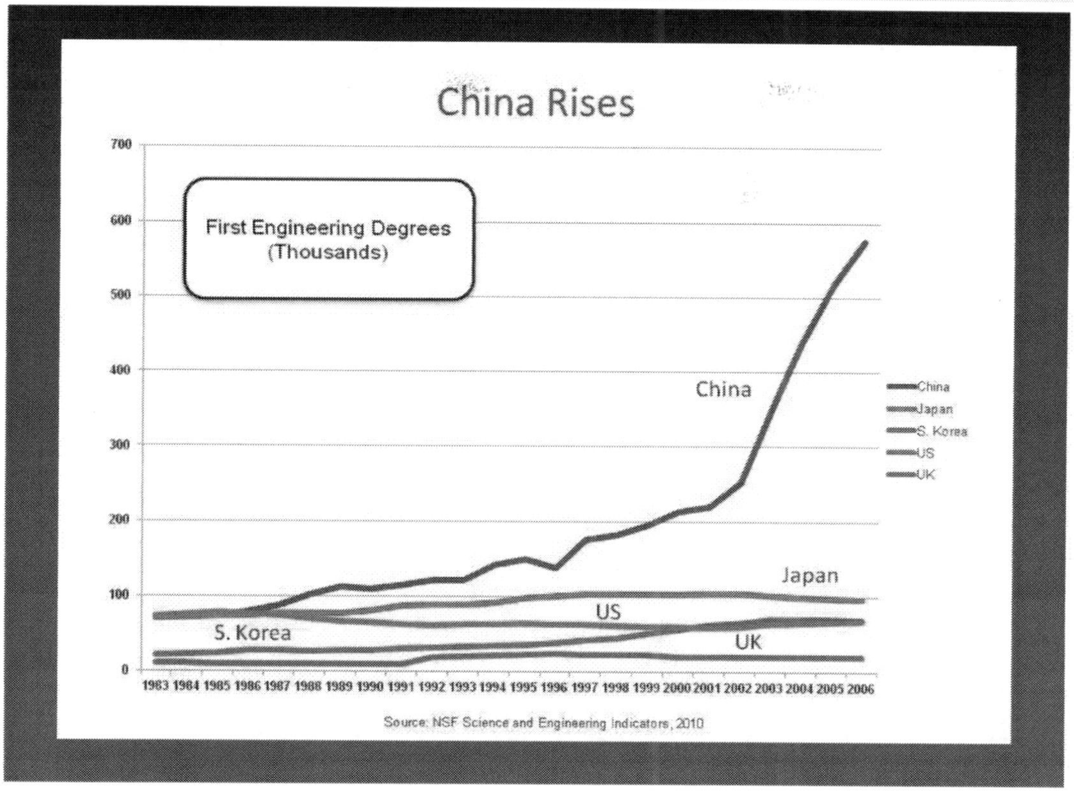
China Rises
First Engineering Degrees
(Thousands)
700
600
500
400
300
200
100
0
China
Japan
US
S. Korea
UK
China
Japan
S. Korea
US
UK
1983 1984 1985 1986 1987 1988 1989 1990 1991 1992 1993 1994 1995 1996 1997 1998 1999 2000 2001 2002 2003 2004 2005 2006
Source: NSF Science and Engineering Indicators, 2010

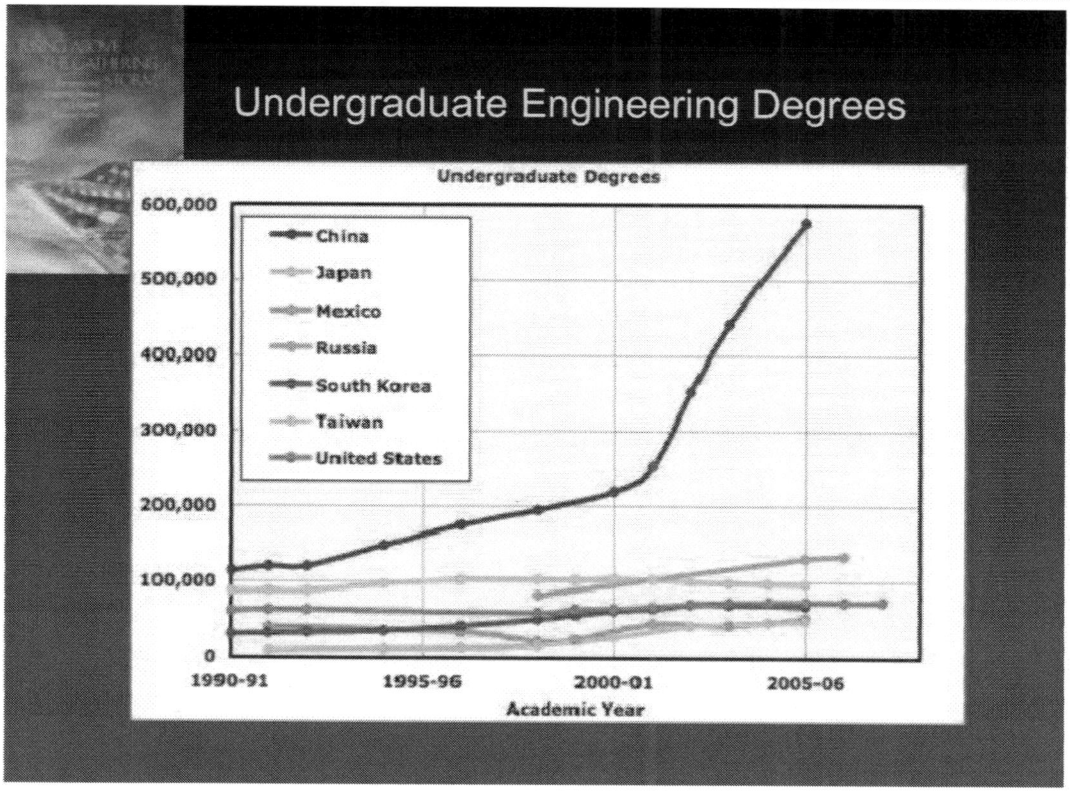
Undergraduate Engineering Degrees
Undergraduate Degrees
600,000
500,000
400,000
300,000
200,000
100,000
0
China
Japan
Mexico
Russia
South Korea
Taiwan
United States
1990-91
1995-96
2000-01
2005-06
Academic Year

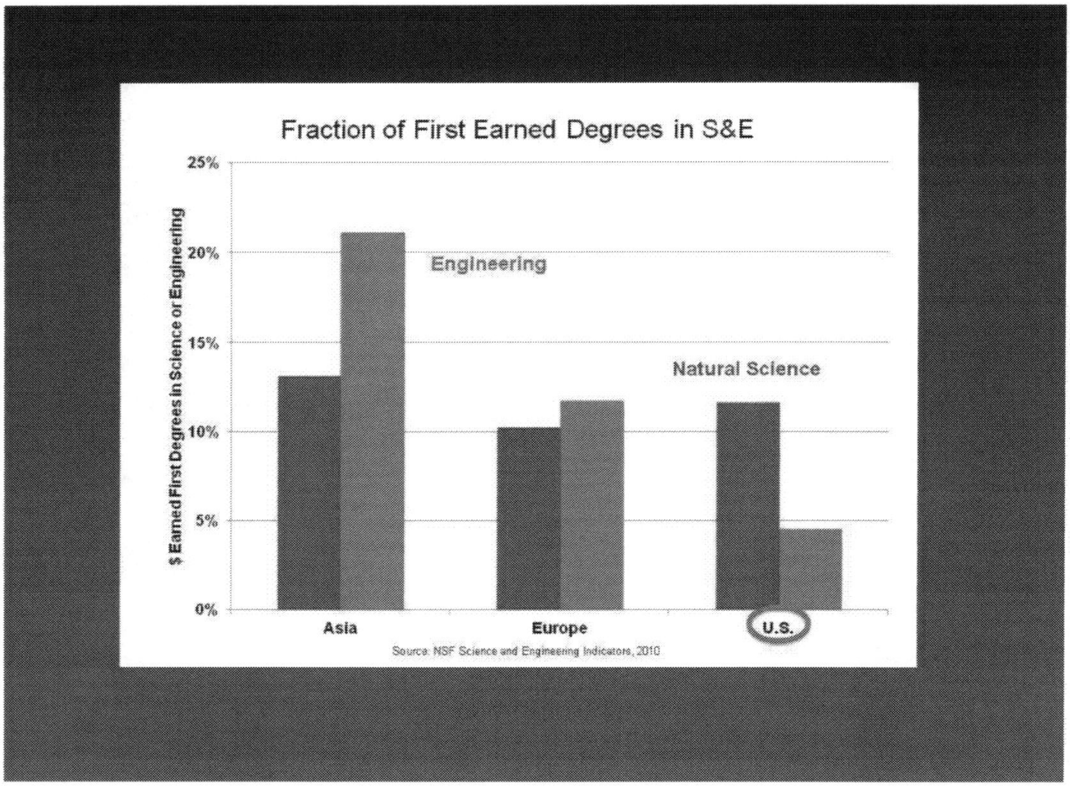
Fraction of First Earned Degrees in S&E
25%
20%
15%
10%
5%
0%
$ Earned First Degrees in Science or Engineering
Engineering
Natural Science
Asia
Europe
U.S.
Source: NSF Science and Engineering Indicators, 2010

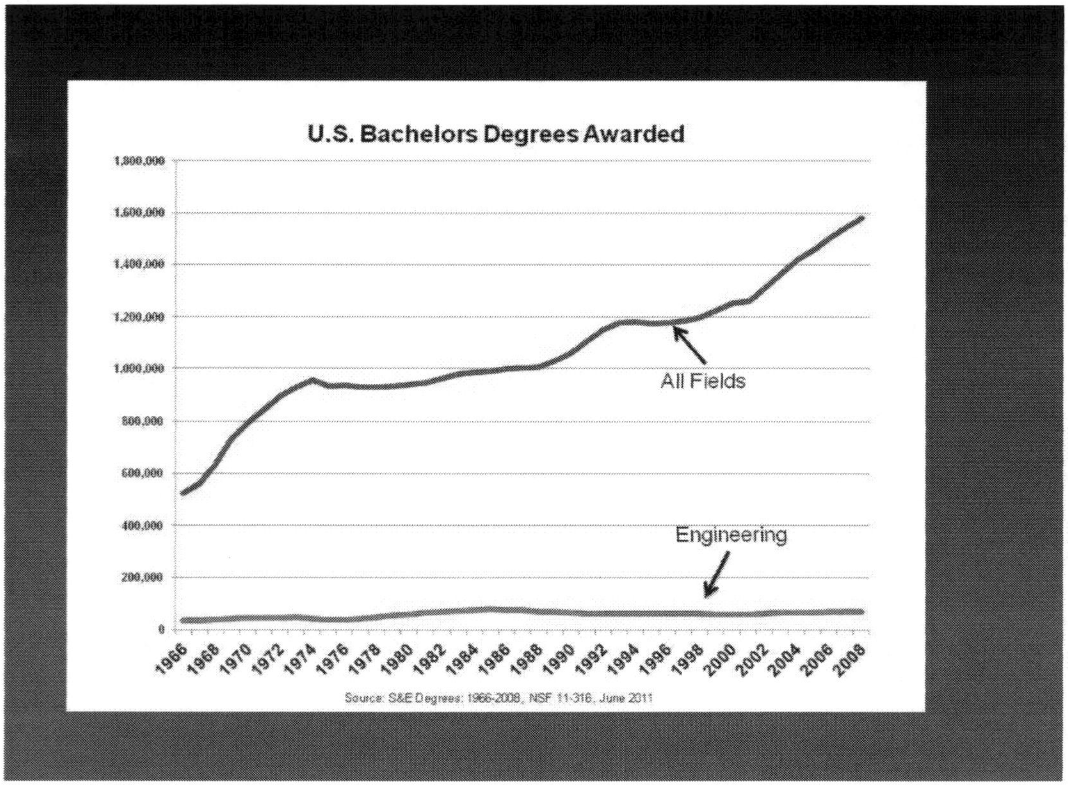
U.S. Bachelors Degrees Awarded
1,800,000
1,600,000
1,400,000
1,200,000
1,000,000
800,000
600,000
400,000
200,000
0
All Fields
Engineering
1966 1968 1970 1972 1974 1976 1978 1980 1982 1984 1986 1988 1990 1992 1994 1996 1998 2000 2002 2004 2006 2008
Source: S&E Degrees: 1966-2008, NSF 11-316, June 2011

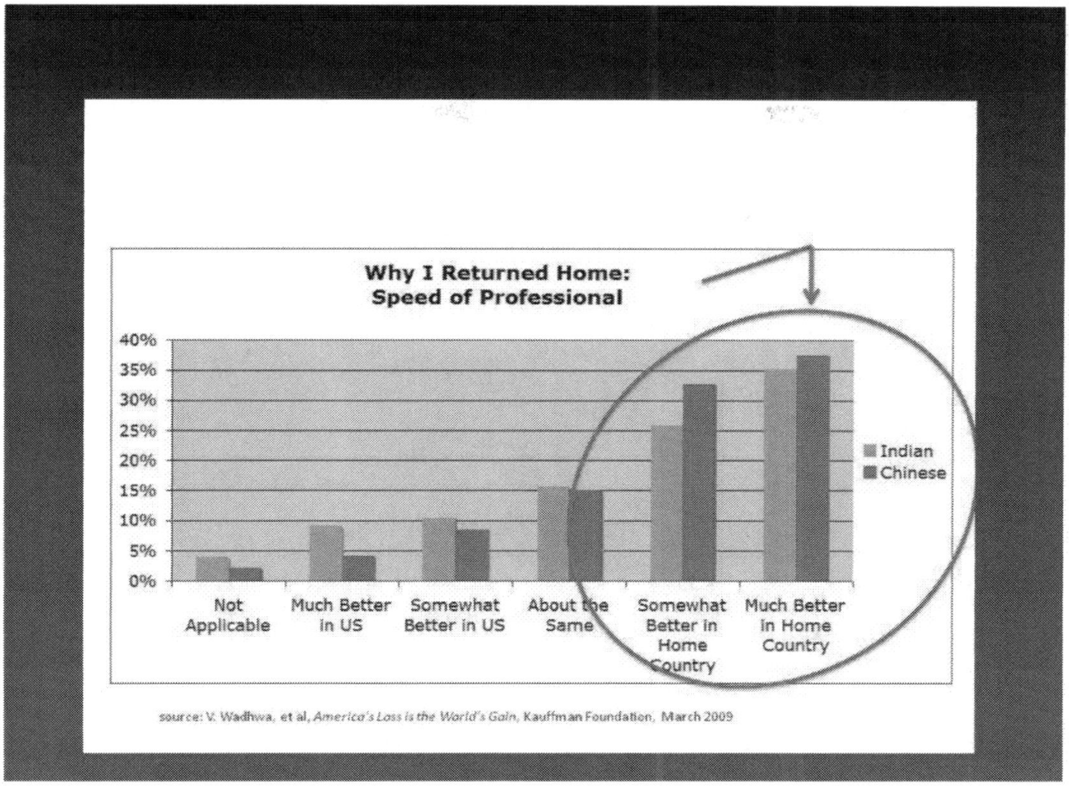
Why I Returned Home:
Speed of Professional
40%
35%
30%
25%
20%
15%
10%
5%
0%
Not Applicable
Much Better in US
Somewhat Better in US
About the Same
Somewhat Better in Home Country
Much Better in Home Country
Indian
Chinese
source: V. Wadhwa, et al, America's Loss is the World's Gain, Kauffman Foundation, March 2009

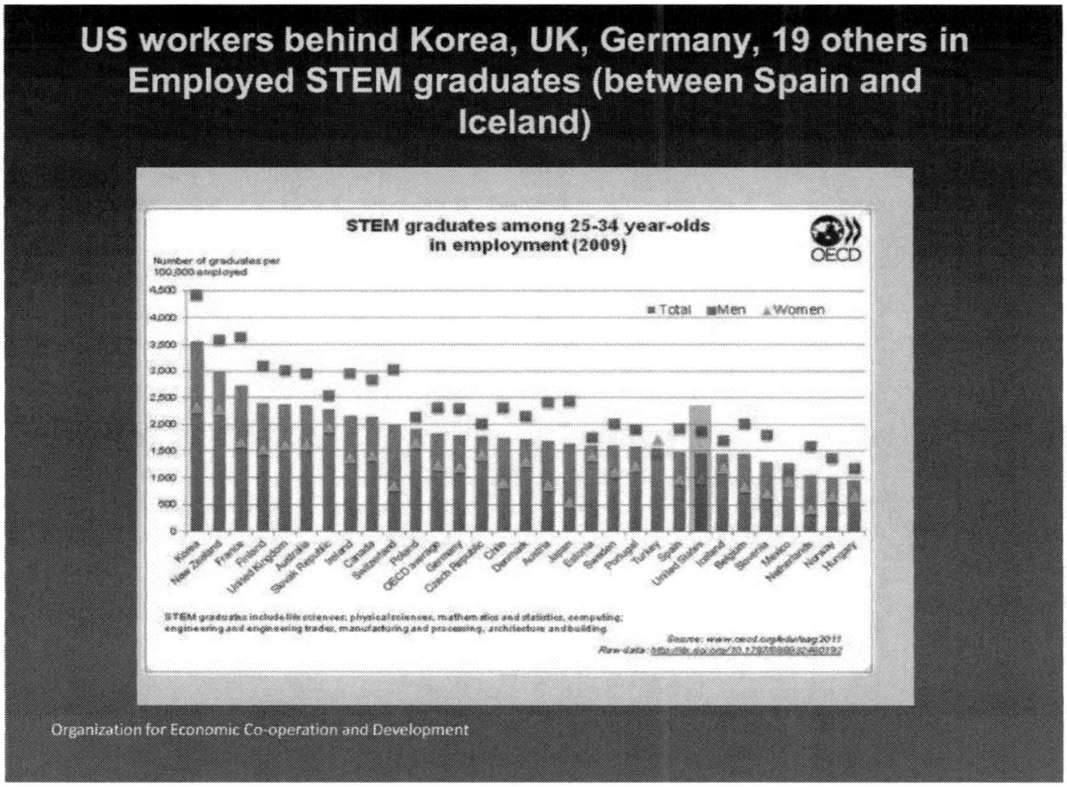
US workers behind Korea, UK, Germany, 19 others in Employed STEM graduates (between Spain and Iceland)
STEM graduates among 25-34 year-olds in employment (2009)
OECD
Total
Men
Women
Organization for Economic Co-operation and Development

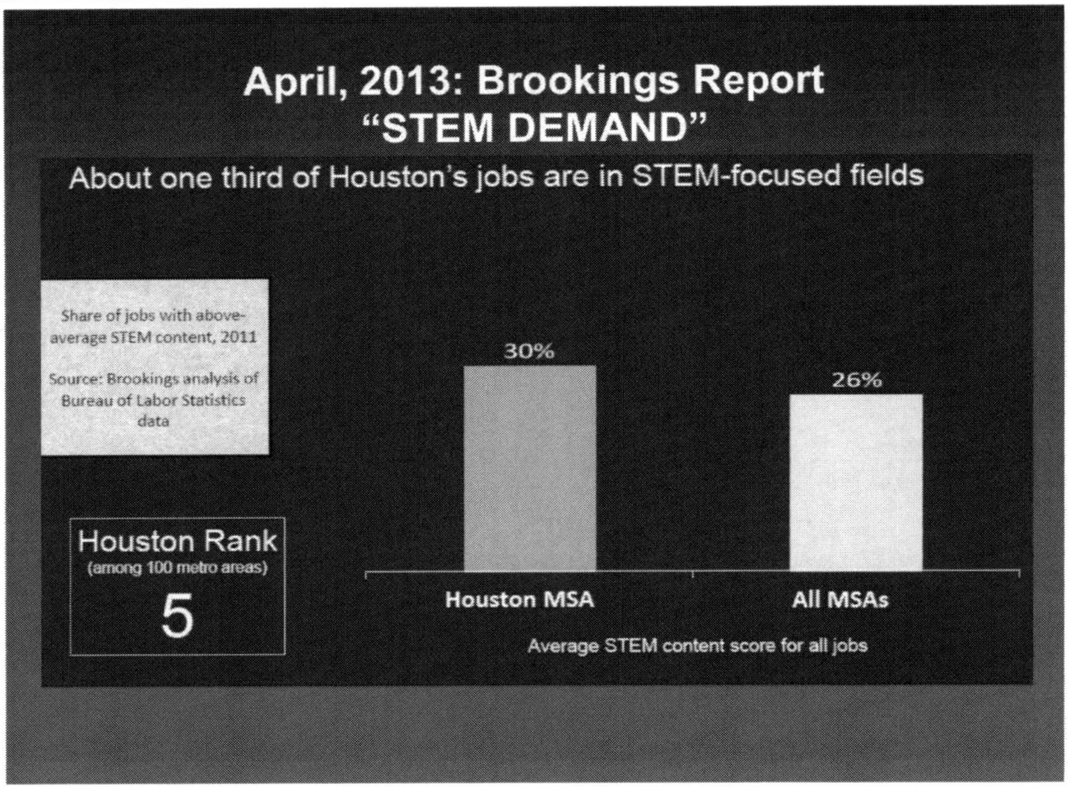
April, 2013: Brookings Report
"STEM DEMAND"
About one third of Houston's jobs are in STEM-focused fields
Share of jobs with above-average STEM content, 2011
Source: Brookings analysis of Bureau of Labor Statistics data
30%
26%
Houston Rank
(among 100 metro areas)
5
Houston MSA
All MSAs
Average STEM content score for all jobs

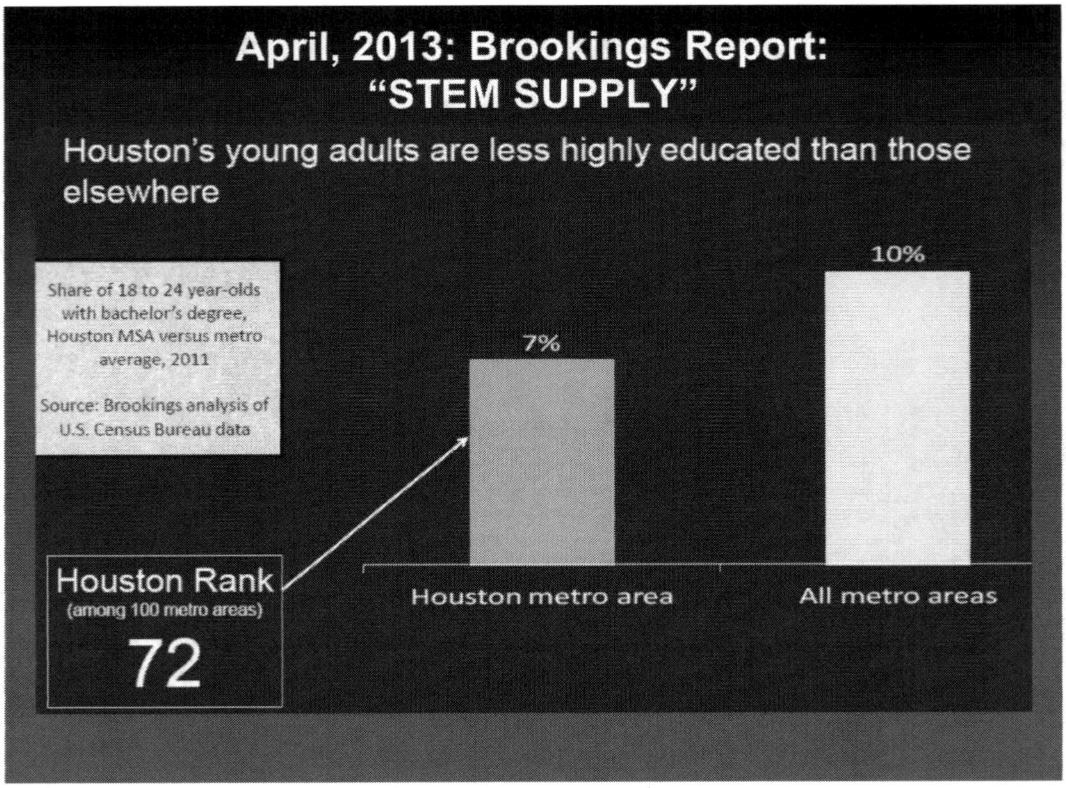
April, 2013: Brookings Report:
"STEM SUPPLY"
Houston's young adults are less highly educated than those elsewhere
Share of 18 to 24 year-olds with bachelor's degree, Houston MSA versus metro average, 2011
Source: Brookings analysis of U.S. Census Bureau data
7%
10%
Houston Rank
(among 100 metro areas)
72
Houston metro area
All metro areas

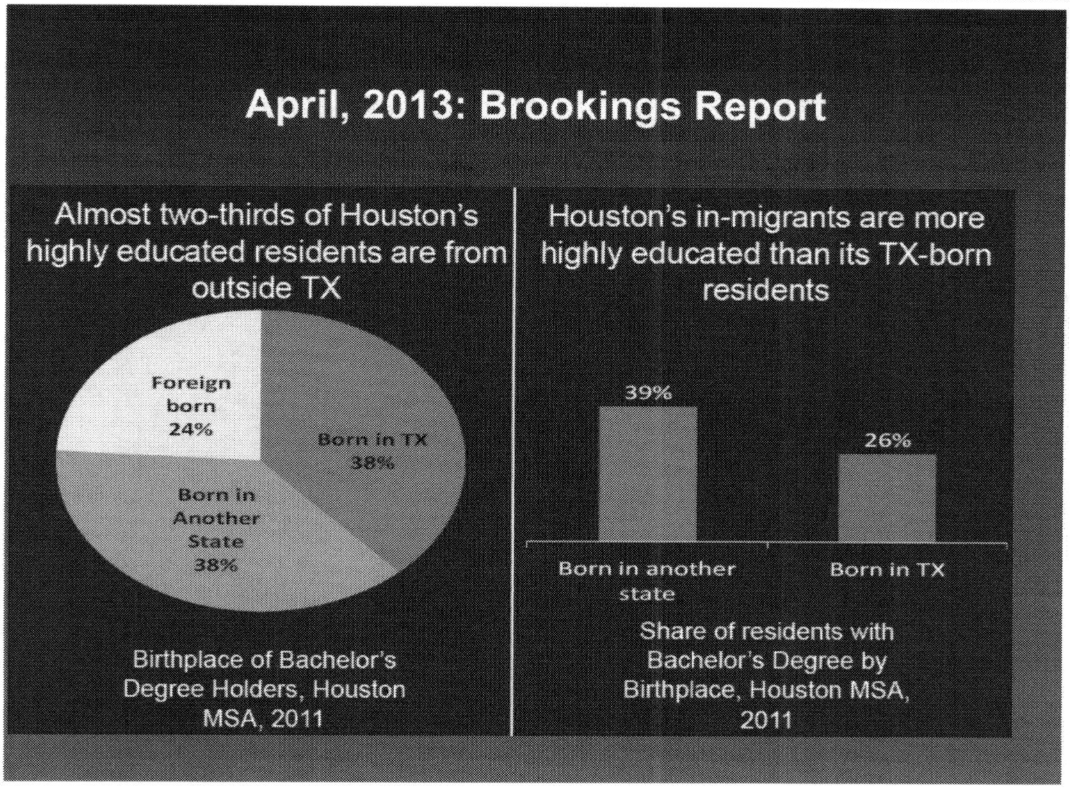
April, 2013: Brookings Report
Almost two-thirds of Houston's highly educated residents are from outside TX
Foreign born 24%
Born in TX 38%
Born in Another State 38%
Birthplace of Bachelor's Degree Holders, Houston MSA, 2011
Houston's in-migrants are more highly educated than its TX-born residents
39%
26%
Born in another state
Born in TX
Share of residents with Bachelor's Degree by Birthplace, Houston MSA, 2011

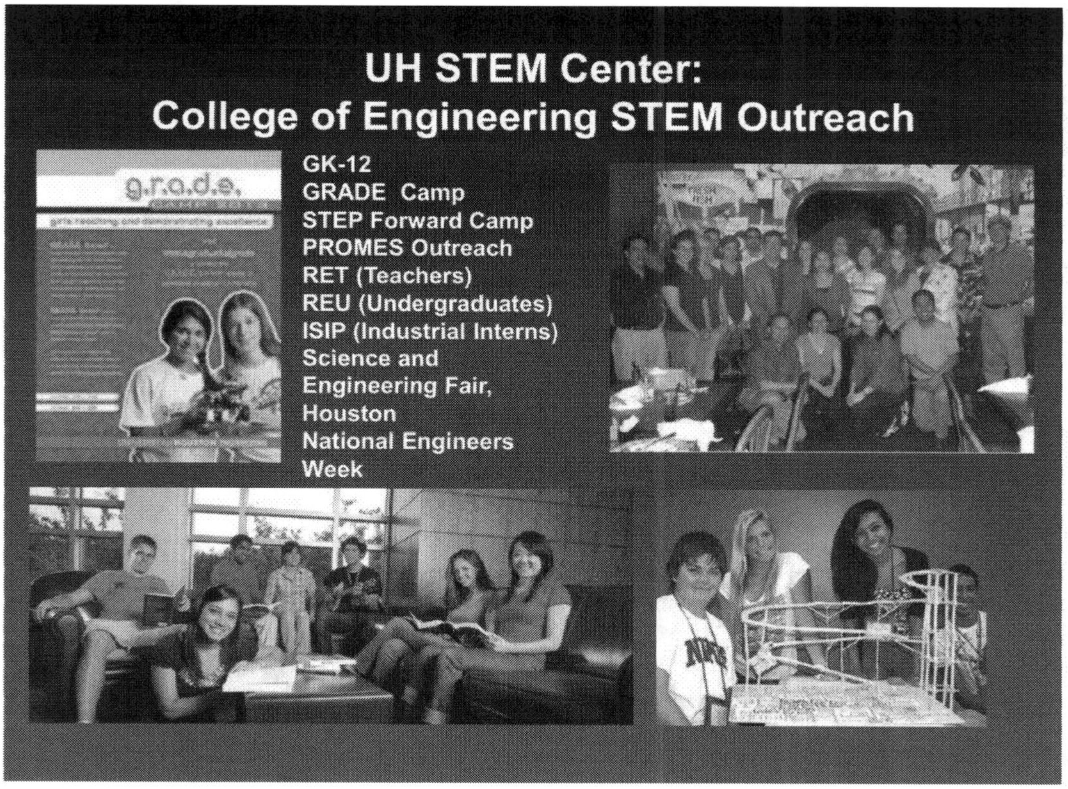
UH STEM Center:
College of Engineering STEM Outreach
g.r.a.d.e.
GK-12
GRADE Camp
STEP Forward Camp
PROMES Outreach
RET (Teachers)
REU (Undergraduates)
ISIP (Industrial Interns)
Science and Engineering Fair, Houston
National Engineers Week

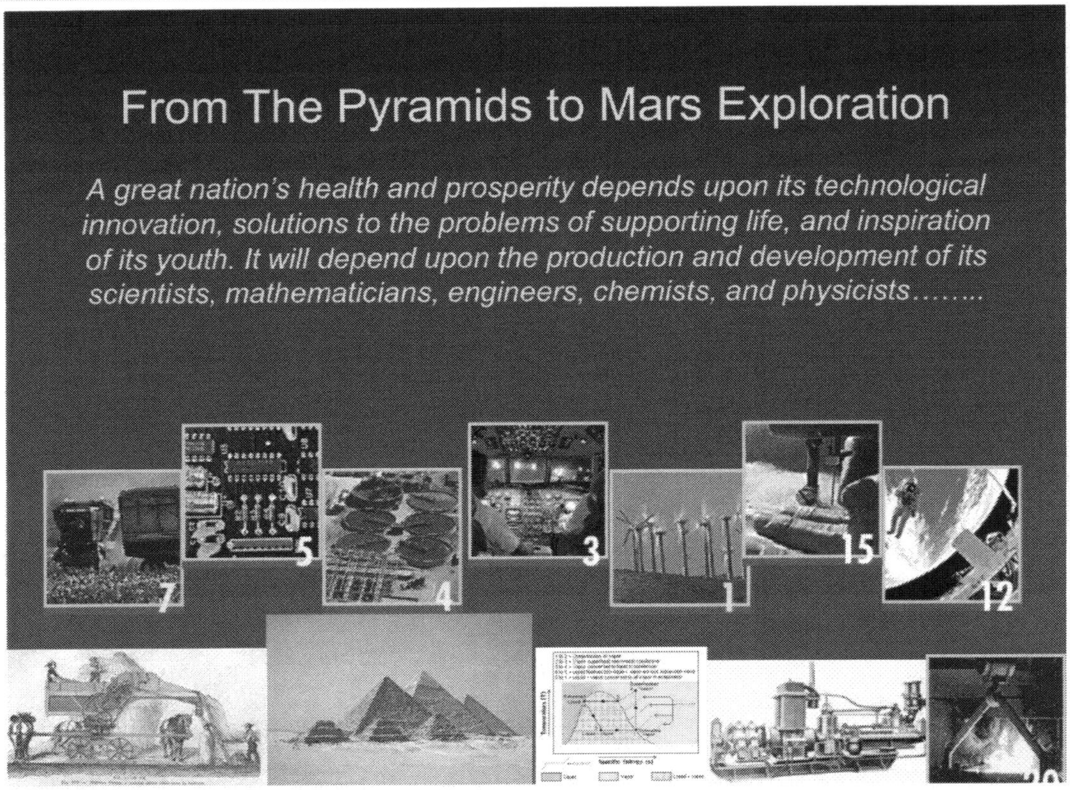
From The Pyramids to Mars Exploration
A great nation's health and prosperity depends upon its technological innovation, solutions to the problems of supporting life, and inspiration of its youth. It will depend upon the production and development of its scientists, mathematicians, engineers, chemists, and physicists........
7
5
4
3
1
15
12

Inspiring and Teaching the Next Generation
Math, Chemistry and Physics are the languages of the Natural World --

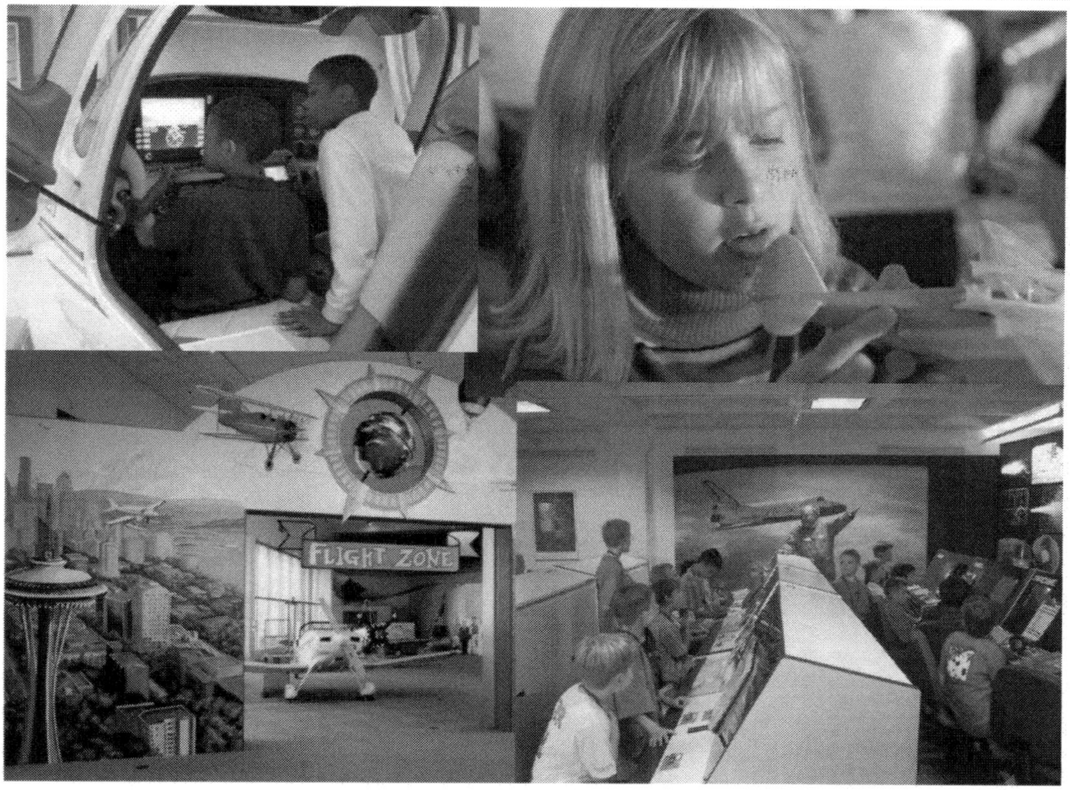
FLIGHT ZONE

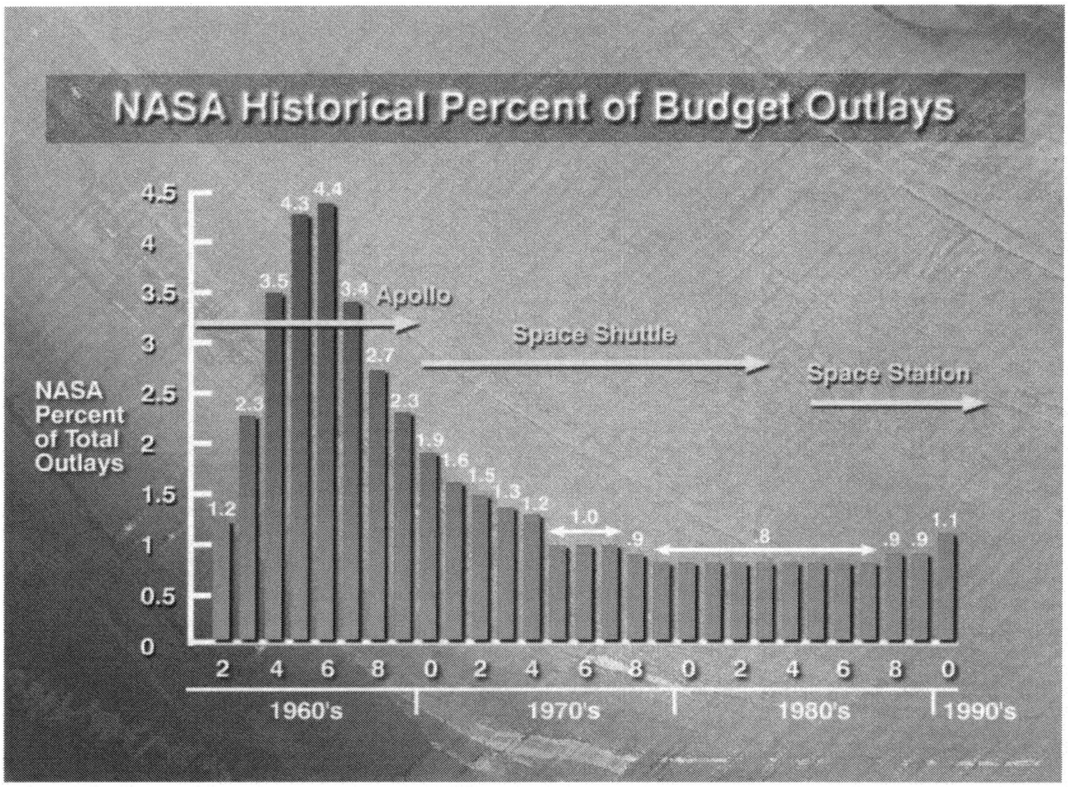
NASA Historical Percent of Budget Outlays
NASA Percent of Total Outlays
4.5
4
3.5
3
2.5
2
1.5
1
0.5
0
1.2
2.3
3.5
4.3
4.4
3.4
2.7
2.3
1.9
1.6
1.5
1.3
1.2
1.0
.9
.8
.9
.9
1.1
Apollo
Space Shuttle
Space Station
2 4 6 8 0 2 4 6 8 0 2 4 6 8 0
1960's
1970's
1980's
1990's

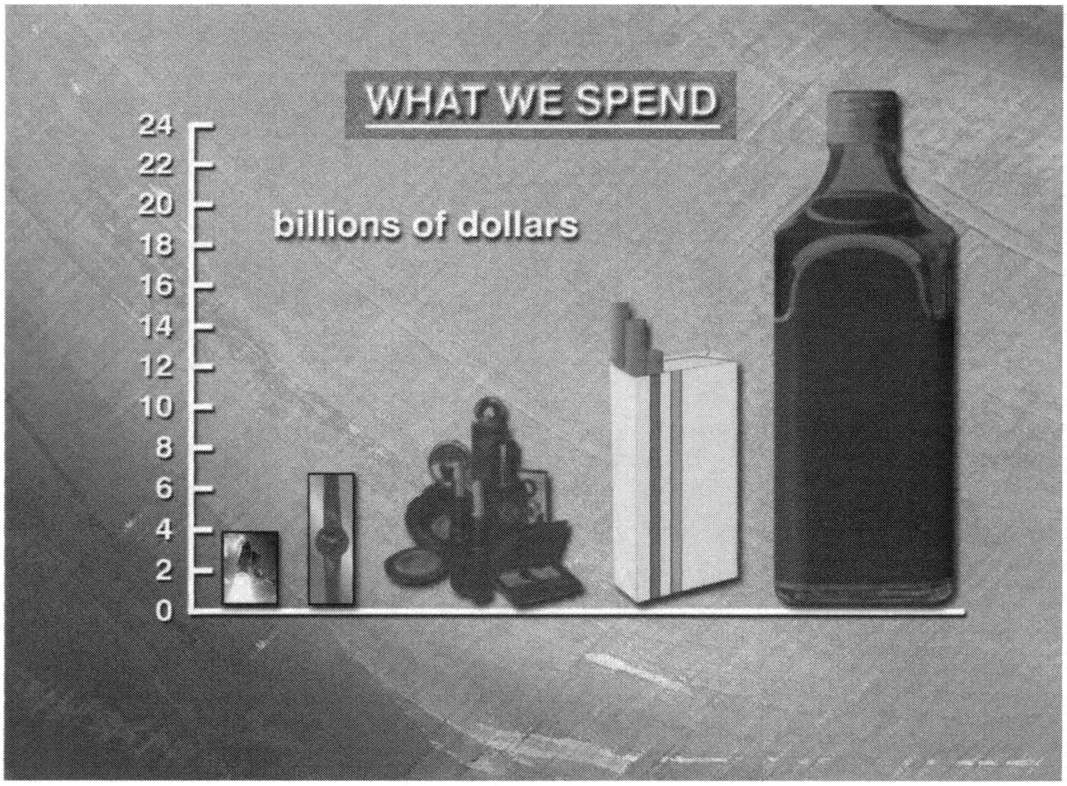
WHAT WE SPEND
billions of dollars
24
22
20
18
16
14
12
10
8
6
4
2
0

Revisiting Viking 1 and II: Not IF--
First Robotic Lander on Mars
United States of America, 1976

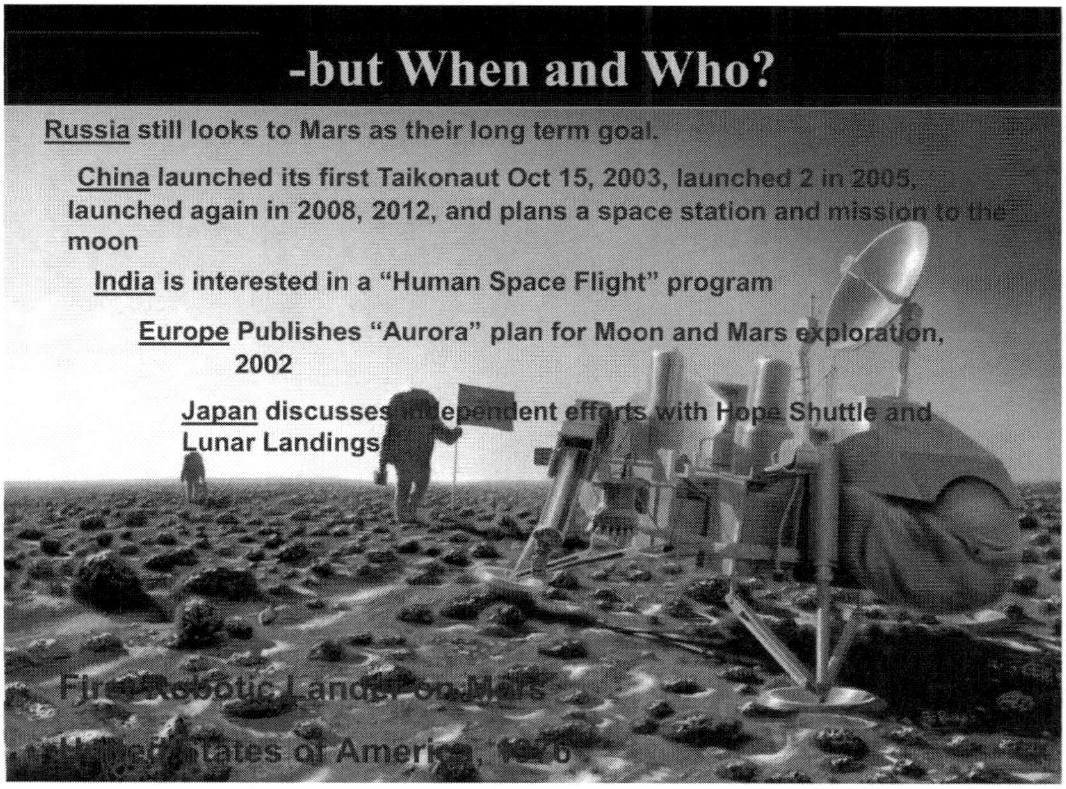

-but When and Who?
Russia still looks to Mars as their long term goal.
China launched its first Taikonaut Oct 15, 2003, launched 2 in 2005, launched again in 2008, 2012, and plans a space station and mission to the moon
India is interested in a "Human Space Flight" program
Europe Publishes "Aurora" plan for Moon and Mars exploration, 2002
Japan discusses independent efforts with Hope Shuttle and Lunar Landings

It is difficult to say what is impossible. For the dreams of yesterday become the hopes of today, and the realities of tomorrow.

Dr. Robert Goddard

Texas Board *of* Professional Engineers
Engineering For A Better Texas

TBPE

Texas Board of Professional Engineers Professional Practice Update / Ethics

C. W. Clark, P.E.
Director of Compliance & Enforcement
October, 2013

Engineering for a better Texas

TBPE

Agenda

- Responsibility / Professionalism and Ethics
- Laws/Rules/Scenarios
- Legislative News
- TBPE Initiatives

Engineering for a better Texas

TBPE

Board Update

- New Chairman – Dr. Daniel Wong, P.E.
- New Board Members
  - Mr. Sina Nejad, P.E. – Beaumont
  - Mr. Kyle Womack, P.E. - Midland

Engineering for a better Texas

TBPE

Expectations of a Licensed Engineer

- Technical Competence
- Professionalism
- Ethical Responsibility
- Understanding of Laws and Rules

Engineering for a better Texas

TBPE

Public Perception of Engineers

- Honesty (Gallup Poll Nov. 2012)
  - Nurses
  - Pharmacists
  - Doctors
  - **Engineers**
  - Dentists
  - Police Officers
  - College Professors
  - Clergy

Engineering for a better Texas

TBPE

Public Perception of Engineers

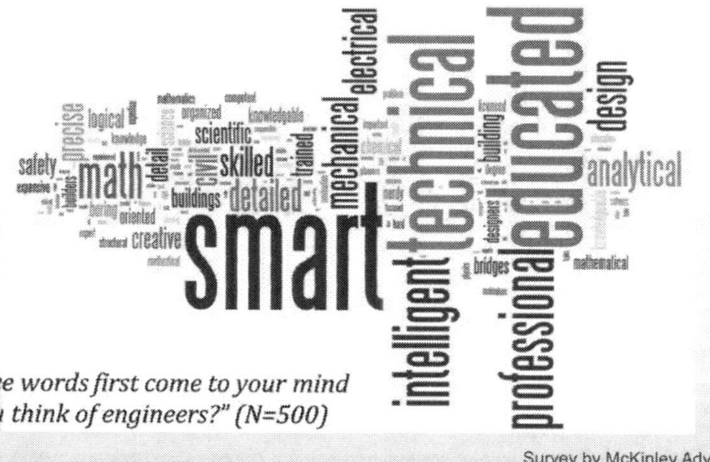

"What three words first come to your mind when you think of engineers?" (N=500)

Survey by McKinley Advisors

Engineering for a better Texas

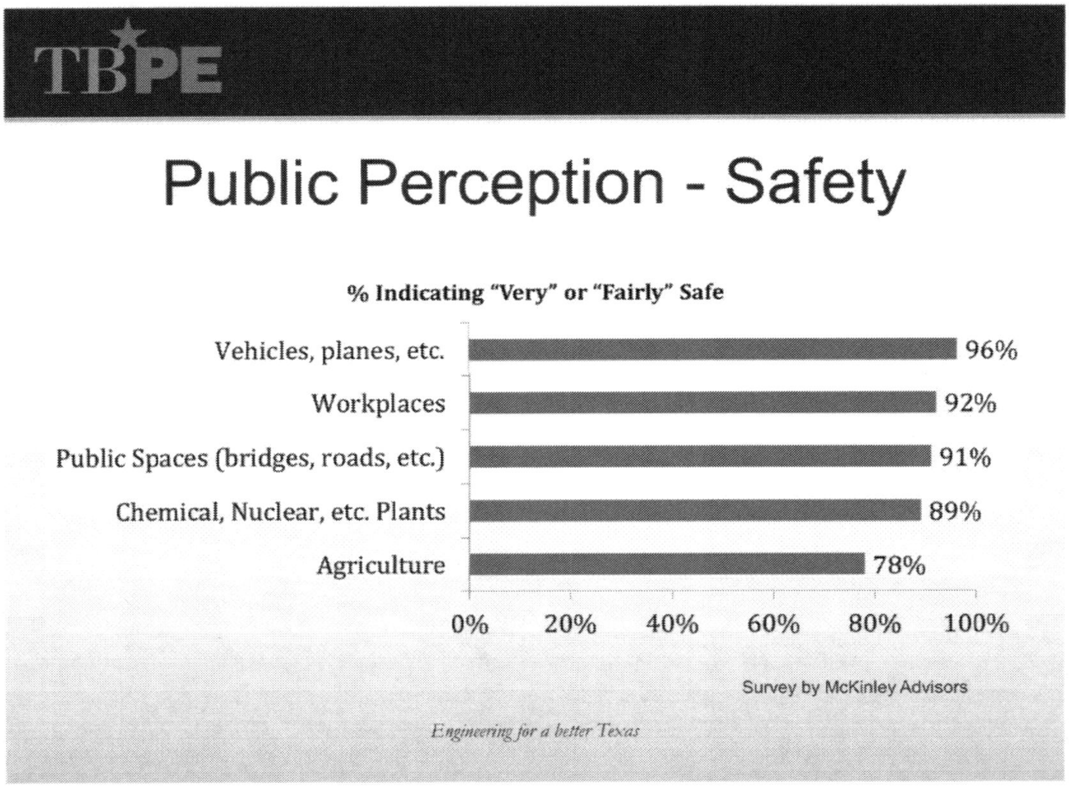
TBPE
Public Perception - Safety
% Indicating "Very" or "Fairly" Safe
Vehicles, planes, etc.
96%
Workplaces
92%
Public Spaces (bridges, roads, etc.)
91%
Chemical, Nuclear, etc. Plants
89%
Agriculture
78%
0%
20%
40%
60%
80%
100%
Survey by McKinley Advisors
Engineering for a better Texas

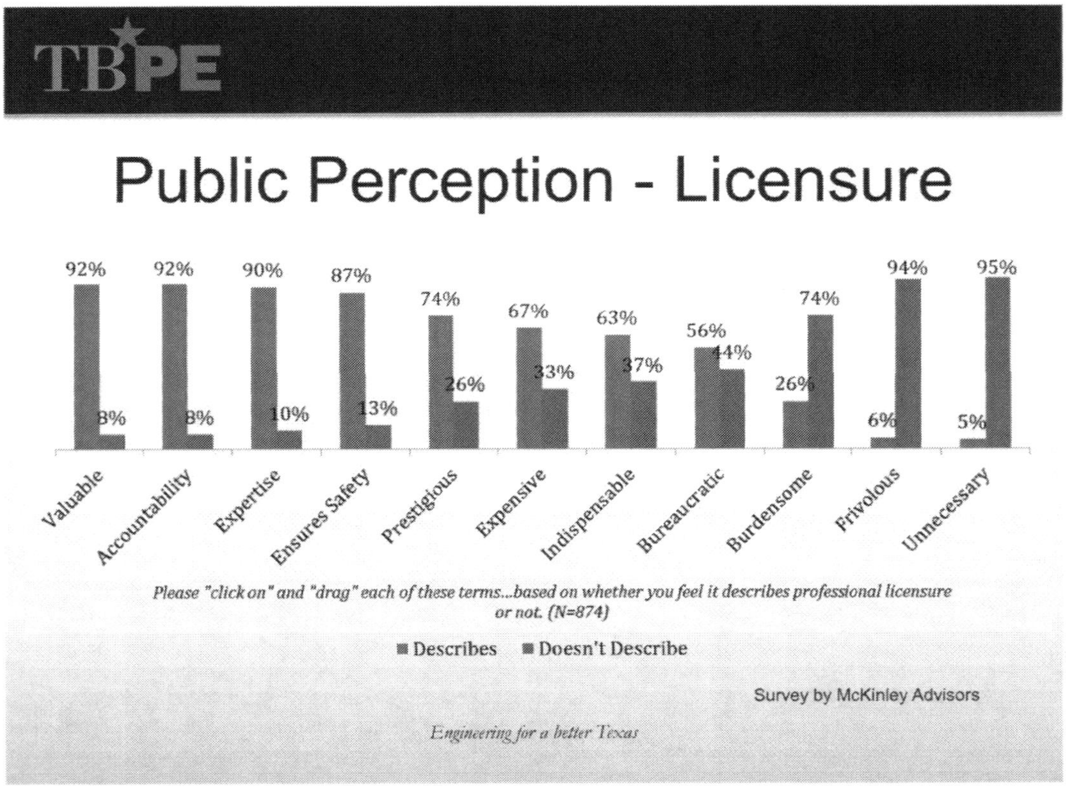
TBPE
Public Perception - Licensure
92%
8%
92%
8%
90%
10%
87%
13%
74%
26%
67%
33%
63%
37%
56%
44%
26%
74%
6%
94%
5%
95%
Valuable
Accountability
Expertise
Ensures Safety
Prestigious
Expensive
Indispensable
Bureaucratic
Burdensome
Frivolous
Unnecessary
Please "click on" and "drag" each of these terms...based on whether you feel it describes professional licensure or not. (N=874)
Describes
Doesn't Describe
Survey by McKinley Advisors
Engineering for a better Texas

TBPE

TBPE

Engineering Competence

- Protection of Public Health, Safety, Welfare
- Professional Engineers must not practice outside of their competence
- Competence is determined by experience, education and examination
- Competence is an expectation of the public

Engineering for a better Texas

TBPE

Engineering Ethics

- Ethical behavior / Professionalism
  - Not just what we do, but how we are expected to treat clients/employers/public/other engineers
  - Most TBPE enforcement is related to ethics rather than technical incompetence
- Laws and Rules

Engineering for a better Texas

TBPE

Law and Rules

- Board is authorized by the Texas Engineering Practice Act
- Other statutes also apply to engineering (PSPA, etc.)
- Board interprets and implements the statute to create Rules
- Texas Professional Engineers are expected to know both

Engineering for a better Texas

TBPE

Scenarios

- Continuing Education
- Professional Services Procurement Act
- Sealing Rules
- Confidentiality
- Supervision/delegation
- Firm Registration

Engineering for a better Texas

TBPE

Continuing Ed

- You have been licensed for years and have always had more than enough hours for each renewal.
- You misplaced your renewal reminder and the expiration date is in a few days.
- You were busy and haven't completed any continuing education activity this year.

Engineering for a better Texas

TBPE

What should a Texas PE do?

A. Go online and renew without completing the Continuing Ed hours and try to take some courses as soon as you can.
B. Keep working, wait to renew until you have the Continuing Ed hours which could be a few weeks and then renew.
C. Check your records from last year to see if you had extra hours.
D. Call the board and ask for an extension.
E. None of the above.

Engineering for a better Texas

TBPE

And the best answer is?

- **C** – You need to have the Continuing Education hours prior to renewing. You can carry forward up to 14 hours from one renewal period to the next, excluding the hour of ethics.
- Self study is limited to 5 hours and can be used for ethics.
- If you don't renew, your license is delinquent and you cannot practice.
- If you do renew and certify that you have completed Continuing Ed but have not, your license is subject to censure.
- Renew in ECHO. It's immediate and you get confirmation.

Engineering for a better Texas

TBPE

Continuing Education

- 15 hours – includes no more than 5 hours of self-study and must include one hour of ethics
- Random audits ongoing
- Keep documentation for 3 years
- Over 65 not exempt
- Fines as high as $5,000; separate violations for claiming Continuing Ed without documentation or not responding to Board.

Engineering for a better Texas

TBPE

Professional Services Procurement Act

- The local Independent School District has properly posted an RFQ and selected an engineering firm to be the prime professional for a new Auditorium/Gym. After negotiations, the ISD and the firm agree on a contract price.
- As the project progresses, the engineering firm contracts with an MEP engineering firm to address the HVAC, electrical and plumbing design.

Engineering for a better Texas

TBPE

PSPA (continued)

- The MEP firm calls you for assistance, because he has worked with you previously. You are a licensed P.E. with HVAC expertise and he asks you to give him a cost estimate to complete the HVAC for the new auditorium/gym.

Engineering for a better Texas

TBPE

What should a Texas PE do?

A. Give him your best estimate of the cost.

B. Wait for a proper authorization to proceed from the prime professional firm who has the contract with the ISD.

C. Call the Board for authorization to provide the estimate.

D. Ask the MEP firm if you have been selected as the HVAC engineering firm prior to providing any cost information.

E. Respectfully decline to offer any cost or price information.

Engineering for a better Texas

TBPE

Best Answer

- **D** – You must follow the Qualification Based Selection process by being selected prior to offering any cost or price information as outlined in the Professional Services Procurement Act (Chapter 2254 of the Texas Government Code) and as identified in the Board Rules, Section 137.53.

Engineering for a better Texas

TBPE

Seals

- Your lead engineer on a project has been overseeing all aspects of the design and construction.
- He has a severe health issue is keeping him from the job for an extended time at the same time that there is a deadline for plan submission.

Engineering for a better Texas

TBPE

What should a Texas PE do?

A. Ask an employee under his supervision to copy his seal and digital signature from the shared drive and use them.

B. Have the documents sent with his seal and assure the authority that he'll come in to sign them later.

C. Mark the plans as "preliminary"

D. Negotiate a new deadline

E. Sign and seal the plans yourself before the deadline.

Engineering for a better Texas

TBPE

Best Answers

- **C or D** – Mark as preliminary or renegotiate the deadline
- 137.33(d): License holders shall take reasonable steps to ensure the security of their physical or electronic seals and electronic signatures;
- 137.33(e): Preliminary documents are not sealed.
- 137.33(f): Sealed documents require a signature.
- Only the licensee may use his or her seal.
- 137.33(a): A PE can only sign and seal their own work or work under direct supervision.

Engineering for a better Texas

TBPE

A Twist

- What happens if the PE never returns to work?

- Another PE can sign and seal the work if the PE performs sufficient calculations to verify the accuracy of the engineering work and make it his own work.

Engineering for a better Texas

TBPE

Confidentiality

- You were the structural design engineer for a new church and that design was released to the church and they contracted the construction to another company. Subsequent to completion, a portion of the church has had a collapse that resulted in some minor injuries.
- It was determined that the failure was a construction problem, as the design had not been followed. You are approached by a good friend of yours whose child had been injured in the collapse and he asks you for a copy of your original design for the project.

Engineering for a better Texas

TBPE

What should a Texas PE do?

A. Since he is a good friend, you give him a copy of your design.
B. Before you give him a copy you make sure that he does not have any lawsuits pending.
C. Remove your seal and signature before you give him a copy.
D. Contact your client, the church, and obtain written approval to release a copy of the design.
E. Refuse to provide a copy since his child was involved.

Engineering for a better Texas

TBPE

Best Answer

- **D -** Rule 137.61 requires that engineers maintain the confidentiality of their clients.
- **137.61 Engineers Shall Maintain Confidentiality of Clients**

(a) The engineer may reveal confidences and private information only with a fully informed client's or employer's consent, or when required by law or court order; or when those confidences, if left undisclosed, would constitute a threat to the health, safety or welfare of the public.

(b) The engineer shall not use a confidence or private information regarding a client or employer to the disadvantage of such client or employer or for the advantage of a third party.

(c) The engineer shall exercise reasonable care to prevent unauthorized disclosure or use of private information or confidences concerning a client or employer by the engineer's employees and associates.

Engineering for a better Texas

TBPE

Supervision

- You are hired by the client to perform an inspection in order for the client to be eligible to procure insurance.
- You are very busy, so you send an employee under your supervision to the property with instructions to check everything out and make sure it meets code.

Engineering for a better Texas

TBPE

Supervision

- Based on his work, you sign an inspection report and send it to the appropriate authority.
- The authority decides to audit this certification and sends you a letter giving you 10 days to respond with all your calculations and product sheets/information to verify that the property meets the required code for the zone in which the property is located.

Engineering for a better Texas

TBPE

Questions

Supervision:

- Can you send an employee to do this inspection?
- What if the employee is an EIT and not a PE?
- Who is responsible?

Engineering for a better Texas

TBPE

Best Answers

- You may put your name and seal on work done by another individual if it is under "direct supervision".
- The work under the supervision of a PE can be done by another PE, an EIT, or any other qualified employee.
- **The PE is responsible for work done under his supervision.**
- 131.81(10) Direct supervision defined.

Engineering for a better Texas

TBPE

Questions

Review:

- Does the authority have the discretion to audit the report?
- Can the authority demand you provide your calculations and product sheets to verify that the property meets pertinent codes?
- Are you required to respond?
- Is 10 days reasonable?

Engineering for a better Texas

TBPE

Best Answers

- Local Officials, TDI, TCEQ, TDLR, etc. have the authority to request additional information and non-PEs may review the work.
- Not responding to a regulatory authority in a timely manner could result in an enforcement action by the requesting agency as well as the TBPE.
- 10 days may be reasonable since this is an audit of existing information, but an extension may be available, if requested.

Engineering for a better Texas

TBPE

Firms

- As a licensed Professional Engineer, you have worked for 20 years performing civil engineering work for your employer, a utility company.
- You have some extra time and want to do some extra work on the side, but haven't done it yet.
- Your church needs some design work and you want to donate your time and talents.

Engineering for a better Texas

TBPE

What should a Texas PE do?

A. Since you are not changing employers or getting paid for the other work, no notification is required.
B. Register yourself as a new engineering firm with the Board after you have been awarded your first paid engineering job for a client.
C. Register the engineering firm you will be operating with the Board before you offer or provide any engineering services.
D. Notify the Board of your secondary employment.

Engineering for a better Texas

TBPE

Best Answer

- **C & D**
- 137.59: An engineer must practice within their area of competence.
- 137.5: The Board must be notified of employment change within 30 days, including secondary employment. Changes made in the ECHO system are official.
- 137.77: Offering or performing engineering services requires registration with the board, even if operating as a sole practitioner (single employee firm). And even if you don't get paid.

Engineering for a better Texas

TBPE

Know the Law

- PEs are required to know the Act, Board Rules, applicable state laws and local codes.
- Windstorm, Architectural Barriers/ADA, etc.
- If you have any questions about interpretation of the Act or Rules, contact us.
- Keep an eye on our emails and newsletter to make sure you know about changes.

Engineering for a better Texas

TBPE

Filing A Complaint

- Online, email, phone, facsimile, mail – all are acceptable for initial contact
- A complaint form or detailed letter/email is needed to cover all the bases
- Provide specific instances of violation
- Provide evidence to show probable cause

Engineering for a better Texas

TBPE

Complaint Process

- If a complaint is filed against a PE or an unlicensed person
- A letter would be sent from the Board explaining the charge in detail
- Written response required
  - Providing as much documentation as possible
  - The person should be open, honest and forthcoming.

Engineering for a better Texas

TBPE

Enforcement Sanctions

- Reprimands (Formal and Informal)
- Suspension (possible probation)
- Refuse to Renew
- Revocation
- $5,000 per violation per day
- Cease and Desist Orders
- Emergency Suspension

Engineering for a better Texas

TBPE

Violations are Published

- By law, all violations, except informal reprimands, must be published
  - On website by Board Meeting Date
  - Added to NCEES Enforcement Exchange (national database)
  - Published in the paper newsletter which is mailed at least annually and quarterly E-newsletter emails

Engineering for a better Texas

TBPE

Preventing Complaints

- CLEAR:
  - Communication (between all parties)
  - Contract (expectations and responsibilities)
  - Calculations (do excellent work)

Most importantly – know the law, and call if you have a question

Engineering for a better Texas

TBPE

Legislative News

Engineering for a better Texas

TBPE

2013 Session

- Regular Legislative Session January – May 2013
  - Tracked any bills filed related to engineering, funding, employees, other agencies, etc.
- SB 204 Sunset Bill for The Texas Engineering Practice Act
  - Effective September 1, 2013

Engineering for a better Texas

TBPE

Legislation / Rule Changes

- Enforcement
  - Cease and Desist for Unlicensed Practice
    - Direct authority
  - Emergency Suspension Orders
    - Immediate action, specific timing for appeal requests
  - Increase maximum penalty from $3k to $5k
    - Suggested Sanctions Table in Rules modified

Engineering for a better Texas

TBPE

Legislation / Rule Changes

- Licensing
  - Application / Professional Fee split
    - Separates application fee from Professional fee
    - Previously collected together with application
    - Implemented August 1, 2013
    - Professional Fee ($200) will be paid when license is issued

Engineering for a better Texas

TBPE

Legislation / Rule Changes

- Licensing (cont)
  - Criminal History Record Checks
    - January 1, 2014
    - All new applicants
    - All Active licensee, 1st renewal
    - Texas Department of Public Safety
      - MorphoTrust USA
      - Fingerprint submittal required
      - Fee

Engineering for a better Texas

TBPE

Legislation / Rule Changes

- Criminal History Record Checks (cont)
  - FAQ on website
  - Detailed written procedures
    - TBPE will provide FAST form to applicant/licensee
    - Unique code for TBPE
    - Schedule and pay through MorphoTrust
    - Electronic submission in Texas
    - Ink card submission outside of Texas

Engineering for a better Texas

TBPE

Legislation / Rule Changes

- Criminal History Record Checks (cont)
  - Submission of fingerprints will allow processing of renewal
  - Records with results will be reviewed on case-by-case basis per Board policies

Engineering for a better Texas

TBPE

Legislation / Rule Changes

- Rule Changes
  - Posted September 20 for comment
  - Legislative Implementation
    - SB 204 – Sunset
    - HB 1865 – self-directed, semi-independent
    - SB 162 – military spouse applications
  - Computer Based Testing Implementation
    - FE exam will be computer-based January 1, 2014

Engineering for a better Texas

TBPE

Update on NCEES

- CBT – Computer Based Testing – FE 2014
- Exam changes
- Industry Exemption discussions
- BS+30/Masters Or Equivalent is a part of NCEES model law to be effective 2020

Engineering for a better Texas

TBPE

TBPE Initiatives

Engineering for a better Texas

TBPE

Journey to Excellence

- Our goal in 2014
- Implementing Malcolm Baldrige Quality Principles
- Pursuing Quality Texas Award for Performance Excellence

Engineering for a better Texas

TBPE

Journey to Excellence

- Strategic Planning
  - Customer Communications
    - Communication Plan / External / Customer Service
  - Organizational Resource Development
    - Staff / Workplace / Internal Communication
    - Innovation / Tools / Performance Measures

Engineering for a better Texas

TBPE

Communication Improvements

- Website redesign
- Quarterly email updates
- Social Media
- More Outreach / Visibility
  - Quarterly Ethics Webinars
  - Increased booth presence at conferences and seminars (TCEQ, TML, K-12 activities, etc).

Engineering for a better Texas

TBPE

Social Media

- Facebook: Texas Board of Professional Engineers
- Twitter: TBPE_Exec
- LinkedIn: Texas Board of Professional Engineers
- RSS Feed available on our website: engineers.texas.gov

Engineering for a better Texas

TBPE

Website Improvements

- Feedback from customer service survey
  - New logo and color scheme
  - Reduced clutter on front page
  - New quick links on side navigation
- Ongoing improvements scheduled for 2014
  - Improve overall navigation and menus
  - More user friendly for mobile devices

Engineering for a better Texas

TBPE

Outreach – continuing excellence

Fiscal Year	Attendees	Presentations
2010	9,145	157
2011	8,954	153
2012	7,332	166
2013	11,559	177

- Quarterly Webinars
- Includes K-12
- E-Week

Engineering for a better Texas

TBPE

We want to hear from you

- How did we do today?
- How are we doing in general?

http://engineers.texas.gov/outreachsurvey
http://engineers.texas.gov/feedback

Engineering for a better Texas

TBPE

Contact TBPE

- 1917 S Interstate 35
- Austin, TX 78741
- Phone: 512-440-7723
- Web: http://engineers.texas.gov
- cw.clark@engineers.texas.gov

Engineering for a better Texas

AIChE
Southwest Process Technology Conference

Systems Interoperability

What is it?
Where does it stand Now?

American Institute of Chemical Engineers
Southwest Process Technology Conference
Galveston, TX
2013-10-04

C.C. (Cliff) Pedersen
CIO (Retired), North West Redwater Partnership
VP & CFO, MIMOSA
Chair, Owner/Operator Leadership Council on Interoperability
ISO TC 184/WG6 Canadian Delegate

MIMOSA™

Pe^x I
Pedersen Enterprises Inc.

AIChE
Southwest Process Technology Conference

C.C. (Cliff) Pedersen

B.Sc., M.A.Sc., P.Eng., P.Mgr.

the open O&M initiative™

Pe^x I
Pedersen Enterprises Inc.

- **North West Redwater Partnership** (3 years) – Chief Information Officer (CIO)
- **Suncor Energy Inc.** (24 years) – Manager, Product Production & Process Control Processes
  - ✓ **(Founded APC, RTO, RWO, led Re-instrumentation Projects, Refinery IT, Operating Plant applications)**
- **Shell Canada Ltd.** – Supervisor, Process Systems Engineering
  - ✓ (Installed/Commissioned SCL's first process computer control system, then all process computers)
- **Imperial Oil Enterprises Ltd.** – Sr. Computer Control Applications Engineer
  - ✓ (Co-Pioneered Multivariable Constrained Supervisory Control at Sarnia Refinery)
- **B.Sc Chemical Engineering, University Alberta**; **MASc** Systems Design (University Waterloo); **P.Eng** (Professional Engineers of Ontario); **P.Mgr** (Canadian Institute of Management)
- Member: PEO, **CSChE**, ISA, CIM, AFPM, MIMOSA
- Boards: AFPM-PADS, ARC-CA, SHARCNET, HW-UAB, HP-HPC, MACC, Petrolia Discovery

MIMOSA™

Pe^x I
Pedersen Enterprises Inc.

AIChE
Southwest Process Technology Conference

Firsts in Refinery Automation

SUNCOR ENERGY

- 1959 - First refinery computer control -Texaco, Port Arthur
- 1975? – First multivariable predictive control (i.e. DMC) aka APC – Shell, Deer Park
- 1986 – Second commercial DMC application – **Sunoco, Sarnia, Hydrotreater Reactor**
- 1991 - First equation oriented unit RTO - **Sunoco, Sarnia, Hydrocracker**
- 2013 - First Refinery-wide Optimization - **Suncor, Sarnia** - [RMO]ProfitMeter™ at daily operating meeting

MIMOSA™

PeˣI
Pedersen Enterprises Inc.

AIChE
Southwest Process Technology Conference

Agenda

- A New Refinery
- Safety Moment
- Integrating Systems & Applications
- Standards Based Interoperability
- Owner/Operator Requirements
- The ISO OGI Pilot
- Another Big Gap – Planning vs Actual
- Your Thoughts?
- Q & A

4

MIMOSA™

PeˣI
Pedersen Enterprises Inc.

NWR A New 'Bitumen' Refinery

NORTH WEST REDWATER PARTNERSHIP

Capacity	3 identical phases Total 150,000 b/d Bitumen feed
Technology	Hydrocracking Resid Hydrocracking Gasification **No Coke, CO_2 Capture (for EOR)**
Capital Cost	About **$5.7 Billion** each phase Produces Diesel & Diluent
Owners	NWU (50%) CNRL (50%)
NWR	325 staff, >1000 contractor staff

BRIK Approved: Feb.16, 2011

Project Sanctioned: Nov.8, 2012

MIMOSA™ **Construction Start: Spring, 2013 (in Progress!!)**

Start-Up: mid 2016

NWU Alberta ENERGY Canadian Natural Pe^{x}I Pedersen Enterprises Inc.

System Integration: What's Missing?

AIChE Southwest Process Technology Conference

Enterprise Business Systems
Enterprise Resource Planning (ERP)

Operations

BIG GAP ? ?

Maintenance

Physical Asset Control
Real-time Systems

MIMOSA

Pe^xI
Pedersen Enterprises Inc.

NWR
NORTH WEST REDWATER PARTNERSHIP

Watch
Out
...

for the
Big
Gaps
!!!

MIMOSA

Pe^xI
Pedersen Enterprises Inc.

AIChE
Southwest Process Technology Conference

A Direct Interface Approach
to Data Exchange within the Enterprise

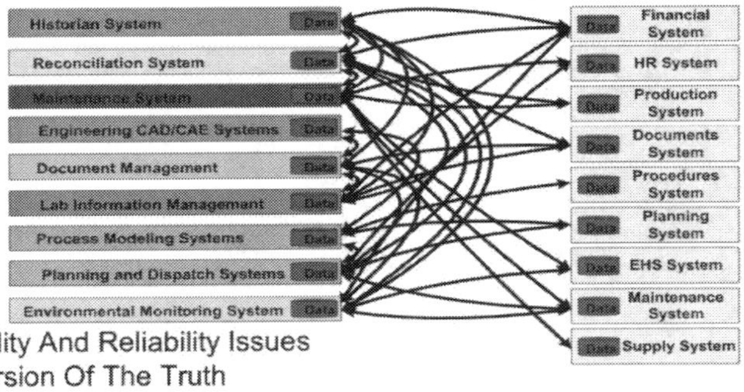

» System Stability And Reliability Issues
» No Single Version Of The Truth
» System Maintenance Issues
» Difficult Access To Multiple Systems/Applications

Don't want to do this anymore!!!

MIMOSA™

(courtesy of Emerson Process Management)

PeχI
Pedersen Enterprises Inc.

AIChE
Southwest Process Technology Conference

A Bottomless Pit ...

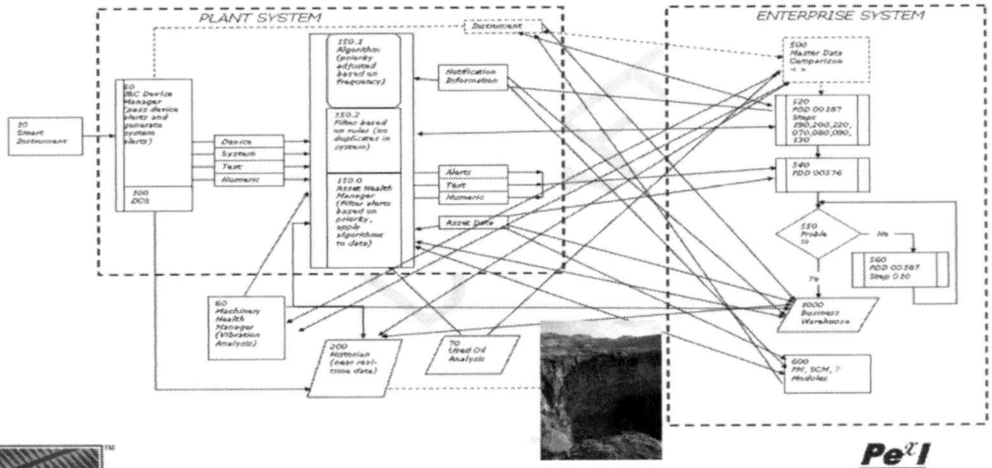

MIMOSA™

PeχI
Pedersen Enterprises Inc.

AIChE
Southwest Process Technology Conference

A Data Warehouse Approach
to Data Exchange within the Enterprise

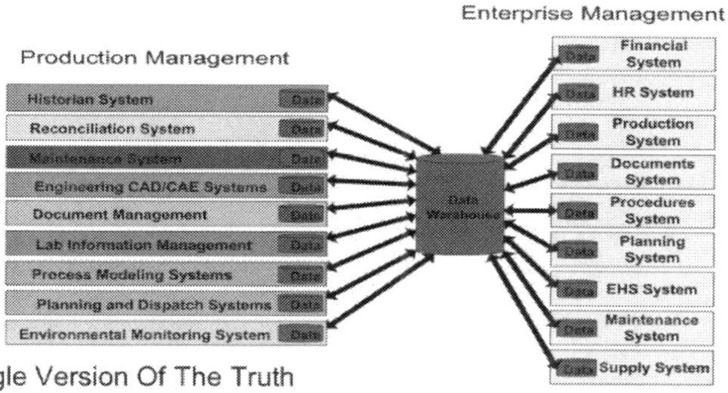

- No Single Version Of The Truth
- Single Point Of Failure
- Difficult Access To Multiple Systems

Don't want to do this either ...

(*courtesy of Emerson Process Management*)

MIMOSA™

Pe²I
Pedersen Enterprises Inc.

AIChE
Southwest Process Technology Conference

What's the Difference?

- **Integration** – project specific solutions process
  - ✓ Use cases, custom code and testing are **all project specific**
  - ✓ **Integration risks** are borne by Owner/Operator and Integrator
  - ✓ **Expensive and high risk to implement and sustain** (20% - 25% annual recurring cost)
  - ✓ **Not repeatable, lower quality** due to project specific code base
- **Interoperability** - industry driven solutions process
  - ✓ Requirements Driven; Guided by Owner-Operators
  - ✓ Industry Foundation Architecture & Compliance
  - ✓ Suppliers develop, maintain, license COTS
  - ✓ Enables repeatable & scalable solutions

MIMOSA™

Pe²I
Pedersen Enterprises Inc.

AIChE Southwest Process Technology Conference

Various Interoperability Definitions

- **IEEE: The capability ...**
  - ✓ of two or more systems or elements to exchange information and to use the information that has been exchanged.
  - ✓ for units of equipment to work together to do useful functions.
  - ✓ that enables heterogeneous equipment, generally built by various vendors, to work together in a network environment.
  - ✓ of two or more systems or components to exchange information in a heterogeneous network and use that information.
- **SEI: The ability of a set of communicating entities to:**
  - ✓ (1) exchange specified state data
  - ✓ (2) operate on that state data according to specified, agreed-upon, operational semantics

➢ **Data/information interoperability is necessary, but only part of the requirement for Interoperable Systems of Systems**

MIMOSA™

Pe^{x}I Pedersen Enterprises Inc. 13

AIChE Southwest Process Technology Conference

What's Interoperability?

- **Interoperability – industry driven solutions process**
  - ✓ *Requirements Driven* – the use case approach by OpenO&M is a good example
  - ✓ *Industry Foundation Architecture* – Open Standards incorporated by reference
  - ✓ *Industry Compliance* data sets are developed and managed by industry
  - ✓ *ISO TC 184 OGI TS* is an example of an activity that sets forward a 'Target Industry Foundation Architecture' and codifies piloted industry solution
  - ✓ *Suppliers develop, maintain and license* compliant adaptors as COTS products
  - ✓ *Enables repeatable and scalable* industry-driven solutions for Oil & Gas shared with other elements of Critical Infrastructure
  - ✓ *Guided by Owner/Operators* with assistance from service companies and standards

MIMOSA™

Pe^{x}I Pedersen Enterprises Inc.

Does It Matter?
(Questions to Consider)

- Does your organization need more ability to share and maintain information important to multiple functional units?
- Will you gain business value from improved sharing of data and information with your key suppliers and business partners?
- Will you gain business value from sharing technology, methods and standards with other related major industry groups including aerospace, defense or military?
- Will technology & best practices evolve quicker or slower in the future than in the past?

Pedersen Enterprises Inc.

Does It Matter? - continued
(Questions to Consider)

- Does the ability to follow evolving best practices matter to your organization?
- Will the future pace of evolution of best practices and technology be likely to result in "orphan" products with painful and/or limited migration options?
- Does the ability to systemically institutionalize organizational knowledge, information and processes matter?
- Does the ability to separate value added knowledge, information and processes from raw technology change matter to your organization?

Pedersen Enterprises Inc.

Oil and Gas Industry Adoption of Standards

AIChE
Southwest Process Technology Conference

Data Models - Standards Positioning

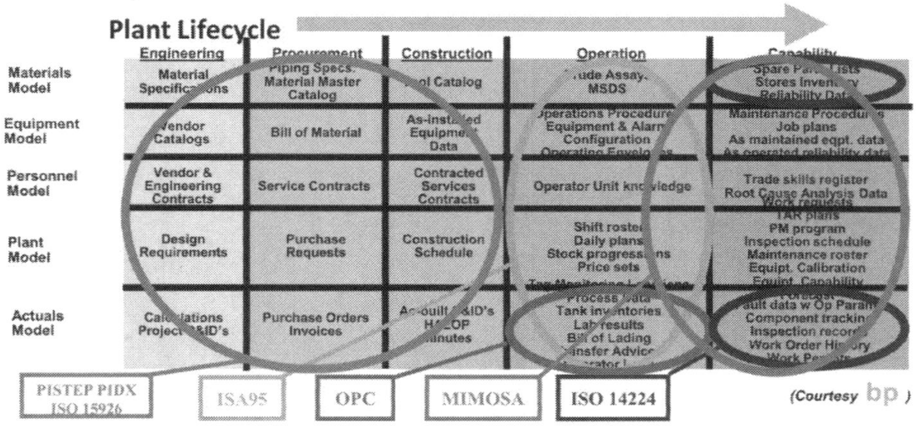

(Courtesy bp)

MIMOSA

Pe^x I
Pedersen Enterprises Inc.

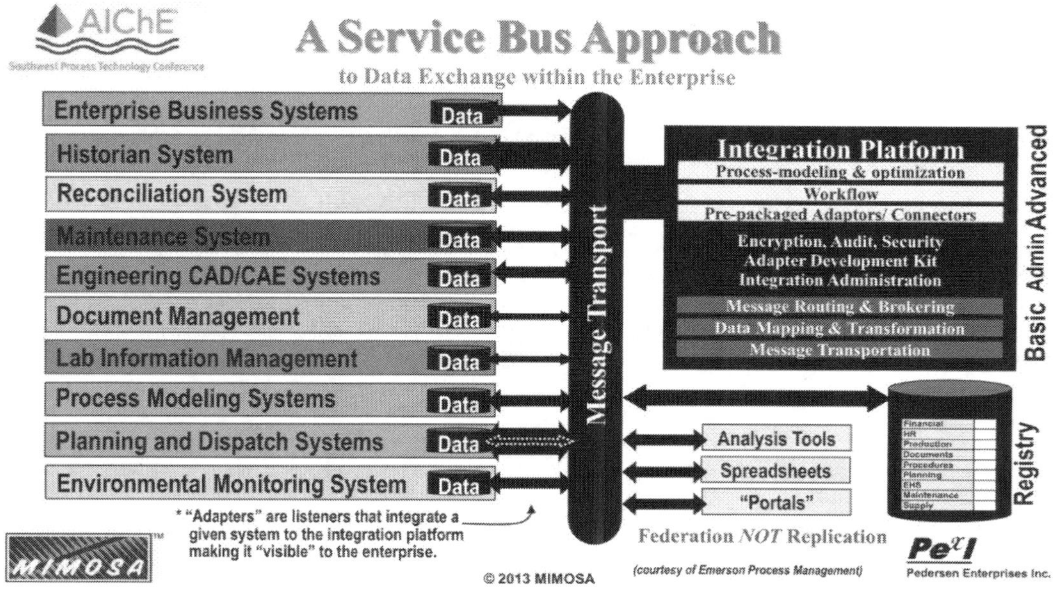
AIChE
Southwest Process Technology Conference
A Service Bus Approach
to Data Exchange within the Enterprise
Enterprise Business Systems
Historian System
Reconciliation System
Maintenance System
Engineering CAD/CAE Systems
Document Management
Lab Information Management
Process Modeling Systems
Planning and Dispatch Systems
Environmental Monitoring System
Data
Message Transport
Integration Platform
Process-modeling & optimization
Workflow
Pre-packaged Adaptors/ Connectors
Encryption, Audit, Security
Adapter Development Kit
Integration Administration
Message Routing & Brokering
Data Mapping & Transformation
Message Transportation
Basic Admin Advanced
Analysis Tools
Spreadsheets
"Portals"
Registry
* "Adapters" are listeners that integrate a given system to the integration platform making it "visible" to the enterprise.
Federation NOT Replication
(courtesy of Emerson Process Management)
© 2013 MIMOSA
Pedersen Enterprises Inc.

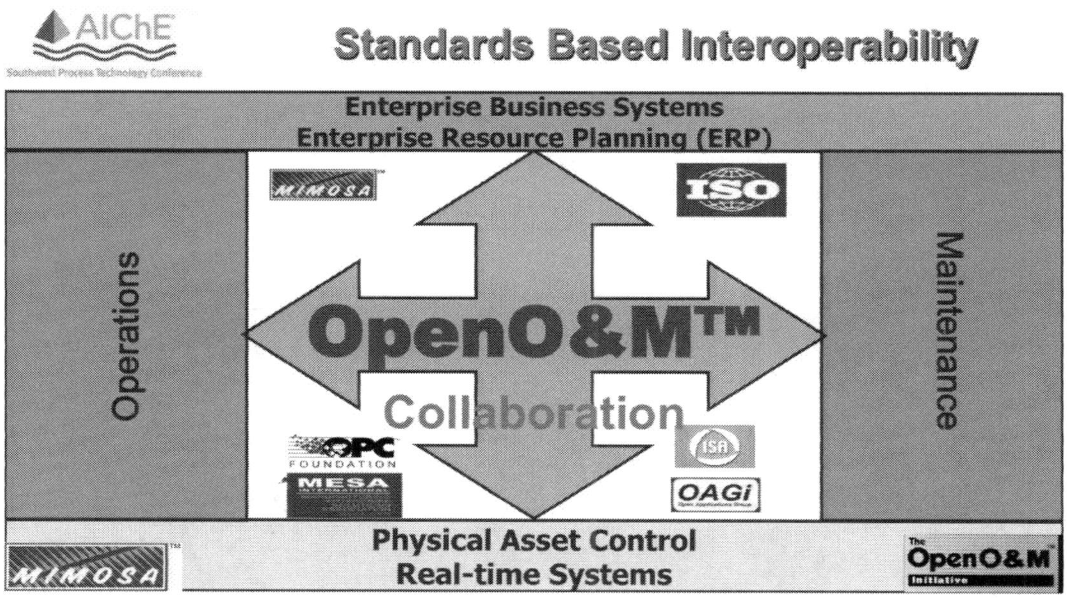
AIChE
Southwest Process Technology Conference
Standards Based Interoperability
Enterprise Business Systems
Enterprise Resource Planning (ERP)
Operations
Maintenance
OpenO&M™
Collaboration
ISO
OPC FOUNDATION
MESA
OAGi
Physical Asset Control
Real-time Systems
The OpenO&M Initiative

What do Owner/Operators Need?

- **The Problem** – Current Oil and Gas and other Asset Intensive industry enterprise solutions are *too complex and too difficult to sustain*.
  - ✓ The current enterprise solutions model is critically dependent on ***large amounts of custom Systems Integration*** and this is a weak link.
    - Expensive to sustain (20% annual recurring maintenance cost)
    - Fragile – Vulnerable to breakage
  - ✓ The current solutions model also forces data to be ***re-entered many times*** rather than managing it on a full life-cycle basis
    - Increases costs
    - Decreases quality
- **Owner/Operators are asking for *a better Solutions Model***

MIMOSA™

Pe²I Pedersen Enterprises Inc.

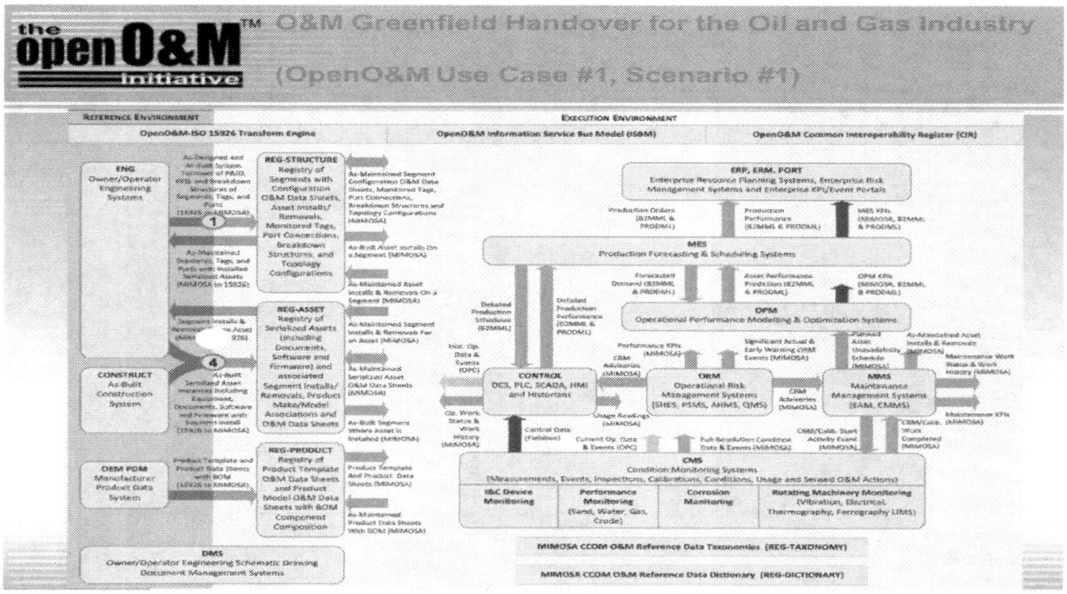

OGI Use Cases

the openO&M™ Initiative — key industry use cases

1. **"digital handover" as-designed/engineered/built O&M information from engineering, procurement, construction phase to O&M phase**
2. recurring updates - send engineering upgrades to O&M systems
3. field engineering changes sent to engineering (bottom up)
4. on-line product data library updated with engineering reference information (asset based data)
5. **operations & maintenance configuration changes (e.g. remove/replace transmitter)**
6. preventive maintenance (PM) triggering
7. condition-based maintenance (CBM) triggering
8. early warning notification
9. incident management – actual & near-miss information captured and escalated along the lines of accountability
10. **O&M systems information provisioning**

+

PCA POSC Caesar Association

Industry Use Cases

1. **Upstream Production Optimization**
2. Drilling Reporting
3. Production Reporting

ISO TC184

Context for Collaboration

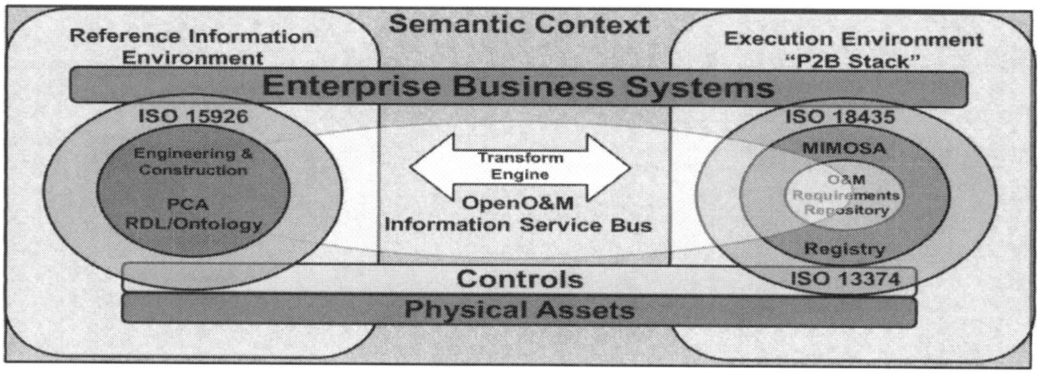

ISO TC 184/WG 6

MIMOSA™

Pe²I Pedersen Enterprises Inc.

Debutanizer Tower

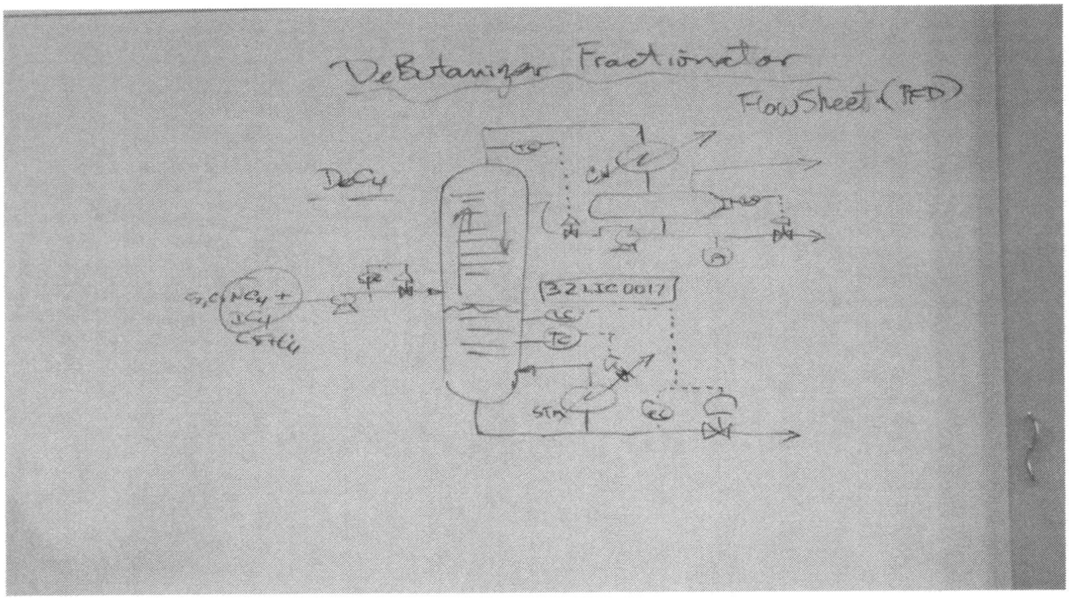

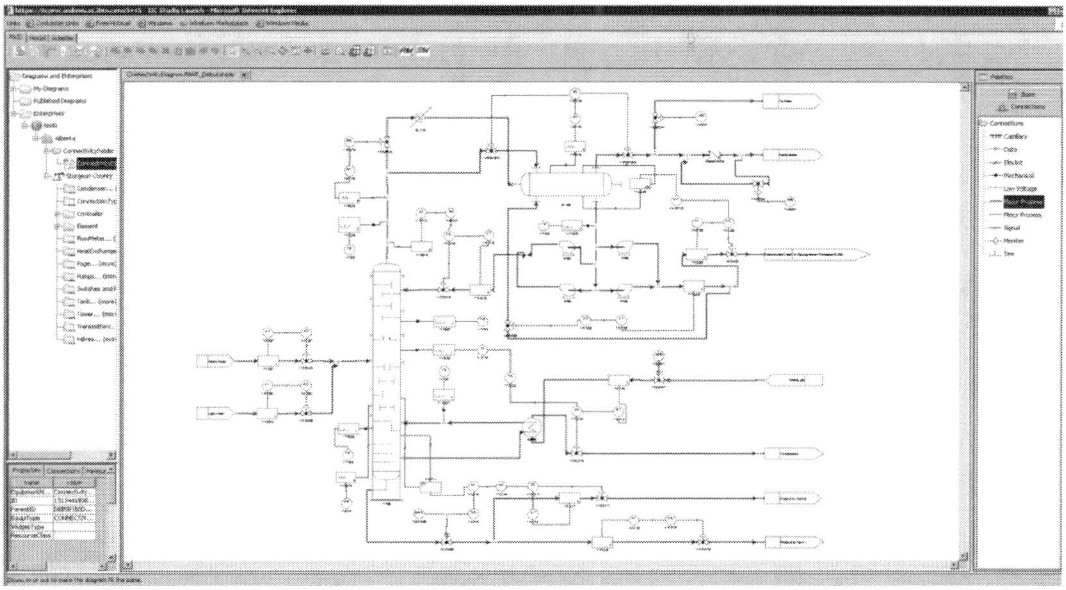

AIChE

Southwest Process Technology Conference

Debutanizer P&ID 001- Worley Parsons

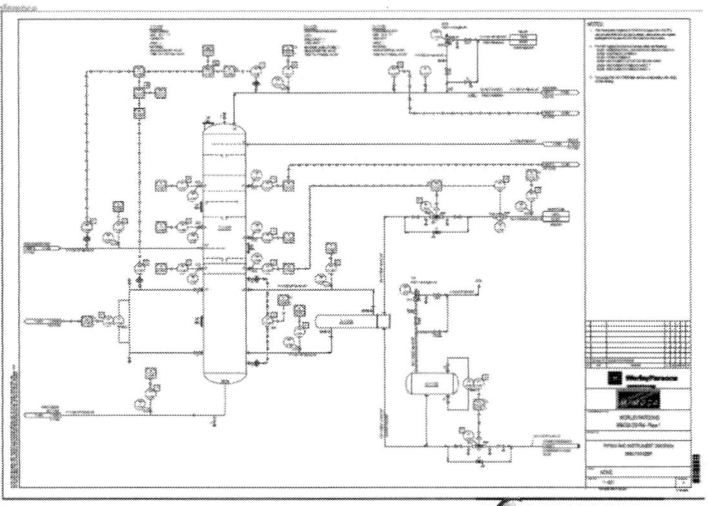

MIMOSA

POSC Caesar Association

Pedersen Enterprises Inc.

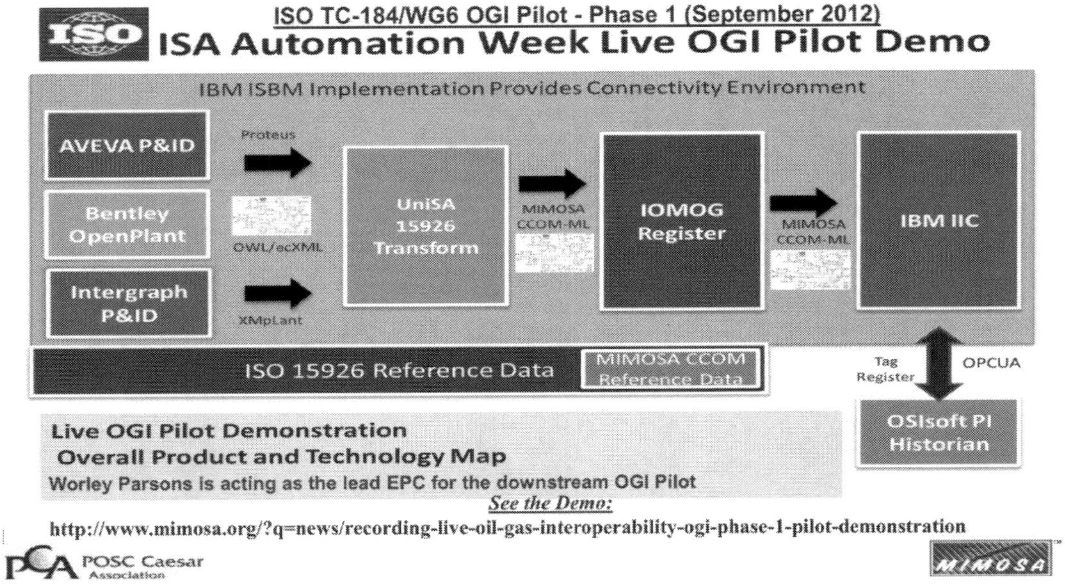
ISO TC-184/WG6 OGI Pilot - Phase 1 (September 2012)
ISA Automation Week Live OGI Pilot Demo
IBM ISBM Implementation Provides Connectivity Environment
AVEVA P&ID
Bentley OpenPlant
Intergraph P&ID
Proteus
OWL/ecXML
XMpLant
UniSA 15926 Transform
MIMOSA CCOM-ML
IOMOG Register
MIMOSA CCOM-ML
IBM IIC
ISO 15926 Reference Data
MIMOSA CCOM Reference Data
Tag Register
OPCUA
OSIsoft PI Historian
Live OGI Pilot Demonstration
Overall Product and Technology Map
Worley Parsons is acting as the lead EPC for the downstream OGI Pilot
See the Demo:
http://www.mimosa.org/?q=news/recording-live-oil-gas-interoperability-ogi-phase-1-pilot-demonstration
POSC Caesar Association
MIMOSA

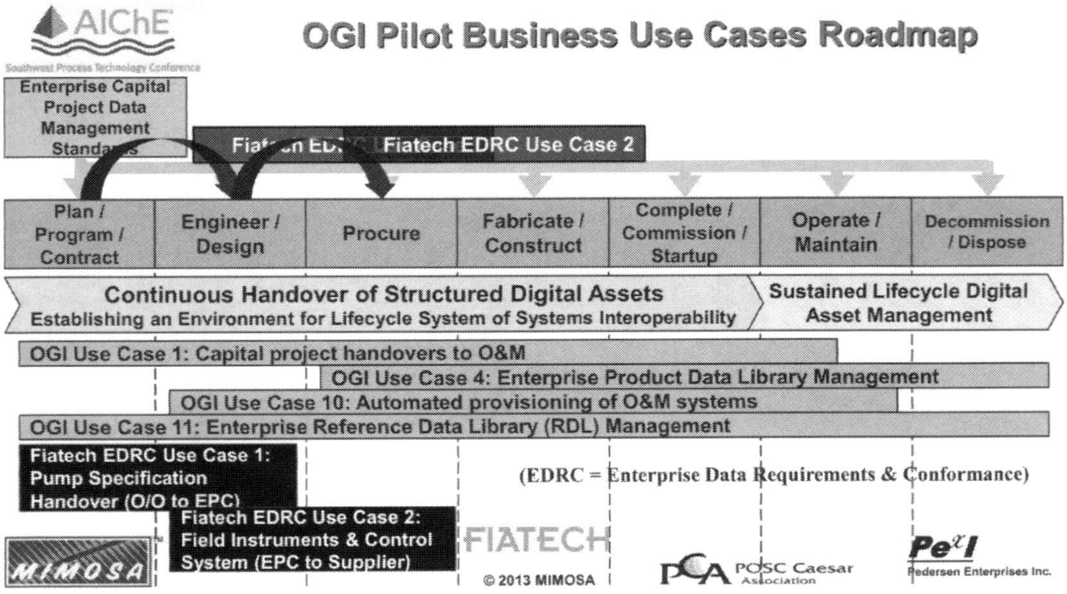
AIChE
OGI Pilot Business Use Cases Roadmap
Enterprise Capital Project Data Management Standards
Fiatech EDRC Use Case 2
Plan / Program / Contract
Engineer / Design
Procure
Fabricate / Construct
Complete / Commission / Startup
Operate / Maintain
Decommission / Dispose
Continuous Handover of Structured Digital Assets
Establishing an Environment for Lifecycle System of Systems Interoperability
Sustained Lifecycle Digital Asset Management
OGI Use Case 1: Capital project handovers to O&M
OGI Use Case 4: Enterprise Product Data Library Management
OGI Use Case 10: Automated provisioning of O&M systems
OGI Use Case 11: Enterprise Reference Data Library (RDL) Management
Fiatech EDRC Use Case 1: Pump Specification Handover (O/O to EPC)
Fiatech EDRC Use Case 2: Field Instruments & Control System (EPC to Supplier)
(EDRC = Enterprise Data Requirements & Conformance)
MIMOSA
FIATECH
© 2013 MIMOSA
POSC Caesar Association
Pedersen Enterprises Inc.

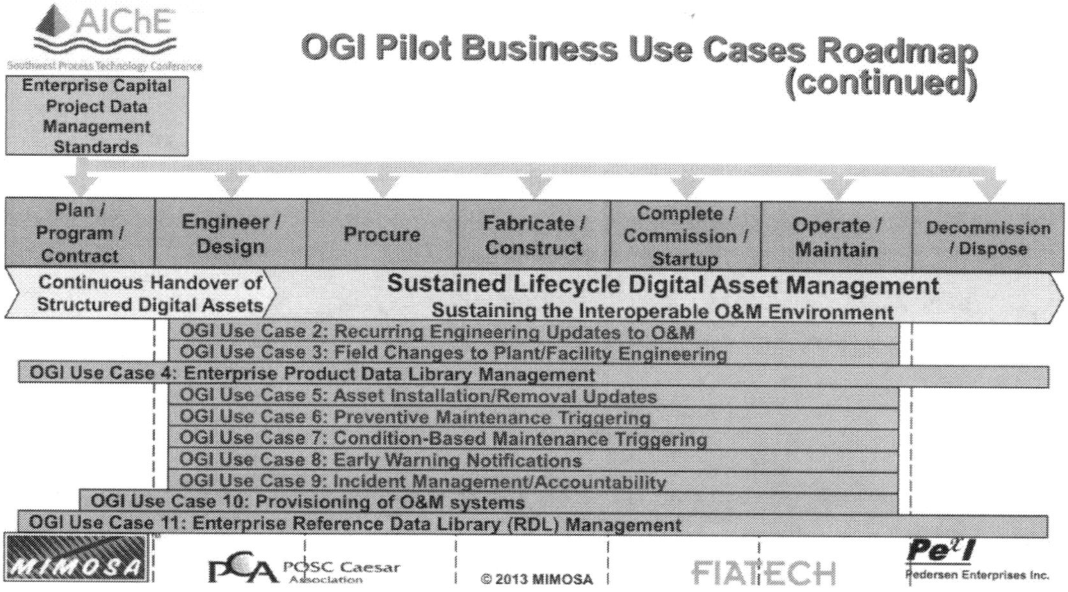
AIChE
Southwest Process Technology Conference
OGI Pilot Business Use Cases Roadmap (continued)
Enterprise Capital Project Data Management Standards
Plan / Program / Contract
Engineer / Design
Procure
Fabricate / Construct
Complete / Commission / Startup
Operate / Maintain
Decommission / Dispose
Continuous Handover of Structured Digital Assets
Sustained Lifecycle Digital Asset Management
Sustaining the Interoperable O&M Environment
OGI Use Case 2: Recurring Engineering Updates to O&M
OGI Use Case 3: Field Changes to Plant/Facility Engineering
OGI Use Case 4: Enterprise Product Data Library Management
OGI Use Case 5: Asset Installation/Removal Updates
OGI Use Case 6: Preventive Maintenance Triggering
OGI Use Case 7: Condition-Based Maintenance Triggering
OGI Use Case 8: Early Warning Notifications
OGI Use Case 9: Incident Management/Accountability
OGI Use Case 10: Provisioning of O&M systems
OGI Use Case 11: Enterprise Reference Data Library (RDL) Management
MIMOSA
POSC Caesar Association
© 2013 MIMOSA
FIATECH
Pedersen Enterprises Inc.

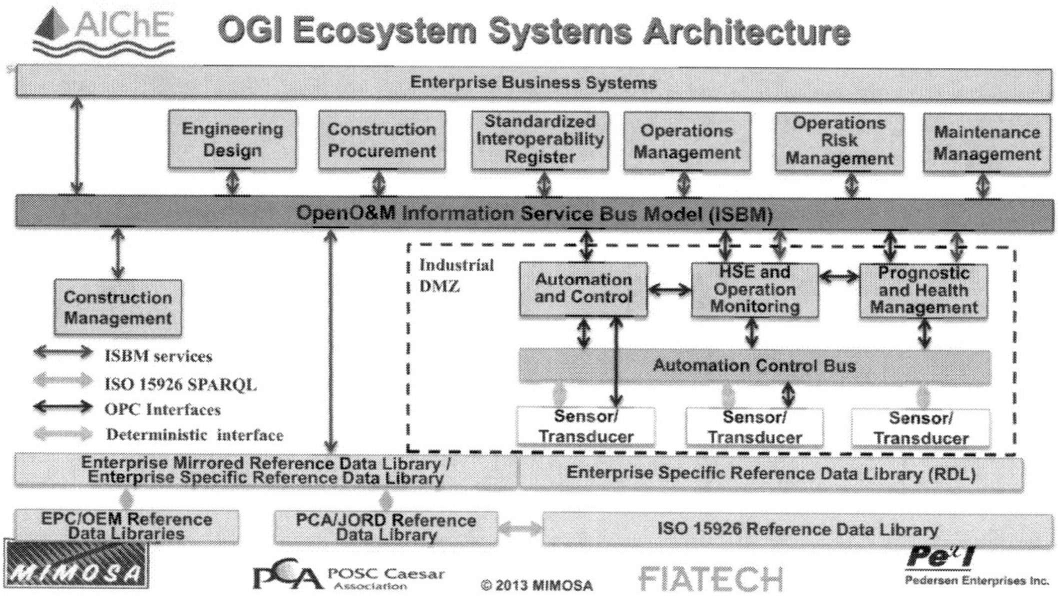
AIChE
OGI Ecosystem Systems Architecture
Enterprise Business Systems
Engineering Design
Construction Procurement
Standardized Interoperability Register
Operations Management
Operations Risk Management
Maintenance Management
OpenO&M Information Service Bus Model (ISBM)
Industrial DMZ
Construction Management
Automation and Control
HSE and Operation Monitoring
Prognostic and Health Management
Automation Control Bus
Sensor/ Transducer
Sensor/ Transducer
Sensor/ Transducer
ISBM services
ISO 15926 SPARQL
OPC Interfaces
Deterministic interface
Enterprise Mirrored Reference Data Library / Enterprise Specific Reference Data Library
Enterprise Specific Reference Data Library (RDL)
EPC/OEM Reference Data Libraries
PCA/JORD Reference Data Library
ISO 15926 Reference Data Library
MIMOSA
POSC Caesar Association
© 2013 MIMOSA
FIATECH
Pedersen Enterprises Inc.

Oil and Gas Interoperability (OGI) Pilot - Methods

- Owner/Operator leadership
- Industry Use Case driven - (OpenO&M, PCA and SPE DSA-TS) Use Cases
- Cooperatively aligned with PCA under **Joint MIMOSA/PCA O&M SIG**
- Overall **Solutions Architecture** under **Joint MIMOSA/PCA IT Architecture SIG**
- Managed like a true capital project- Worley Parsons-Lead EPC for downstream pilot
- Pragmatic focus on Commercial Off The Shelf (COTS) products
- Suppliers assume responsibility for compliance of their own products

> Publication – All working documents and results are on the MIMOSA website at www.mimosa.org

> **Proven OGI Pilot output provides basis for an ISO Technical Specification (TS)**

Oil and Gas Interoperability (OGI) Summary

- The **OGI Ecosystem**
  - ✓ Is the open industrial ecosystem for related industries
  - ✓ Is fully based on published supplier neutral standards
  - ✓ Is driven by owner/operator defined, fully dressed use cases
  - ✓ Provides the interoperability test-bed for defined use cases
- The **OGI Pilot** is the OGI Ecosystem development program
  - ✓ Collaboratively managed under the Joint MIMOSA PCA O&M SIG
  - ✓ All OGI Pilot Artifacts are published on **MIMOSA.org**
- The **ISO OGI Technical Specification**
  - ✓ Is under development by ISO TC 184/WG 6
  - ✓ Is based fully on proven output from the OGI Pilot

AIChE
Southwest Process Technology Conference

MESA White Paper

MESA INTERNATIONAL
MANUFACTURING ENTERPRISE SOLUTIONS ASSOCIATION

ISA-95 for the Continuous & Process Industries (CPI)

Interpreting and Applying the Standards in Specific Case Studies

1. Migration Planning and Organizational Structure
2. Life Cycle Management
3. Business Process Globalization
4. System architecture for a new plant
5. Production accounting in mining

The Automation Conference
Chicago, IL
May 14, 2013

MESA White Paper #41 (www.mesa.org/resourcelibrary)
Available to MESA members; nominal fee for others

MIMOSA

Pe^{x}I
Pedersen Enterprises Inc.

AIChE
Southwest Process Technology Conference

Another Big GAP
Financial Planning & Operational Decision Making

Monthly Plan
↓
Weekly Schedule
↓
Daily Orders
↓
Actual

The GAP

The actionable GAP

(courtesy of Manufacturing Technology Network)

MIMOSA

Pe^{x}I
Pedersen Enterprises Inc.

Opportunity Cost Visualization Today

"A Day Late and a Dollar Short"

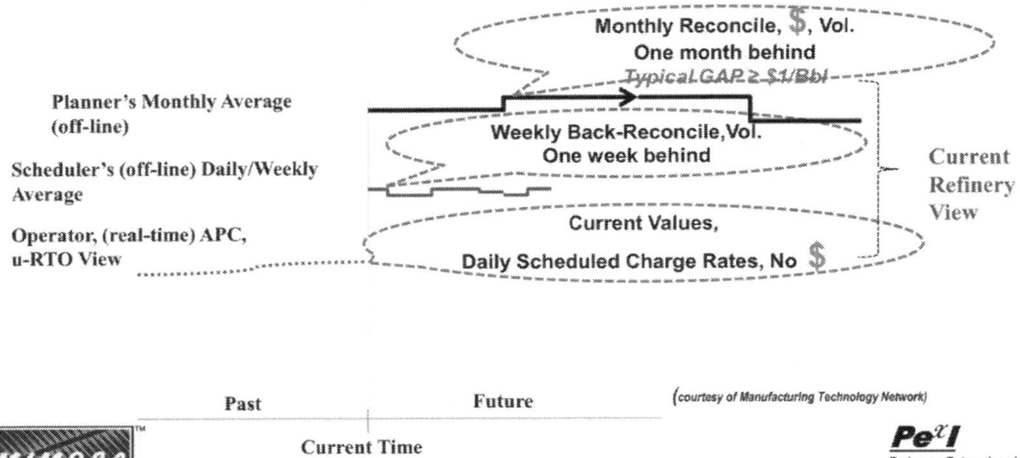

(courtesy of Manufacturing Technology Network)

Pedersen Enterprises Inc.

Opportunity Cost Visualization

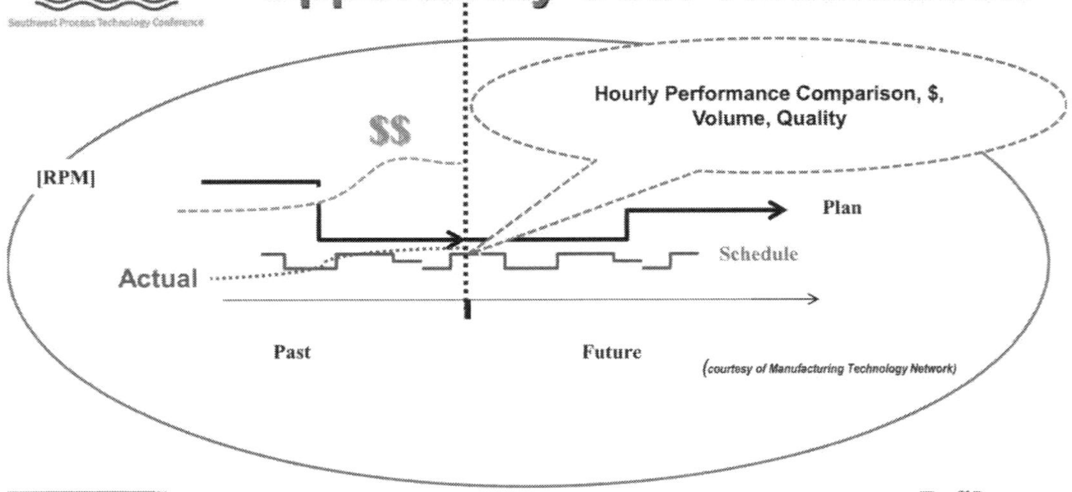

(courtesy of Manufacturing Technology Network)

Pedersen Enterprises Inc.

What are Your Thoughts?

Does Interoperability Matter?

- Yes: Get Involved & Support:
  - ✓ Join the Owner/Operator Leadership Council (www.mimosa.org)
    - MIMOSA/PCA O&M SIG
    - MIMOSA/PCA IT Architecture SIG
  - ✓ Contribute to industry standards (like ISA-95)
  - ✓ Join the MESA CPI SIG (www.mesa.org)
- No: Keep on connecting systems the old ways!!

Pedersen Enterprises Inc.

Owner/Operator Leadership Council

- provide input, direction, oversight & prioritization of use-cases & scenarios
- use-cases address high-value added activities
- scenarios combine use cases to depict system interactions

Chair & Industry Facilitator – Cliff Pedersen

BP

Chevron

North West Redwater Partnership

Origin Energy

Southern Company

Suncor Energy

Woodside Energy

Pedersen Enterprises Inc.

AIChE
Southwest Process Technology Conference

For More Information:

Cliff Pedersen
cliffpedersen@sympatico.ca
519-312-5256

Collaborating with:

Alan Johnston
atjohn@mimosa.org
205-553-1804

Questions ???

Ken Dunn, BP
Bruce Taylor, Suncor
Matt Rettagliata, Chevron
Steve Treiber, MTN

MIMOSA™

Pe^xI
Pedersen Enterprises Inc.

Advances in Automation: Wireless the Enabler

Tim Olsen
Emerson Process Management

AIChE
Southwest Process Technology Conference

EMERSON.
Process Management

Agenda

- Wireless
  - Then and Now
- Culture
  - Solomon Associates
- Challenges
- Examples: Wireless the Enabler

2008 Post Spring Meeting Topical 7 Survey

1. New Technologies in Refining
2. Biodiesel and Renewable Fuels
3. Gas and Coal to Liquids
4. Energy Efficiency Tutorial
5. Bottom of the Barrel Processing
6. Upgrading and Processing of Opportunity Crudes
7. BioRefining
8. Energy and Conservation
9. CO2 Reduction Tutorial
10. CO2 Conversion
11. Advances in Hydroprocessing
12. Hydrogen Production
13. Simulation, Engineering and Design
14. Gas Separation and Conversion Technologies
15. Environmental Solutions
16. Refining and Petrochemical Integration
17. Hydrogen Management
18. Sustainable Developments for Petrochemical Refineries
19. Expansion, Revamps, and Debottlenecking
20. Advances in FCC
21. Training: New Generation Engineers and Operators
22. Regulations and Sulfur Removal
23. Fuel Cells
24. LNG
25. Process Control and Optimization
26. Fouling
27. Advances in Aromatics
28. Oil Water Separation
29. Water Conservation Design Methods
30. Sour Water Stripper and Amine Systems
31. Wireless Technology

Global Automation Research

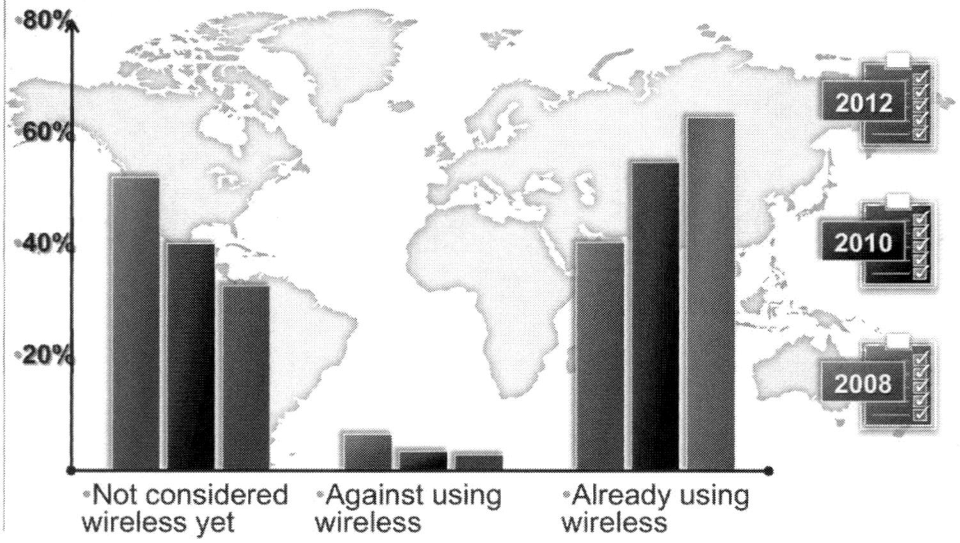

Latest Numbers for WirelessHART®

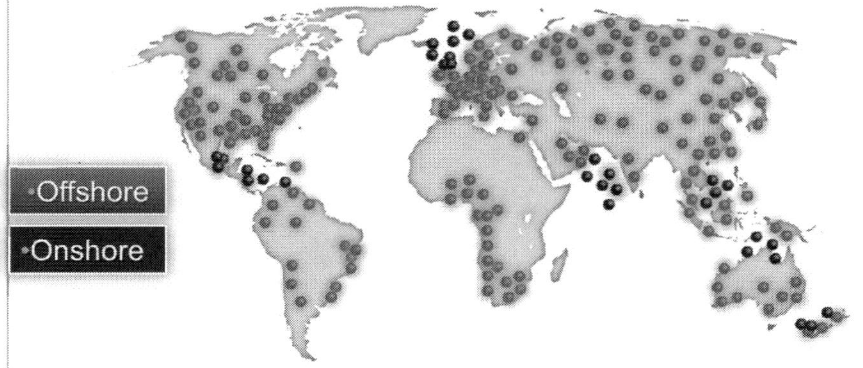

1.7+ billion hours operation

Over 14,000 wireless field networks

IEC 62591 International Standard (March 2010)

www.HARTCOMM.org

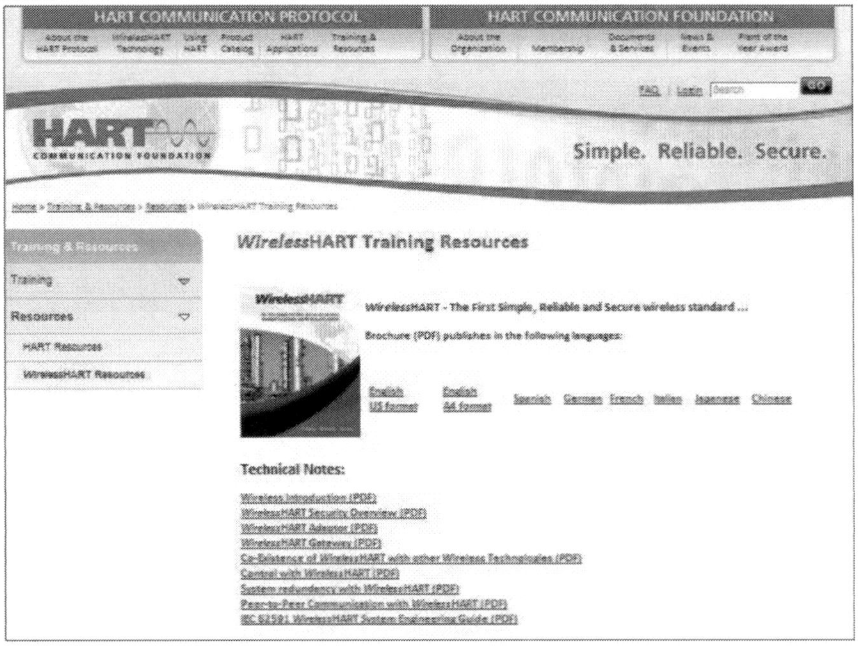

Culture and Challenges

Reliability, Availability, Maintenance, and Safety

AIChE
Southwest Process Technology Conference

EMERSON.
Process Management

Solomon Associates

Recent AFPM Event

- Culture
  - Traditional vs.
  - Progressive
- Top quartile refiners with highest availability actually spend less on Maintenance
  - Reactive vs.
  - Planned

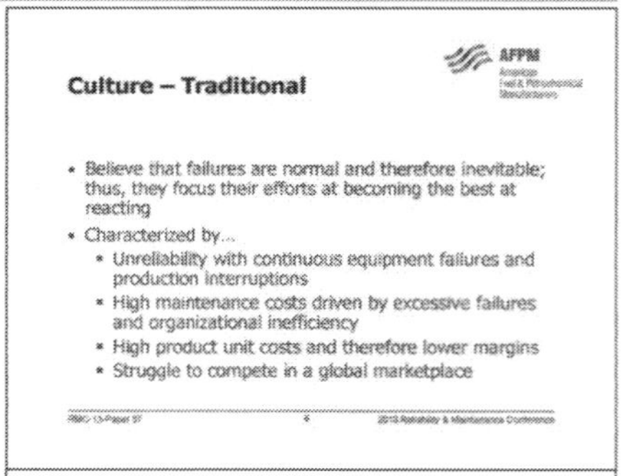

Culture – Traditional

- Believe that failures are normal and therefore inevitable; thus, they focus their efforts at becoming the best at reacting
- Characterized by...
  - Unreliability with continuous equipment failures and production interruptions
  - High maintenance costs driven by excessive failures and organizational inefficiency
  - High product unit costs and therefore lower margins
  - Struggle to compete in a global marketplace

Culture – Progressive

- Believe that failure-free operation is normal and therefore focus their efforts on failure avoidance
- Characterized by...
  - Highly reliable equipment
  - Nearly failure-free operation
  - Low maintenance costs
  - Consistently high quality product
  - Low product unit cost
  - Generous margins
  - Very competitive in a global marketplace

Al Poling from Solomon Associates
2013 AFPM (NPRA) Reliability and Maintenance Conference

Challenge for Older Facilities

Most process units were designed and built with the ***minimal*** amount of instrumentation.

Only that which is required for:

- Safe operations
- Validate process design

Route-based manual data collection for non-critical assets and some process measurements

Route based data collection does not prevent all asset failures

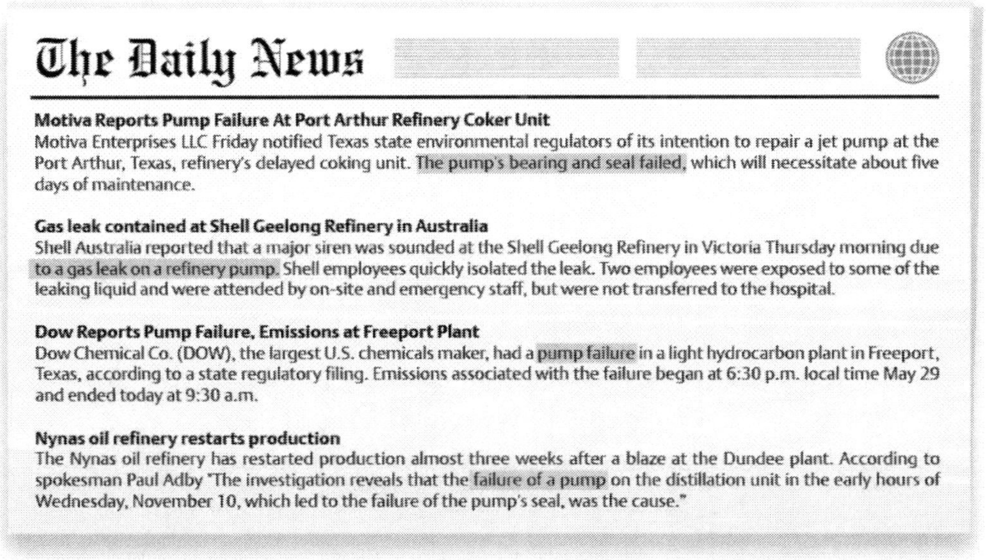

The Daily News

Motiva Reports Pump Failure At Port Arthur Refinery Coker Unit
Motiva Enterprises LLC Friday notified Texas state environmental regulators of its intention to repair a jet pump at the Port Arthur, Texas, refinery's delayed coking unit. The pump's bearing and seal failed, which will necessitate about five days of maintenance.

Gas leak contained at Shell Geelong Refinery in Australia
Shell Australia reported that a major siren was sounded at the Shell Geelong Refinery in Victoria Thursday morning due to a gas leak on a refinery pump. Shell employees quickly isolated the leak. Two employees were exposed to some of the leaking liquid and were attended by on-site and emergency staff, but were not transferred to the hospital.

Dow Reports Pump Failure, Emissions at Freeport Plant
Dow Chemical Co. (DOW), the largest U.S. chemicals maker, had a pump failure in a light hydrocarbon plant in Freeport, Texas, according to a state regulatory filing. Emissions associated with the failure began at 6:30 p.m. local time May 29 and ended today at 9:30 a.m.

Nynas oil refinery restarts production
The Nynas oil refinery has restarted production almost three weeks after a blaze at the Dundee plant. According to spokesman Paul Adby "The investigation reveals that the failure of a pump on the distillation unit in the early hours of Wednesday, November 10, which led to the failure of the pump's seal, was the cause."

In Extreme Cases Reliability Can Impact HSE

Marsh – The 100 largest losses in Hydrocarbon Industries since 1972

Piping failures, start-up and shut-down events are a majority of significant incidents

Motivation for Reliability

- More available operations
- Lower probability for safety incident
- Lower maintenance costs

However, route-based manual data collection does not prevent all asset failures

What is the barrier to upgrade manual checks to online measurements?

Wiring costs!

Examples

Wireless Applications

AIChE
Southwest Process Technology Conference

EMERSON.
Process Management

Missing Online Temperatures Around Heat Exchangers

- New opportunity crudes and new normal with blending different combinations of crudes
- Difficult to take measurements
  - Hot process fluids
  - Stacked exchangers
  - Other access constraints
- Lack of temperature measurements between bundles
- Time consuming to take measurements manually

Up to 50% of the heat delivered to the crude unit comes from the crude pre-heat exchangers

Process Pumps and Motors

- Vibration
- Suction Strainer dP
- Auxiliary Seal Flush Level
- Hydrocarbon Leak Detection

Think about a pump prior to a fired heater, or the overhead reflux pump on a distillation column.

Vacuum bottoms and LPG with insufficient NPSH.

Cooling Towers

- Supply pump
- Fans
- Water Chemistry

A crane is required to repair a fan on top of a cooling tower. Chemical costs are linked to blowdown rates and water quality measurements.

Blowers / Fans

- Fired Heater ID and FD fans

What are the consequences of a fired heater shutting down? How long to bring process temperatures back to operating conditions?

Safety Shower and Eye Wash

- Typically no indication to operations when safety shower or eye wash being used.

Is it worth knowing *immediately* when a safety or environmental incident occurred in the field?

"Critical" Measurements

- Frequent earthquake zone in Japan
- Wires burned, wireless still functioning after earthquake, tsunami flooding, and fire
- Add second measurements (wireless) on process "critical" needed in an emergency

Fires caused by the earthquake and flood water from the subsequent tsunami devastated the facility just 5 months after the wireless network was commissioned

PRV and Steam Trap Leak Detection

- Acoustic technology 'listening' for leaks

Difficult to identify small pressure relief valve leaks.

Air-Cooled Heat Exchangers

- Hot climates with cooling limited operation (peak summer) – loss of air cooled exchanger may result in relief valve lifting or process slowdown with less cooling available

Distillation Column Temperature and dP Profile

- Additional temperature measurements
- Differential pressure across tower to detect flooding

Tank Farm Level & Valve Position Indicator

- Secondary tank level measurements
- Mitigate human error with valve position indicator for correct line-up
- Intermittent pump use and challenging to capture manual data collection

Wastewater Treatment Plant

- Similar to cooling tower water management for chemical additives

Stranded Diagnostics on Valves and Transmitters

- Older systems not able to get HART diagnostics "online"
- Critical valves like hydrocracker hydrogen quench valves or separator letdown valves should be closely monitored

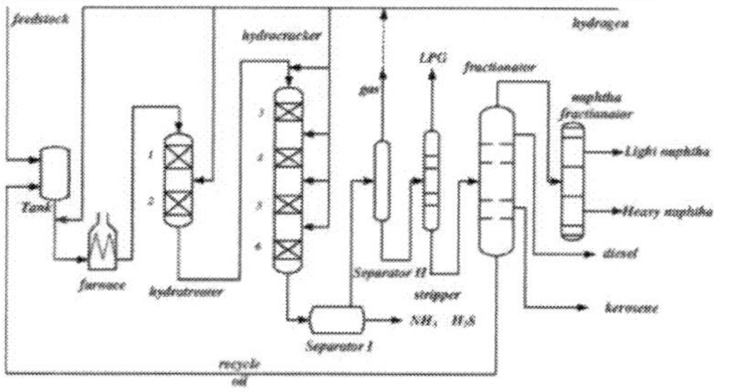

Good for not only day to day operations, but also for planning upcoming turnarounds

Wireless the Enabler

Conclusion

AIChE
Southwest Process Technology Conference

EMERSON. Process Management

The Manufacturing Balance Has Changed

It is time to revisit technology justifications

YESTERDAY:

- Cheap Energy
- Cheap Water
- Cheap Waste
- 'Free' Emissions
- Predictable Raw Materials
- Available Talent Pool

- Expensive Controls & Automation Technology
- Expensive Software and IT

TODAY:

- Cheap Controls & Automation Technology
- Cheap Software and IT

- Energy cost swings
- Water cost rising
- Expensive Waste
- Expensive Emissions
- Unpredictable Raw Materials
- Limited Talent Pool

ARC Advisory Group

THOUGHT LEADERS FOR MANUFACTURING & SUPPLY CHAIN

Ref: A. Chatha; "Current Business Drivers and Trends;" ARC 2009 Orlando Conference; February, 2009

Technology is Not the End Game

Wireless ***enables*** low cost implementation to upgrade process and asset health measurements

However, there is a ***culture*** shift also required:

- Measure
- Analyze
- Alert
- Take Action

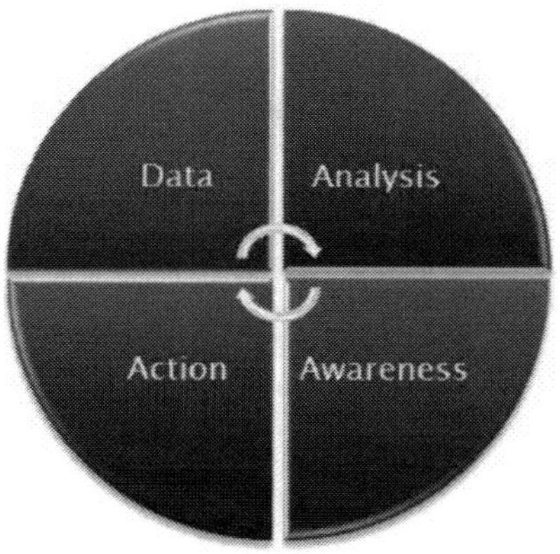

Questions?

Tim.Olsen@Emerson.com

AIChE
Southwest Process Technology Conference

EMERSON
Process Management

The Human Reliability Company

Human Reliability

Optimizing the Human-Automation Relationship to Help Humans Make Better Decision

AICHE Southwest Process Technology Conference

Eddie Habibi, PAS Founder and CEO
Galveston, Texas
October 4, 2013

Human Ingenuity & Unprecedented Prosperity

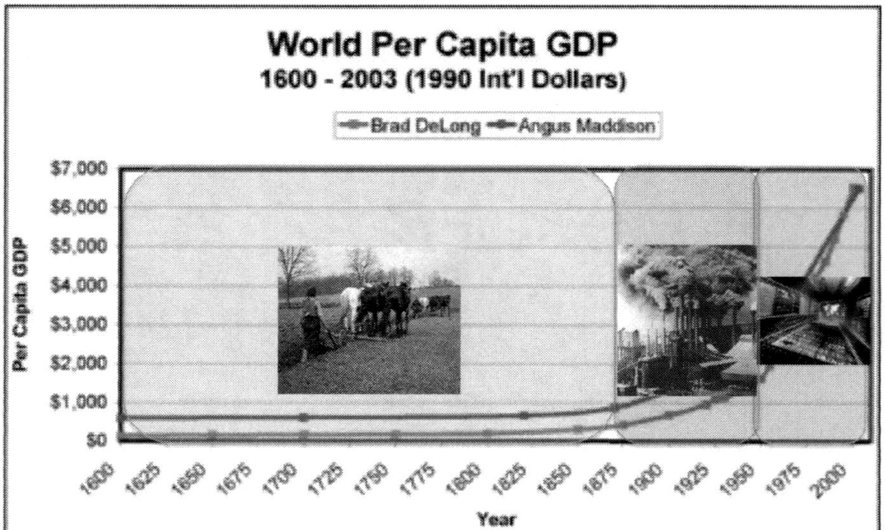

Source: J. Bradford DeLong, "Estimating World GDP, One Million B.C. - Present" http://www.j-bradford-delong.net/TCEH/1998_Draft/World_GDP/Estimating_World_GDP.html. Accessed Mar 5, 2008. Angus Maddison. "Contours of the World Economy, 1-2030 AD. Essays in Macro-Economic History." New York: Oxford University Press, 2007. 382.

The Industrial Automation Revolution

Innovations

- DCS, PLC, Historian
- Model-based Optimization
- Fault Tolerant SIS

Benefits

- Production Optimization
- Extended Asset Runtime
- Significant Financial Returns

Faster! Better! Cheaper!

PAS

Complexity and Interdependency

- Best-in-breed, heterogeneous
- Highly complex configuration
- Limited documentation
- Platform for improvement / change

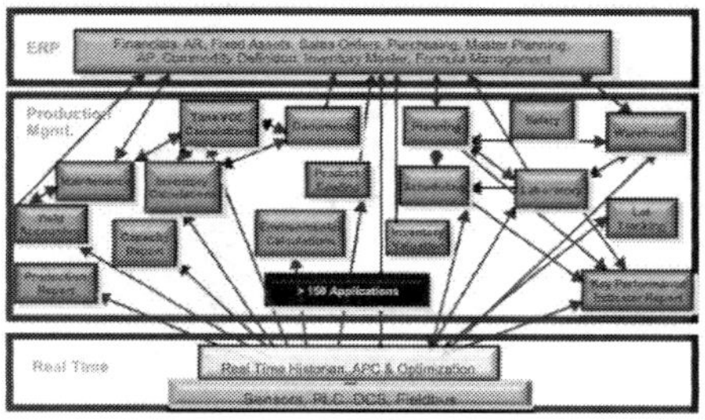

PAS

Unintended Consequences

A Closer Look…

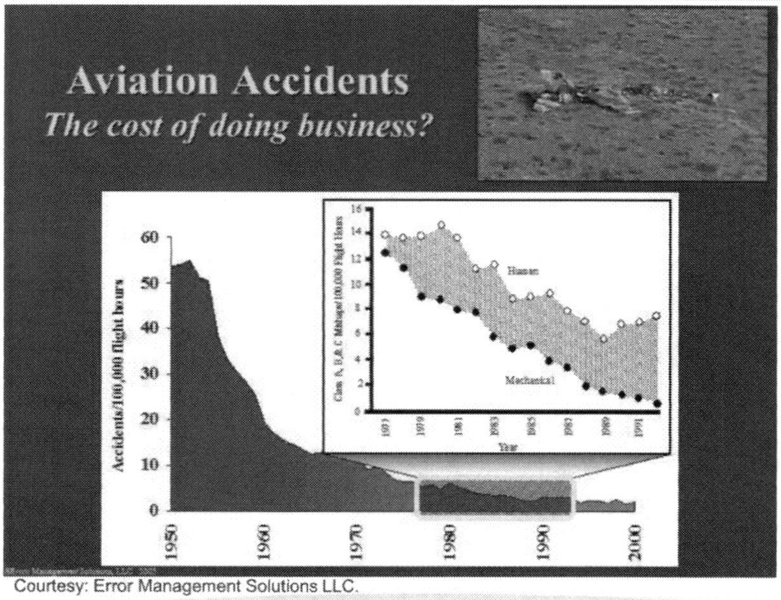

Courtesy: Error Management Solutions LLC.

Preventing Industrial Accidents

The time has come to focus on preventing *Human Error*.

PAS

Types of Human Error

Unintentional Errors

- Unwittingly doing the wrong thing
- "Fat Finger error"
- Missing an important alarm

Intentional Errors

- Deliberate action
- Mostly not malicious
- Shortcuts, misperceptions

PAS

What is Human Reliability

Human Reliability Goals

- Goal is to prevent human error
- Similar to physical Asset Reliability
- Focuses on the human element

Situation Awareness (SA)

- Essence of *human reliability*
- Critical requirement in complex operations
- Essential to preventing human error

Situation Awareness: 3 Essential Elements

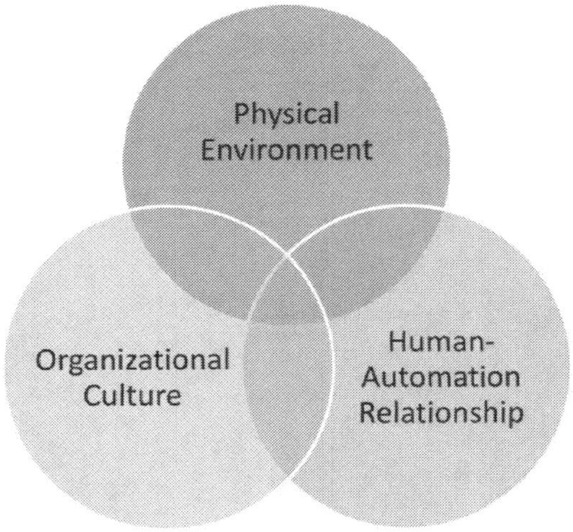

Situation Awareness : 3 Essential Elements

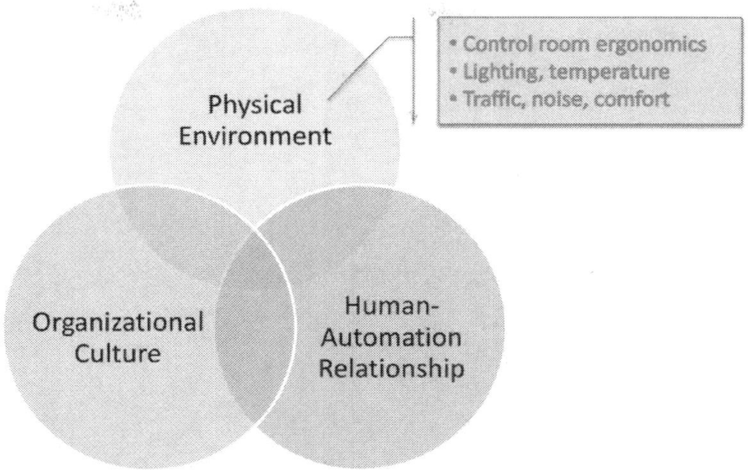

Situation Awareness: 3 Essential Elements

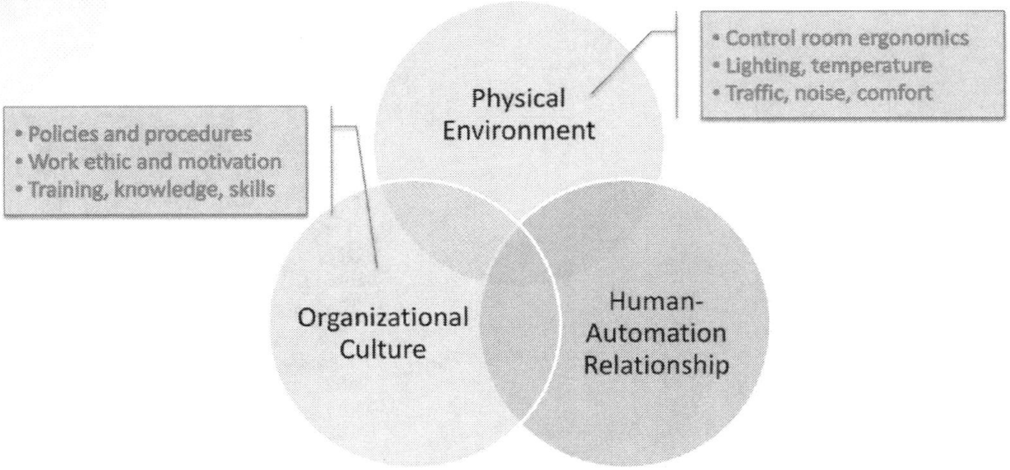

Human-Automation Relationship:

Decision Support – Operations & Engineering

- Operator HMI
- Alarm Management
- Boundary Management
- "Right-Click" Access to Information in Context
- Procedural Automation
- Integrated Checklists
- Operator Logbook

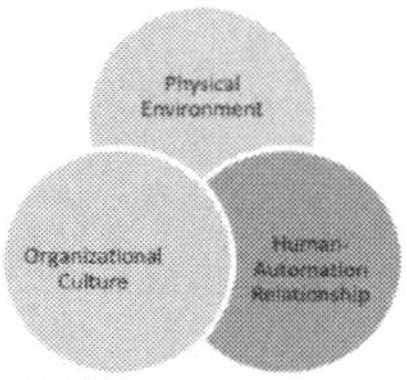

PAS

#ptc2013 | 12

Human-Automation Relationship:

Automation Asset Management

- Automation Asset Inventory Management
- Management of Change
- Configuration Management
- Cyber Security
- Automated Documentation
- Simple Visualization
- Control Loop Performance Management

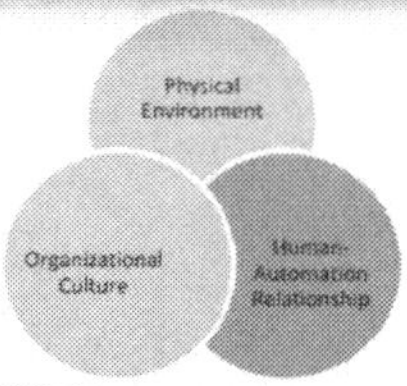

PAS

#ptc2013 | 13

The Human Reliability Company
Thank You!

VALERO

VALERO

Advanced Process Control (APC) Program at Valero – An Overview

Sriram Ramaganesan
Director – APC Technology

Valero – Company Background

- World's largest independent refiner
  - 16 Refineries
  - 3 Million barrels per day (BPD) of throughput capacity
- More than 7,300 branded marketing sites
  - Nearly 1,900 CST Brands Sites (Valero former retail business that was spun off (80%) in May 2013)
  - Brands include: Valero, Ultramar, Texaco, Shamrock, Diamond Shamrock & Beacon
- One of the largest renewable fuels companies
  - 10 Corn ethanol plants with total of 1.2 Billion gallons/year of nameplate capacity
  - Diamond Green Diesel JV plant started up and operating
- Approximately 10,500 employees

Valero – Company Background (contd.)

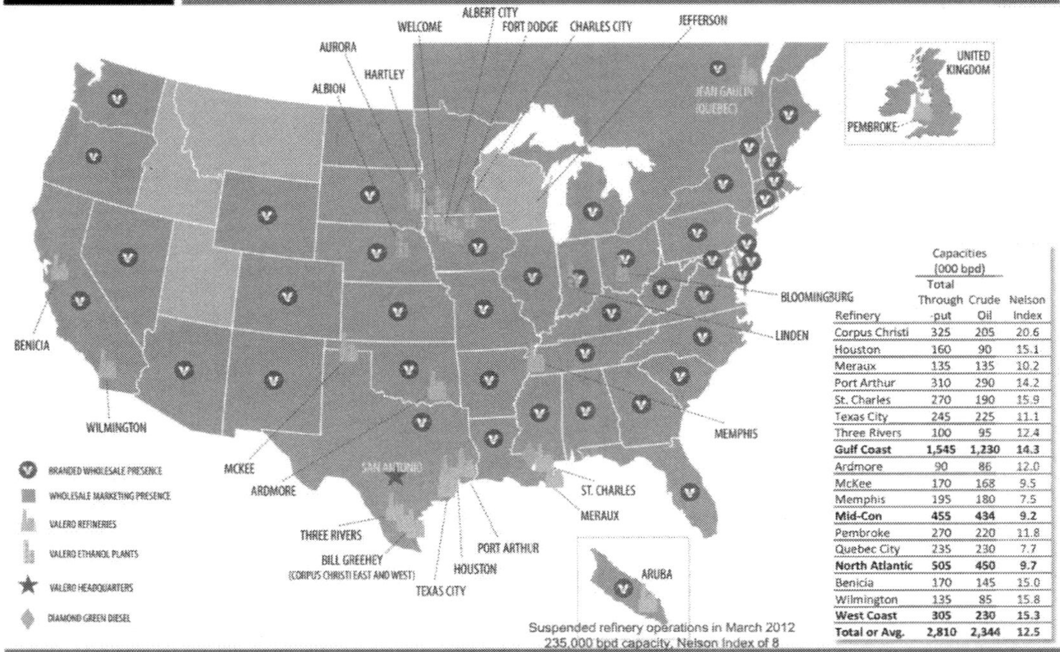

Refinery	Capacities (000 bpd) Total Through-put	Crude Oil	Nelson Index
Corpus Christi	325	205	20.6
Houston	160	90	15.1
Meraux	135	135	10.2
Port Arthur	310	290	14.2
St. Charles	270	190	15.9
Texas City	245	225	11.1
Three Rivers	100	95	12.4
Gulf Coast	**1,545**	**1,230**	**14.3**
Ardmore	90	86	12.0
McKee	170	168	9.5
Memphis	195	180	7.5
Mid-Con	**455**	**434**	**9.2**
Pembroke	270	220	11.8
Quebec City	235	230	7.7
North Atlantic	**505**	**450**	**9.7**
Benicia	170	145	15.0
Wilmington	135	85	15.8
West Coast	**305**	**230**	**15.3**
Total or Avg.	**2,810**	**2,344**	**12.5**

Business Problem – Process Variability

- Process variability results in lost $$
  - Poor quality products / Poor yield
  - Reduced Throughput

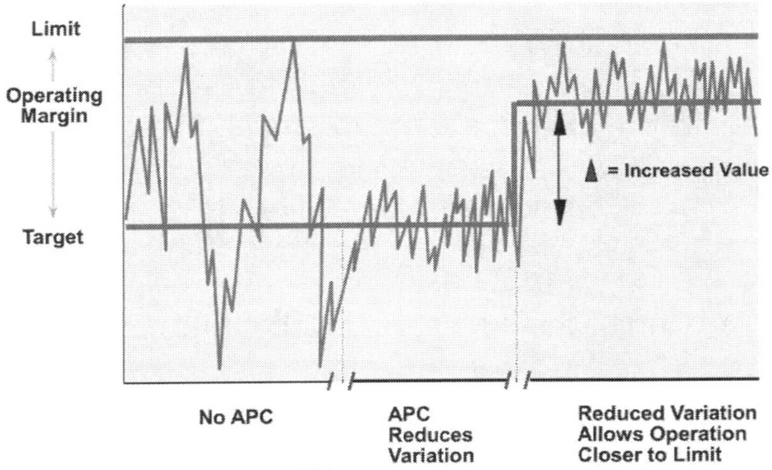

- Addressing process variability results in increased $$
  - Better quality products / Better yield of more valuable products
  - Increased throughput

Business Problem – Operating at Constraints

- APC uses linear empirical process models to optimize process unit operation
  - These cause & effect relationships allow the unit to be moved pro-actively to the appropriate operating point and allows operation close to profitable process constraints

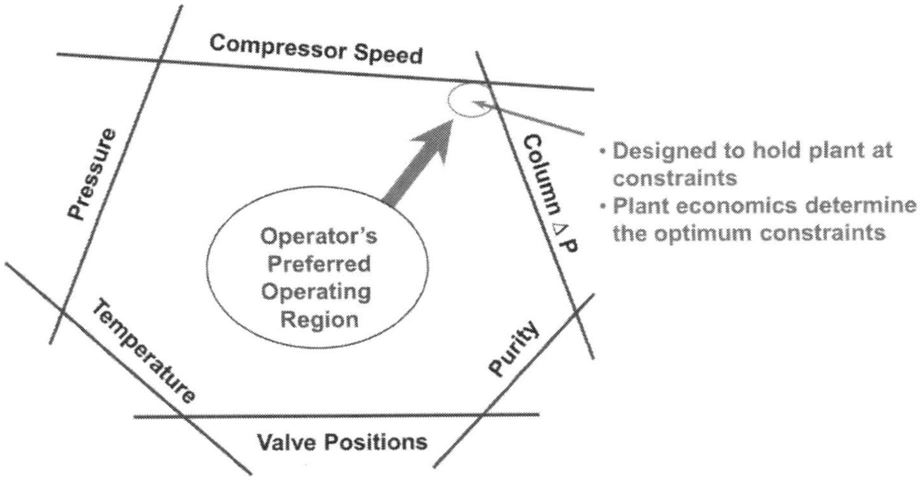

Typical APC Objectives

- Minimize Variability in key operating variables
- Achieve Energy Minimization objectives
- Better Product Quality Control by developing inferentials and using analyzer/lab signals for rapid/reliable feedback
- Immediate response to deviations from production / quality targets and problems of compliance with environmental / safety regulations
- Better disturbance rejection by using process information and models to take proactive corrective action

APC Adoption Initiative – Key Elements

- **People**
  - Trained, Competent, Onsite resources (Engineering and Operations)
  - Central Support team for implementation and support
  - Senior Leadership buy-in, support
- **Technology**
  - Standard supportable technology
  - Tools to speed up implementation (Automated Step testing, Adaptive Modeling etc)
  - Tools to make Monitoring/Sustainment easier (Easy to implement/monitor KPIs, drilldown etc)
- **Work Processes**
  - Documented Best practices for enable easy implementation and long term support
  - Flexible Support model (In-house, Contract, Remote support etc)
  - Technical network (bi-quarterly meetings to facilitate knowledge transfer)
  - Regular Training classes for both Engineering and Operations personnel
  - ***Publish benefits to key decision makers on a monthly basis***
- **Process Equipment Maintenance**
  - Valves, Analyzers etc
  - PID loop tuning

APC Adoption Initiative – Results

- APC Adoption Program has made significant progress
  - Over 60 new process units with APC in ~5 years
  - More than doubled the number of units with APC in ~5 years
  - Improved utilization of applications from ~55% to over 80%
  - Several best practices documents published or currently working (Implementation, Inferential deployment, Value Sustainment)
  - Technical network functioning well to enable knowledge sharing

- Significant post audit verified $$ benefits – over $100MM cumulative benefits in the last 5 years
  - KPIs monitored and published on a monthly basis to refinery leadership
  - Benefits information published on a monthly basis to senior leadership

APC Adoption Initiative – Future

- Continue to accelerate Adoption till APC is implemented on 100% of units that can benefit including the Ethanol Plants

- Utilize newly available technology to :
  - Preserve application benefits over time by keeping models evergreen
  - Help overcome challenge posed by personnel constraints at the refineries

- Select Real Time Optimization (RTO) applications, for example Diesel Maximization, Hydrogen Management etc

Q & A

- Questions / Comments ?

Making Digital Technologies Work For You

Gary Mintchell
The Manufacturing Connection
www.themanufacturingconnection.com

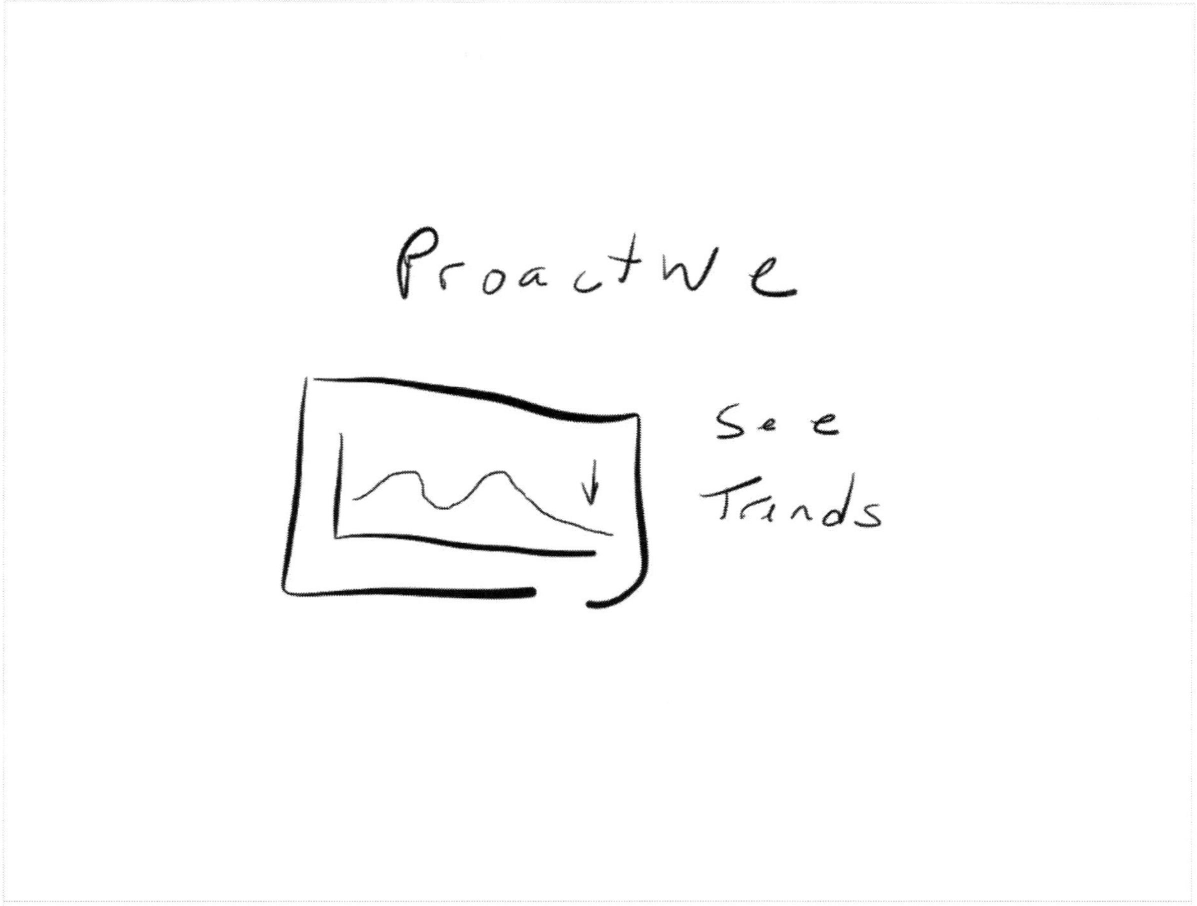
Proactive
See Trends

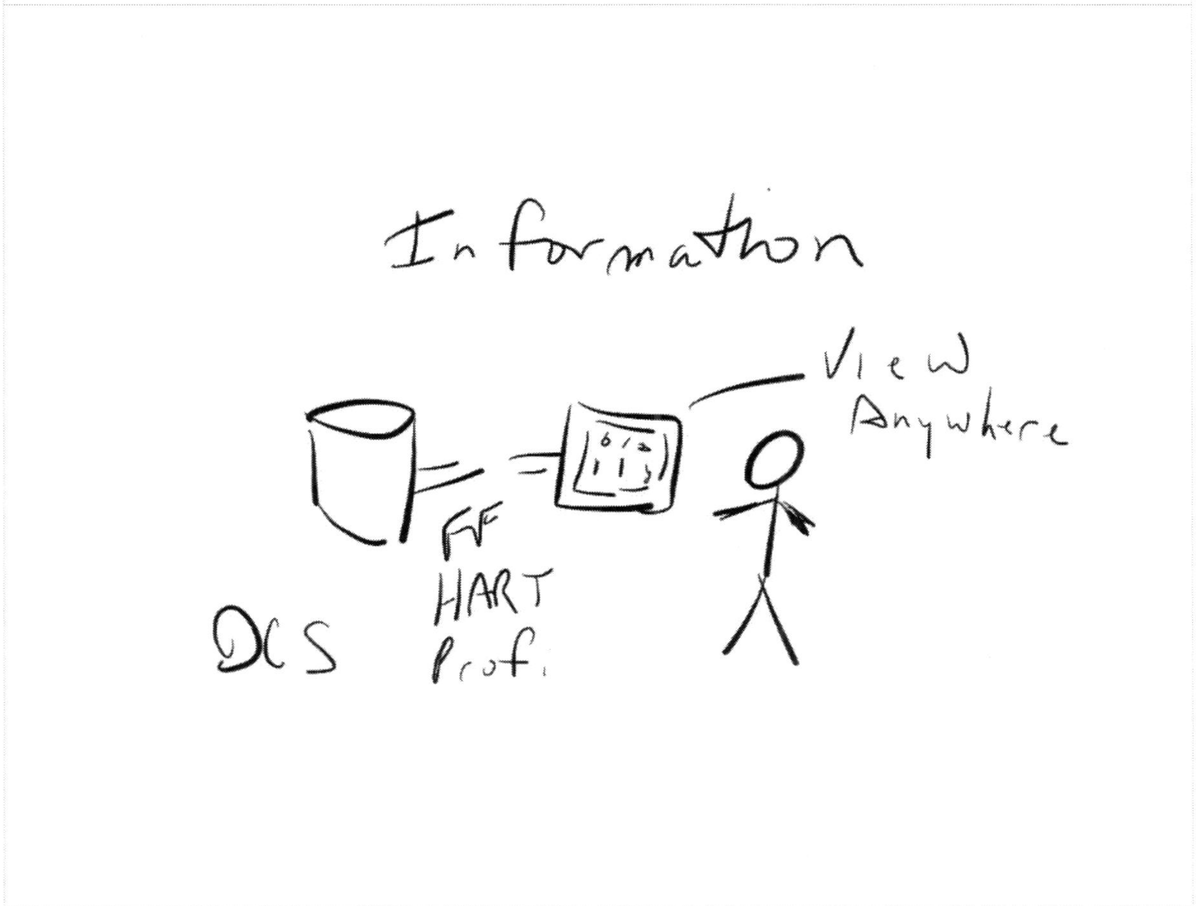
Information
View Anywhere
FF
HART
Profi
DCS

Calibration

Better
Information To
Operators So
Fewer Wasted
Alerts

Mobility—
Technician or
Engineer Can See
and speak to
expert, photo,
video

Collaboration

Right Tools to the Site First Trip

Engineer
v
ops
v
maintenance

But

All went
success,
good job
plant that
lasts

Working Together!

Gary Mintchell
Founder, The Manufacturing Connection
www.TheManufacturingConnection.com
gary@TheManufacturingConnection.com

2013 AICHE Southwest Process Technology Conference

Oct 3 – 4, 2013

Burns & McDonnell
1700 West Loop South, Suite 300
Houston, TX 77027

Friday, Morning Session, Oct 4; Track 2: Chemical Process Technology Session

Expanded Abstract:

"Shale Gas Boom – A New Era of Technology for Ethylene and Ethylene Derivatives"

There is a new era in the U.S. with the rise of the Shale Gas Boom that has emerged in this country. Large production increases are occurring from shale gas plays ranging from C1's – C4's, C5's/condensate and "tight oil". The plentiful supply of C1's – C4's versus current demand is driving prices for these molecules down. The recent result is directionally lower prices for methane, ethane, propane and butane as well as heavier products.

The plentiful supply and low prices for these building block molecules is having a marked effect on downstream industries that use the molecules either as fuel for energy or for raw materials as feeds. Steam crackers are moving to maximally change feedstock from naphtha's and gas oils to ethane to make ethylene and recently experiencing very attractive margins as a result.

Ethylene yield is very high when using ethane as feed which results in very favorable economics given the low price of ethane that has been prevalent over recent years. Ethylene producers have already made moves to feed as much ethane as possible within the limits of their existing production facilities. Ethylene producers are also moving to debottleneck existing plants including various debottlenecks that have already been announced. Additionally, as many as seven (7) new grass-roots facilities have been announced or are under consideration primarily along the gulf coast. Although, at many as two new ethylene plant projects are under review to be located in the North East to enable taking advantage of ethane production from the Marcellus Shale Play located in the states of Ohio, Pennsylvania and West Virginia.

Ethylene capacity totaling over 7 mm MTons/yr have been announced or are under study between debottlenecking of existing plants and new grass-roots ethane crackers. The debottlenecking projects can be accomplished in a shorter timeframe with the capacity increases continuing through 2013 to 2015. The new grass-roots capacities are slated to come on-line through 2016 and 2017. Several of the new ethane crackers will also install additional polyethylene capacity as part of the new facilities.

Historically, naphtha and gas oils have been used as feed to steam crackers. These feeds typically come from processing crude oils in refineries and, as such, the pricing for naphtha and consequently ethylene is directly related to the price of crude. Ethane production from shale gas plays provides crackers a new source of feed that is not related to the refining of crude oil or world-wide oil markets. This represents a significant paradigm shift in the traditional economics for ethylene production.

Ethylene is a major building block for numerous chemicals and petrochemicals. Consequently, the shale gas renaissance is affecting a broad spectrum of the chemical industry.

Shifting ethylene production from naphtha and gas oil to ethane as feed has a dramatic effect on the production of other byproducts that traditionally come from steam crackers. The yields of propylene, C4's, C5's and heavier products that are produced when feeding ethane are very low.

The reduction in propylene production is affecting the economics of producing propylene through the "on purpose" route. Numerous propane dehydrogenation projects are already in design with others being announced. This is just one example of the changing markets and the necessary transition that is taking place in the U.S. as the markets are in transition.

The total spending for the chemical/petrochemical industry for the period 2013 to 2030 is now estimated at approximately $120 billion. There is an anticipated sharp increase in spending starting in 2014 to peak in 2016/2017 at over $13 billion. The total spending for the period 2013 to 2017 totals nearly $48 billion.

With the advent of the announced ethylene plants slated to come on line by 2017, ethylene will be in greater supply. The economics of supply and demand suggest an oversupply of ethylene produced in the U.S. which will result in lower ethylene prices. A lower ethylene price has the effect of swinging economics in favor of producing more derivatives made from ethylene which will have to be exported from the U.S. to the global market place in order to maintain a supply and demand balance.

It is estimated that along with ethylene production various ethylene derivatives including polyethylene, ethylene oxide, ethylene glycol, EDC, VCM, PVC, among others will also increase. This has the effect of increasing the capital spending through 2020 - and beyond.

Where will all this production go? It is forecasted that the U.S. will be increasing net exports of the ethylene derivatives.

Lower energy cost along with cheap ethane has caused a marked shift in where U.S. stands on the global ethylene cost of production curves. An impressive result of the Shale Gas Boom and consequent lower cost of energy and cheap ethane has been the effect on U.S. cost of production for ethylene on the global cost curve. Comparing the 2007 and 2012 cost of production curves it is clear that the shift has now brought the U.S. much closer to the lowest cost producers including Canada and the Middle East.

Ethylene Dichloride (EDC) is one of the derivatives of ethylene that is forecasted to experience capacity increases in the U.S. EDC is used to produce Vinyl Chloride Monomer (VCM) which in turn is used to produce PVC. The U.S. is likely to increase export of PVC for use in home building and plumbing products world-wide. Burns & McDonnell has chosen to highlight the EDC -> VCM value chain in this presentation.

The process to produce chlorine is very energy intensive at 20,200 Btu/lb. Reduced U.S. energy costs represent a significant advantage in lowering the cost of production for chlorine which is another advantage in the EDC, VCM and PVC value chain. There are two widely accepted methods for producing EDC: direct chlorination and oxychlorination.

In direct chlorination, chlorine and ethylene are reacted to form the EDC molecule such that two chlorine molecules are used for each molecule of ethylene. The typical reaction is between the dissolved gases in liquid over an iron chloride catalyst at 15 to 75 psig pressure and 200 to 300 deg F. The use of high purity reactants is typical to avoid the formation of a broad range of impurities which may further complicate EDC purification. An excess of chlorine is typical to ensure complete ethylene conversion.

In oxychlorination, the reaction occurs over a copper chloride impregnated alumina catalyst in the liquid phase at a temperature of approximately 300 deg F and a pressure of up to 75 psig. The typical fluid bed technique offers heat transfer, prevents the occurrence of hot spots and allows more efficient catalyst regeneration. Ethylene conversion of 93 to 97% can be achieved with selectivity in EDC of 91 to 96%.

The EDC purification section removes trace amounts of impurities in the furnace feed which can inhibit the cracking reactions and cause fouling. In addition, EDC must be dry to prevent excessive corrosion in equipment and piping downstream of the reactor. Therefore the combined EDC from three different sources must be purified to greater than 99% EDC from direct chlorination, EDC from oxychlorination and EDC recovered and recycled from the cracking step. This combined EDC is washed with water and scrubbed with caustic to remove chlorine, HCl and other water soluble impurities including chloral. The crude EDC and unconverted EDC from the VCM column are combined and dried and purified in two distillation columns. In the first "Lights Column", water and lower boiling point impurities are separated as the overhead product. The bottoms product is fed to a "Heavies Column". Pure EDC of approximately 99% purity is taken as the overhead product while the heavy impurities leave the column as the bottoms product.

The pure EDC is forwarded to a furnace for thermal cracking of EDC to vinyl chloride. A typical furnace consists of a radiant section, a convection section, and a stack. The furnace fire box operates at 950 deg F. The convection utilizes combustion gases to preheat the feed before entering the firebox.

The vinyl purification section consists of two distillation columns in series which are used to separate vinyl chloride from EDC, HCl and remaining by-products. The first column ("HCL Column") distills the HCL mixture to a pure overhead product which is recycled to the oxychlorination reactor. The HCL Column is a trayed column and operates at a pressure of 135 psig. The bottoms product of the HCL Column is fed to the second column ("VCM Column") which operates at 65 psig. A VCM product of greater than 99 wt% is produced as the overhead product of the VCM Column. The bottom of the VCM Column is recycled to the Lights Column for re-purification.

A brief overview of the economic analysis for EDC/VCM plant investment indicates that the total installed cost (TIC) for a 300,000 metric ton/yr facility is in the range of $450MM to $500MM. With current EDC pricing of $245-275/ton and assuming an ethylene and chlorine cost of $1320-1362/ton and $275-310/ton, respectively and assuming the energy, labor, overhead and tax contributions are in the range of $70-85/ton results in a payback of less than five years. These rough economics do not include cash costs for land, start-up nor the capital investment required for utilities and infrastructure. Constructing a new EDC plant at an existing facility where existing infrastructure and utilities are present will make the investment more attractive than a pure green field site. Debottlenecking an existing EDC facility is likely yield an ROR that represents another attractive investment.

Furthermore, installing or debottlenecking existing VCM and PVC facilities make this value chain one of keen interest in the coming low cost ethylene environment that is predicted.

In conclusion:

- The Shale Gas Boom is resulting in high production for C1's – C4's, which has resulted in reduced prices for these chemical and petrochemical building blocks.
- Several new grass-roots ethane crackers and ethylene plant debottlenecks to increase ethane feed capacity have been announced.
- Increased ethylene production in the U.S. is forecasted to lower ethylene prices in the long run.
- Ethylene is a central building block in producing various chemicals and petrochemicals.
- Therefore, a large sector of the U.S. chemical industry will be affected.
- Ethylene derivatives benefit from reduced ethylene prices.
- U.S. exports of ethylene derivatives are forecasted to increase to supply the world-wide market.
- The U.S. can compete well in the world chemical and petrochemical market place with the advent of the Shale Gas Plays.
- Chlorine cost of production is and will continue to benefit from the low energy prices in the U.S.
- EDC and VCM are examples chemicals that are forecasted to benefit as more ethylene production comes on line along with consequent lower ethylene pricing.
- The economics of producing EDC and VCM in a low cost ethylene environment is compelling.
- The EDC->VCM->PVC value chain will benefit. Export of these chemical products to the world-wide market is forecasted to increase in the future.

The Shale Gas Boom has shifted the U.S. energy picture as well as the outlook for chemical and petrochemical production. The U.S. is once again a key actor and leader in the world chemicals stage.

This truly is in a new era of Technology!

Mark W. Lockhart, P.E., Process Technology Manager, Chemicals and Petrochemicals
Burns & McDonnell, Houston, TX
832-214-1975

Elliott Robertson, P.E., Senior Process Engineer, Chemicals and Petrochemicals
Burns & McDonnell, Houston, TX
832-214-2916

Shale Gas Boom – A New Era of Technology for Ethylene and Ethylene Derivatives

AICHE – Southwest Process Technology Conference
Galveston, Texas
October 4, 2013

Mark W. Lockhart, P.E., Process Technology Manager, Chemicals and Petrochemicals
Elliott Robertson, P.E., Sr. Process Engineer, Chemicals and Petrochemicals

1

Introduction / Overview

- Shale Gas Boom → C1 – C4's, Condensate, "Tight Oil"
- Supply / Demand → Pricing Effect
- Ethane based Ethylene → Plant Announcements
- Naphtha/Gas Oil vs Ethane Feeds → SHIFT IN THINKING!
- Ethylene Key Building Block → Much of the Chem / Petrochem Industry is affected
- Chemicals Capacity and Spend → Forecasts

2

Introduction / Overview

- Increase in Ethylene Supply → Lower pricing
- Ethane → Ethylene → Ethylene Derivatives
- Ethylene in High Supply → Need to Export!!
- Value Chain Focus → EDC, VCM, PVC
- Conclusions → Future looks very bright!!

3

Shale Gas – Products in High Supply

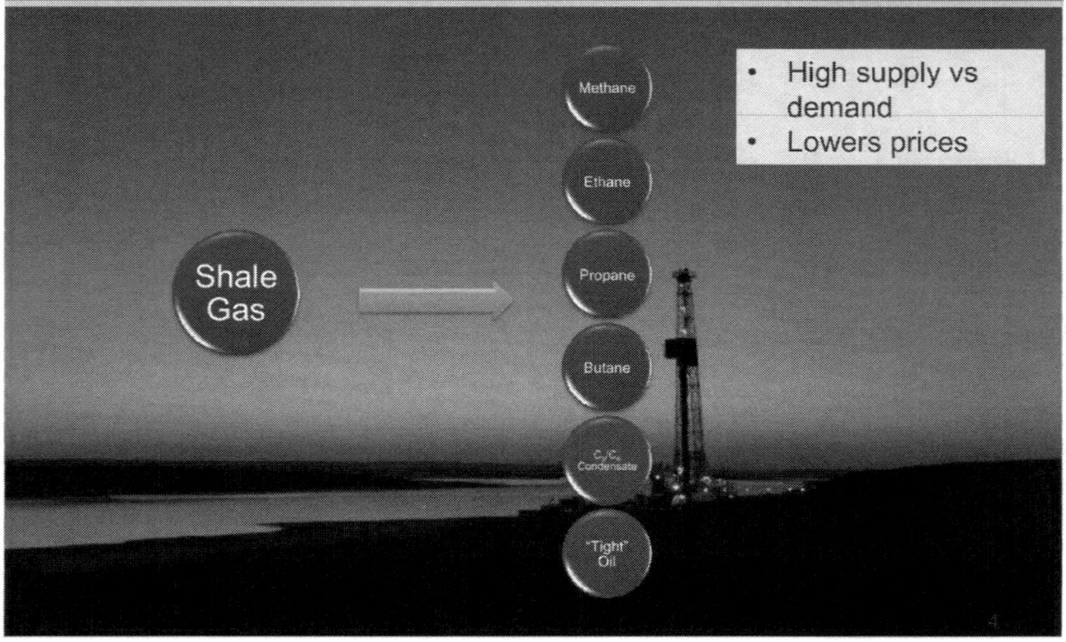

Burns & McDonnell SINCE 1898

Shale Gas - Potential Products

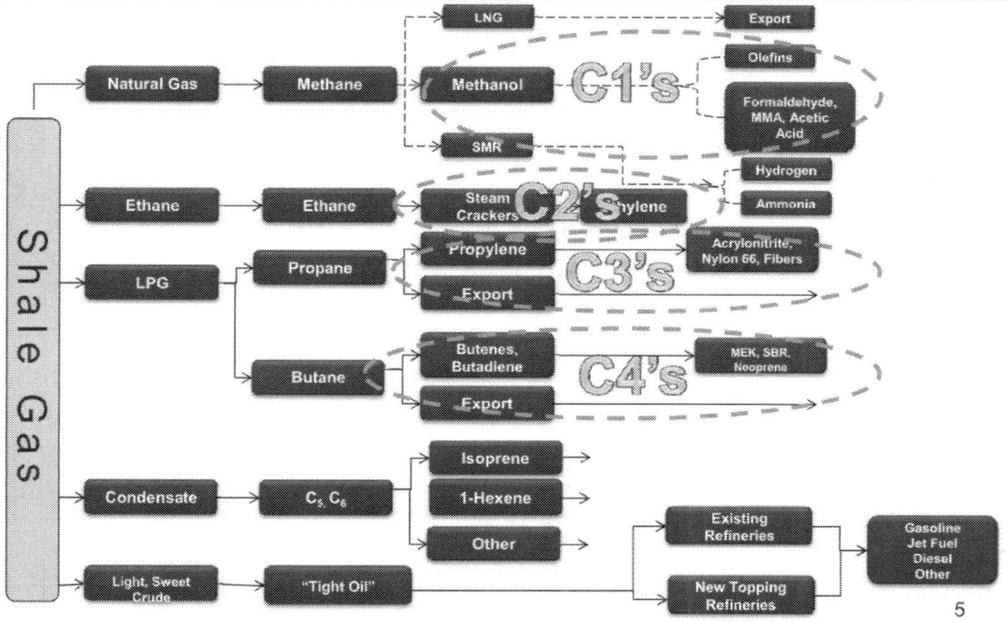

5

Burns & McDonnell SINCE 1898

Petrochemical Industry – Ethylene Producers

- Shale Gas Having Major Effect on Petrochemical Industry!
  - Lower Energy Costs
  - Lower Feedstock Costs
- U.S. Ethylene Producers:
  - Maximally displacing: Naphtha and Gas Oil → Ethane
  - Currently realizing high profitability
  - Quite a change from earlier last decade
- Ethylene Manufactures Making Moves:
  - Maximizing C2 feed to existing plants → Now
  - Debottlenecking Existing Facilities → 1 to 2 year cycle
  - Building New Grass-roots Ethane-based Crackers → 3 to 4 year cycle

6

Burns & McDonnell SINCE 1898

Ethylene Plant Announcements

Announced/Under Study

Company	Project	Capacity, tonnes/year	Location	Cost	Startup
ExxonMobil Chemical	New Cracker	1.5 m	Baytown, TX	NA	2016
Chevron Phillips Chemical	New Cracker	1.5 m	Cedar Bayou, TX	NA	Q1 2017
Dow Chemical	New Cracker	World-scale (>800,000m)	U.S. Gulf Coast	NA	2016-2017
Shell Chemicals	New Cracker	World-scale	U.S. Northeast	NA	2016-2017
Formosa Plastics	New Cracker	800,000	Point Comfort, TX	$1.7B	2016
Dow Chemical	Restart	390,000	St. Charles, LA	NA	End 2012
Westlake Chemical	Expansion	108,863	Lake Charles, LA	NA	Q2 2012
Williams	Expansion	272,158	Geismar, LA	$350M-$400M	Q3 2013
INEOS	Debottleneck	115,000	Chocolate Bayou, TX	NA	End 2013
Westlake Chemical	Expansion	113,399	Lake Charles, LA	NA	2014
LyondellBasell	Expansion	386,000	LA Porte, TX	NA	2014
Westlake Chemical	Expansion	82,000	Calvert City, KY	NA	2014
Dow Chemical	Expansion	NA	Plaquemine, LA	NA	2015
LyondellBasell	Expansion	363,000	Corpus Christi, TX	NA	2015

Expansions

7

Burns & McDonnell SINCE 1898

Ethylene Plants – Under Consideration

Considered Expansions

Company	Project	Capacity, tonnes/year	Location	Cost	Startup
Sasol	New Cracker	1.0M-1.4M	Lake Charles, LA	$3.5B-$4.5B	NA
Indorama Ventures	New Cracker	World-scale	NA	NA	2017
LyondellBasell	Expansion	NA	Channelview, TX	NA	NA
SABIC	New Cracker	World-scale	U.S.	NA	NA
Braskem	New Cracker	World-scale	U.S.	NA	NA
Occidental Chemical	New Cracker	NA	Ingleside, TX	NA	NA
Aither Chemicals, Renewable Manufacturing Gateway	New Cracker	NA	U.S. Northeast	$750M	2016
BASF, Total Petrochemicals LLC	New Cracker	NA	Port Arthur, TX	NA	NA
BASF, Total Petrochemicals LLC	Expansion (Add 10th C_2 furnace)	NA (increase by 15%)	Port Arthur, TX	NA	2014

Many producers will include polyethylene plants

8

Ethane Feed – A Paradigm Shift

- Historically Ethylene Feedstocks:
  - Naphtha / Gas Oil Based
  - Feeds supplied from Refineries
  - Tied to the price of Crude Oil
- Ethane from Shale Gas:
  - New and plentiful source of cracker feed
  - Independent of the price of Crude Oil

THIS IS A MAJOR PARADIGM SHIFT!!!

- Much of the Global Market Steam Cracking will Remain with Naphtha Based Feeds
  - **This represents a distinct advantage for the U.S.!**

9

North America Energy Trends

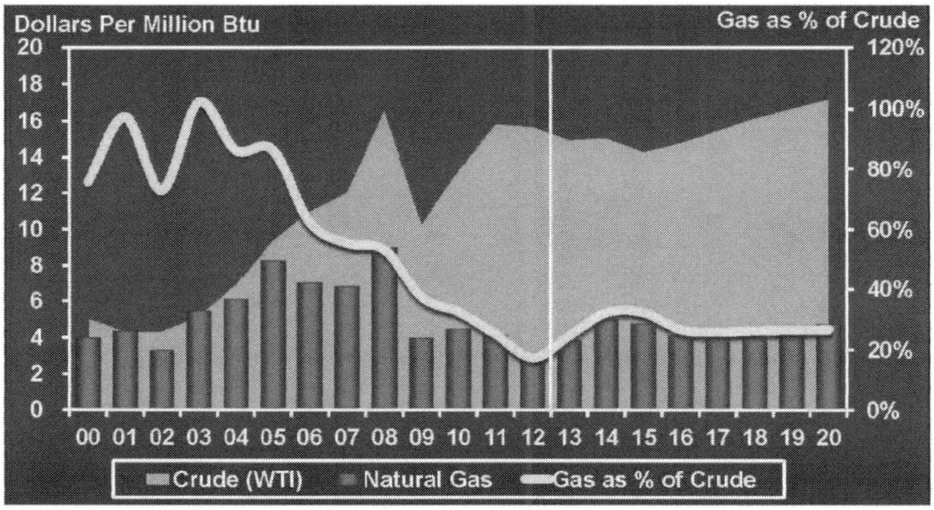

Crude Oil versus Natural Gas

Source IHS, Inc.

US Ethylene Industry

Both natural gas and crude prices have contributed to differential performance in the US ethylene industry

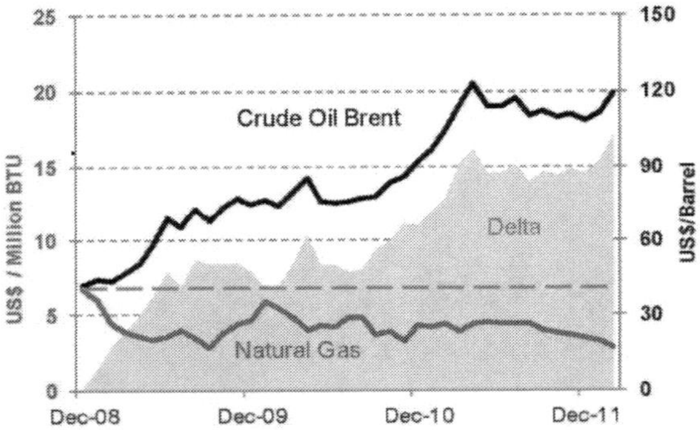

Cost of Ethylene Production
(¢/lbs)

	1H'09	Q1'12	Delta
Ethane-based (US)	20	23	3
Naphtha-based (US)	35	54	19
Ethylene price (NEA)	34	61	27

Source CMAI

Ethylene – A Major Building Block

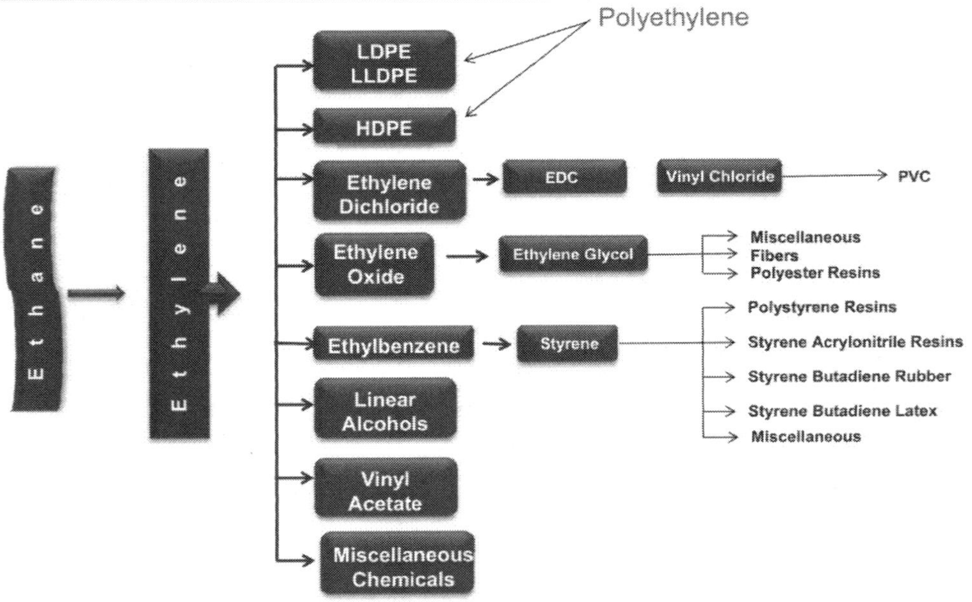

12

Olefins Flow Diagram

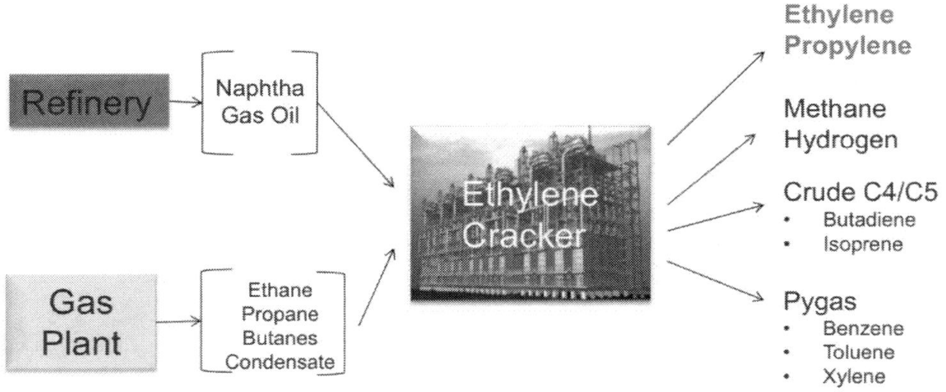

- The purpose of the cracker is to make light olefins (Ethylene and Propylene)
- Shifting to Ethane Feedstock dramatically affects olefin yield

13

Cracker Co-product Yields

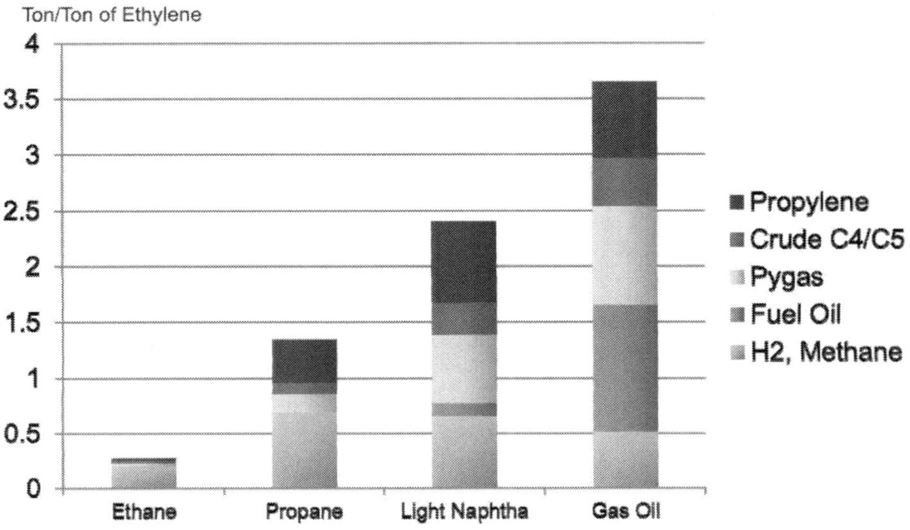

14

Rise of the USGC

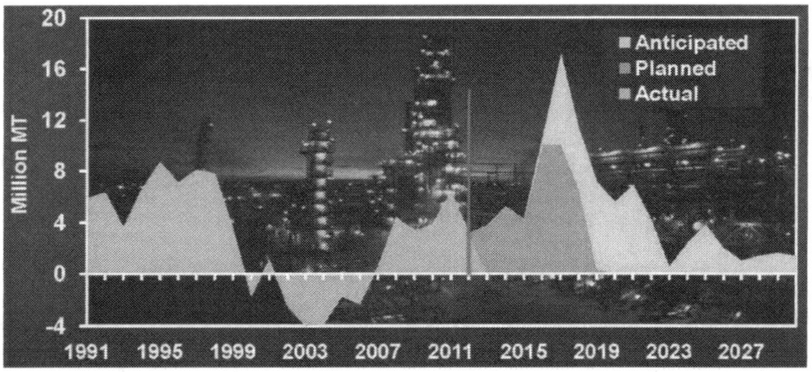

- 13 MM MT of capacity added since 2010
- Planned addition currently add 40 MM MT by 2018
- Over 45 MM MT of additional capacity anticipated by 2030

Source IHS, Inc.

15

New Plants

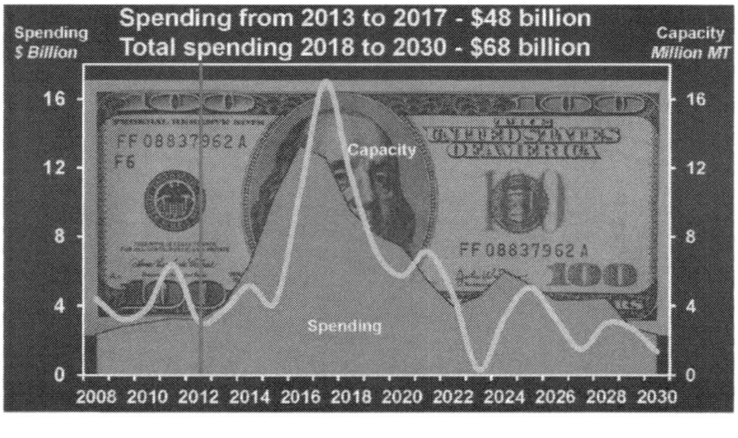

- Total spending from 2013-2030 estimates ~$120B
- Anticipated sharp increase in spending starting in 2014 to peak in 2016 / 2017 at over $13B
- Total spending between 2013-2017 totals nearly $48B

Source IHS, Inc.

16

Impact by Product Chains

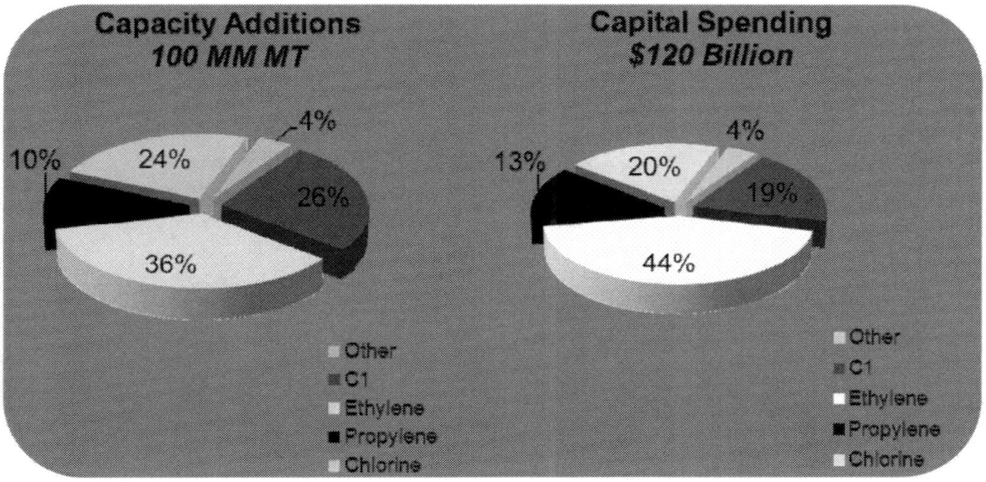

Source IHS, Inc.

17

U.S. Exports to the Global Market Place

- Where will all of this ethylene production go?
  - Increase in U.S. exports already in progress!!
  - Exports will have to increase significantly as Ethane based capacity comes on line in 2016 / 2017 and beyond
  - Need outlet for balanced supply / demand
  - Ethylene Derivatives much more easily exported
- Ethylene Derivatives forecasted for increased export:
  - Polyethylene
  - Ethylene oxide / Ethylene glycol
  - EDC, VCM, PVC
  - Other

EXPORTS!!

18

Global Capacity Cost Curve

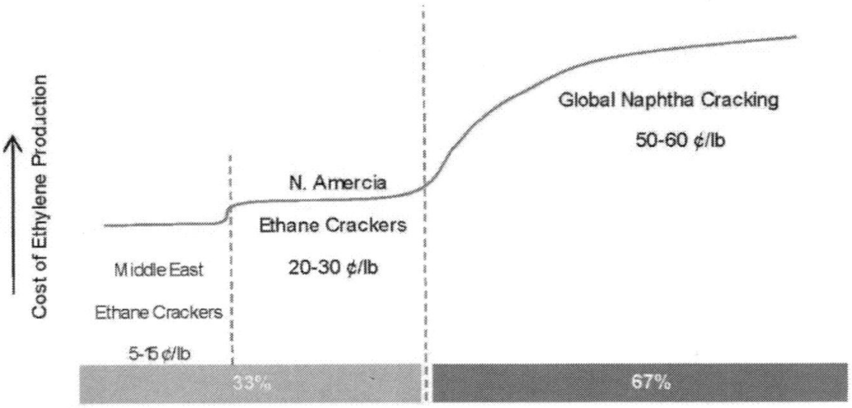

Raw material factors define regional competitiveness

Source CMAI

Global Ethylene Cost Curve

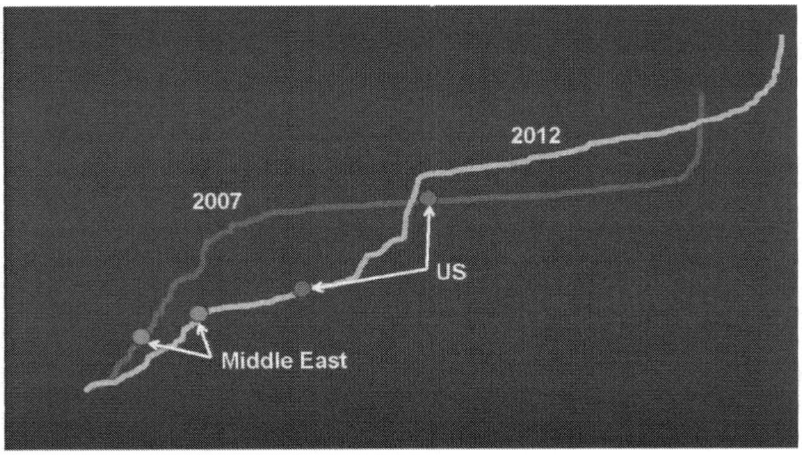

Source IHS, Inc.

20

Ethylene Derivatives

North America Will Become A Major Exporter Of Ethylene Derivatives

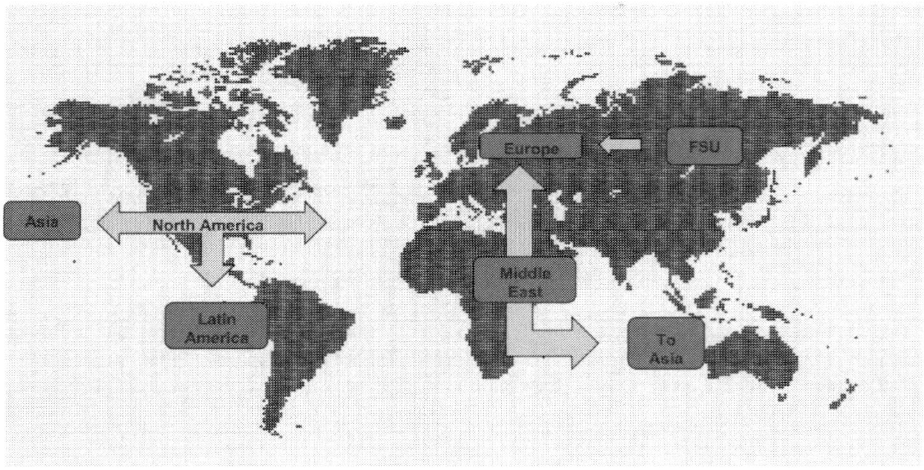

Chlorine Derivatives Expand

Due to low cost of ethylene & chlorine:

- Value chain through PVC forecast to expand
  - New EDC/VCM/PVC capacity
  - By 2017 – 4.1 million MT
  - By 2025 – 9.5 million MT
- Low cost of energy additionally benefits chlorine

Source IHS, Inc.

North America Spending by Product

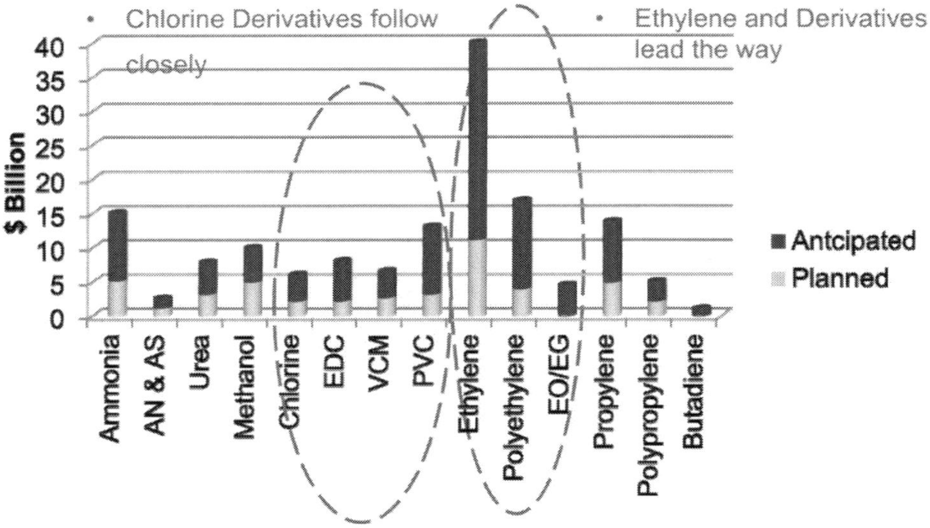

Source IHS, Inc.

23

VCM as Building Block

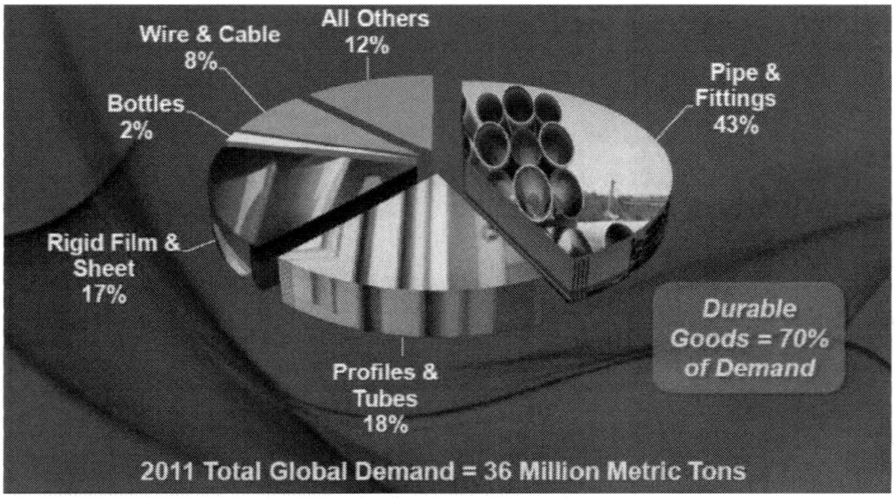

Source IHS, Inc.

24

VCM and Export

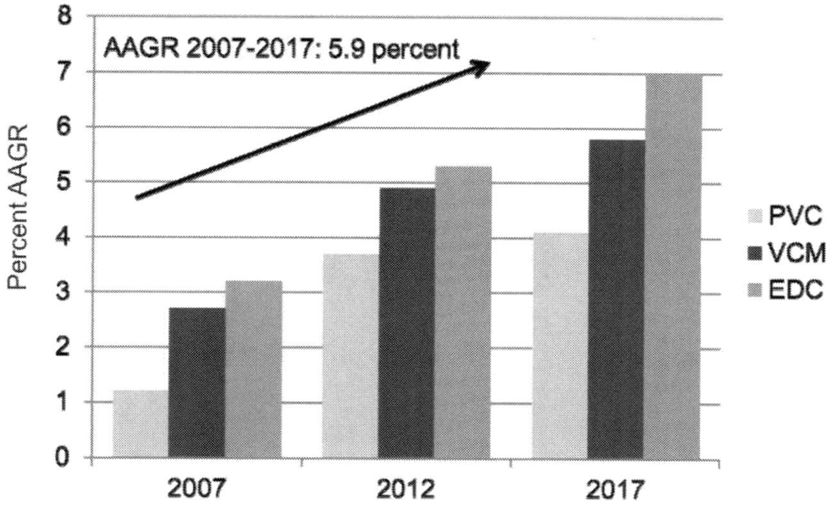

Source IHS, Inc.

25

Shale Impact on Vinyl Chain

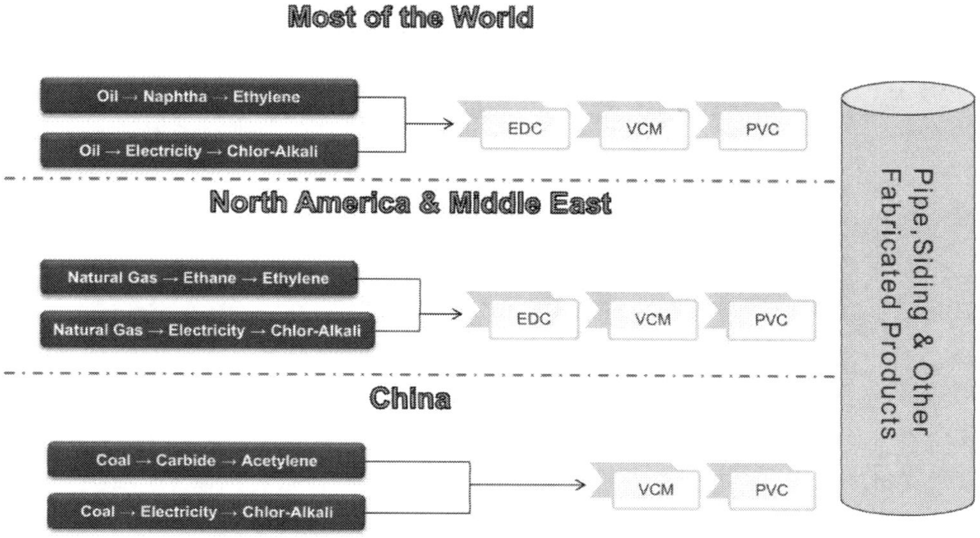

Source IHS, Inc.

26

EDC VCM Process

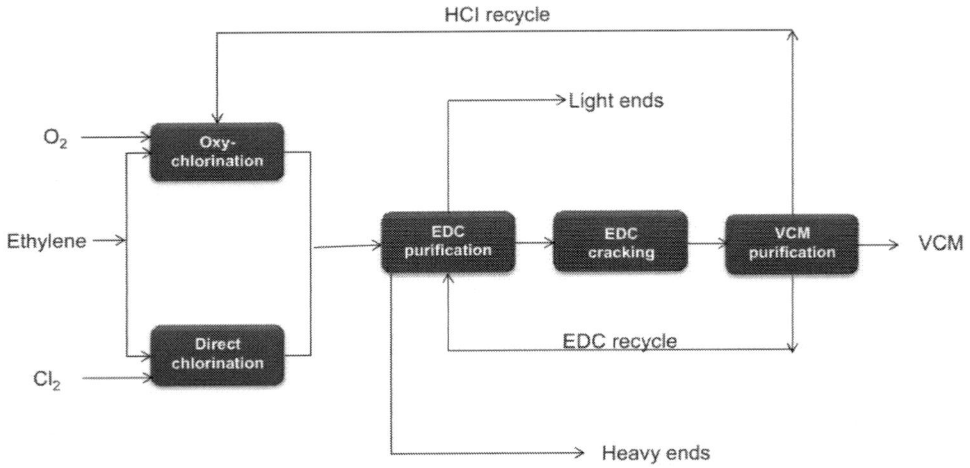

27

EDC VCM Operating Economics

		(ton/ton-VCM)	($/ton)	
Ethylene		462	1320-1362	Raw Materials & Products
Chlorine		578	275-310	
Oxygen		131	40	
EDC			245-275	
VCM			677-820	
Electricity	kWh	105		Utilities per ton VCM
Steam	Tons	0.15		
Cooling Water	Ft3	6200		
Fuel	Mcal	700		

28

Burns & McDonnell SINCE 1898

EDC VCM Capital Economics

Capacity MT/year 300,000

Service Life	Years	25		
Working Capital	$MM	20		
Direct Production Cost	$MM	189		
Plant Capital Cost	$MM	350	400	450
Fixed Costs	$MM	19	22	25
Gross Income	$MM	80	76	72
Net Profit	$MM	63	56	50
Pay out Period, Years	Years	4.9	6.4	8.2

29

Burns & McDonnell SINCE 1898

EDC VCM Outlook

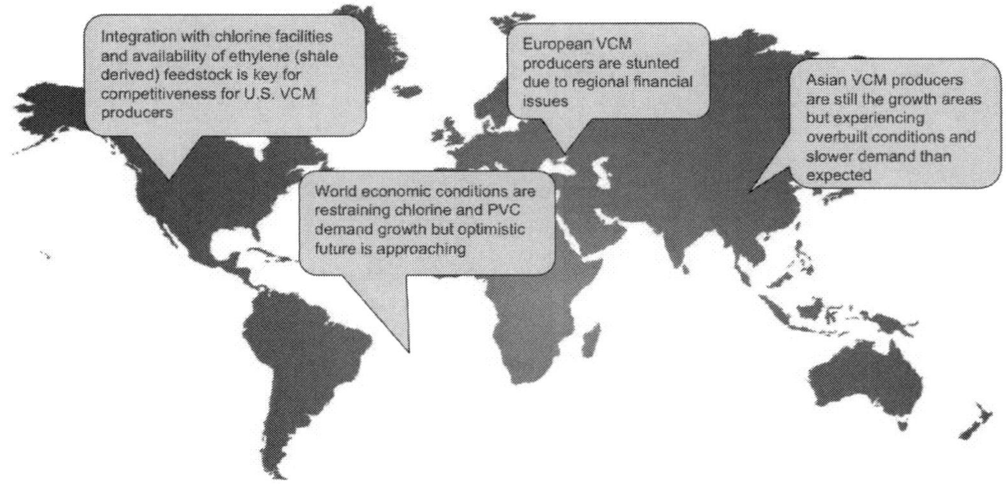

Source IHS, Inc.

30

Conclusions

- Shale Gas Boom resulting in high production of C1's-C4's → **Higher supply lowers prices**
- Ethane-based Ethylene Plant Announcements → **Grass-roots and debottlenecks!!**
- Ethylene → Central building block petrochemical → **Large sector of U.S. chemical industry affected**
- Lower Ethylene prices coming in 2017 ....?
- Ethylene derivatives will benefit → **New Capacity**

31

Conclusions

- Where will the increased production go? → **EXPORTS!!**
- One Example of Ethylene Derivative value chain that is forecasted to benefit: **EDC → VCM → PVC**
- Also leveraged by Chlorine COP → **Low energy prices**
- EDC and VCM Economics → **Compelling!!**

Shale Gas Boom has created a New Era of Technology in the U.S.!!!

32

Q&A

Burns & McDonnell – 100% Employee Owned!!!

Mark W. Lockhart
mlockhart@burnsmcd.com
832-214-1975

Elliott Robertson
erobertson@burnsmcd.com
832-214-2916

33

Impact on Chemicals

- The competitive landscape has changed
- Many product chains directly effected
- Ripples through product chains are becoming evident
- Impact is transformational for the US

C1
- Ammonia
- Ammonium Nitrate
- Ammonium Sulfate
- Urea
- Methanol

Chlorine
- Chlorine
- EDC
- PVC
- VCM

Ethylene
- Ethylene
- Ethylene Glycol
- Ethylene Oxide
- Polyethylene

Propylene
- Propylene
- Polypropylene

Other
- Butadiene

Source IHS, Inc.

34

Sebastian HOLZ
Business Development Manager
Linde Engineering North America Inc.
3700 West Sam Houston Parkway South, Suite #425 | Houston, TX 77042 | USA
Mobile: +1-832-560-1851 | Office: +1-281-717-9082 | Fax: +1-281-717-9091
E-mail: sebastian.holz@linde-le.com

Title: Mega Crackers – Where is the limit?

Abstract: The size of steam cracking plants has increased significantly over the last two decades. Currently the biggest plants built have around 1,500 kta* ethylene capacity. Looking at this trend, Linde has studied its design to find any constrain in building a 2,000 kta* ethylene cracker. Starting with some basic design criteria, this presentation is going through the major process units and key equipment and is giving some insight to the actual limits of furnaces, rotating equipment, columns, heat exchangers and piping design. The results are promising and surely in a few years we will see the next generation of Mega Crackers.

*kta = one thousand metric tons per year

Paper: The size of steam cracking plants has increased significantly over the last two decades. Until the mid-1990 the size of crackers was between 200 and 500 kta and then boosted quickly above the 1,000 kta line. Currently the biggest grass root plants built have around 1,500 kta ethylene capacity plus valuable side products. The question is now: Is this already the limit?

The trend is supported by the growing demand for plastics and specialized chemicals. Resins producers worldwide are competing for growing markets and try to cut cost by taking advantage of the "economy of scale". This means that in most cases it is cost beneficial building one large single train plant instead of two units with half the capacity. While in the recent years many new plants were built in Middle East, the abundance of low cost ethane feed is currently boosting big investments in the United States. And a fair portion of it in Mega Crackers.

Carrying forward this trend Linde has studied its technology to find any constrain in building a single train 2,000 kta gas or liquid cracker. Here single train refers to the cracked gas compressor. Building parallel could be wise and beneficial for other equipment and should be evaluated on a case by case basis. Major equipment, such as the cracking furnaces, compressors, columns and heat exchangers have been analyzed and respective vendors interviewed. To minimize risk the study adhered to proven concepts.

The development of furnaces mirrors the trend one for the steam crackers. After the twin cell design has been developed in the 1990s the size of the heaters has grown from around 75 kta up to 250 kta. And this is not the end. By adding more coils to each row of the furnace the throughput of each furnace could be increased. And of course there is always the option to simply build more furnaces.

The cracked gas compressor is the bottleneck of a steam cracker plant. It is the most costly single equipment and has a high complexity on site. Current installations have a power rating of up to 75 MW. For a 2,000 kta plant even more power is needed. With modifications to the casing arrangement and the latest developments in the steam turbines in mind, machinery vendors agree that we have not reached the limits of a single compressor train yet.

Taking advantage of the economy of scale is especially important when looking at the columns. The increased capacity would lead to either shell diameters that are difficult to transport or CAPEX intense parallel equipment. By using high capacity trays tower diameters will remain within limits that are easy to handle. Sophisticated internals are more expensive but have overall cost advantages because of the savings on the shell material, isolation and transportation. However, not all towers can be equipped with cap trays. Another possibility to safe energy and invest are smart process solutions like distributive distillation in case parallel columns are needed.

Heat exchangers are also critical. Depending on the service a good maintainability is crucial. Simple scaling the standard design would bring this journey to a quick end as tube bundles of more 12 meters and a mass of more than 50 metric tons are everything but very easy to maintain. The solution could either be to use several smaller exchangers in a row or to use "block-in-kettle" type exchangers. The latter is a plate fin exchanger mounted in a kettle and thereby combining the benefits of a high surface per volume ratio and easy accessibility. Above all, this concept also saves invest cost.

And even piping can become a concern in chemical plants of this magnitude. The cracked gas line in today's crackers is already impressive. Diameters above 100 inches have already been realized and can be further increased. Handling pipes this size is an enormous challenge. Constraints and challenges will be with instrumentation and required plot space due to piping stress. In some cases parallelization of piping may be more economical.

After all, have crackers reached their capacity limit? No, they have not.

Linde has developed the technical know-how to build liquid feed crackers with more than 2,000 kta and gas feed crackers with more than 2,500 kta ethylene capacity. There are definitely more criteria to consider before we will see the next generation of Mega Crackers but from a technical perspective it is feasible.

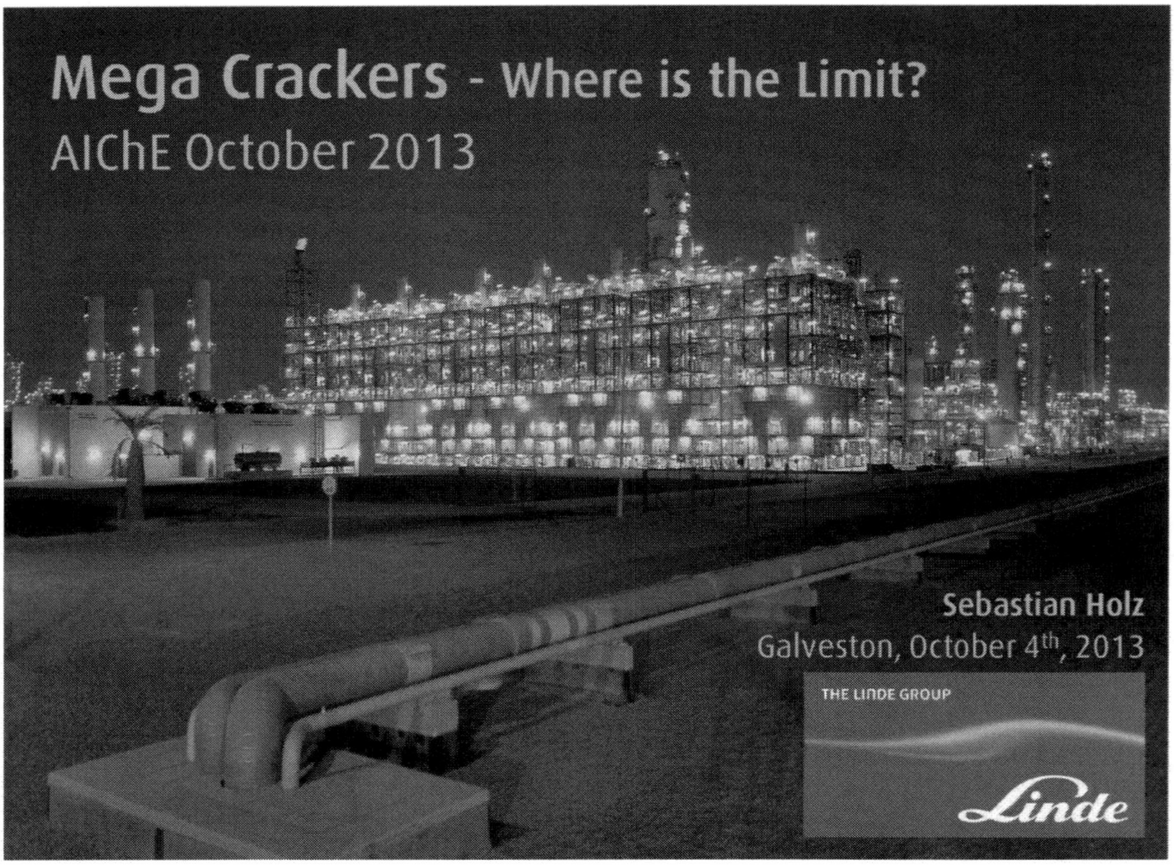

Steam crackers are the heart of a petrochemical complex

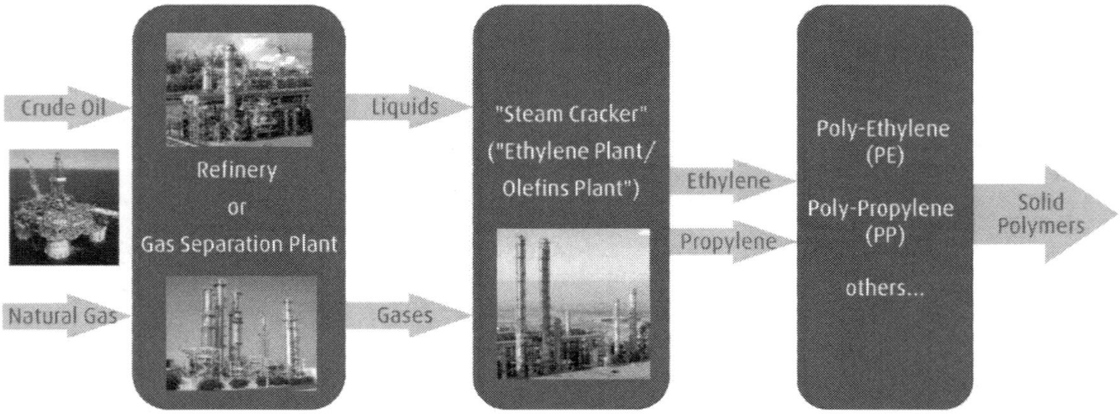

The size of steam crackers has increased significantly in the last decades

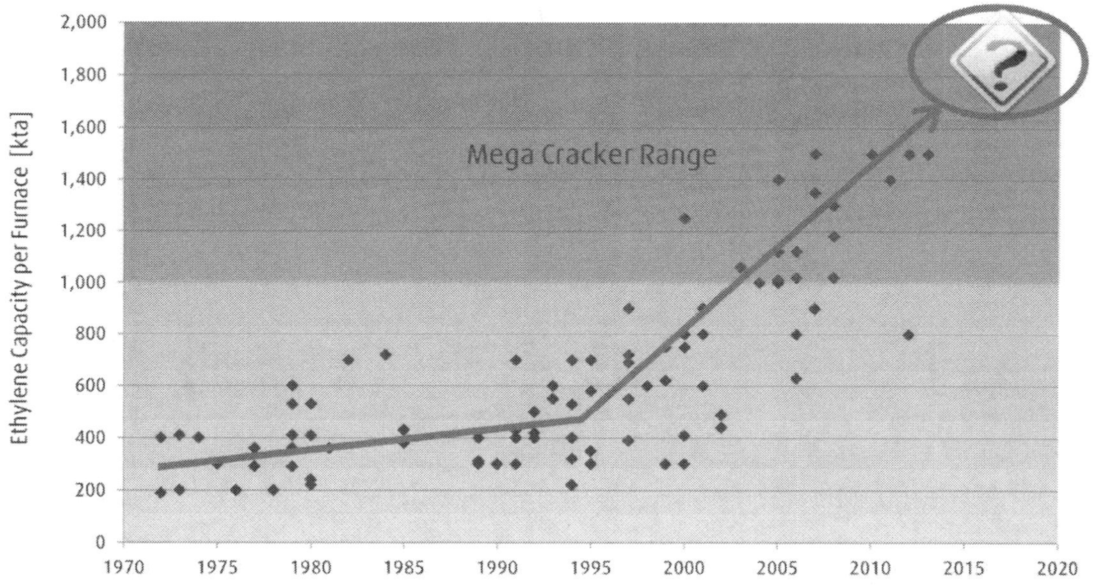

The decission to build a MEGA CRACKER is to be made on a case by case decission

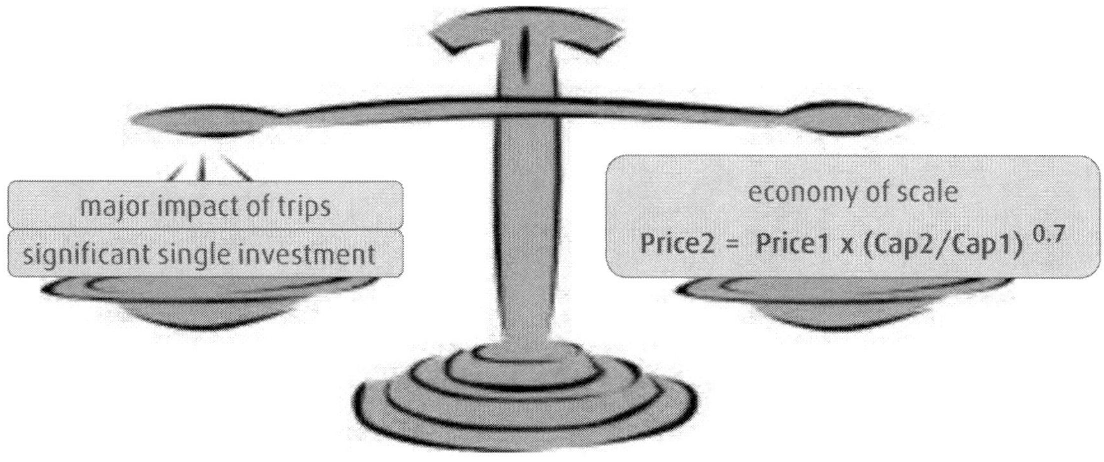

- ✓ less specific operational cost
- ✓ increased profit margins
- ✓ faster return of investment

MEGA CRACKERs are designed using new concepts based on existing guidelines to fullfill all major goals

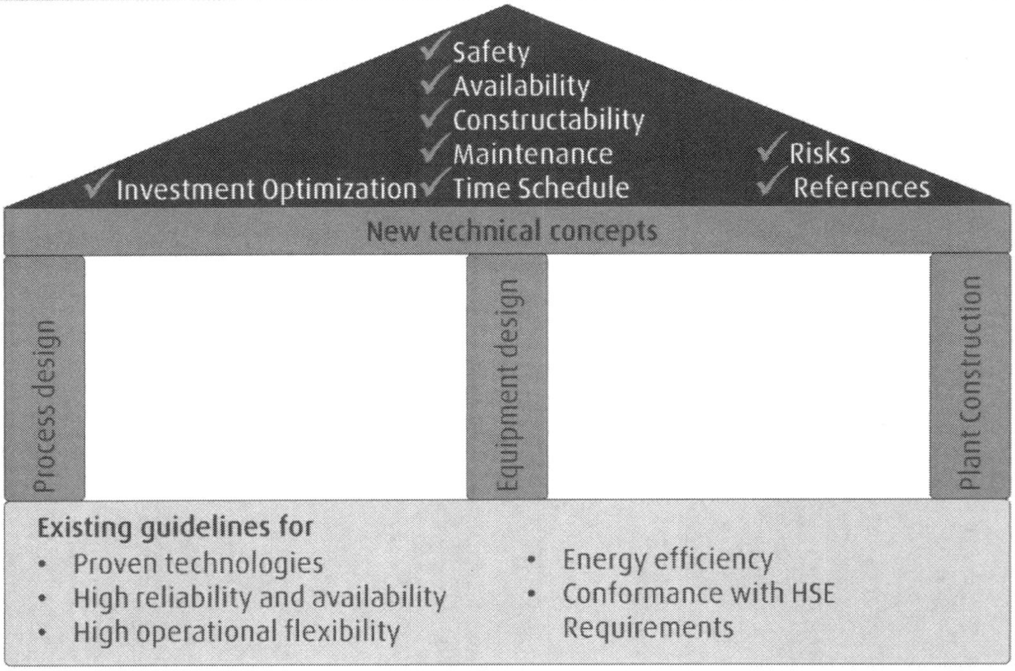

The size of furnaces has increased in the last decades

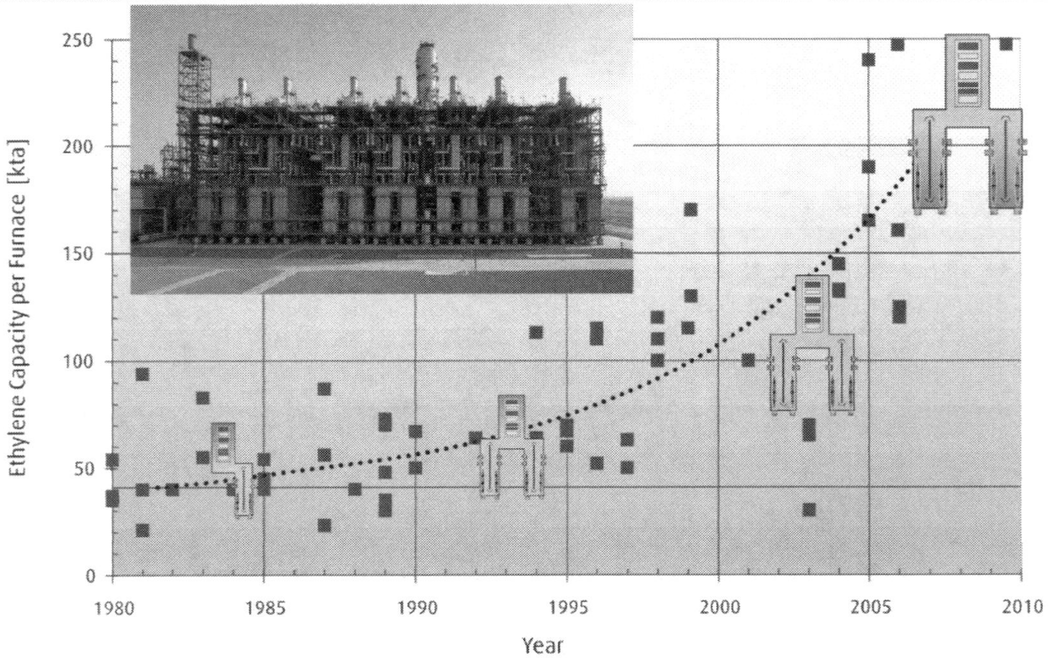

Mega Cracker – Furnaces
Individual Furnace Capacities

THE LINDE GROUP
Linde

Increasing capacity vs. increasing number

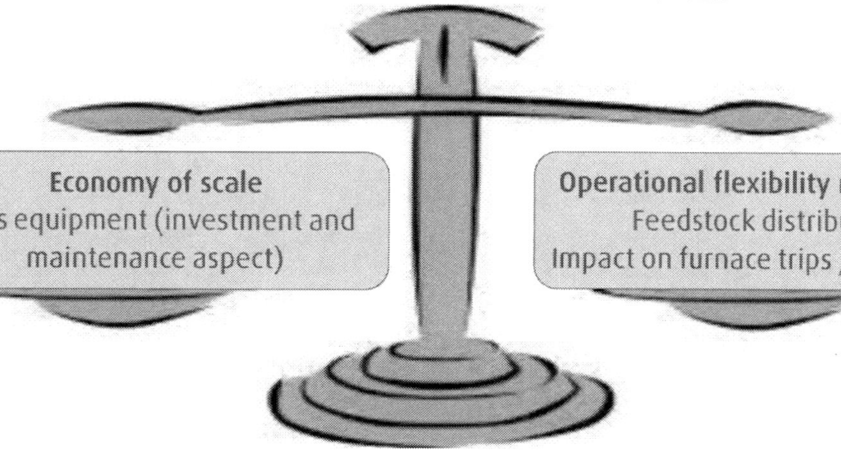

Economy of scale
Less equipment (investment and maintenance aspect)

Operational flexibility restrictions
Feedstock distribution
Impact on furnace trips / shutdown

- ✓ Minimum number of furnaces
- ✓ Individual furnace capacity is increasing

Two solutions to tackle increasing column dimensions: High capacity trays & distributive distilling

THE LINDE GROUP
Linde

High Capacity Trays

- easy scale up of size
- no risk in column design
- no froth gradient
- no stagnant area
- Overall capital savings
  Increased tray costs, but reduced costs for column shell + isolation transport / installation/erection

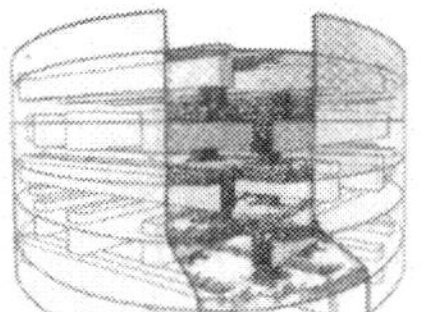

Distributive Distilling

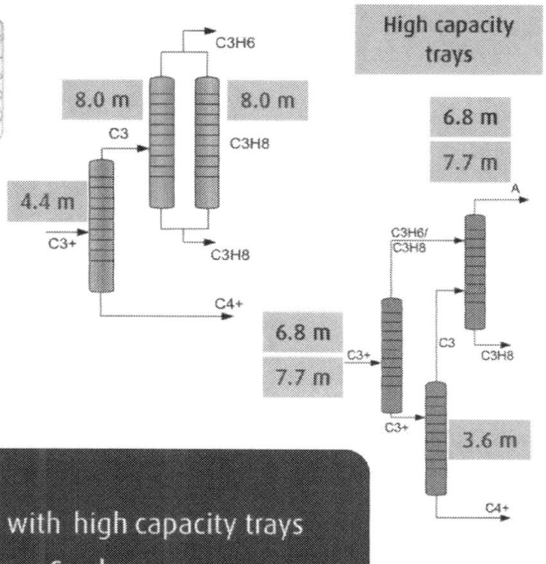

- ✓ approx. 10% energy reduction
- ✓ approx. 30% capacity increase with high capacity trays
- ✓ Manageable dimensions for Mega Crackers

Rotating equipment is key in a cracker

THE LINDE GROUP

Linde

Typical cracked gas compressor arrangement

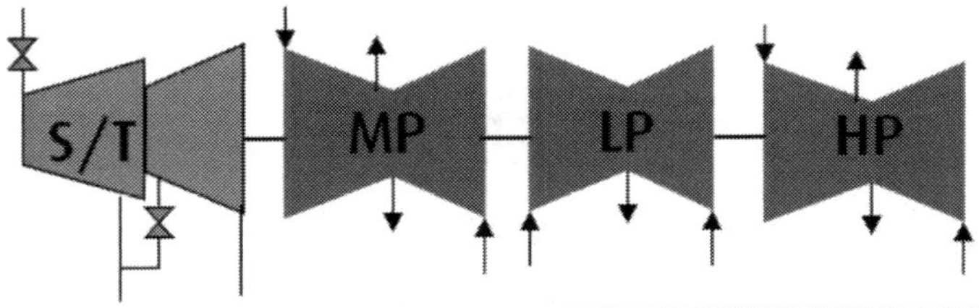

Proven Design for 1500 kta	2000 kta Mega Cracker
• $350.000\ m^3/hr$ suction flow	• $730.000\ m^3/hr$ suction flow
• 54 inch suction nozzle	• 84 inch suction nozzle
• 1500 mm impeller diameter	• 1800 mm impeller diameter
• 73 MW power rating	• 110 MW power rating
• 770 t/h life steam	• 1000 t/h life steam
• 24 x 5 x 5 m	• 33 x 6 x 6 m
• 370 tons package weight	• 520 tons package weight

Alternative solutions for rotating equipment in MEGA CRACKERs are available

THE LINDE GROUP

Linde

Different cracked gas compressor arrangements for Mega Crackers

Variant 1

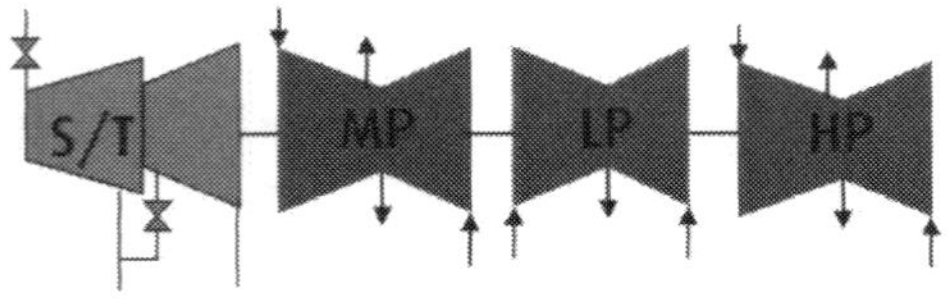

33 m x 6m x 6m
520 tons

Variant 2

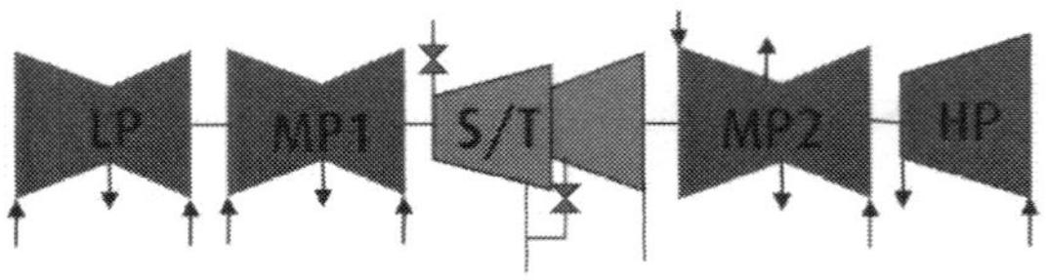

38 m x 6m x 6m
620 tons

✓ classical single train setup
✓ reasonable train dimension

✓ lower torque transmission
✓ smaller bearing &gas seals diameter
✓ HP casing smaller and easier to seal

New concepts for heat exchangers reduce size, weight and cost

THE LINDE GROUP

Linde

Current Design Limits

- case by case decision depending on type, service, transport restrictions, plot space
- Dimensions (length, diameter)
- Max. weight
- Maintainability
- Accesability

Block in Kettle - Concept

Brazed platefin exchanger in a conventional shell

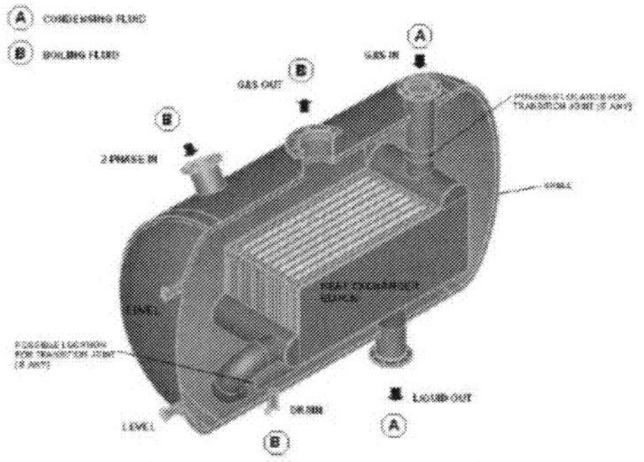

✓ Reduced equipment size & weight
✓ 30% cost savings
✓ Removable cores

Various other aspects have to be considered planing a MEGA CRACKER

THE LINDE GROUP

Linde

Pre-commissioning and Commissioning

⚠ Tightness of large size flanges
⚠ Flushing of lube oil systems main machinery
⚠ Cleaning, steam/air blowing of large diameter piping

Plate Fin Heat Exchangers

⚠ Increased heat transferred per volume is more critical to load temperature changes

Vessels, Reactors and Piping

⚠ CFD calculations can be required to prove reasonable flow distributions

Main machinery and piping

⚠ Dynamic simulation required to ensure a sound design of kick-back layout

THE LINDE GROUP Linde

MEGA CRACKERs will be built eventually

Linde Engineering North America Inc. 13 AIChE Conference 2013 / Sebastian Holz

THE UNIQUENESS AND CHALLENGES OF TECHNOLOGY BASED BUSINESSES

By Dennis W. McCullough, InnovCore Technology Management Consultants LLC

Technology-Based Businesses (TBBs) are a growing contributor to the profitability of mature industries as well as the primary basis for the emergence of small entrepreneurial enterprises that drive the dynamic growth of our economy. However, the unique characteristics of such enterprises are not widely recognized in strategy development, asset management, human capital deployment and growth planning. When faced with the overall responsibility for guiding such entities, it is imperative that the special nature of TBB enterprises be addressed in order to preserve their vitality and their long-term impact on our society.

A TBB is defined as a commercial entity that utilizes intellectual assets (e.g. patents, know-how, trade secrets, etc.) as the core asset of its business pursuit and from which its profits are derived. A TBB may be a small start-up company with only its R&D resources built around its patent assets without prior commercial know-how or manufacturing assets; or a TBB may be an entity with considerable manufacturing and market channel assets but which relies on its technology position as the principle market differentiator for its commercial success. In all cases, the TBB is considered by both customers and competitors to have established itself in its market through leveraging its technology assets.

A TBB often is thought of as a pure licensing entity in which access to technology assets are exchanged under license in return for license fees. While this certainly is one type of TBB, the licensing business is a rather narrow model that is far too constrained to be representative of TBBs in general. Rather the TBB classification provides great flexibility under the umbrella definition above to encompass entities that do not license out their technology at all. The one essential ingredient is that technology be the defining core of its business, without which its business could not be viable.

What is "Technology" in this context?

The term "Technology" in this context is considered to apply to assets that are characterized by the existence of an advanced level of understanding and/or practice of scientific and technical principles, embodied in both tangible and intangible forms that are essential to the financial success of a commercial entity. That sounds more complicated than it is in reality. Think of it diagrammatically as depicted in Figure 1.

Figure 1: Science, Technology & Knowledge Based

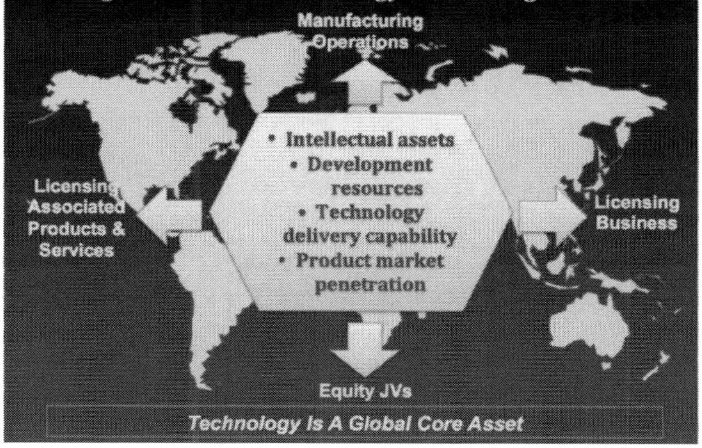

This illustrates a form of TBB that relies on its broad technology assets to extract value as a multi-faceted business where its strategic options are defined by its ability to convert or utilize the technology in various commercial initiatives.

TBB Business Model Basics

The conceptual model for TBBs can vary all over the map depending, of course, on the corporate environment into which the business is born. Startup TBBs often face the two-path option decision (Figure 2):

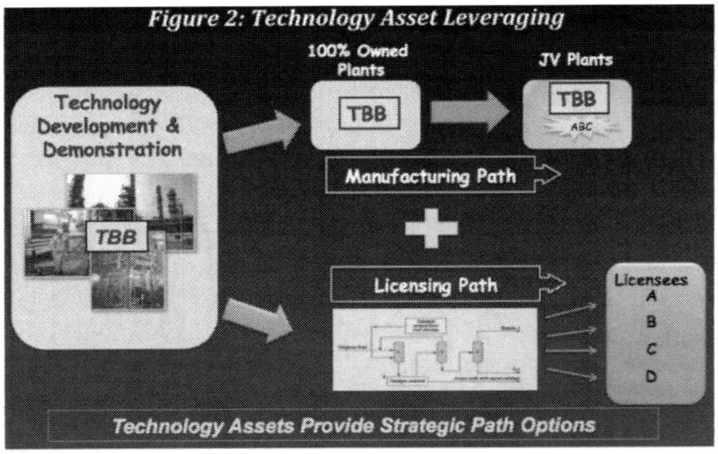

Figure 2: Technology Asset Leveraging

1. **Manufacturing Path** --- Investing in plant and equipment based on its technology to produce products for sale directly to customer markets. This can either be in its wholly owned plants or plants where the TBB is part of a manufacturing JV. This path can include both options in sequence or parallel.
2. **Licensing Path** --- Proceeding to license the technology to third party licensees who build their own manufacturing plants, with the TBB commercial profits deriving from license fees, services and license associated products (e.g. equipment, catalysts, materials, etc.)

TBBs that are formed within an existing larger entity face the same choices but have greater flexibility to pursue its options. Some TBBs stay in one mode or the other without ever entering the manufacturing or licensing paths, respectively. This decision may be shaped by a variety of commercial factors such as competitive pressures, capital limitations, risk appetite, profitability objectives and core strengths, to name a few.

TBB Strategies

A common strategy is to use all TBB options in constructing a sequence of commercial moves to maximize the impact of the differentiated technology position. Assuming largely three options:

A. Wholly-owned manufacturing;
B. JV manufacturing
C. Licensing & technology product sales

The various combinations could lead to the following illustrative scenarios:

<u>Scenario 1</u>: A => B => C

The TBB first builds its own plants to demonstrate the technology and establish its products in the TBBs home market. This initial strategy is followed by entertaining selected JVs, where feasible, to penetrate foreign markets under equity investment constraints. Then, where equity investment may be high risk, selectively license the technology with potential product off-take or lucrative licensing associated product options.

<u>Scenario 2</u>: C => B => A

The TBB first licenses the core technology to a first adopter to establish the technology at large scale. The next move may be to license the technology to others to build up the wider licensing market and generate sufficient capital and product demand to allow the TBB to feasibly consider participating in the manufacturing path. However, the TBB may need partners to fulfill capital and risk mitigation requirements and so chooses next to selectively enter manufacturing JVs in various world areas. After gaining manufacturing and product know-how as well as building up capital reserves, the TBB then may decide to build its own plants based on the now fully demonstrated and mature technology.

Of course, other combinations could be followed either in sequence or parallel. The idea is that the TBB is able to leverage the core technology in various ways to optimize its return. The TBB is not bound by some fixed idea of what sequence is appropriate but instead uses the technology core asset in flexible scenarios to profitable strategies (Figure 3).

Figure 3: Combination of Licensed, Owned & JV Plants

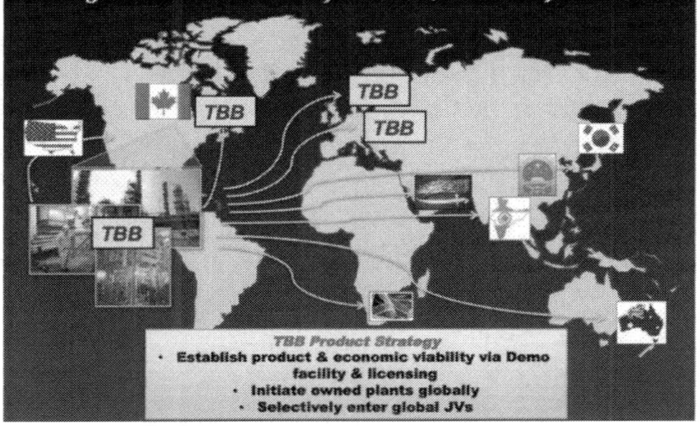

Key Industry Examples

Classically, one of the most dramatic examples of a TBB was the 1970s launch by Union Carbide Corporation (UCC) of a new technology for producing polyethylene. Traditionally, in the Chemical Industry, such innovative developments would be held closely, to be exploited through manufacturing with licensing exposure only after the technology reached maturity and its competitive advantage had waned. UCC decided to monetize its breakthrough technology in a unique way; through the worldwide licensing program designed to dominate the rapidly expanding polymers industry and simultaneously in its own manufacturing operations. The technology took the industry by storm (although not without some product penetration struggles) and at one time, UCC was participating in profits from 25% of the polyethylene sold in the world.

Today, over thirty years later, the UCC PE technology still forms the basis of a strong and vibrant business and, like the early days, both manufacturing and licensing businesses are pursued with vigor, respectively, by Dow Chemical (the acquirer of UCC) and in Univation (the Dow licensing JV with ExxonMobil).

There are many other examples of TBBs. In the chemical industry, the UOP division of Honeywell and the Lummus Technology Division of CB&I, both are examples of highly profitable TBBs based on an expanded licensing business model. In heavy equipment manufacturing, Austrian based Ebner Industries emphasizes R&D and considers its technical knowhow as its core asset. Perhaps the most popular current example of a TBB is Apple, not based on a licensing business component, but certainly characterized by a core that is technology intensive. These are only a few such entities and all change somewhat over the life of the company.

The most prevalent current example of TBBs are the biofuels and biochemicals startup companies who continue to spring up in an attempt to make an impact on our energy and chemicals markets that are now served primarily by carbon-based products. There are hundreds of these companies: Elevance, Gevo, Kior, Rennovia and Cobalt Technologies to name a handful. Some of these types of companies are following a "capital lite" strategy outlined above whereby the initial commercialization is via a licensee or JV partner who has the capital and resource base to build these capital-intensive plants. Once the early commercialization risks are bridged, these companies may then begin to enter the manufacturing phase with greater confidence of survival.

TBBs Are Unique

The most important point to recognize is that special strategies are needed for a company that considers itself a TBB. The first is that the technology must be nurtured above all else. This will be reflected in arranging priorities on capital allocation, IP protection, organization structure, hiring and firing, salary structure, partnering, R&D focus and market branding. Most importantly, it will affect short and long term strategies for survival and growth.

This isn't to say that key supportive manufacturing, financial, commercial and marketing initiatives are not essential. It is just that the entity can never forget the true source of its competitive advantage and the preservation of the technology asset that is crucial for its existence in the first place.

The Uniqueness & Challenges of Technology Based Businesses

Impact On The Chemical Industry

Speaker: Dr. Dennis McCullough
InnovCore Technology Management Consultants

Friday October 4th , 2013

AIChE American Institute of Chemical Engineers

Southwest Process Technology Conference

Topics

- TBB Business Basics
- CPI and HPI Components
- HPI Technology Sources
- TBB Business Model Strategic Options
- Technology Licensor & Service Provider Strategies
- Final Thoughts

2

TBB Business Basics

3

Technology Based Business

"A *commercial entity* that utilizes *intellectual assets* (e.g. patents, know-how, trade secrets, etc.) as ***the core asset*** of its business pursuit and from which its profits are derived."

4

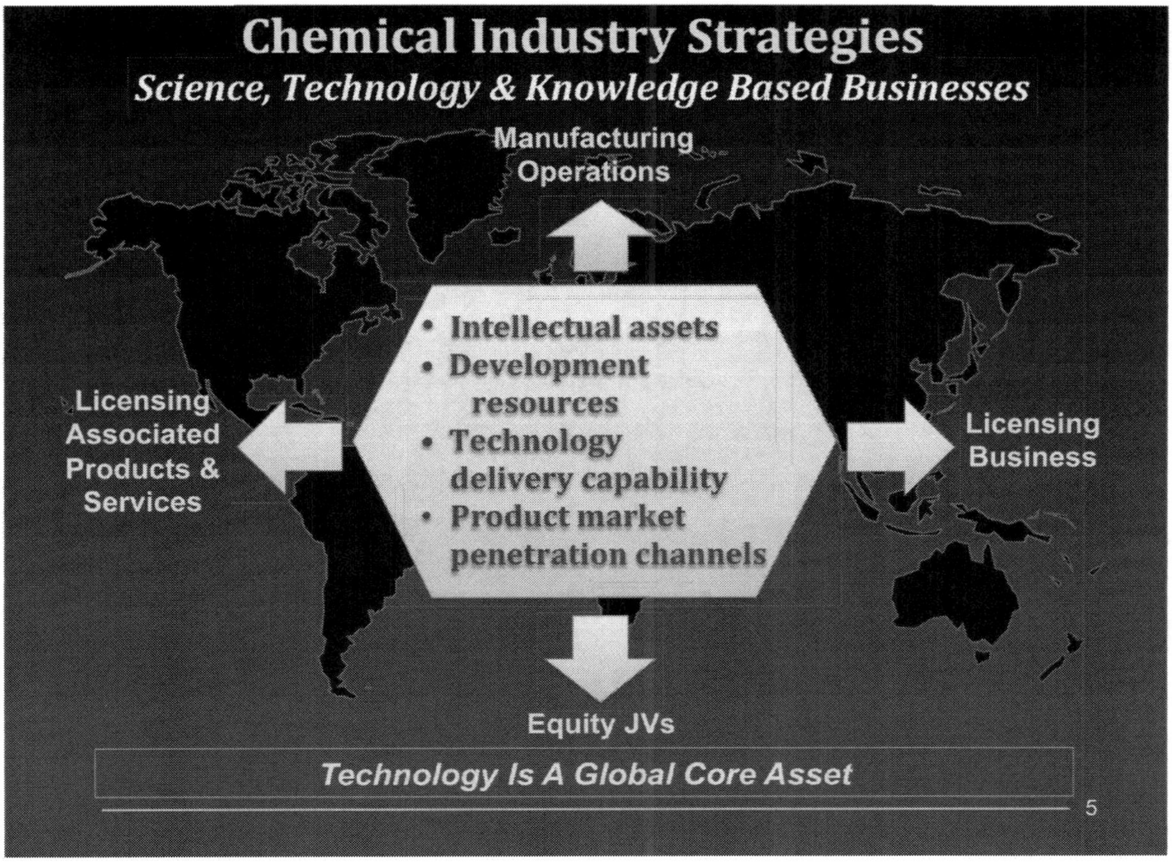
Chemical Industry Strategies
Science, Technology & Knowledge Based Businesses
Manufacturing Operations
• Intellectual assets
• Development resources
• Technology delivery capability
• Product market penetration channels
Licensing Associated Products & Services
Licensing Business
Equity JVs
Technology Is A Global Core Asset
5

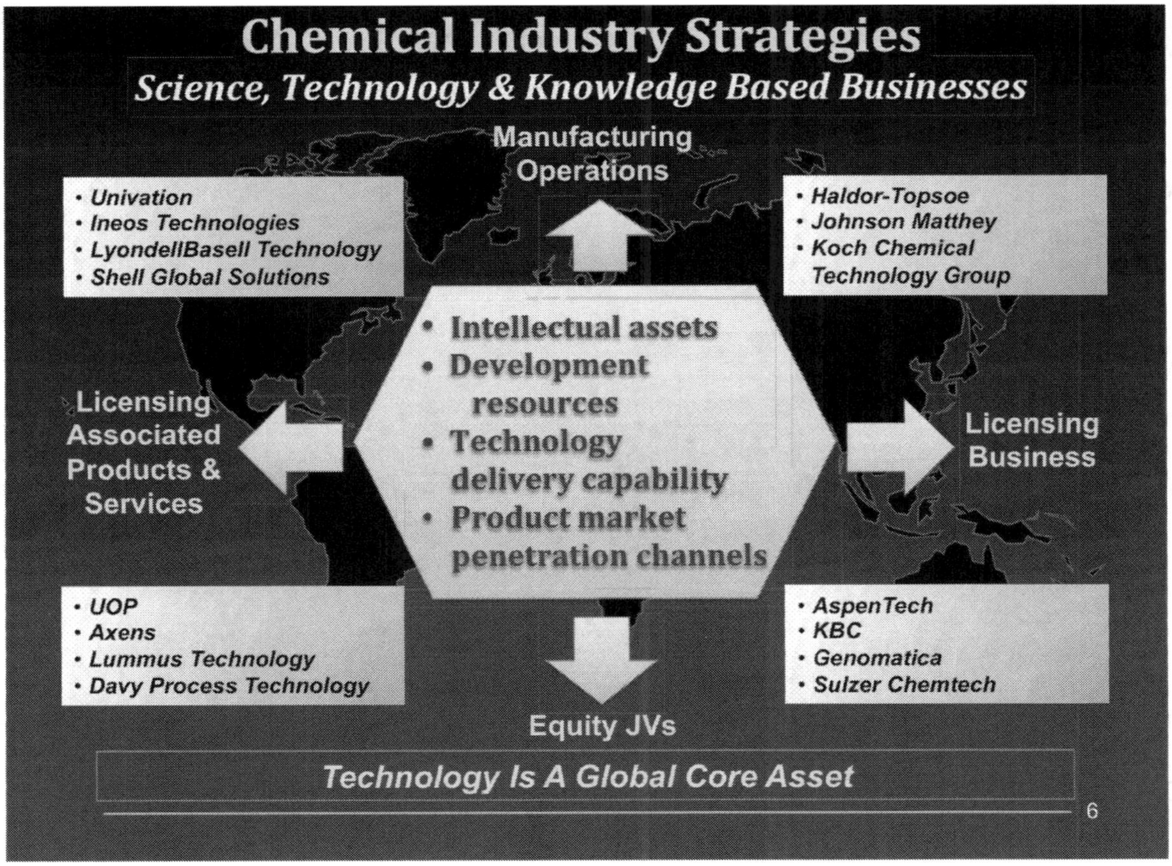
Chemical Industry Strategies
Science, Technology & Knowledge Based Businesses
Manufacturing Operations
• Univation
• Ineos Technologies
• LyondellBasell Technology
• Shell Global Solutions
• Haldor-Topsoe
• Johnson Matthey
• Koch Chemical Technology Group
• Intellectual assets
• Development resources
• Technology delivery capability
• Product market penetration channels
Licensing Associated Products & Services
Licensing Business
• UOP
• Axens
• Lummus Technology
• Davy Process Technology
• AspenTech
• KBC
• Genomatica
• Sulzer Chemtech
Equity JVs
Technology Is A Global Core Asset
6

Chemical Processing Industry & Hydrocarbon Processing Industry

Industry Components

7

Overall HPI Processing

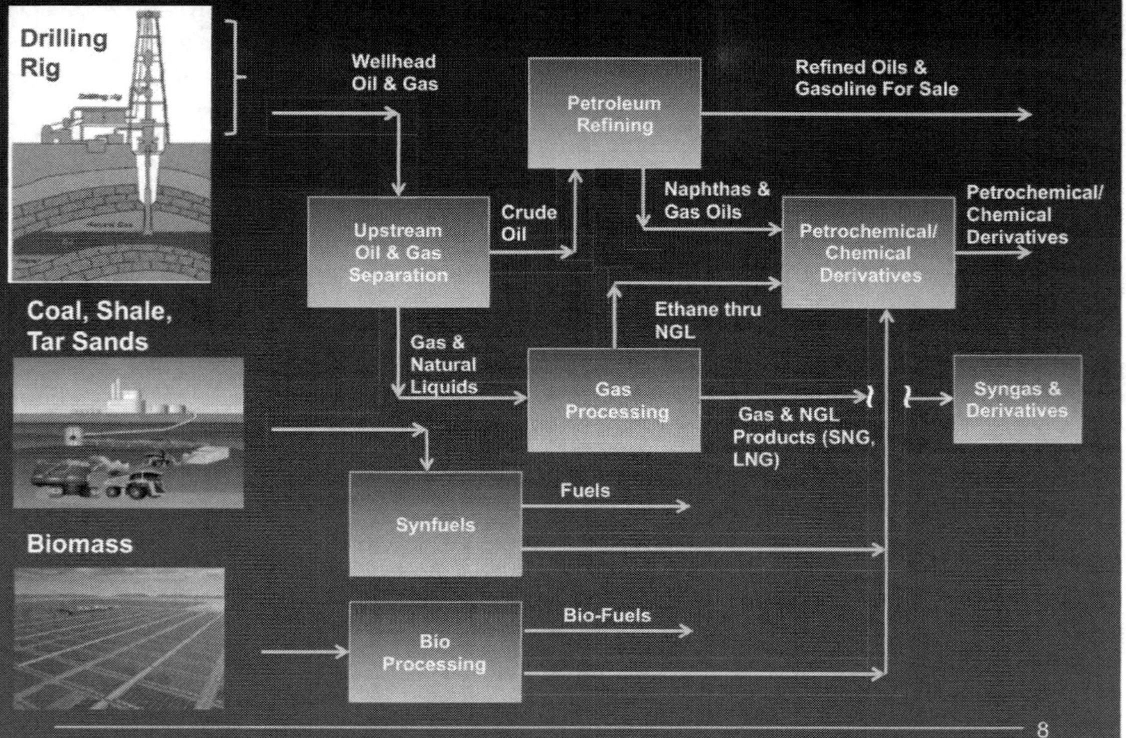

8

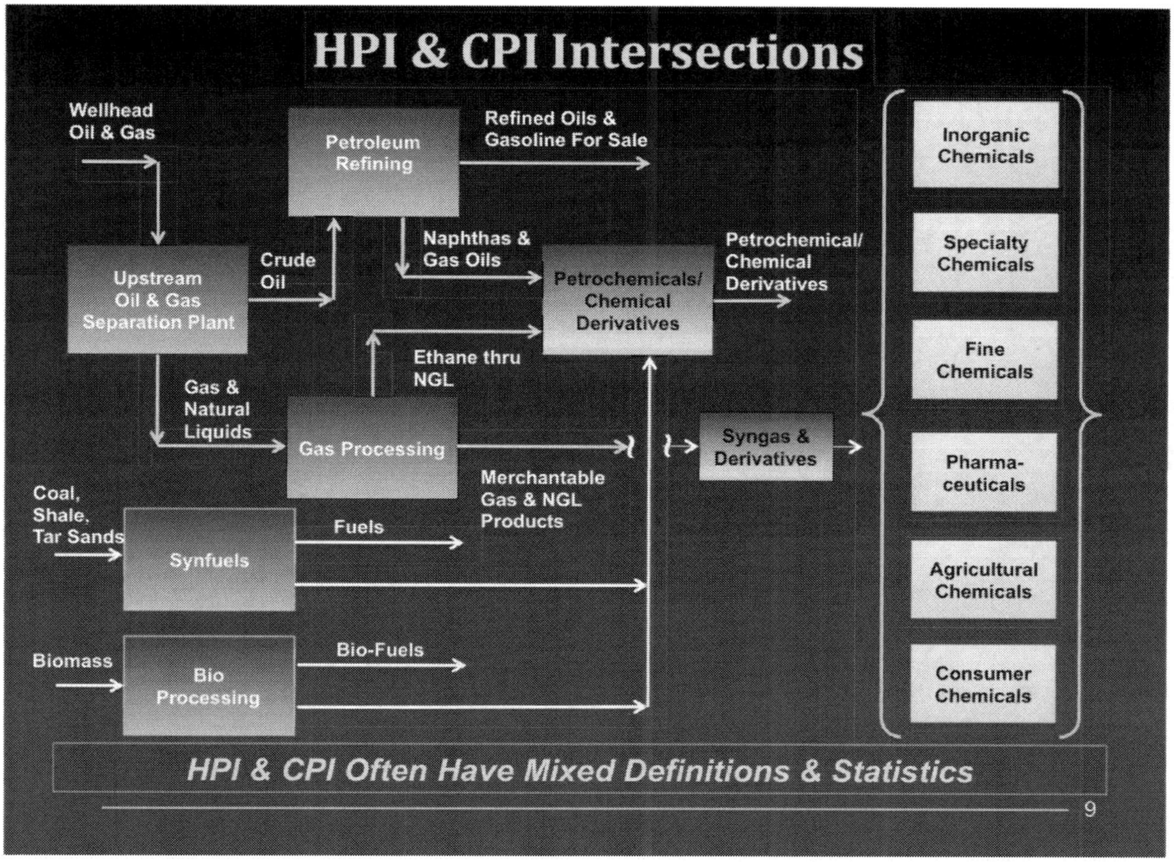
HPI & CPI Intersections
Wellhead Oil & Gas
Petroleum Refining
Refined Oils & Gasoline For Sale
Upstream Oil & Gas Separation Plant
Crude Oil
Naphthas & Gas Oils
Petrochemicals/ Chemical Derivatives
Petrochemical/ Chemical Derivatives
Ethane thru NGL
Gas & Natural Liquids
Gas Processing
Syngas & Derivatives
Merchantable Gas & NGL Products
Coal, Shale, Tar Sands
Synfuels
Fuels
Biomass
Bio Processing
Bio-Fuels
Inorganic Chemicals
Specialty Chemicals
Fine Chemicals
Pharma-ceuticals
Agricultural Chemicals
Consumer Chemicals
HPI & CPI Often Have Mixed Definitions & Statistics
9

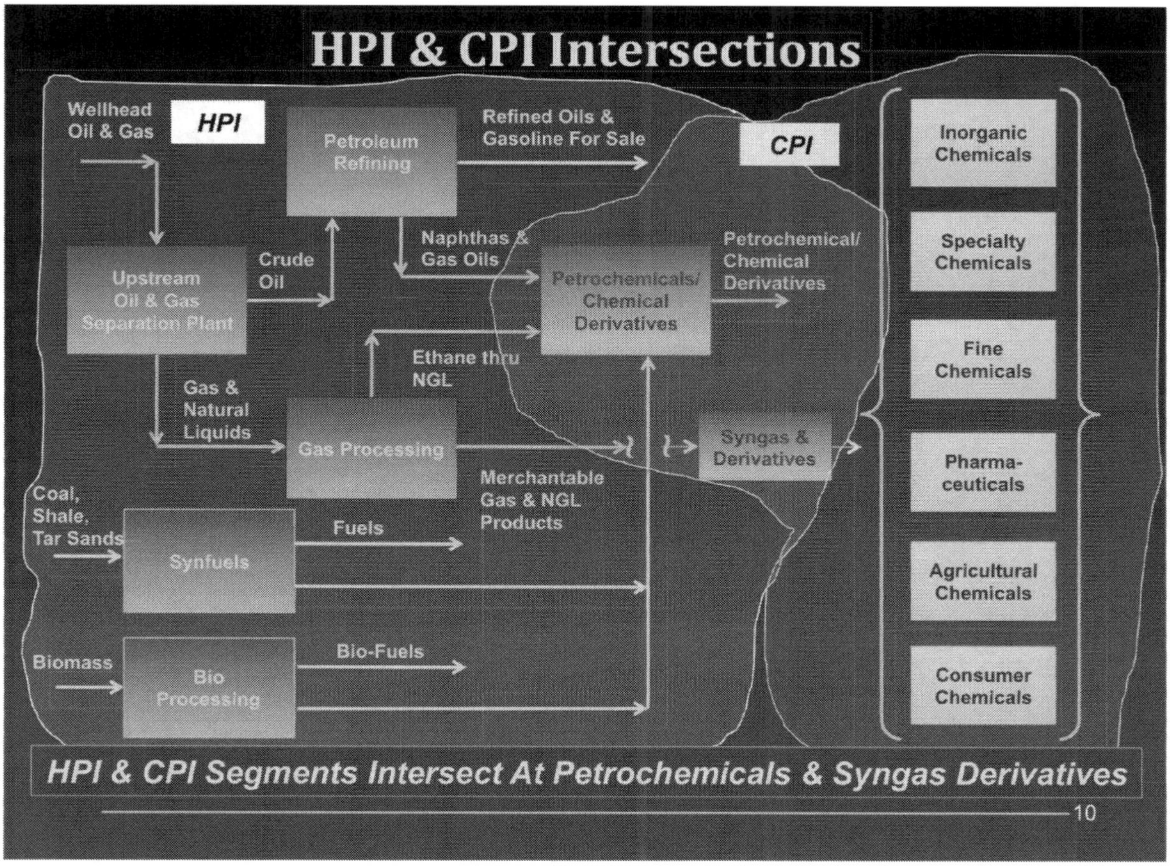
HPI & CPI Intersections
HPI
CPI
Wellhead Oil & Gas
Petroleum Refining
Refined Oils & Gasoline For Sale
Upstream Oil & Gas Separation Plant
Crude Oil
Naphthas & Gas Oils
Petrochemicals/ Chemical Derivatives
Petrochemical/ Chemical Derivatives
Ethane thru NGL
Gas & Natural Liquids
Gas Processing
Syngas & Derivatives
Merchantable Gas & NGL Products
Coal, Shale, Tar Sands
Synfuels
Fuels
Biomass
Bio Processing
Bio-Fuels
Inorganic Chemicals
Specialty Chemicals
Fine Chemicals
Pharma-ceuticals
Agricultural Chemicals
Consumer Chemicals
HPI & CPI Segments Intersect At Petrochemicals & Syngas Derivatives
10

HPI Technology Sources
11

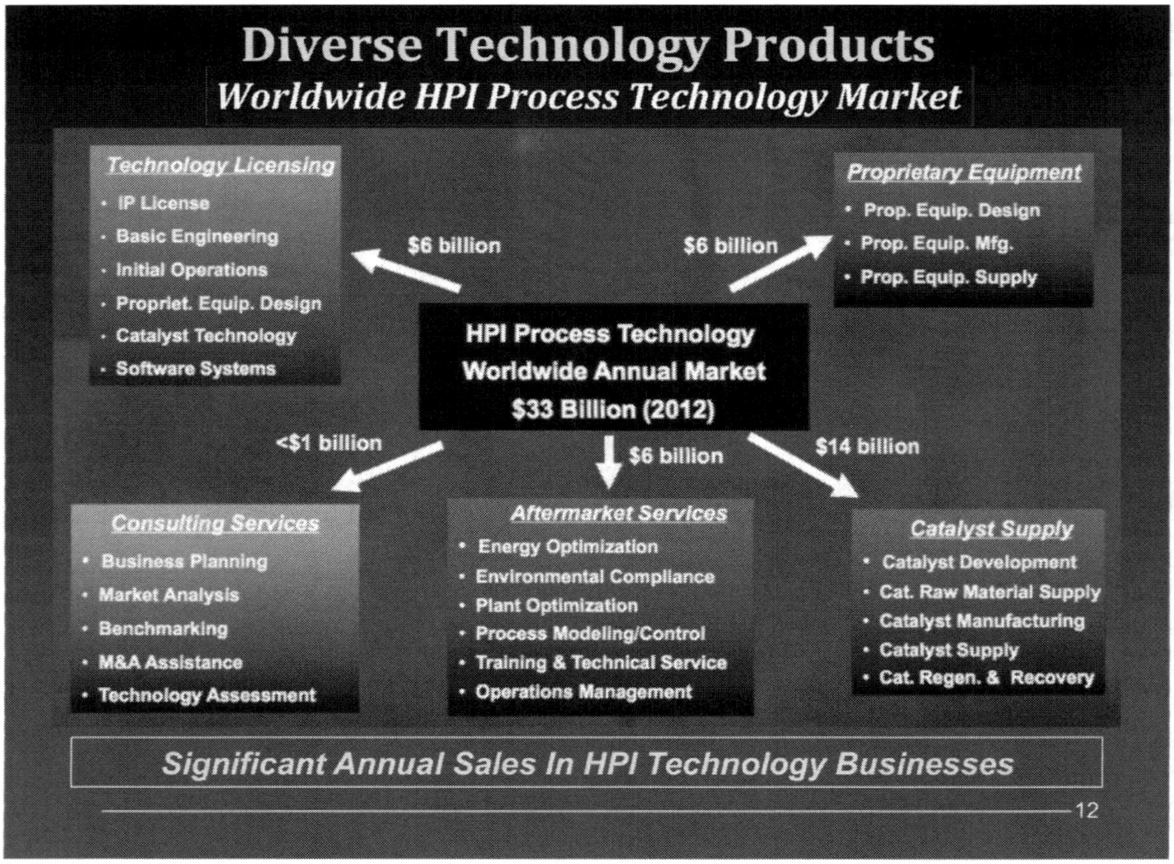
Diverse Technology Products
Worldwide HPI Process Technology Market
Technology Licensing
• IP License
• Basic Engineering
• Initial Operations
• Propriet. Equip. Design
• Catalyst Technology
• Software Systems
$6 billion
$6 billion
Proprietary Equipment
• Prop. Equip. Design
• Prop. Equip. Mfg.
• Prop. Equip. Supply
HPI Process Technology
Worldwide Annual Market
$33 Billion (2012)
<$1 billion
$6 billion
$14 billion
Consulting Services
• Business Planning
• Market Analysis
• Benchmarking
• M&A Assistance
• Technology Assessment
Aftermarket Services
• Energy Optimization
• Environmental Compliance
• Plant Optimization
• Process Modeling/Control
• Training & Technical Service
• Operations Management
Catalyst Supply
• Catalyst Development
• Cat. Raw Material Supply
• Catalyst Manufacturing
• Catalyst Supply
• Cat. Regen. & Recovery
Significant Annual Sales In HPI Technology Businesses
12

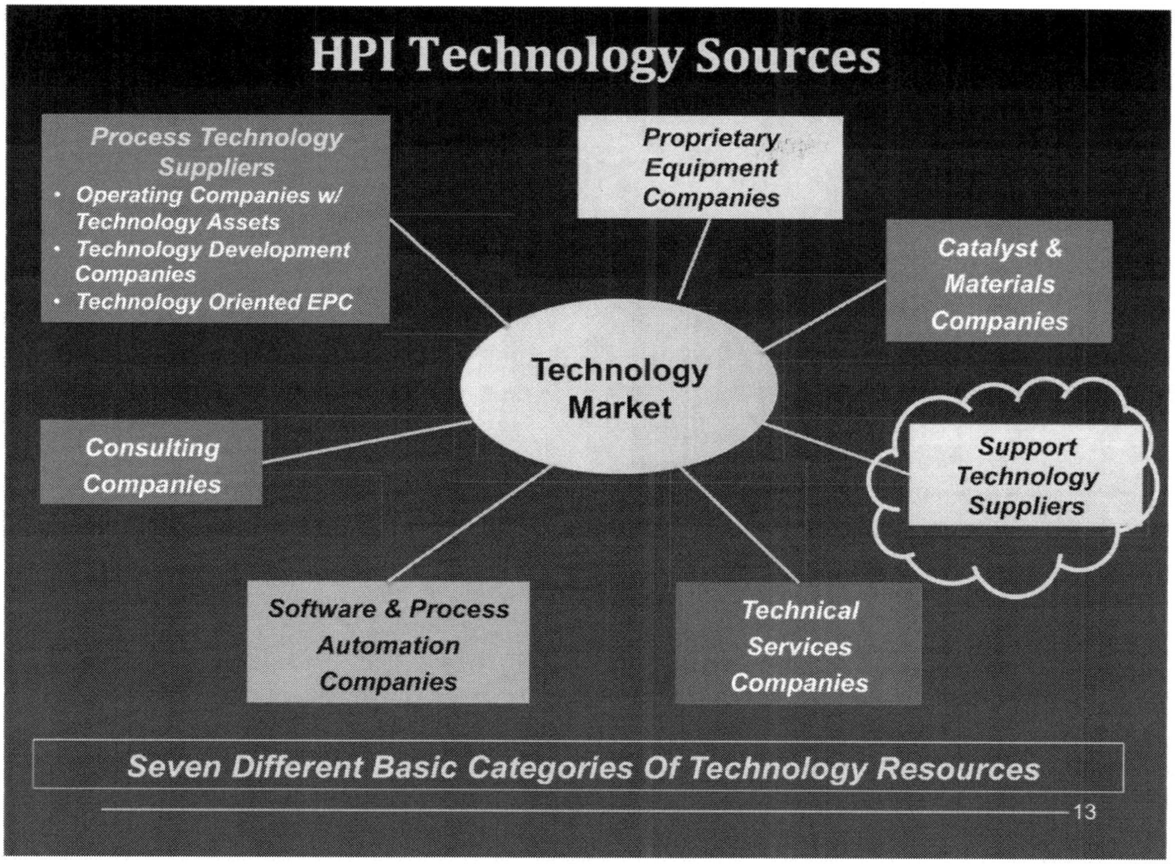
HPI Technology Sources
Process Technology Suppliers
• Operating Companies w/ Technology Assets
• Technology Development Companies
• Technology Oriented EPC
Proprietary Equipment Companies
Catalyst & Materials Companies
Technology Market
Consulting Companies
Support Technology Suppliers
Software & Process Automation Companies
Technical Services Companies
Seven Different Basic Categories Of Technology Resources
13

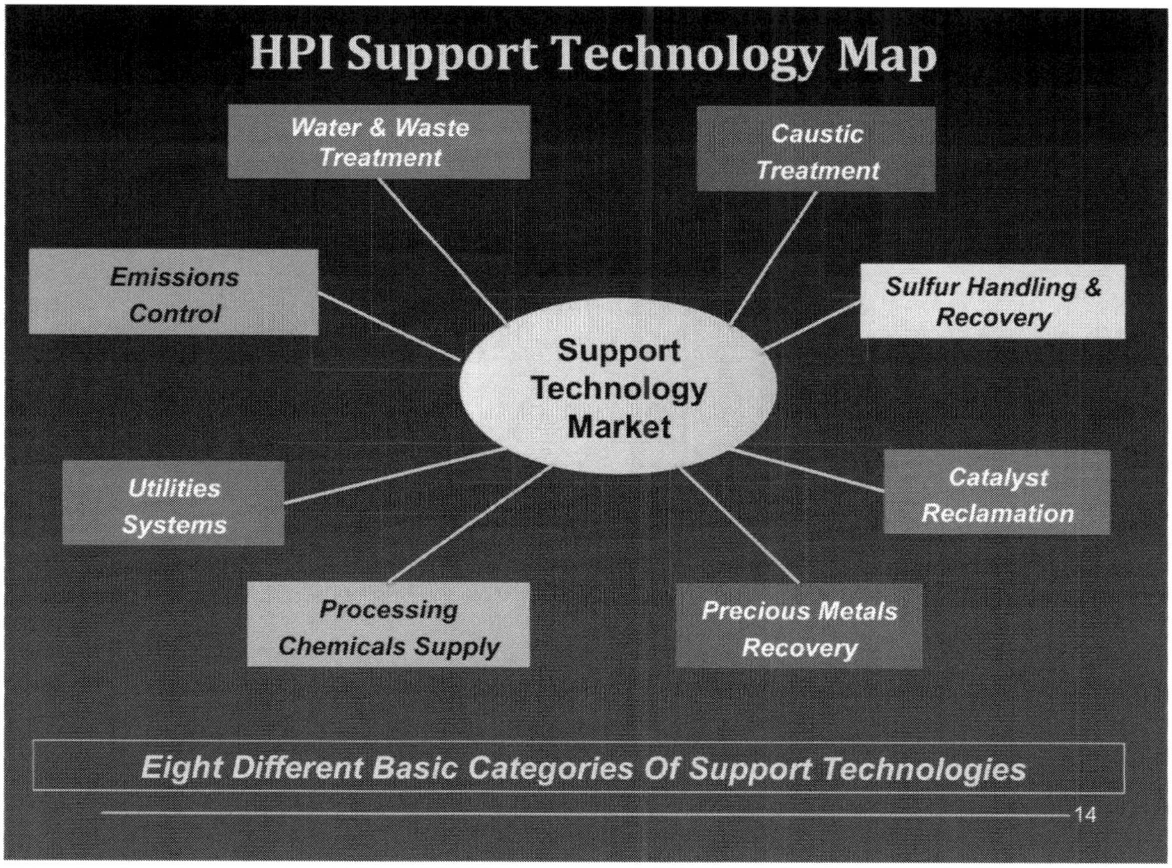
HPI Support Technology Map
Water & Waste Treatment
Caustic Treatment
Emissions Control
Sulfur Handling & Recovery
Support Technology Market
Utilities Systems
Catalyst Reclamation
Processing Chemicals Supply
Precious Metals Recovery
Eight Different Basic Categories Of Support Technologies
14

TBB Business Models

Strategic Options

15

Example: Entity Potential Strategic Paths

Technology Asset Leveraging

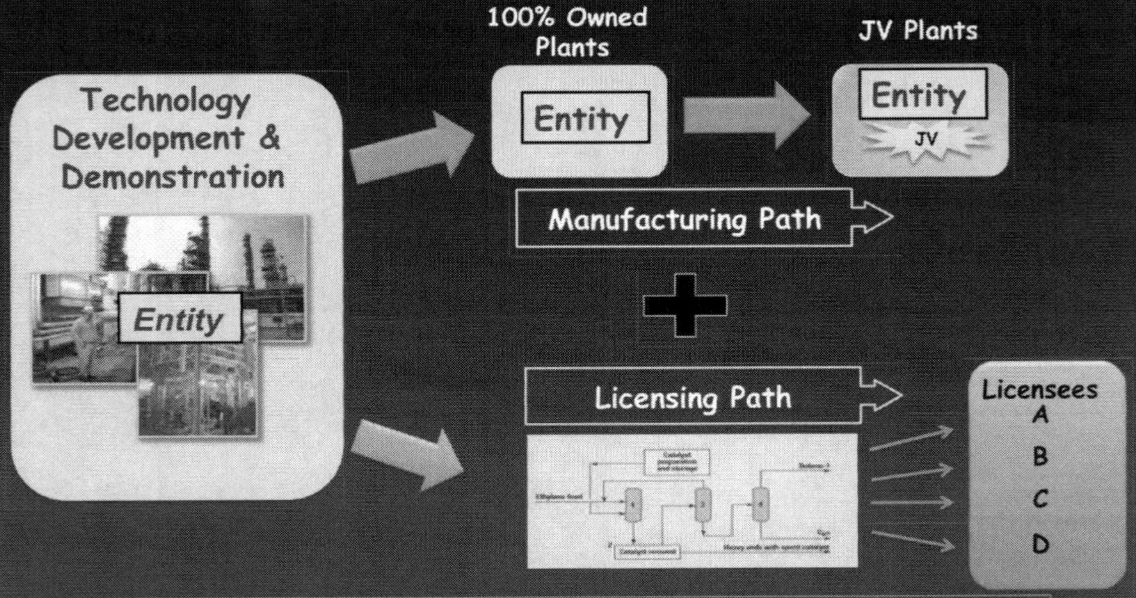

Technology Assets Provide Strategic Path Options

16

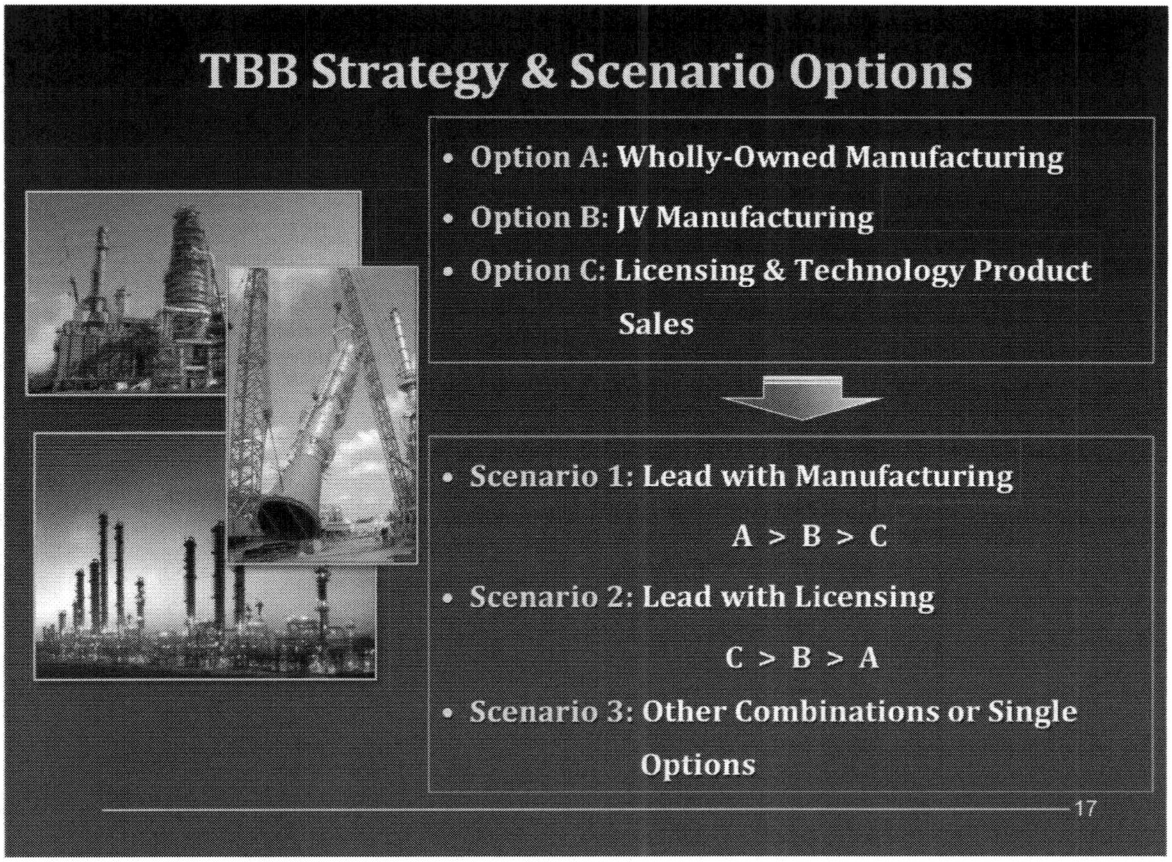
TBB Strategy & Scenario Options
• Option A: Wholly-Owned Manufacturing
• Option B: JV Manufacturing
• Option C: Licensing & Technology Product Sales
• Scenario 1: Lead with Manufacturing
A > B > C
• Scenario 2: Lead with Licensing
C > B > A
• Scenario 3: Other Combinations or Single Options
17

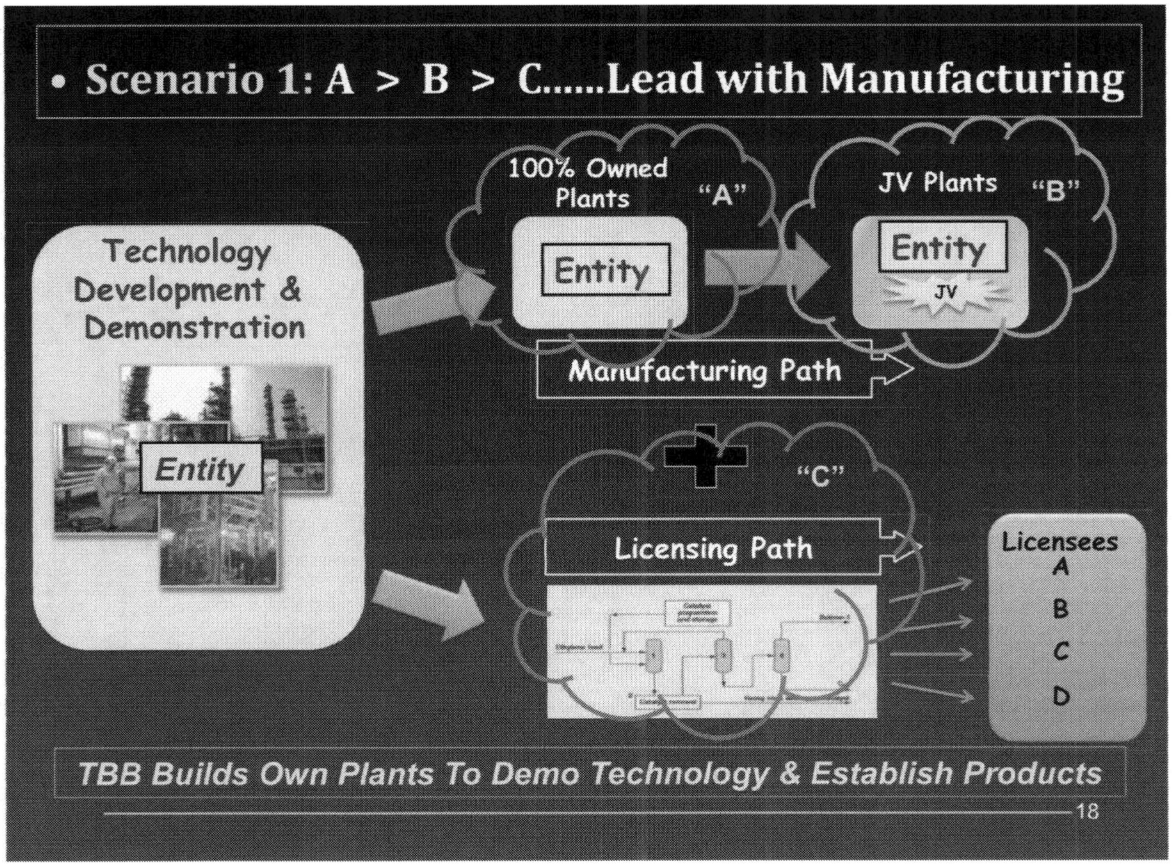
• Scenario 1: A > B > C......Lead with Manufacturing
Technology Development & Demonstration
Entity
100% Owned Plants
"A"
Entity
JV Plants
"B"
Entity
JV
Manufacturing Path
"C"
Licensing Path
Licensees
A
B
C
D
TBB Builds Own Plants To Demo Technology & Establish Products
18

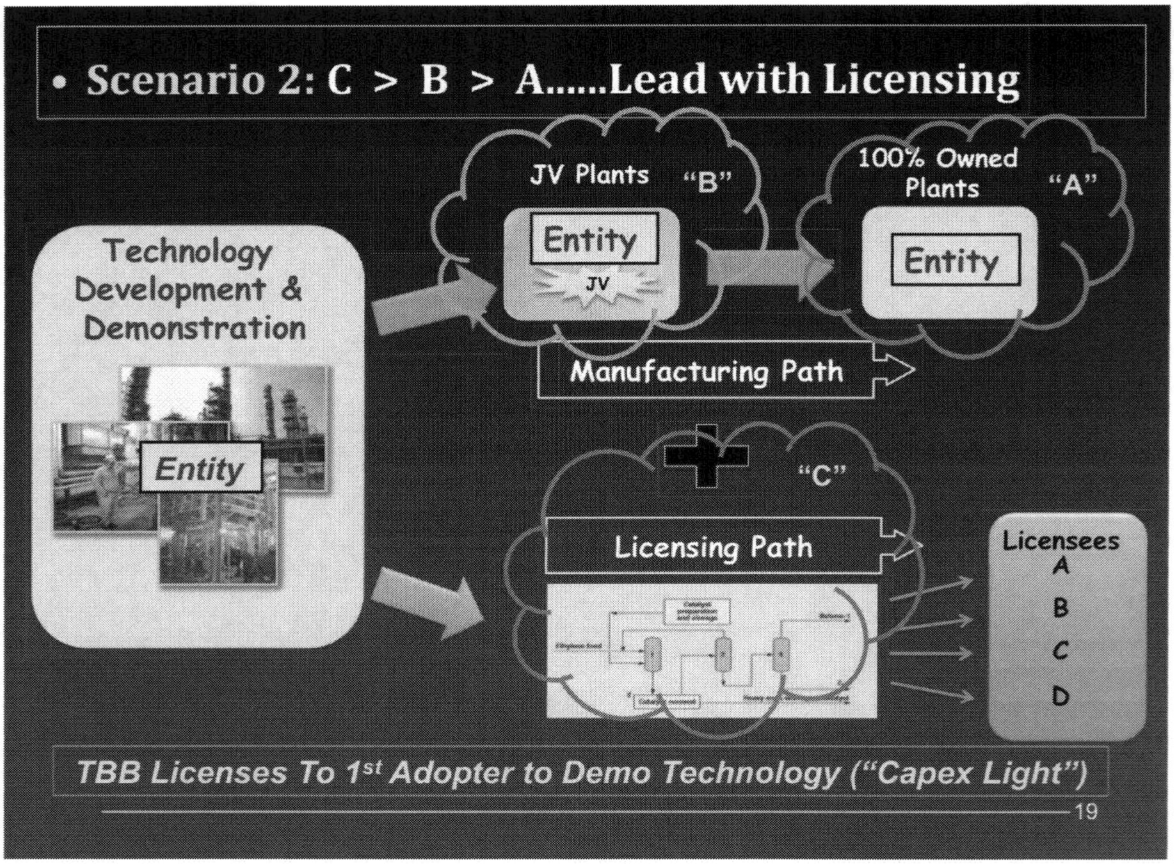
• Scenario 2: C > B > A......Lead with Licensing
JV Plants "B"
Entity
JV
100% Owned Plants "A"
Entity
Technology Development & Demonstration
Entity
Manufacturing Path
"C"
Licensing Path
Licensees
A
B
C
D
TBB Licenses To 1st Adopter to Demo Technology ("Capex Light")
19

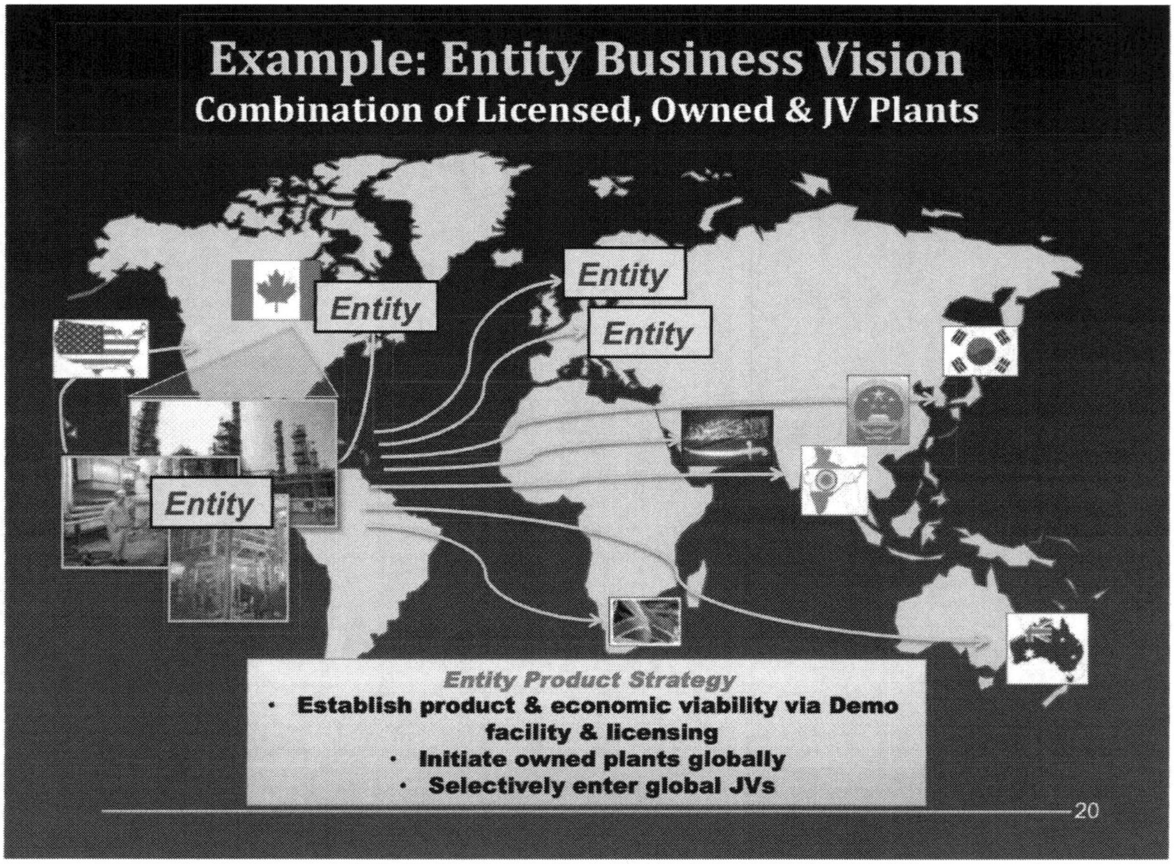
Example: Entity Business Vision
Combination of Licensed, Owned & JV Plants
Entity
Entity
Entity
Entity
Entity Product Strategy
• Establish product & economic viability via Demo facility & licensing
• Initiate owned plants globally
• Selectively enter global JVs
20

Technology Licensor & Service Provider Strategies

21

Licensor Expanded Product Strategies

Process & Associated Products

HPI Process License Royalties

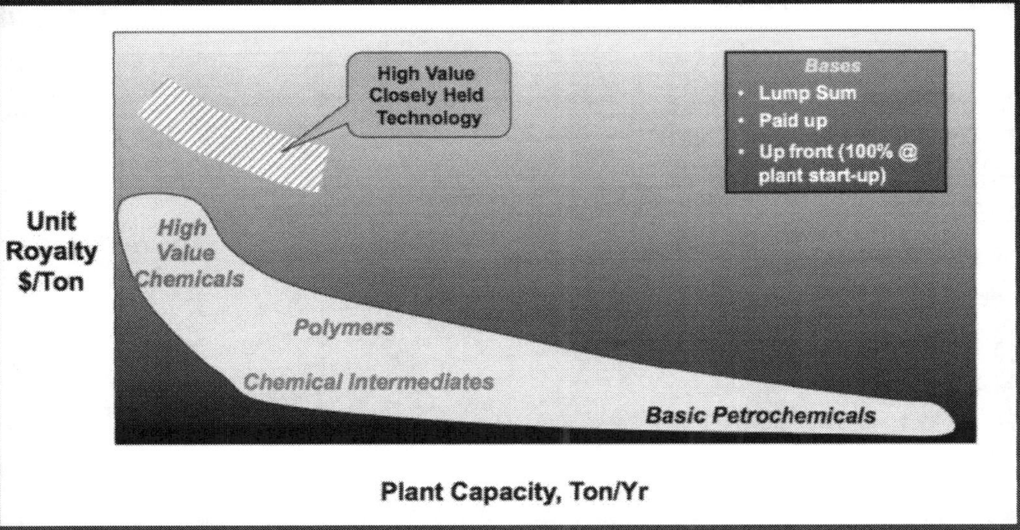

HPI Industry Provides Strong Royalty Framework

22

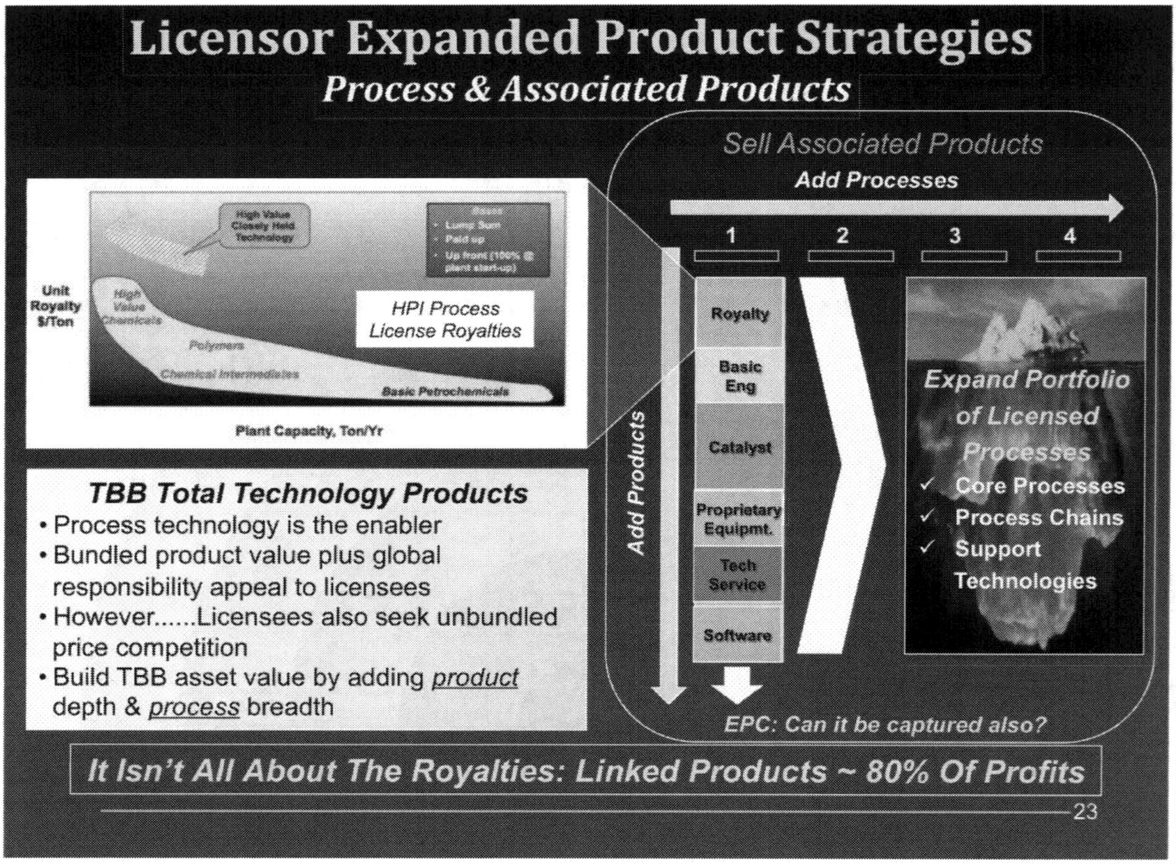

Licensor Expanded Product Strategies
Process & Associated Products
Sell Associated Products
Add Processes
1
2
3
4
High Value Closely Held Technology
Unit Royalty $/Ton
HPI Process License Royalties
Polymers
Chemical Intermediates
Basic Petrochemicals
Plant Capacity, Ton/Yr
TBB Total Technology Products
• Process technology is the enabler
• Bundled product value plus global responsibility appeal to licensees
• However......Licensees also seek unbundled price competition
• Build TBB asset value by adding product depth & process breadth
Add Products
Royalty
Basic Eng
Catalyst
Proprietary Equipmt.
Tech Service
Software
Expand Portfolio of Licensed Processes
✓ Core Processes
✓ Process Chains
✓ Support Technologies
EPC: Can it be captured also?
It Isn't All About The Royalties: Linked Products ~ 80% Of Profits
23

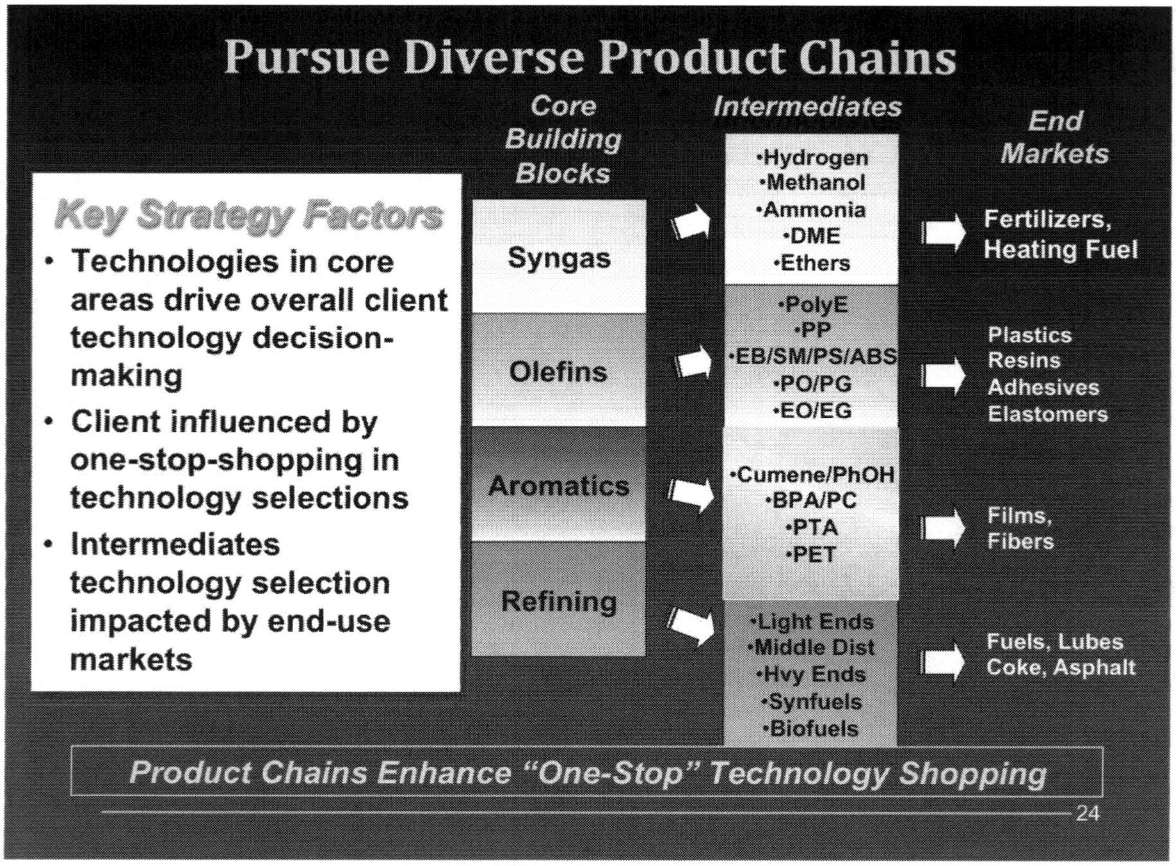

Pursue Diverse Product Chains
Core Building Blocks
Intermediates
End Markets
Key Strategy Factors
• Technologies in core areas drive overall client technology decision-making
• Client influenced by one-stop-shopping in technology selections
• Intermediates technology selection impacted by end-use markets
Syngas
Olefins
Aromatics
Refining
•Hydrogen
•Methanol
•Ammonia
•DME
•Ethers
•PolyE
•PP
•EB/SM/PS/ABS
•PO/PG
•EO/EG
•Cumene/PhOH
•BPA/PC
•PTA
•PET
•Light Ends
•Middle Dist
•Hvy Ends
•Synfuels
•Biofuels
Fertilizers, Heating Fuel
Plastics
Resins
Adhesives
Elastomers
Films,
Fibers
Fuels, Lubes
Coke, Asphalt
Product Chains Enhance "One-Stop" Technology Shopping
24

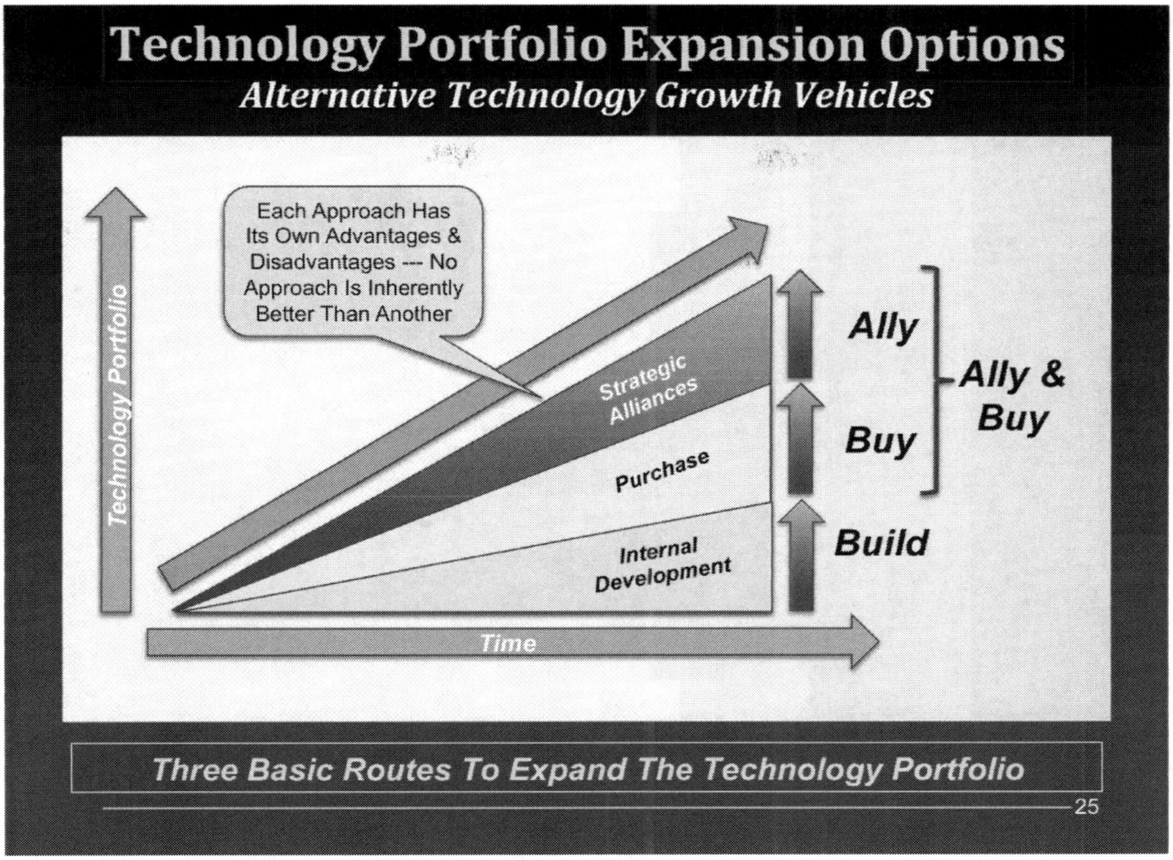
Technology Portfolio Expansion Options
Alternative Technology Growth Vehicles
Each Approach Has Its Own Advantages & Disadvantages --- No Approach Is Inherently Better Than Another
Technology Portfolio
Strategic Alliances
Purchase
Internal Development
Time
Ally
Buy
Build
Ally & Buy
Three Basic Routes To Expand The Technology Portfolio
25

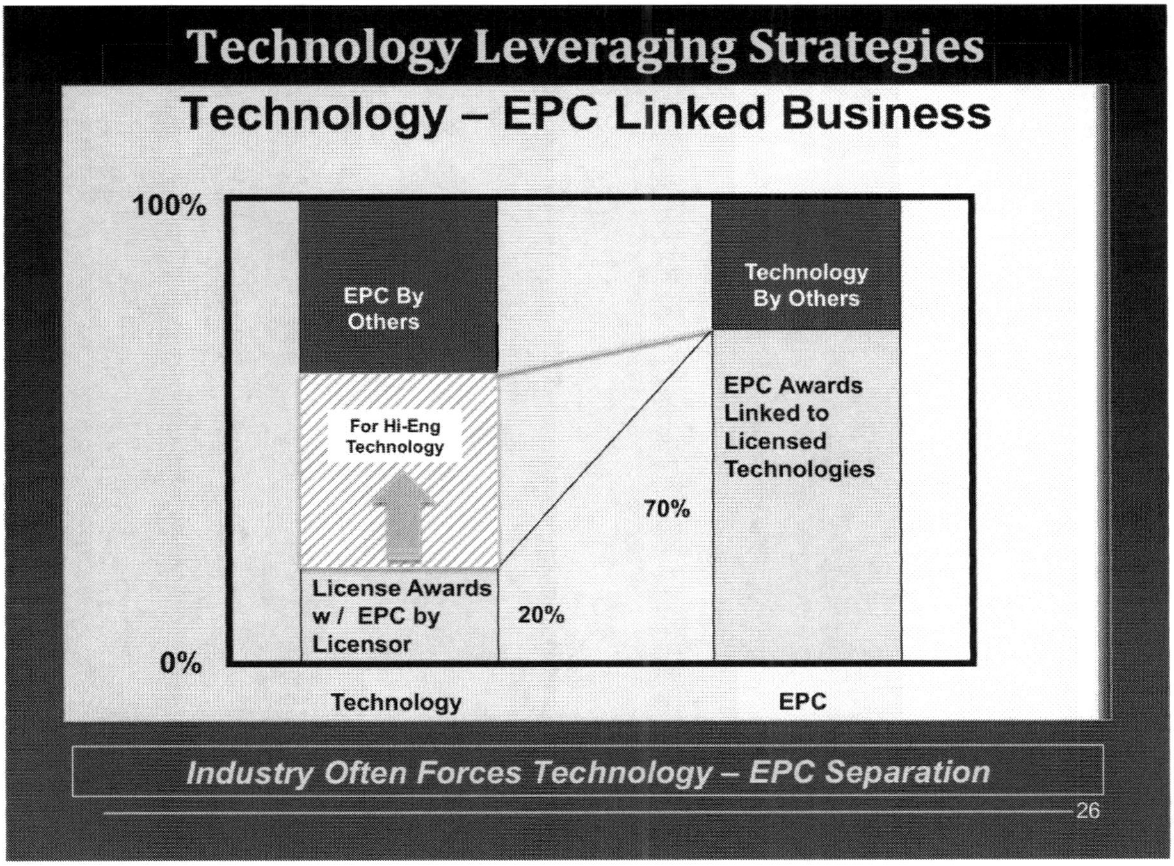
Technology Leveraging Strategies
Technology – EPC Linked Business
100%
0%
EPC By Others
For Hi-Eng Technology
License Awards w / EPC by Licensor
20%
70%
Technology By Others
EPC Awards Linked to Licensed Technologies
Technology
EPC
Industry Often Forces Technology – EPC Separation
26

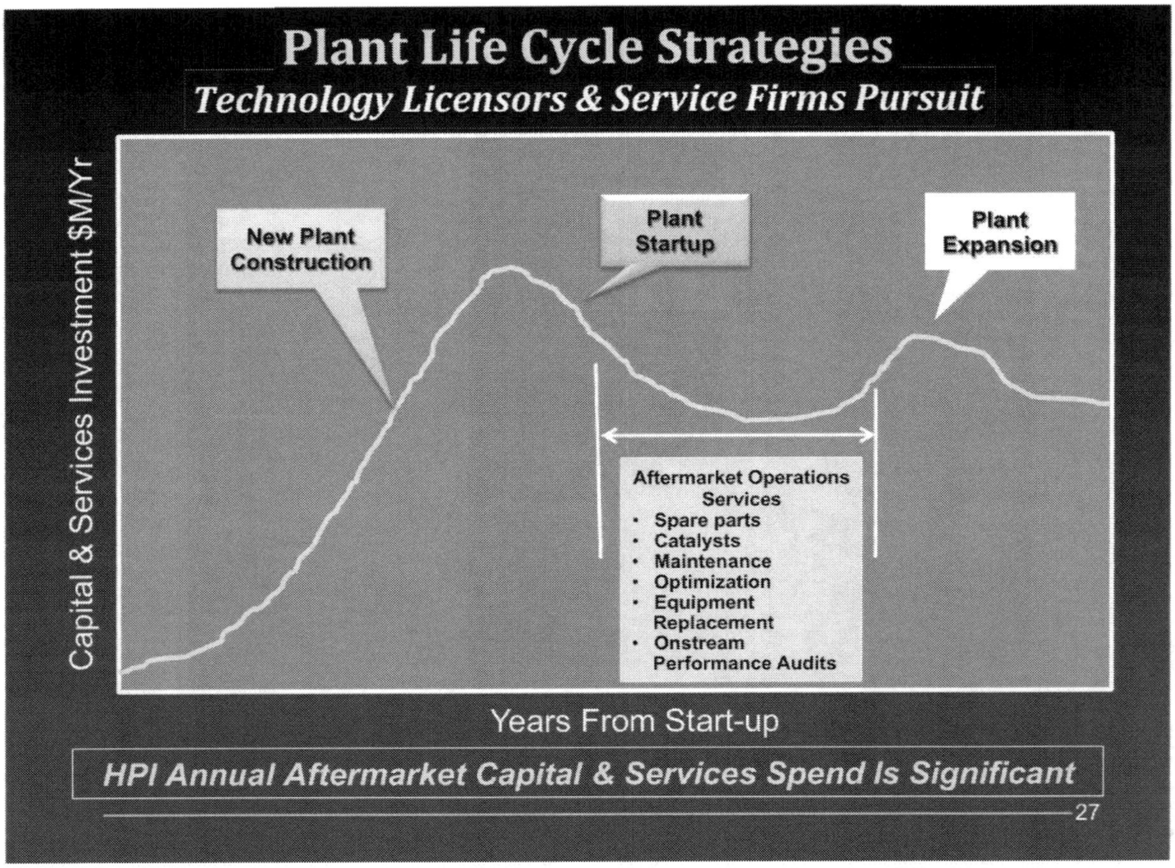
Plant Life Cycle Strategies
Technology Licensors & Service Firms Pursuit
Capital & Services Investment $M/Yr
New Plant Construction
Plant Startup
Plant Expansion
Aftermarket Operations Services
• Spare parts
• Catalysts
• Maintenance
• Optimization
• Equipment Replacement
• Onstream Performance Audits
Years From Start-up
HPI Annual Aftermarket Capital & Services Spend Is Significant
27

Licensor Expanded Product Strategies
New Technology Development Strategies
Refinery
Natural gas
Crude oil
Gasoline
Diesel
Clean fuels
Ethylene/ Petrochemicals
Plastics/ Polymer
HPI Technology Development Characteristics
$1MM - $10MM R&D Projects
$100MM - $1Bn Plant Capex (w/ $30MM Demo Unit)
8 -10 Yr Development-to-Commercialization Cycle
$1MM+/Day Plant Down Time Revenue Loss
HPI Characteristics Place High Barriers On New Technology Entry
28

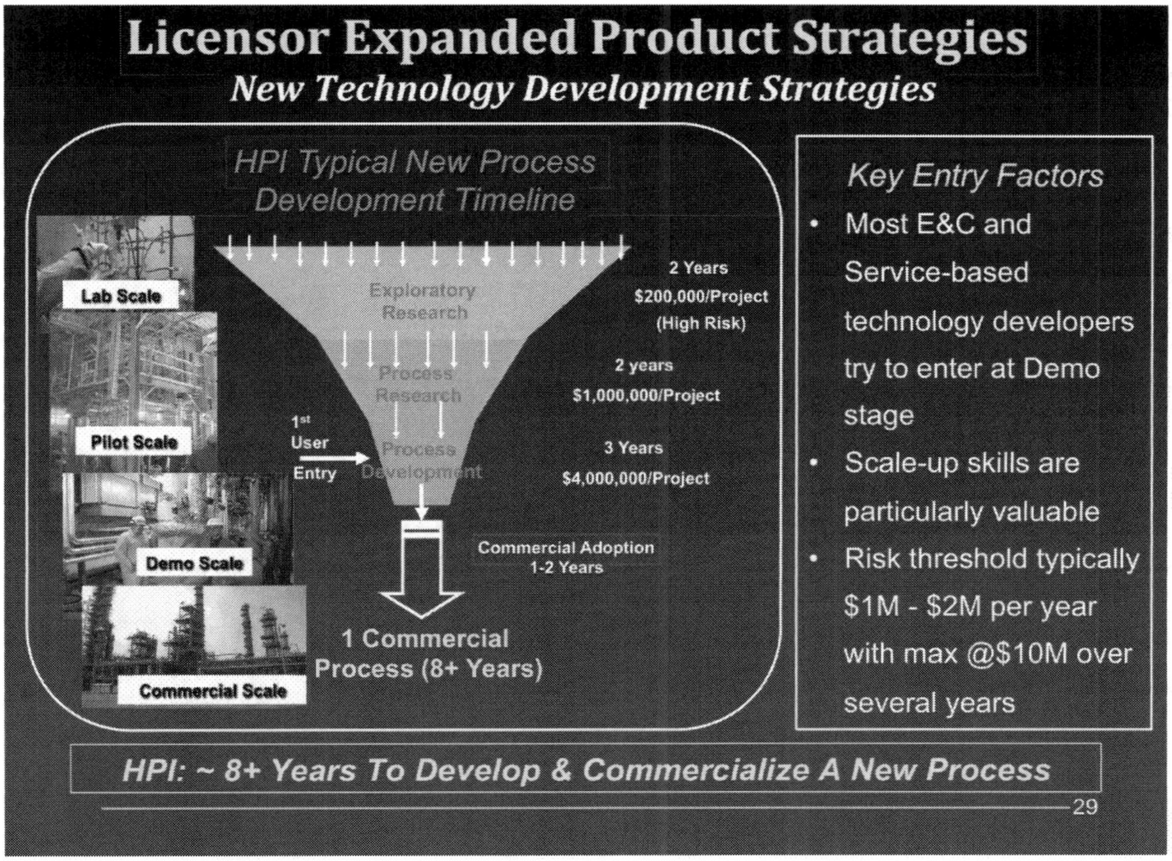
Licensor Expanded Product Strategies
New Technology Development Strategies
HPI Typical New Process Development Timeline
Lab Scale
Pilot Scale
Demo Scale
Commercial Scale
Exploratory Research
Process Research
Process Development
1st User Entry
2 Years
$200,000/Project
(High Risk)
2 years
$1,000,000/Project
3 Years
$4,000,000/Project
Commercial Adoption
1-2 Years
1 Commercial Process (8+ Years)
Key Entry Factors
• Most E&C and Service-based technology developers try to enter at Demo stage
• Scale-up skills are particularly valuable
• Risk threshold typically $1M - $2M per year with max @$10M over several years
HPI: ~ 8+ Years To Develop & Commercialize A New Process
29

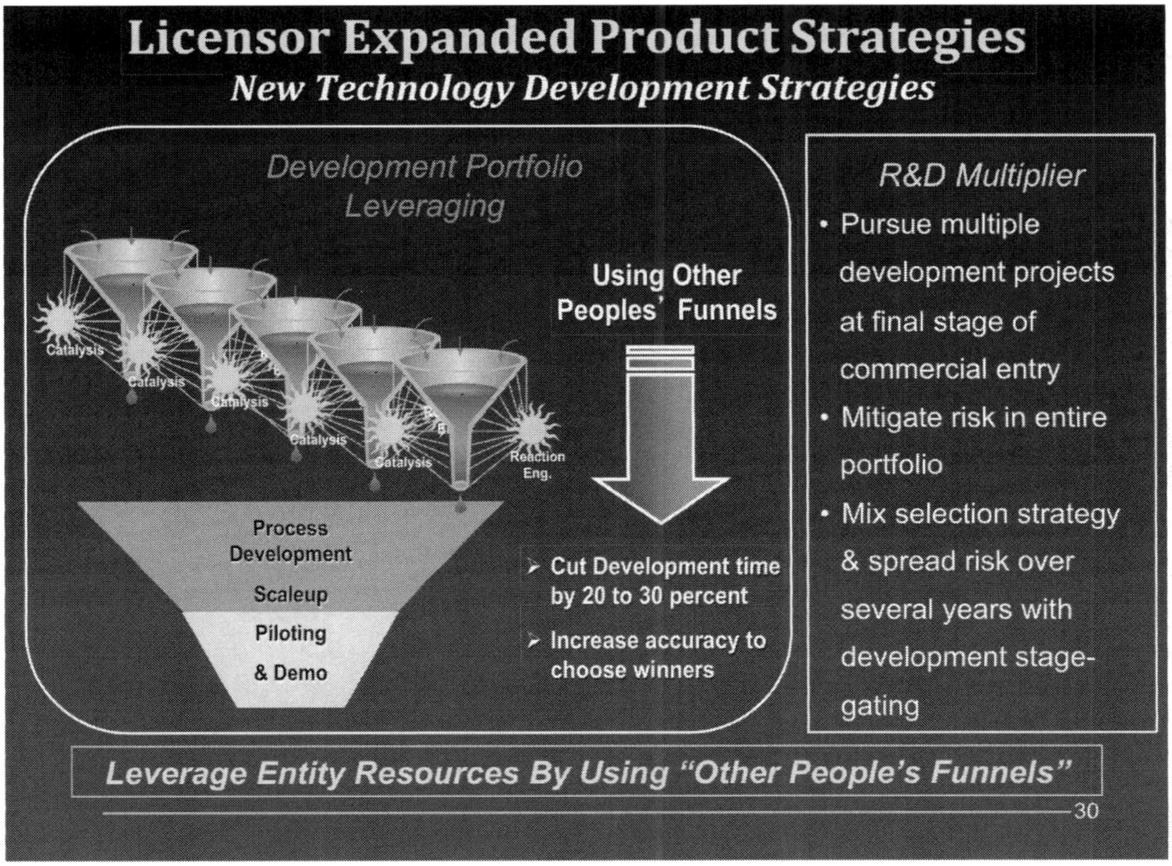
Licensor Expanded Product Strategies
New Technology Development Strategies
Development Portfolio Leveraging
Catalysis
Reaction Eng.
Using Other Peoples' Funnels
Process Development
Scaleup
Piloting
& Demo
➢ Cut Development time by 20 to 30 percent
➢ Increase accuracy to choose winners
R&D Multiplier
• Pursue multiple development projects at final stage of commercial entry
• Mitigate risk in entire portfolio
• Mix selection strategy & spread risk over several years with development stage-gating
Leverage Entity Resources By Using "Other People's Funnels"
30

Final Thoughts

31

Technology Based Businesses Overview

Uniqueness	*Challenges*
• Starts with a differentiated asset base • Global by its nature • Core asset legally protected worldwide • Licensing mode provides high margins/ low risk to begin • Enabling technology provides high asset equity value in JV manufacturing entry mode • Provides business model flexibility	• Significant part of asset goes out the door every night • Often stolen by others despite legal protection • Takes significant capital to exploit commercially • Long time-line from start to commercialization • Asset difficult to differentiate

32

"Buy land. They' ve stopped making it!"
Mark Twain
Questions?
33

Understanding Membrane Separation of Gases in the Transition Flow Region Through Multi-physics Modeling

James Sturnfield, Kishori Deshpande, Alexia Finotello, John Pendergast
Dow Chemical Company, Freeport TX

October 3&4, 2013
5th AIChE Southwest Process Technology Conference
Galveston, TX

Outline

- Why transition flows through membranes?
- Literature work
- Tools and techniques for multi-physics modeling
- Case Studies
  - Flow profile – pressure dependence (first order and second order slip approximation)
  - Validating the model through standard flow (helium and nitrogen)
  - Variation of Knudsen flow along the pore
  - Example of binary flow
- Summary

Background and Motivation

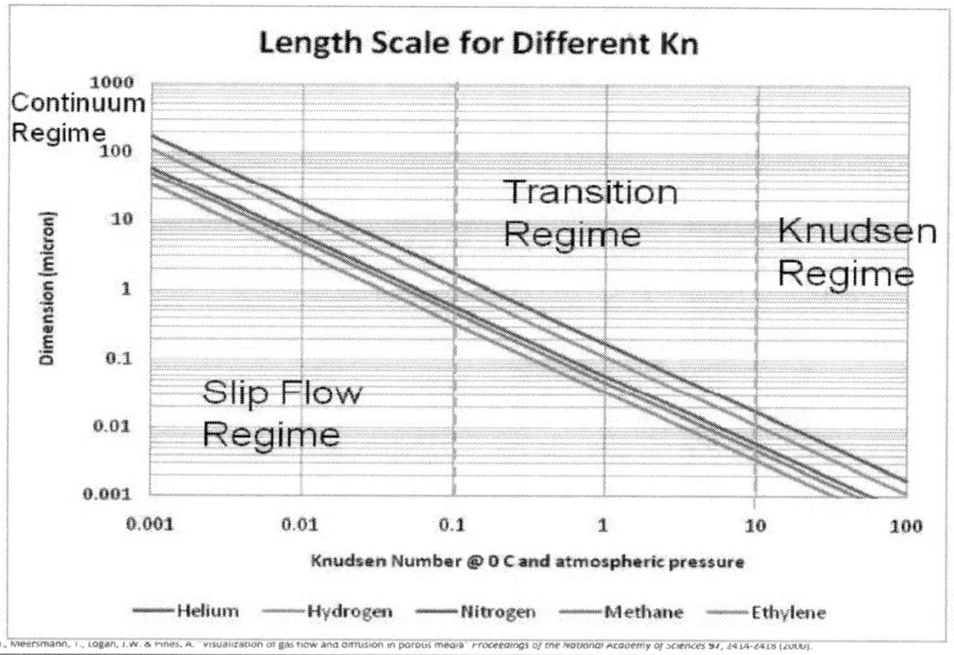

0.

Kaiser L.G., Meersmann, T., Logan, J.W. & Pines, A. "Visualization of gas flow and diffusion in porous media" Proceedings of the National Academy of Sciences 97, 2414-2418 (2000).

Momentum Transfer

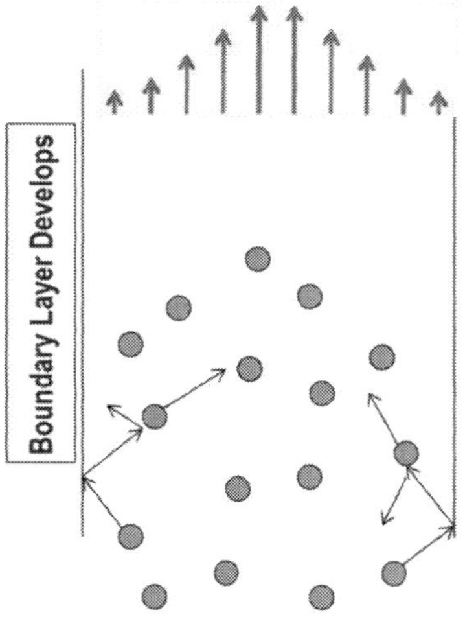

In **continuum regime**, the momentum reflected off the wall is transferred towards the center of the fluid by molecular collisions and reducing the forward momentum near the wall resulting in a parabolic forward momentum distribution

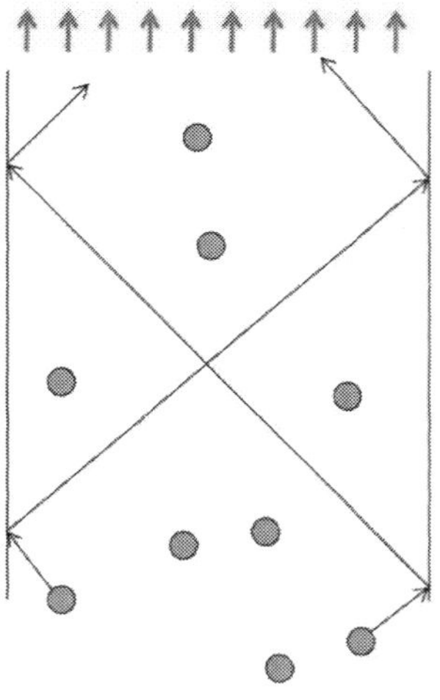

In **Knudsen regime**, the momentum reflected off the wall is transferred to the other wall keeping the forward momentum uniform across the channel diameter

For perfectly reflecting wall, velocity at wall is the amount of velocity change which can be moved to the wall in the mean free path distance $\lambda\ \partial v/\partial n$

Gas Flow through Pore

- Hagen-Poiseuille Equation
  - Q_{HP} volumetric flow rate
  - μ is viscosity
- Slip flow
  - L_s slip length
- Knudsen Regime
  - V_m Molar Volume
  - σ areal pore density
  - A total membrane area
  - Assumptions –
    - » Hard sphere, ideal gas
    - » Viscosity, molecular diameter will modify the relationship

$$Q_{HP} = \frac{\pi (d/2)^4}{8\mu} \frac{\Delta p}{L}$$

$$Q_{SLIP} = \frac{\pi \left[(d/2)^4 + 4(d/2)^3 L_s \right]}{8\mu} \frac{\Delta p}{L}$$

$$Q_{gas} = \frac{2}{3} \sqrt{\frac{8\pi}{MRT}} (d/2)^3 V_m \frac{\Delta p}{L} \sigma A$$

Do not photocopy Page 5

Multi-physics Problems

- Tools like COMSOL, CFD-ACE, ANSYS, Abaqus
- Convection, pressure driven flow, facilitated transport (reaction)
- Numerical techniques based on finite element method for spatial discretization
- Platform combines different physical processes and allows predictions in presence of interdependent and competing effects
  - Heat transfer, diffusion, electrochemistry, fluid flow, and membrane separation and allows coupling of these effects to gain better understanding of the system

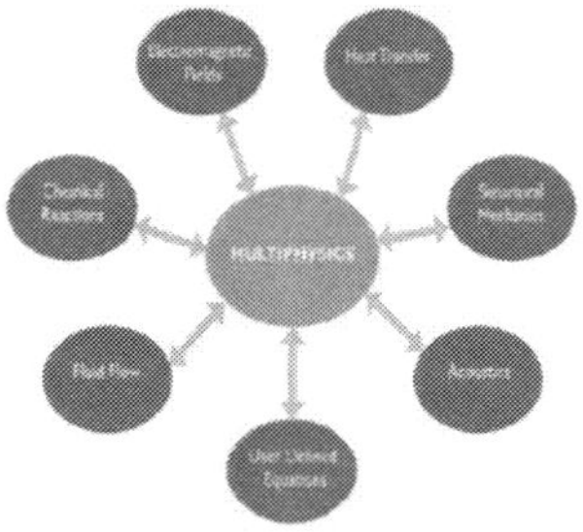

Do not photocopy Page 6

Case 1: Impact of Pressure on Pure Gases

DOW

As the average pressure goes down, the Knudsen number goes up and molecules are able to recoil further from the walls. This results in less drag on the wall, so the velocity profile flattens.

Helium has higher Knudsen number than Nitrogen so for the same conditions it has more slip on the wall

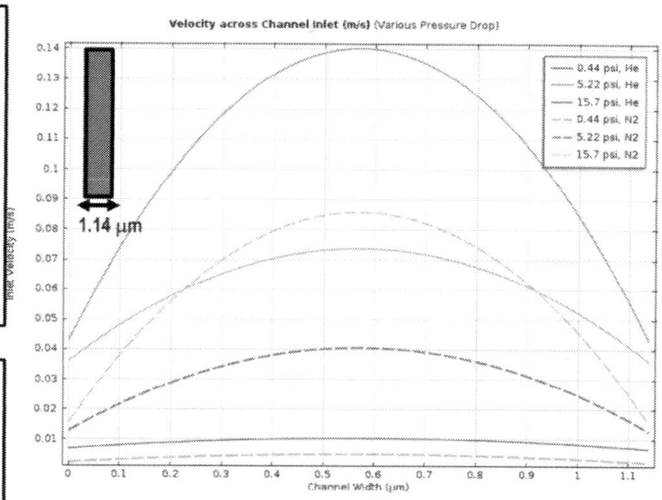

Channel width: 1.14 µm
Outlet pressure: 4.35 psia

Local Knudsen Number

DOW

- Knudsen Number is Pressure Dependent
  - Knudsen Number at Average Pressure is Often Used
  - Amount of Slip at the Wall depends on the local Pressure

Local Knudsen number varies inversely with the Local Pressure

The greater the pressure drop, the greater the variation of the Slip

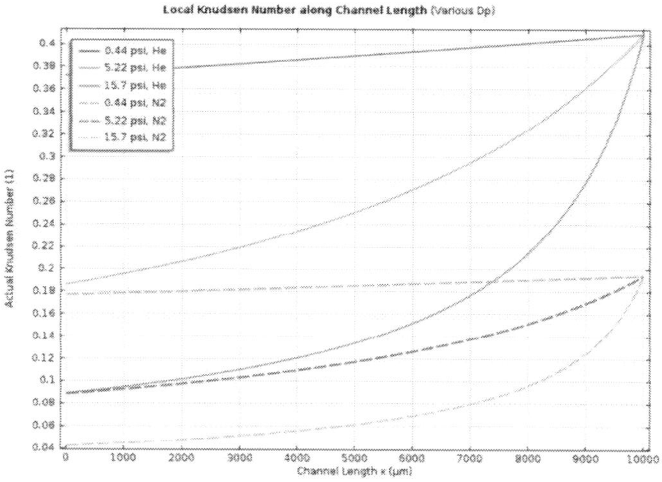

Slip Order and Slip Coefficient

➢ 2nd Order Slip factor increases for larger Knudsen #

$$S \approx 1 + 6\,A_1 K_n + 12\,A_2\,K_n^{\,2}\ ^{*}$$

Nitrogen A_1= 1.3, A_2 =0.26

Helium A_1= 1.2, A_2 =0.23

A_1 determined by molecular spectroscopy techniques for glass and silicon
A_2 determined by fitting data

The Slip Coefficient is significantly increased when the Local Kn (instead of the Kn at average pressure) is used to solve for Flow

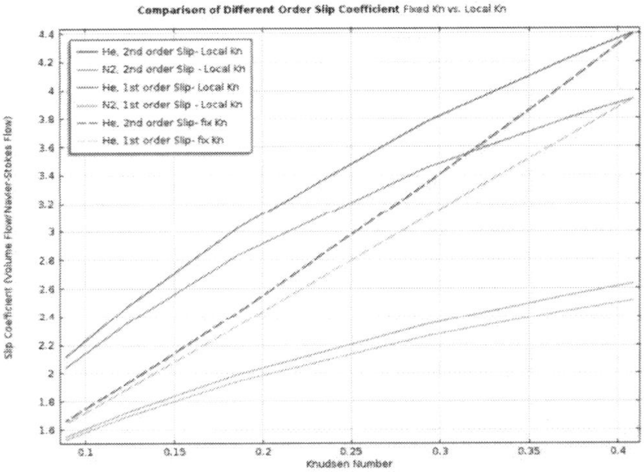

Channel width: 1.14 μm
Outlet pressure: 4.35 psia

*Maurer, J., Tabeling, P., Joseph, P. & Willaime, H. "Second-order slip laws in microchannels for helium and nitrogen" *Physics of Fluids* **15**, 2613-2621 (2003).

Do not photocopy

Case Study 2: Flow Through 5.6 nm Pores

For Helium, pore density was adjusted to fit the extreme experimental data. The 2nd order slip model is a little high in the middle of experiment range. The model parameter could be adjusted to improve the fit.

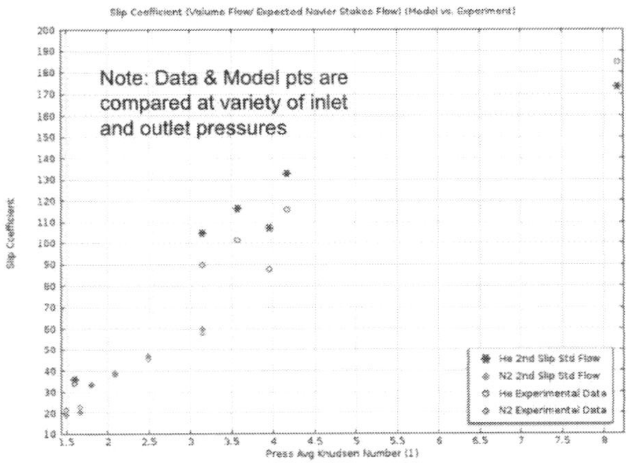

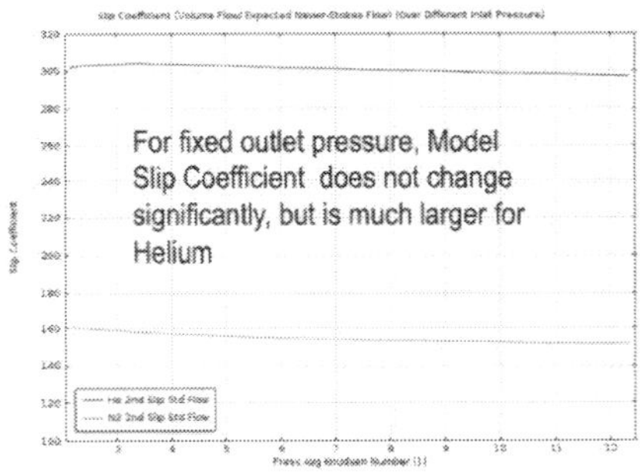

For Nitrogen, the pore density calculated for Helium Data is used. The 2nd order slip model matches this experimental data extremely well.

The 2nd order slip model is using σ and A2 from Micro-channel study.
These parameter were from fits of data in range of 0.1 to 0.4 Knudsen number

Do not photocopy

Case study 3: Binary Molecular Flow

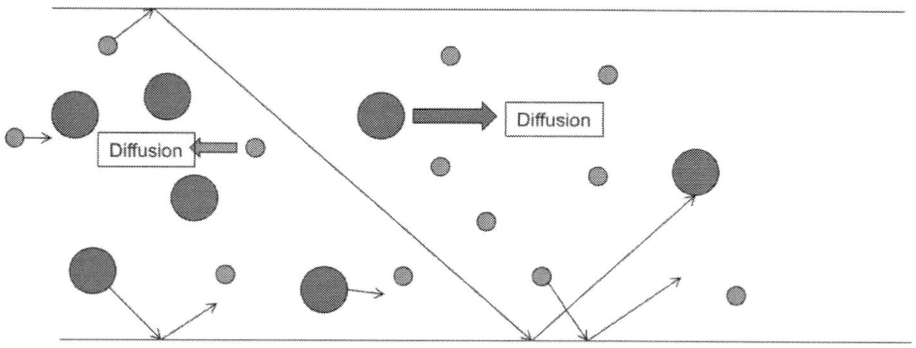

The larger particles are more likely to collide with other particles and transfer their momentum to other particles. The momentum transfer from the wall is more likely to be moved towards the center

Molecular diffusion will move the species toward the lower concentration, dragging some of the larger molecules forward and holding some of the smaller molecules back

Binary Flow through Pore

Dusty Gas Model

Flux of species "a" is given by

Fickian Diffusion — Binary Diffusion

$$N_a = -D_a \nabla c - \frac{x_a D_a}{D_{ab}} \left(\frac{D_a \nabla c_a + D_b \nabla c_b}{x_a \gamma_a + x_b \gamma_b} \right) + x_a N''.$$

Knudsen Diffusion — Molecular Diffusion — Viscous Flux

where c is the molar density of the gas mixture, the subscript a and b indicates the restrict to the particular species, x is the fraction, D is a diffusion coefficient, and γ are ratio of diffusion coefficients.

Fickian Diffusion is partial pressure driven
Knudsen Diffusion adds a total pressure component

Modeling Binary Flow Model

- Each specie separately modeled with their own CFD physics using the same mixture viscosity, but the gas densities are based on the individual molecular weight and the slip constants depend on the specie.
- Flow of the gas mixture is weighted average of the flow of these individual components.
- Mole fraction approximated from relative velocity changes for each specie (a PDE model of the mole fraction can be included for molecular diffusion effects)

With Slip Conditions, Mole Fraction varies across pore

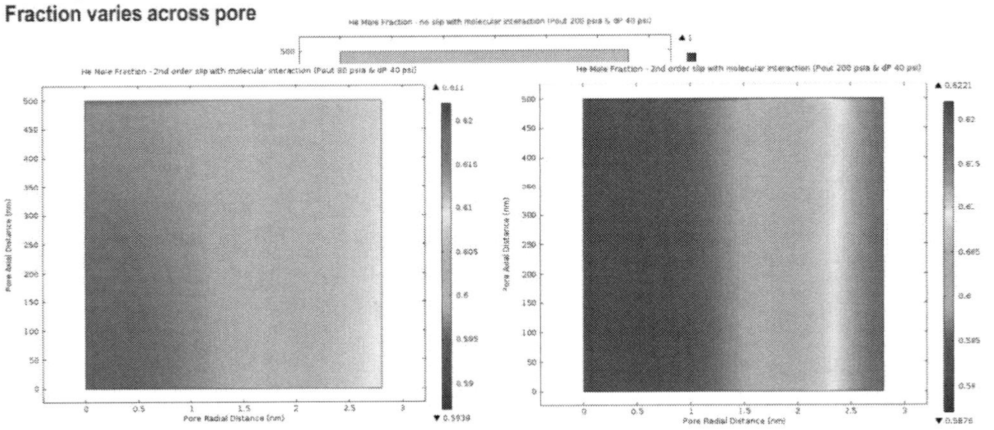

Impact of Binary Mixture on Knudsen Number

Mean Free Path of Gas Component in Mixture

For a molecule in a binary gas mixture, the mean free path is given by

$$\lambda_i = \frac{1}{n_1 \pi \sigma_{11}^2 \sqrt{2} + n_2 \pi \sigma_{12}^2 \sqrt{1 + m_1/m_2}}$$

where n_i is the number density, σ_{ij} is the collision diameter, and m_i is the molecular weight.

In the Binary Mixture, the mean free path used in the wall boundary condition needs to be modified for the specie interactions

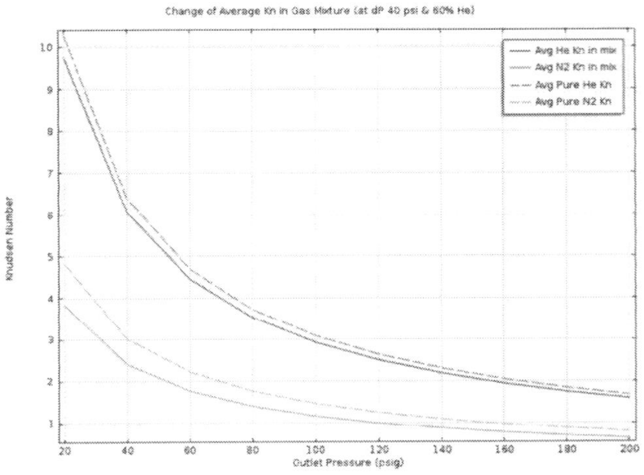

Impact of Molecular Interaction on Velocity

- The molecular interactions reduces the velocity of each specie through the pore
    - Impacted by change in mean free path
- Higher Knudsen have higher speeds
    - Molecular interaction has greater impact for higher Knudsen Number
- Permeance of Gas Mixture is lower than weighted average permeance of species

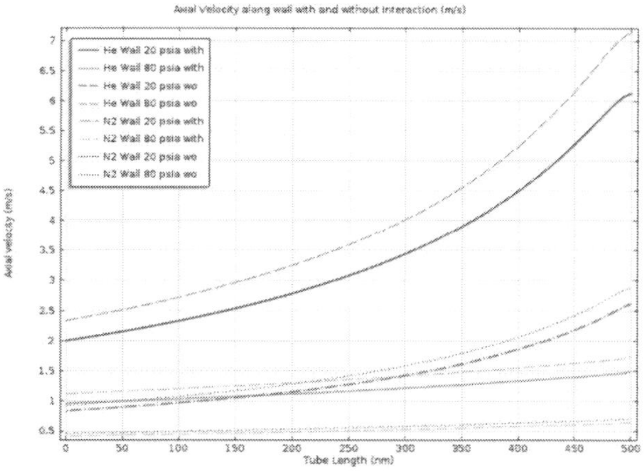

5.6 nm pore with 40 psi pressure drop and 60 mole% Helium

Summary and Conclusions

- Transition flow through porous media important for many applications including catalysis and water purification
- Developed multi-physics model to understand influence of slip coefficient on single component flow
    - Investigated the influence of pressure on local Kn
    - Investigated the relation between slip order and slip coefficient for helium and nitrogen flow
    - Model predictions compare well with experimental data for nitrogen
- Binary flow modeling
    - Coupled the flow models of the individual specie to get binary flow
    - Composition varies in pore with slip
    - Molecular interaction reduces velocity compared to individual species

Back Up Slides

References

- Holt, J.K., Park, H.G., Wang, Y., Stadermann, M., Artyukhin, A.B., Grigoropoulos, C.P., Noy, A. & Bakajin, O. "Fast Mass Transport Through Sub-2-Nanometer Carbon Nanotubes" *Science* **312**, 1034-1037 (2006).
- Maurer, J., Tabeling, P., Joseph, P. & Willaime, H. "Second-order slip laws in microchannels for helium and nitrogen" *Physics of Fluids* **15**, 2613-2621 (2003).
- Dongari, N. & Agrawal, A. "Modeling of Navier–Stokes equations for high Knudsen number gas flows" *International Journal of Heat and Mass Transfer* **55**, 4352-4358 (2012)
- Agarwal, R.K., Yun, K.-Y. & Balakrishnan, R. "Beyond Navier--Stokes: Burnett equations for flows in the continuum--transition regime" *Physics of Fluids* **13**, 3061-3085 (2001).

Case Study: Single Component Diffusion

➢ Approach

- Laminar Flow Physics in 2 dimensions assuming an infinitely wide channel
- Isothermal conditions, locally fully developed steady flow, and ideal gas behavior
- Maxwell's slip velocity equation

Parameters

Parameters

Name	Expression	Value	Description
hchannel	1.14[um]	1.14E-6 m	Channel Height
Lchannel	10000[um]	0.01 m	Channel Length
wchannel	200[um]	2.0E-4 m	Channel Width
Pin	.33[bar]	33000.0 Pa	Inlet Pressure
Pout	.3[bar]	30000.0 Pa	Outlet Pressure
Temp	293.15[K]	293.2 K	Temperature
Mn	4.0029[g/mol]	0.004003 kg/mol	Molecular Weight
Mdiam	210[pm]	2.1E-10 m	Molecular Diameter
mfpath	k_B_const*Temp/(sqrt(2)*pi*(Mdiam^2)*(Pin+Pout)/2)	6.558E-7 m	Mean Free Path
KnudsenN	mfpath/hchannel	0.5752	Knudsen Number
sigma	.91	0.91	Wall reflect fraction
A2	.23	0.23	2nd order slip coeff
PsqDiff	(Pin^2-Pout^2)/2	9.45E7 $kg^2/(m^2s^4)$	Average Difference of Press Square

Single Component Diffusion: 1st order Slip

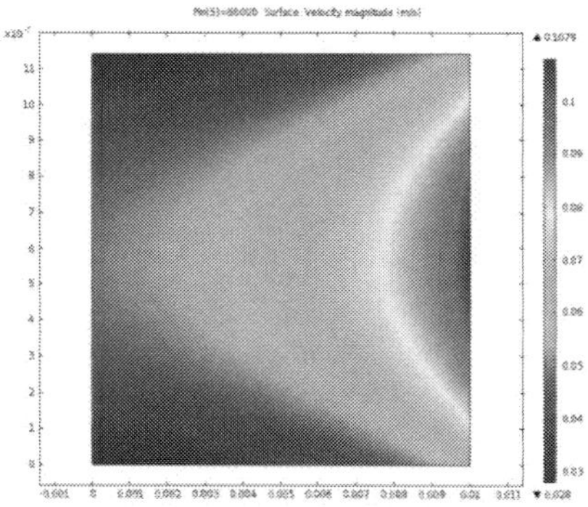

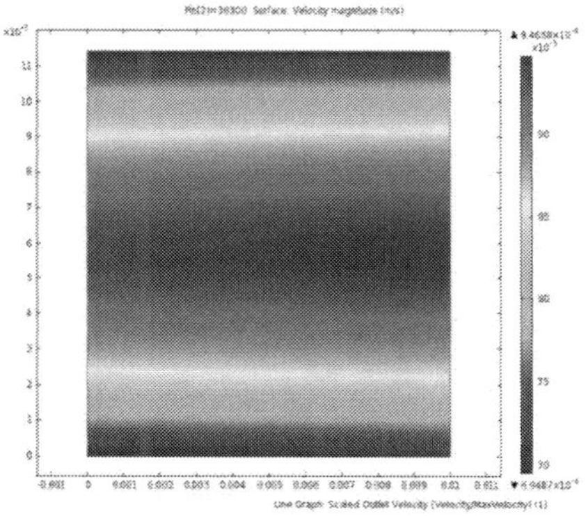

Figure 17: Velocity Profile with Slip at Pin= .66 bar (L) and =.3(

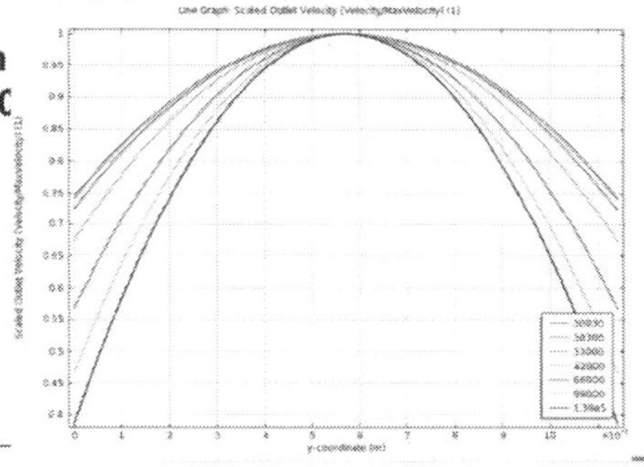

Single Component Diffusion: 2nd order Slip

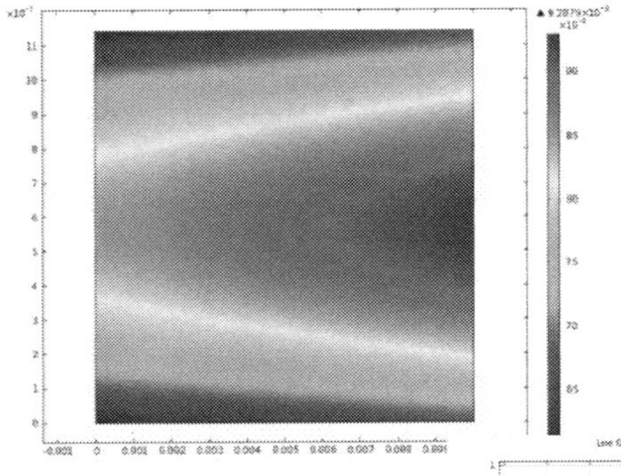

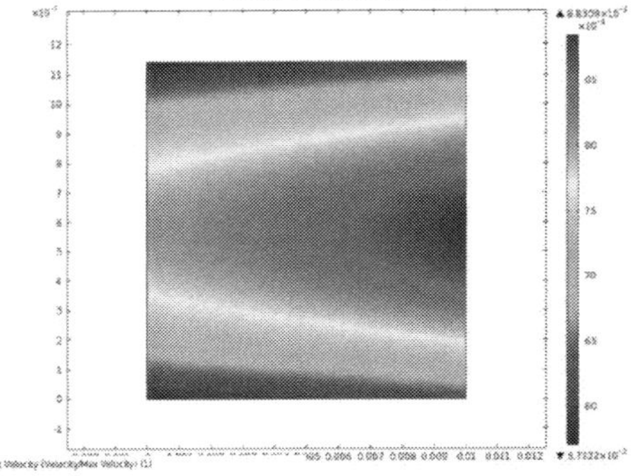

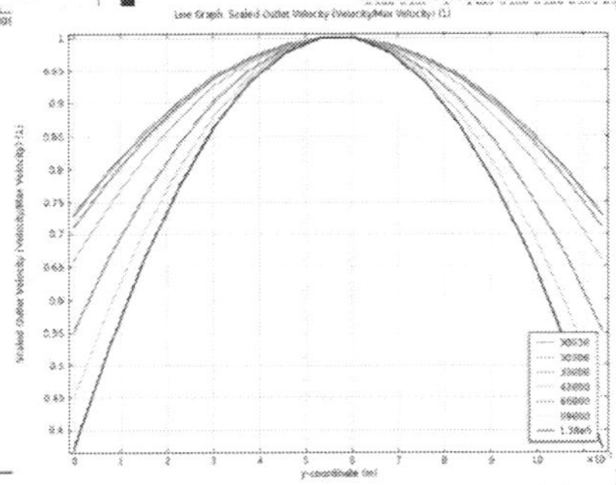

Do not photocopy

Knudsen Number Variation down the Pore

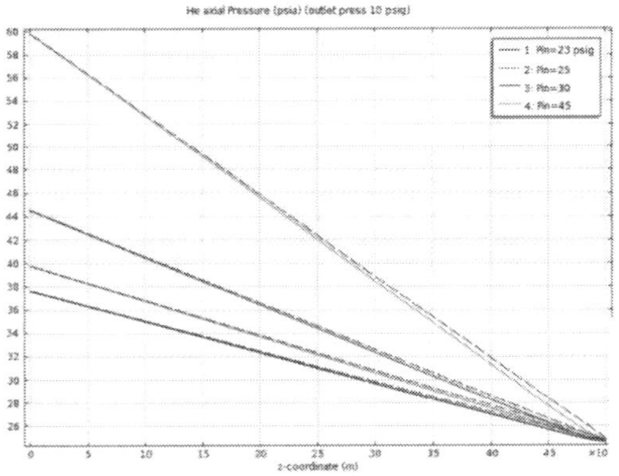

As pressure drops along the length of pore, the Knudsen number changes. For high pressure drop, the flow can range from the slip flow regime to the Knudsen regime

With gas mixture, one component could be in the Knudsen regime while another is still in the transition regime

There are a number of competing models in the literature, so experimentation is critical

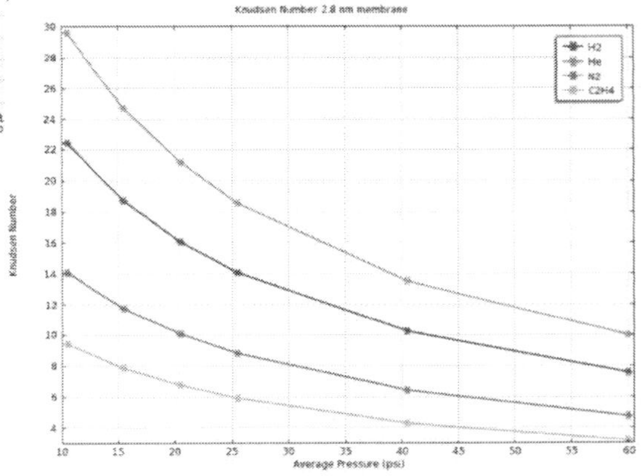

Do not photocopy

Realities of Utilizing a Reference Cracker Design for Grass Roots Projects

Brian Bahr, Technip

Technip is a world leader in designing steam crackers, from concept stage through construction and commissioning, for both grassroots plants (including mega-crackers) and plant expansions. Since its 2012 acquisition of Stone & Webster process technologies, Technip is able to propose either former Stone & Webster ethylene technology and/or Technip ethylene technology. With respectively 70 years of expertise for Stone & Webster and 40 years for Technip and a total of 150 grassroots plants, Technip is the largest ethylene licensor and contractor in the world.

Ethylene is one of the basic building blocks of the petrochemical industry. It is primarily used as the feedstock for polymerization (polyethylene), oxidation (ethylene oxide), halogenation (ethylene dichloride), alkylation (ethylbenzene, a precursor to styrene) and other compounds. The majority of these intermediate chemicals are further processed to create many consumer products utilized on a daily basis.

A key step in the production of ethylene is the cracker process. The name is derived from the reactor system which constitutes a large furnace that heats up hydrocarbon molecules until its bonds begin to crack, creating double-bonded olefins molecules. The remainder and larger portion of the plant separates the various components into ultra-pure products. The feedstock can range from ethane and propane to a full range of naphtha derived from crude oil or refinery processes.

Within the U.S., fracking has led to a shale gas boom which has increased natural gas production. One of the by-products of natural gas production is ethane, a new supply of cheap feedstock for the ethylene industry.

Technip, as a key supplier of the ethylene cracker technology, is participating in this renaissance of the petrochemical business. As clients express interest in a new cracker, the conversation usually begins with a discussion about existing plants with a similar feed type.

As the conversation progresses, Technip often turns to its "Reference Cracker Design." This design captures the company's cumulative knowledge about a multitude of plants into a single document that can serve as the starting point for a new project, bringing advantages for the client and contractor. For Technip, utilizing a Reference Cracker Design allows us to offer:

- A defined capacity, 1500 metric tonnes per annum ethylene plant
- Completed and optimized heat and material balances
- Preliminary process flow diagrams with major control systems
- Preliminary equipment sizing leading to better cost estimates
- Lower front-end engineering cost
- Shorter project implementation schedule

The benefits of using a reference design is not limited to grassroots projects, but extends even to smaller projects, as starting with a base plan or a reference piece of equipment expedites the overall design process.

Utilizing the term reference design has created a false expectation from some clients that the cracker will be a duplicate of an existing facility operating somewhere in the world. In reality, there is no such thing as a duplicate plant as Technip's reference design reflects our cumulative knowledge about a multitude of plants. It is only by continually applying new technology ideas and lessons learned from many projects that the reference design provides the most efficient base from which to optimize the production process.

Three factors – environmental and safety requirements, location of the new plant, and client preferences – can have a significant impact to the process design for a new project.

The first hurdle, environmental and permitting requirements, tends to be driven by the location of the plant site, and safety tends to be driven by the client's experience and risk tolerance. The cracker is designed to meet the regulatory requirement for its proposed location. If a client wants to go beyond the site-specific regulations and apply additional environmental and safety measures, then changes to the process design are required. Examples will be presented to highlight these considerations.

The second hurdle, location of the new plant, becomes the next biggest area of change to the design. Location specific issues include climate and weather, available plot space for the existing cracker, and coordinating location of existing or new facilities including utilities and utility conditions. Examples will be presented to highlight how these concerns can impact the process design.

The third hurdle, client preferences, plays a major role in changing a reference design. Each client approaches a project with a difference set of expectations, backgrounds and requirements. Key items that influence the process design are feedstock flexibility, product specifications and conditions, by-product recovery, lessons learned from other crackers within the client's company, and specifications to design the equipment and integration with existing and new plants. Again, examples will be presented to highlight key changes and impacts to the reference design.

In conclusion, utilizing a reference design brings potential added value to any project. However, a reference design is not a duplicate design as there are many key questions and issues that need to be considered early in a project's life.

This paper and presentation will present key items that should be considered as the design basis is finalized and a project proceeds to implementation. With the increased availability of ethane from shale gas, only ethane feedstock will be used as the reference.

The Realities of Utilizing a Reference Cracker Design for Grass Root

Brian Bahr, Ethylene Technology Manager
Technip Stone & Webster Process Technology
October 4, 2013

Why Talk Ethylene?

We Offer a Wider, Enhanced Portfolio of Downstream Technologies and Alliances

Business Domains

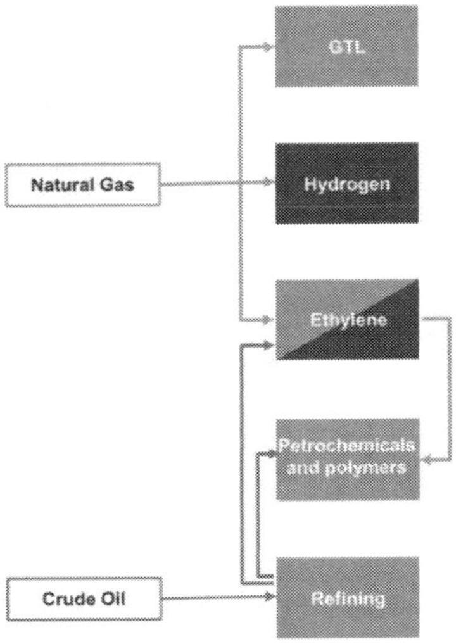

Technologies and Skills

- Exclusive co-developer of Sasol Fischer Tropsch reactor technology
- Steam reformer/ proprietary technology/ complete Hydrogen plant
- Complementary proprietary technologies with different clients & geographic bases
- Phenolics and styrenics
- ABS and Polystyrene
- BenzOUT™
- Resid Fluid Catalytic Cracking
- Deep Catalytic Cracking

■ Technip

■ Stone & Webster process technologies and associated oil and gas engineering capabilities

Technip

Ethylene: *What is it?*

- **Ethylene –** Double bonded molecule that can be easily reacted to make other petrochemical products
- **Ethylene -** key feedstock to many down-stream plants (i.e. - derivative plants)
- **Ethylene Plant –** Heart of most chemical complexes
- **Olefins –** Often used synonymous with ethylene & propylene
- **Ethylene –** A base or commodity chemical
- **Steam Cracker…Olefins Plant….Ethylene Plant** – all used synonymously

Technip

DERIVATIVES COMMERCIALLY PRODUCED FROM ETHYLENE

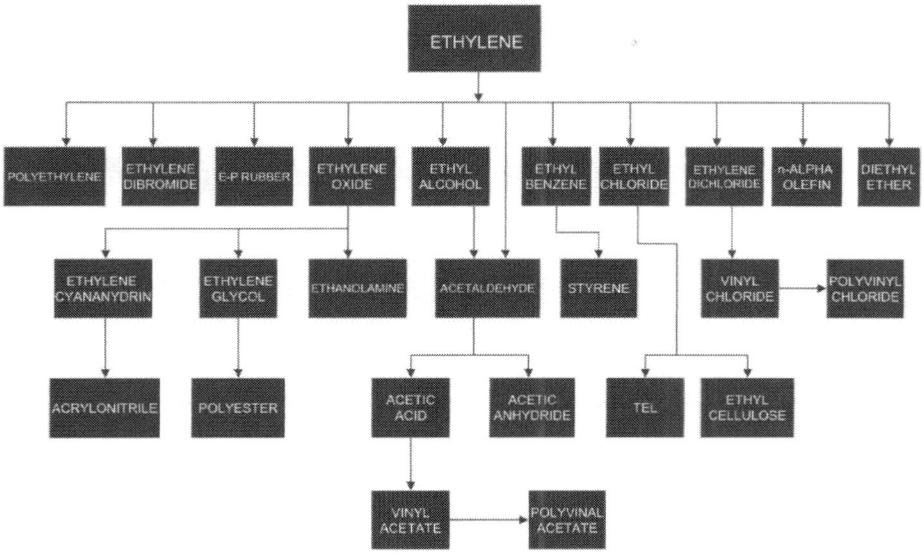

Technip

The Impact of Shale Gas

US ETHYLENE EXPANSIONS BASED ON SHALE GAS

Company	Project	Capacity, tonnes/year	Location	Cost	Start-up
Chevron Phillips Chemical	New cracker	1.5m	Cedar Bayou, Texas	NA	Q1 2017
Dow Chemical	New cracker	World-scale	US Gulf Coast	NA	2016-2017
Shell Chemicals	New cracker	World-scale	US Northeast	NA	2016-2017
Formosa Plastics	New cracker	800,000	Point Comfort, Texas	$1.7bn	2016
Dow Chemical	Restart	390,000	St. Charles, Louisiana	NA	end 2012
Westlake Chemical	Expansion	108,863	Lake Charles, Louisiana	NA	H2 2012
Williams	Expansion	272,158	Geismar, Louisiana	$350m-$400m	Q3 2013
INEOS	Debottleneck	115,000	Chocolate Bayou, Texas	NA	end 2013
Westlake Chemical	Expansion	113,399	Lake Charles, Louisiana	NA	2014
LyondellBasell	Expansion	386,000	Laporte, Texas	NA	2014
Considered expansions					
Sasol	New cracker*	1.0m-1.4m	Lake Charles, Louisiana	$3.5bn-$4.5bn	NA
Indorama Ventures	New cracker**	World-scale	NA	NA	2017
LyondellBasell	Expansion	NA	Channelview, Texas	NA	NA
SABIC	New cracker	World-scale	US	NA	NA
Braskem	New cracker	NA	US	NA	NA
Occidental Chemical	New cracker***	NA	Ingleside, Texas	NA	NA
Aither Chemicals, Renewable Manufacturing Gateway	New cracker	NA	US Northeast	$750m	2016

NOTE: *Feasibility study to be complete by H2 2013 **Feasibility study to be complete by H1 2013 ***Company to "explore various options for future supply of ethylene" SOURCE: ICIS

Technip

What is a "Reference Design"?

Technip

What is a "Reference Design"?

- An optimized plant design based on proven technology
- Plant design documentation that is readily available
- Major equipment requisitions ready for inquiry / purchase
- Design that conforms to the relevant codes and standards
- Plant that is operable, maintainable, reliable and safe
- Environmentally sound

[The design captures cumulative knowledge about a multitude of plants into a single document.]

Technip

Reference Design Flow Scheme

- **For a gas cracker, Technip Stone & Webster can offer three proven flow schemes**
  - **Front End Demethanizer and Back End Hydrogenation**
  - **Two Front End Deethanizer and Front End Hydrogenation options**

9

Technip

Technip's Process Flow Schemes Gas Cracker

Front End Demethanizer And Back End Hydrogenation

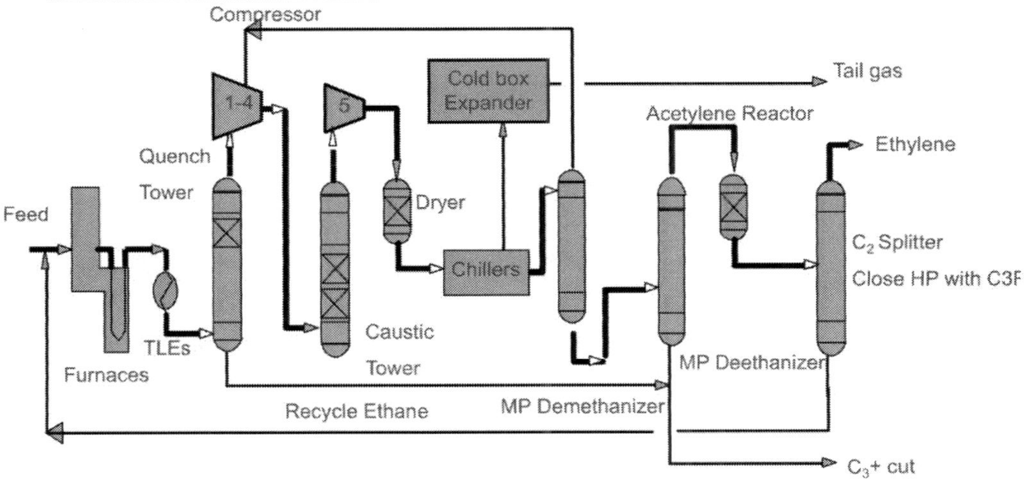

10

Technip

Technip's Process Flow Schemes Gas Cracker

Front End Deethanizer And Front End Hydrogenation

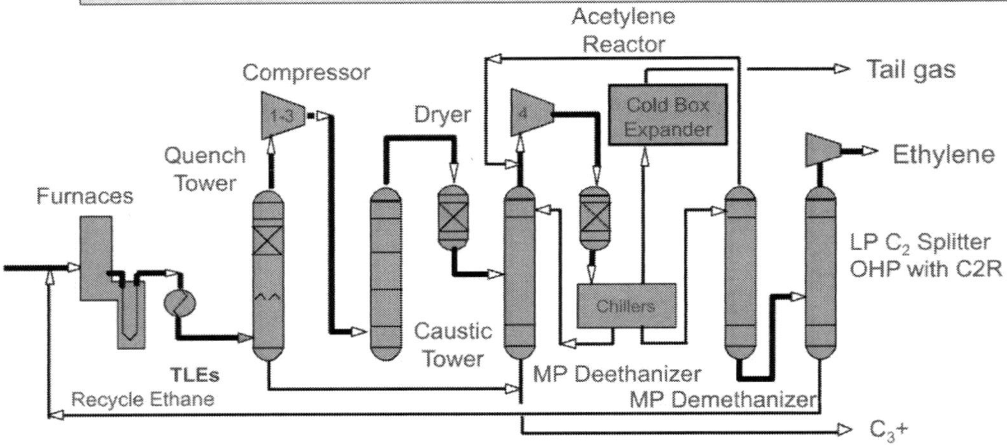

11

Technip

S&W's Process Flow Scheme Gas Cracker

Front End Deethanizer And Front End Hydrogenation

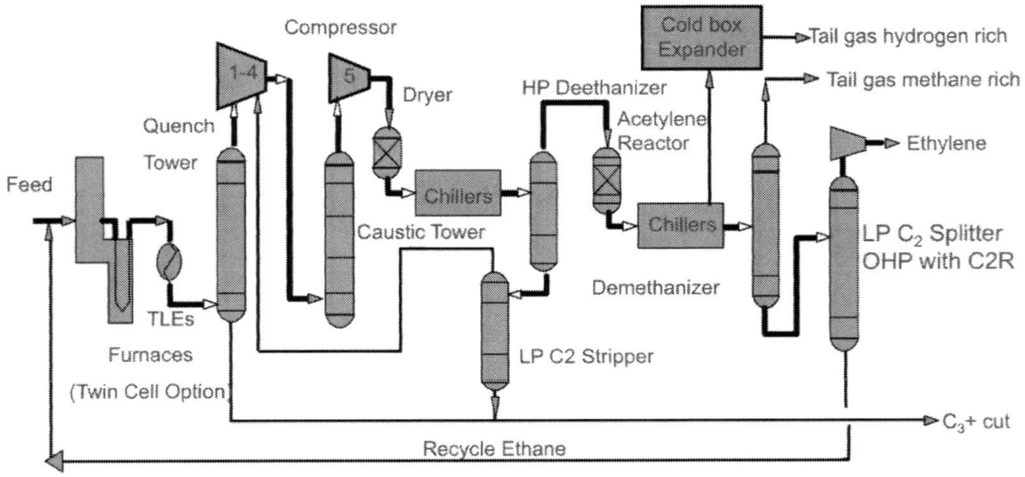

12

Technip

How Does The Design Benefit Clients?

- A proven ethylene plant capacity
- Preliminary process flow diagrams with major control systems
- Preliminary equipment sizing leading to better project cost estimates
- Equipment arrangement and preliminary plot plan
- Lower front-end engineering cost
- Shorter project implementation schedule

Technip

The Dilemma: The "Reference Cracker" is Not a Duplicate Design

Technip

Three Key Areas Where Every Design Gets Customized

- Environmental and safety requirements
- Location of the new plant
- Client preferences

15 Technip

Environmental and Safety Considerations

- Environmental issues are driven by plant location
- Governmental permitting organization continue to add additional regulations
- Plants are design to meet industrial accepted norms and practices
- Each client has own experiences and norm's

[Applying new requirements results in changes to the "Reference Design"

16 PT External Technip

Plant Locations

- Climate and weather concerns
- Available plot space
- Coordination with other units and/or existing ethylene plant
- Availability and access to utility requirements

[Conditions not matching the "Reference Design" will add changes

Technip

Reference Design Cracker – T-shape Layout

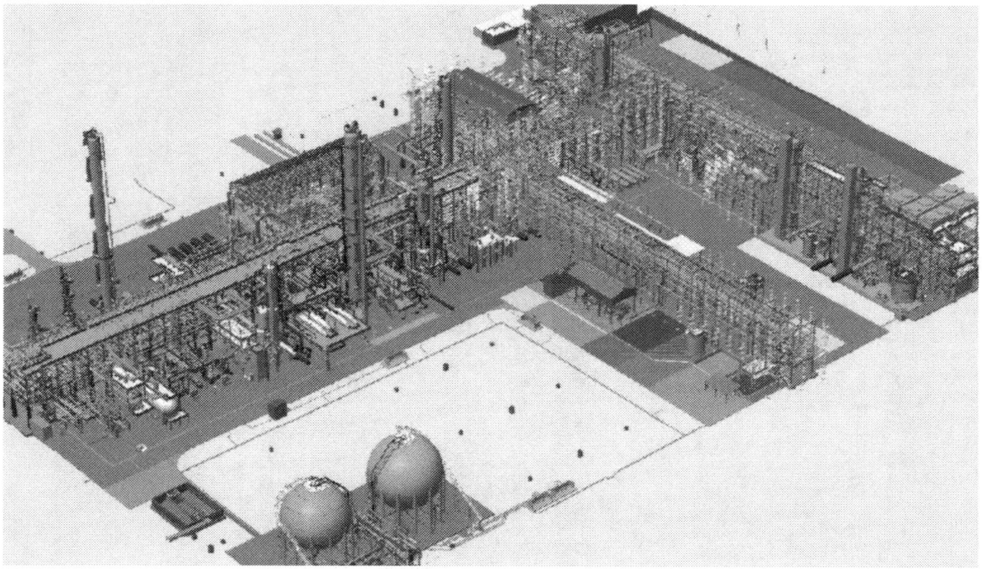

Technip

Reference Design Cracker – T-shape Layout

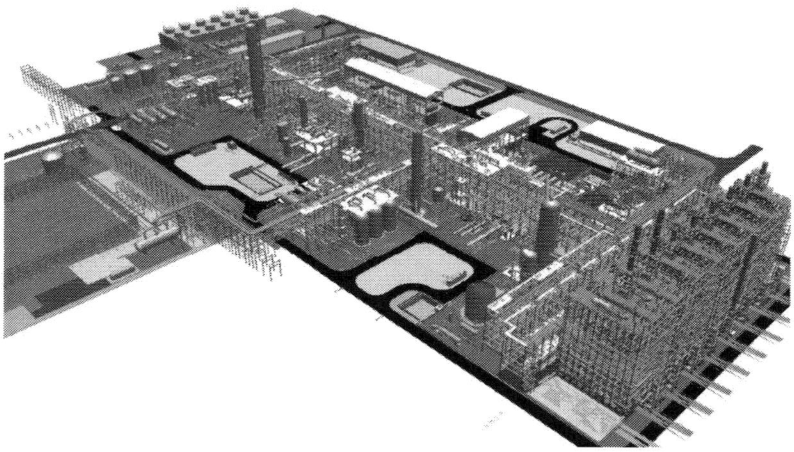

Technip

Client Preferences

- Feedstock flexibility
- Product specifications and conditions
- By-product recovery
- Lesson learned from other crackers within the client's company,
- Design specifications for piping and equipment
- Integration with existing and/or new plants.

[A difference set of expectations, backgrounds and requirements result in potential modifications to the "Reference Design"]

Technip

Reference Design Codes and Standards

- **ASME I & VIII**
  - Pressure vessels, steam drums, heat exchangers, package boilers
- **TEMA Heat exchangers**
- **API**
  - Pumps, compressors, steam turbines, cracking furnaces, gears, oil systems, flare, tanks
- **ANSI**
  - Piping, pumps
- **NEC, IEEE, UL**
  - Motors, electrical equipment & materials

21

Technip

Conclusion

- Using the reference design adds value for the client and Technip.
- It is not a duplicate of an existing plant.
- Some customization will occur.
- Technip has significant recent experience in ethane cracker designs.

22

Technip

The Realities of Utilizing a Reference Cracker Design for Grass Root

Thank you for your attention!

Technip

Effective Interviewing Tips

AIChE
Southwest Process Technology Conference

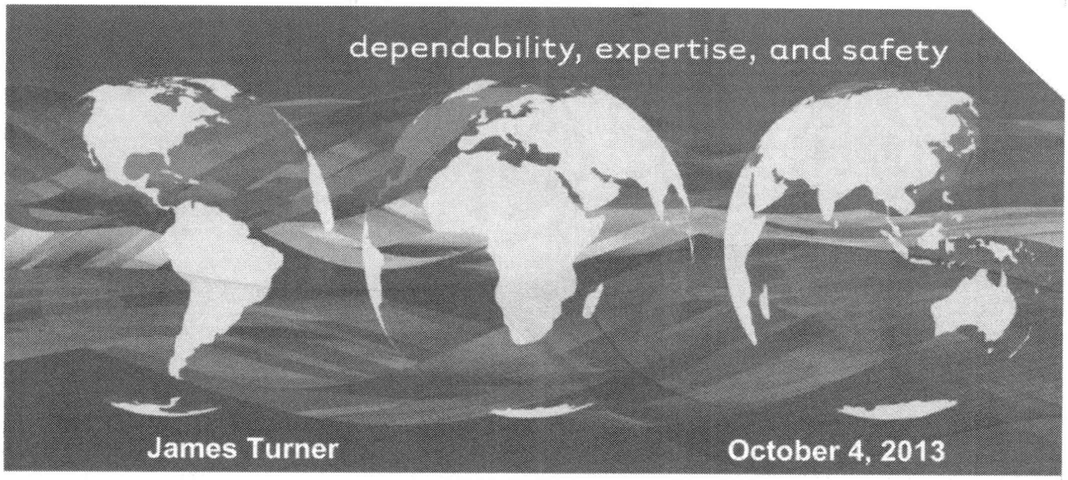

FLUOR®

Agenda

- Introduction
- Preparing in Advance of the Interview
- At the Interview
- Additional Comments
- Q&A

©Johnny Sajem * www.ClipartOf.com/1089280

FLUOR®

James Turner

- Celebrated 25 years at Fluor Houston Office in May
- Process Technology Department Manager since 2011
- Knowledgeable in refining, hydrotreating, clean fuels, project execution
- Active in AIChE, AFPM, Texas A&M Chemical Engineering
- Authored 20+ Papers
- Two patents

FLUOR

Fluor Overview

One of world's largest publicly-traded engineering, procurement, construction, maintenance, and project management companies

- **#110** in the FORTUNE 500 in 2013
- 2012 Results:
  - Revenue: **$27.6 billion**
  - New awards: **$27.1 billion**
  - Backlog: **$38.2 billion**
- Over **1,000** projects annually, serving more than **600** clients in **66** countries
- Workforce of over **42,000** men and women executing projects globally
- Offices in **28** countries on **6** continents
- Celebrated **100** years in **2012**

Fluor Corporate Headquarters
Dallas, Texas

FLUOR

Who is Hiring?

- Operating Companies
  - Upstream
  - Downstream
  - Chemicals
- EPC Companies
- Technology Vendors
- Service Companies

Companies can range from very large to less than 10 employees

FLUOR®

What Employers Look For

Leadership

Experience

Aptitude

Enthusiasm

Flexibility

Communications Skills

FLUOR®

Get Organized and Be Prepared!

- Research the company in advance
  - Understand the company's business
  - Understand what they are looking for
- Prepare a good resume
  - Emphasize previous experience and skills
  - Consider customizing resume specific for each company
  - Where applicable, highlight benefits and value added work experience compared to skills
- Analyze Yourself
  - Strengths and Weaknesses
  - What are you really looking for?
- Prepare good questions

FLUOR

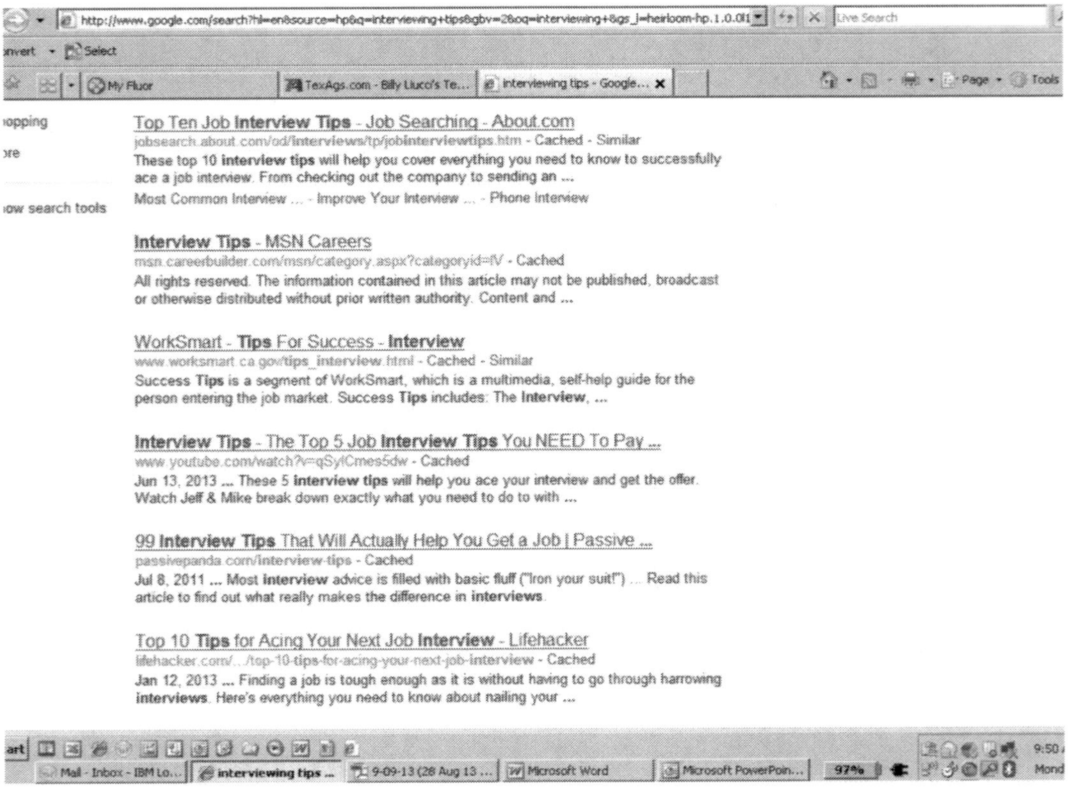

At the Interview

- Dress for Success
  - If not told in advance, suit and tie is best bet
- Show up 10-15 minutes early
- Bring resume, folder, paper & pen
- Silence all cell phones and devices in advance
- Use positive body language
- Act (and be) enthusiastic and open
- Expect a two way conversation
  - You are interviewing the company as well
  - Be prepared to ask questions/take the lead when appropriate

FLUOR®

At the Interview

- Try to direct the conversation about your qualifications, previous accomplishments, and interest in the job
  - Highlight how previous experiences prepared you for this position
  - Many companies are more interested in attitude than skills
- Don't be afraid to highlight areas that company may not ask about
  - Language skills
  - Interest in travel
  - Long term career goals
  - Your innate desire for team success (i.e. to "do what it takes" to accomplish the team goal)

FLUOR®

Differentiate Yourself

I know MS Office, same as everyone else…

I utilized MS Excel to uncover mass balance error, and discovered leak in exchanger

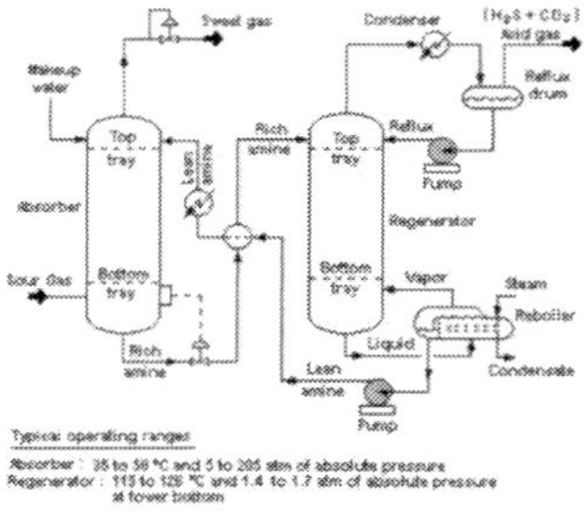

FLUOR

At the Interview

- **Types of questions you may be asked:**
  - **What interests you about this job?**
  - **Why do you want this position?**
  - **What types of activities do you enjoy?**
  - **What do you know about this company?**
  - **What are your Strengths? Weaknesses?**
  - **Where do you see yourself in 5 years?**
  - **What was your biggest accomplishment?**
  - **Tell us about a time when things did not go as planned. What did you do?**
  - **How do you deal with difficult co-workers?**

- **Plus (maybe) technical questions! Examples:**
  - **Fluid Dynamics**
  - **Heat Transfer**
  - **Mass and Energy Balances**

FLUOR

Ask Good Questions!

- Where do you make most of your revenue?
- What are your most profitable products or services?
- How many of your executive leaders are ChE's?
- What is a typical career path for a ChE?
- What opportunities are there for training, travel, continuing education?
- How do you help employees with work-life balance?
- What is your strategy for globalization?
- What is your strategy for growth?
- Do you see my experience and skills as a good fit for this position?
- Is there any reason why you don't think I am right for this opportunity?

FLUOR®

Post Interview

- **Before you leave, ask "what the next step is"**
  - **When will you hear from company?**
- **Thank everyone that you talk to as you leave**
- **Send a follow-up thank you note or email**
- **Analyze your performance**
  - **What should I do differently next interview?**
- **Follow up if you have not heard from the company**

FLUOR®

Build A Network

- Your first job most likely will not be your last
  - Remain on good terms with employers and past-employers
- Interview widely to get to know many companies, see different business cultures, and get practice interviewing
- Keep contact with classmates
- Be Professional – LinkedIn
- Be careful about content – Facebook, Twitter – use social media wisely!
- Stay involved with AIChE
- Local Sections
- Technical Divisions, such as Fuels & Petrochemicals
- Be willing to offer assistance-best way to network

FLUOR

Effective Communication

You are judged on your *perceived* capabilities and qualifications, not your *actual* capabilities and qualifications

FLUOR

- Thank You – to the AIChE Fuels and Petrochemicals Division, for providing the starting point for this presentation!

FLUOR.

Questions?

FLUOR.

5th AIChE
Southwest Process Technology Conference

Implications of the Evolving North American Crude Supply Outlook

MUSE STANCIL

Presented by:

Neil K. Earnest
President

October 3-4, 2013

Overview

- NORTH AMERICAN CRUDE OIL SUPPLY OUTLOOK – CURRENT AND FUTURE
- NORTH AMERICAN CRUDE OIL MARKET FORECAST
  - DISPOSITION OF NORTH AMERICAN CRUDE OIL SUPPLY
  - PROJECTED CRUDE OIL PRICES
- CONCLUSIONS

MUSE STANCIL

-2-

GLOBAL CRUDE OIL SUPPLY LANDSCAPE

AIChE

- The U.S. is the third largest crude oil producer in the world, and North America production rivals that of the very biggest global producers
- The forecast increase in North America production, if a separate country would rank fourth in the world

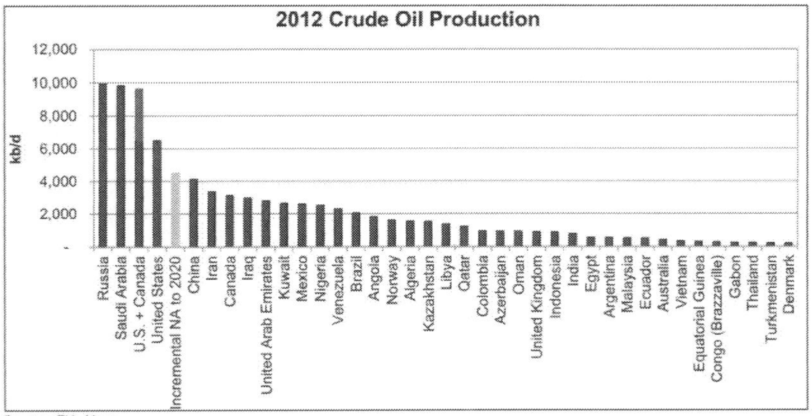

Source: EIA, Muse

MUSE STANCIL

NORTH AMERICAN TOTAL CRUDE OIL SUPPLY

AIChE

- At a high level, the international border can be ignored and total crude oil supply can be partitioned into two key supply regions
  - **Southern Supply Region**: West Texas; Eagle Ford; Gulf of Mexico offshore, plus onshore Gulf Coast
  - **Northern Supply Region**: Western Canada; Bakken; and, Rockies
- The U.S. crude oil supply outlook has been generated via a joint effort by Muse and Crane Energy LLC
  - Crane Energy is a boutique reservoir engineering firm
- The Canadian crude oil outlook is the very recent CAPP 2013 forecast

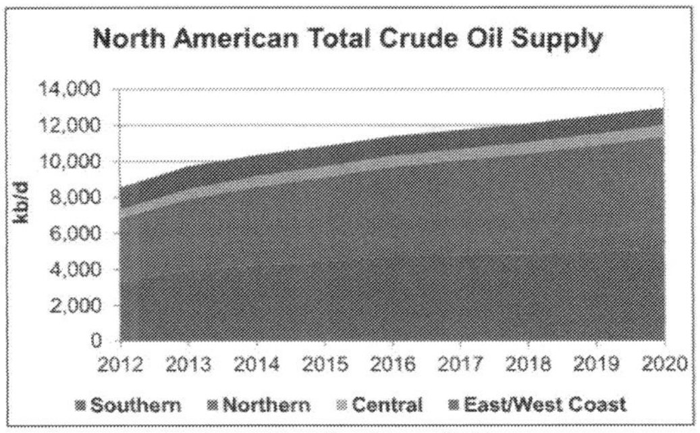

Source: Muse, Crane Energy, CAPP

MUSE STANCIL

North American Light Crude Oil Supply

- **Of the ~2,300 kb/d of incremental light crude oil supply, essentially all of it is in the U.S.**
  - The Bakken (+~600 kb/d), Eagle Ford (+~600 kb/d), and the Permian Basin (+~300 kb/d) are the key North American supply regions for light crude oil
  - The new CAPP 2013 forecast has light synthetic supply (not production) falling over time, although conventional light is now growing by ~100 kb/d to 2020
  - With additional contributions from the Gulf Coast onshore and offshore, Midcontinent, and the Rockies
- **About 60 percent of the light crude supply growth is in the "Southern Supply Region"**
- **The Gulf of Mexico offshore will also add a significant contribution to medium sour crude oil supply**
- **The situation for the heavy crude oil supply is quite different**
  - Little of the incremental North American supply outside of Western Canada is heavy

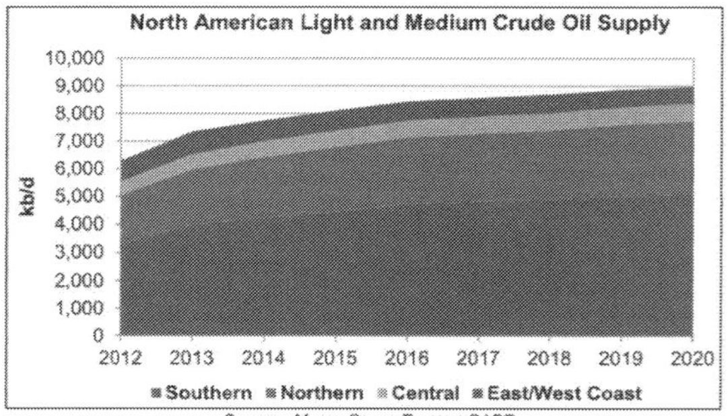

Source: Muse, Crane Energy, CAPP

MUSE STANCIL

-5-

Forecasting the Future – Muse Approach

- **Muse uses a highly detailed, linear programming (LP)-based distribution model of the North America crude oil markets to assess the implications of various forward scenarios**
- **The Muse Crude Market Optimization Model can be used to quantify the impact of differing supply volumes, new pipeline and rail routes, and changing refinery capacities and capabilities over time**
- **The optimization model is seeking to maximize the aggregate netback for all Western Canadian crudes (conventional, synthetic, and heavy) and U.S. crudes**
  - Refining values are generated outside the optimization model for various grades of Canadian synthetic and heavy sour crudes for the individual demand nodes
  - Muse uses the PIMS™ LP modeling system to develop the refining values for the various refineries accessible from Western Canada
  - PIMS™ is the LP modeling system used by most North American refiners
- **The market scenario presented herein is regarded as a reasonable Base Case**
  - But there are many forward possibilities

MUSE STANCIL

-6-

CANADIAN HEAVY CRUDE DISPOSITION

AIChE

- **The U.S. Gulf Coast and Northeast Asia will join the Midwest to become key outlets for Canadian heavy crude**
- **Absent Northern Gateway and the Trans Mountain expansion, most of the Asia-bound volumes would be redirected to the U.S. Gulf Coast**
  - If the west-bound pipelines are not built, the required heavy crude market share on the Gulf Coast will reach very high levels by the middle of the next decade
  - Raising substantive issues of a competitive response by the existing heavy crude suppliers

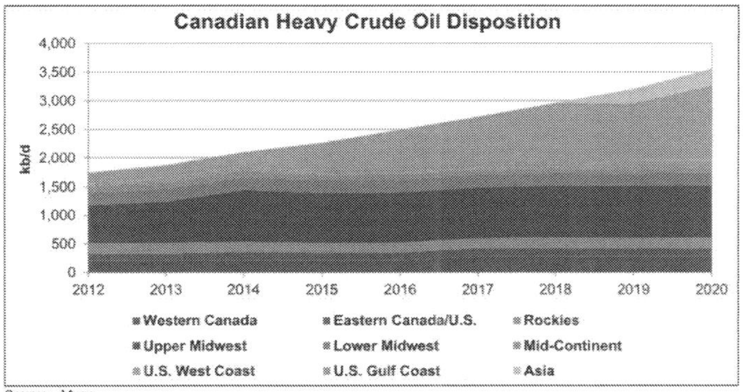

Source: Muse

MUSE STANCIL

-7-

DISPOSITION OF NORTHERN LIGHT CRUDE OIL SUPPLY

AIChE

- **The incremental markets for light crude oil producers in the Northern Supply Region are on the East and West Coast of North America**
  - Shipments to the large Gulf Coast market are projected to be fairly static
- **The Line 9B re-reversal and rail are the key transportation modes for accessing the eastern markets**
  - And likely the TransCanada Energy East project
- **However, Northern light crude producers will encounter competition from Eagle Ford crude in Atlantic Canada and the U.S. East Coast**

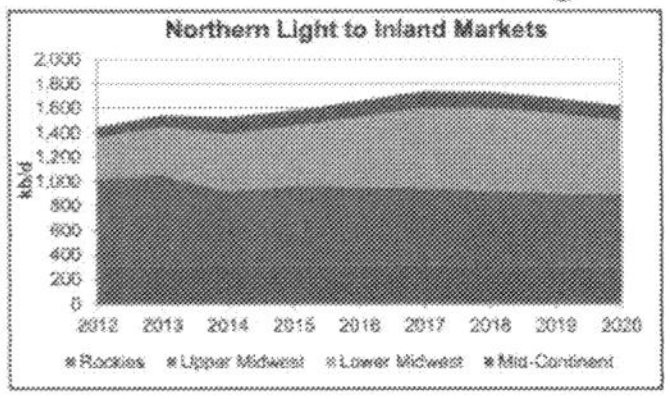

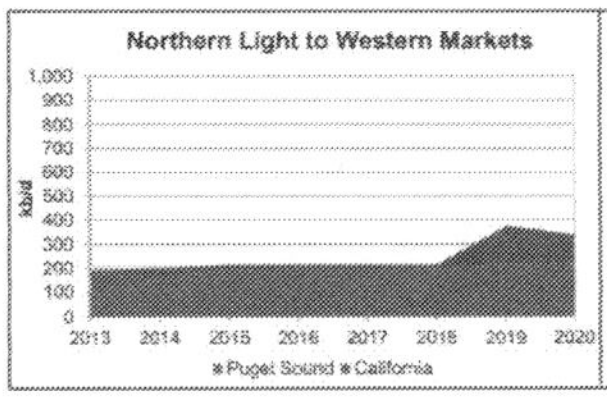

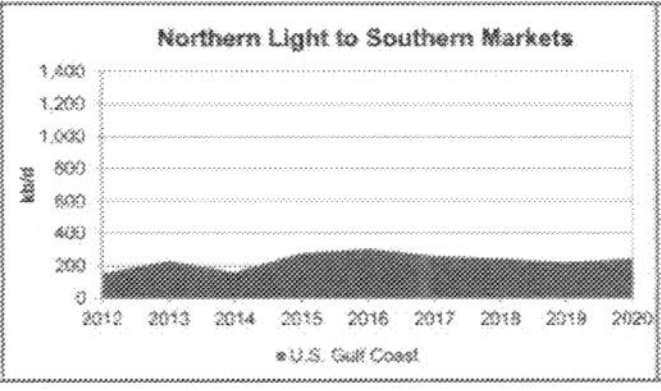

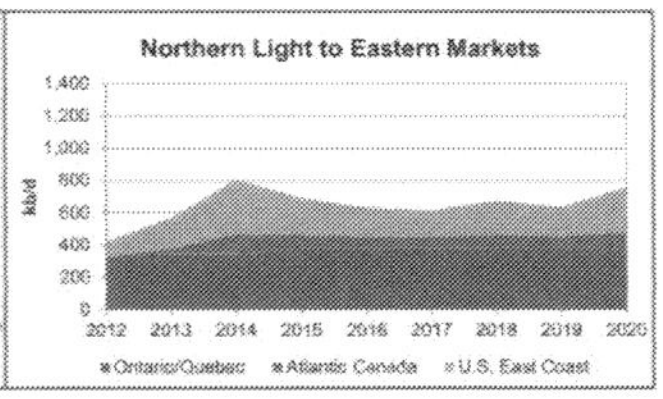

Source: Muse

MUSE STANCIL

-8-

Tension in the Midwest

- **Total light crude oil supply from the Northern Supply Region to the inland markets is forecast to continue to increase**
  - The Midwest is the largest component of the inland market
  - Upper right figure
- **However, this masks a significant shift in the precise source of supply to the inland market**

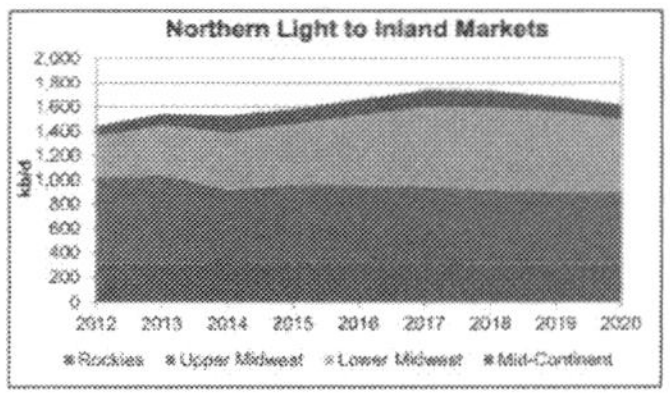

- **The Canadian crude oil producers are projected to lose market share to the inland U.S. producers**
  - Creating a need to access other markets to the east and west

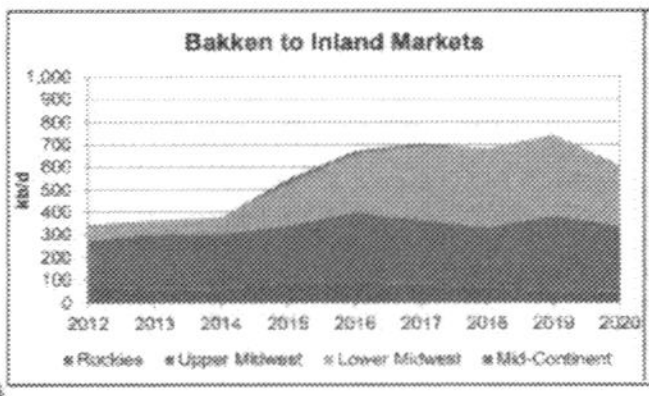

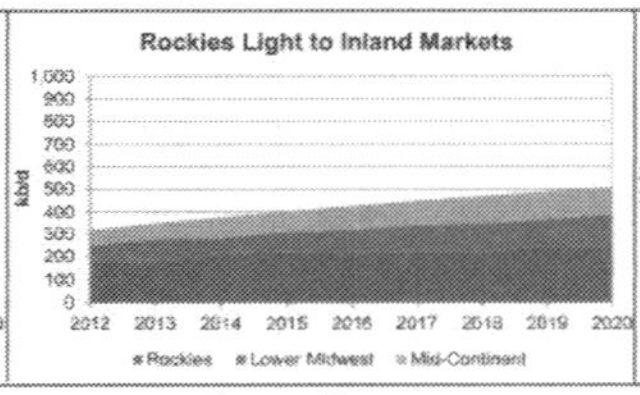

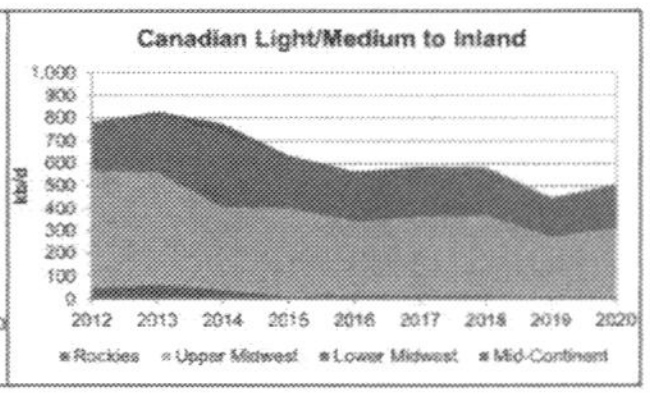

Source: Muse

MUSE STANCIL

Non-Traditional Transportation Modes

- **Rail will remain an important transportation mode for the Williston Basin**
  - Rail shipments of Bakken crude are projected for the entire forecast period
  - If the major outbound pipeline projects are further delayed, Canadian rail shipments will continue
- **A very recent development is the waterborne shipment of Eagle Ford crude to the U.S. East Coast and Atlantic Canada**
  - Much of the Eagle Ford production is pipelined into Corpus Christi, Texas, from which it must be further transshipped by barge or tanker to other markets
  - Eagle Ford shipments to Atlantic Canada can be made on a lower cost, foreign-flagged tanker

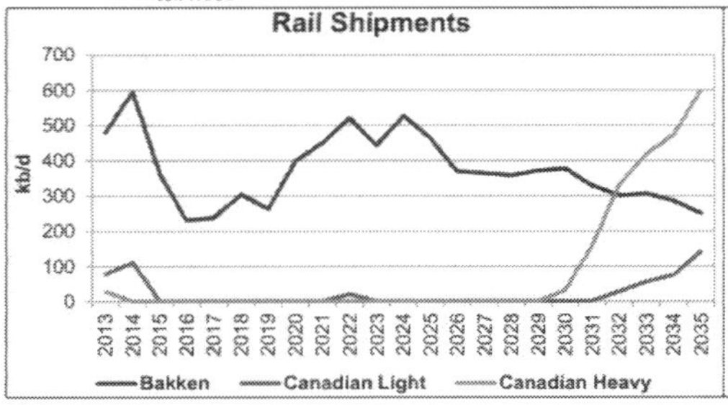

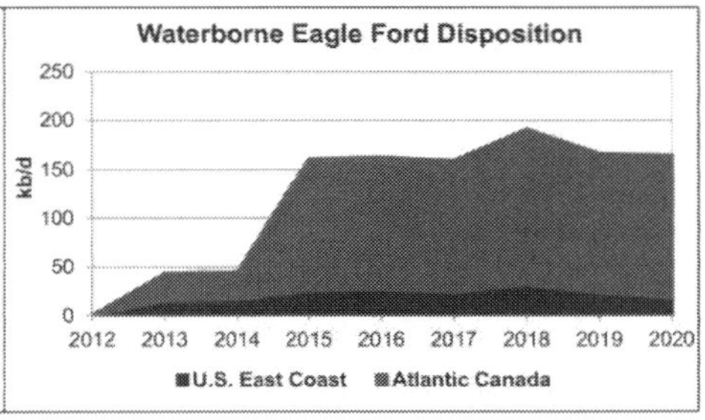

Source: Muse

MUSE STANCIL

CONVENTIONAL CRUDE PRICE FORECAST – LLS AND WTI

AIChE

- **As southbound pipeline capacity is added, WTI is expected to stabilize at about $7.00/barrel under LLS**
  - This is a combination of about a $4.50/barrel Cushing-Gulf Coast pipeline tariff and a $2.50/barrel refining quality differential between LLS and WTI
  - The 2013 WTI price is overstated due to the use of annual average capacity for new pipelines
  - The 2012 values are a backcast using the model
- **The refining quality differential is a function of the crude oil assays used, the gasoline-diesel price differentials, and RIN costs**

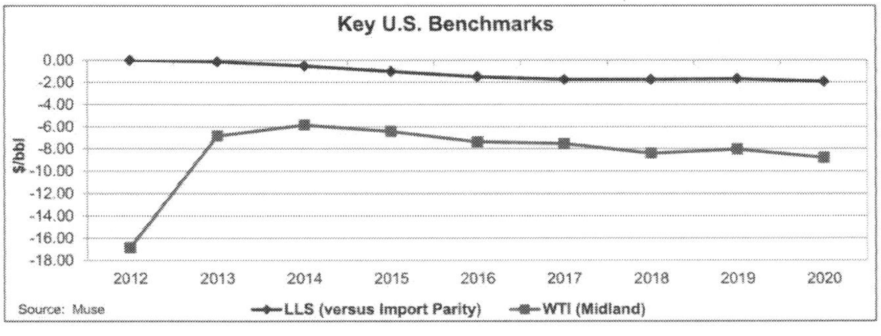

MUSE STANCIL

-12-

AIChE

Gulf Coast Crude Oil Price Forecasting

- **Once a straightforward exercise, quantifying the future price for Gulf Coast light sweet crude oil versus Dated North Sea is now a forecasting challenge**
- **Domestic light sweet crude oil will have to push its way into the more complex Gulf Coast refineries designed to process medium sour and heavy sour crude oil to clear the market**
  - The potential markets on the North American East Coast are not large enough, and the inland markets are blocked by rising Northern Supply
- **Quantifying the "degree of discounting" required is an active area of research by Muse**
  - The below figures are illustrative of the implications of pushing increasing volumes of light sweet crude oil into a medium sour coking configuration
  - The below data were generated by Muse's detailed refinery LP model of the aggregate Gulf Coast medium sour coking configurations

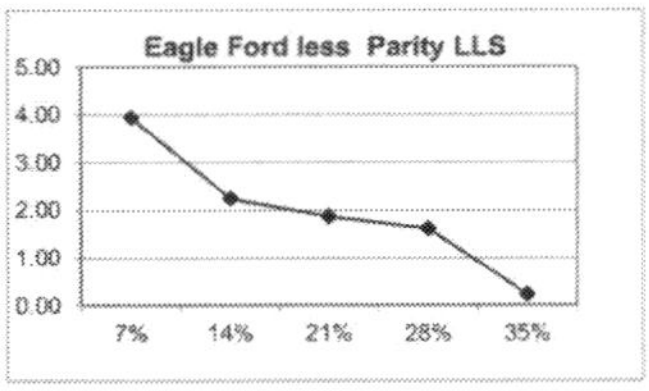

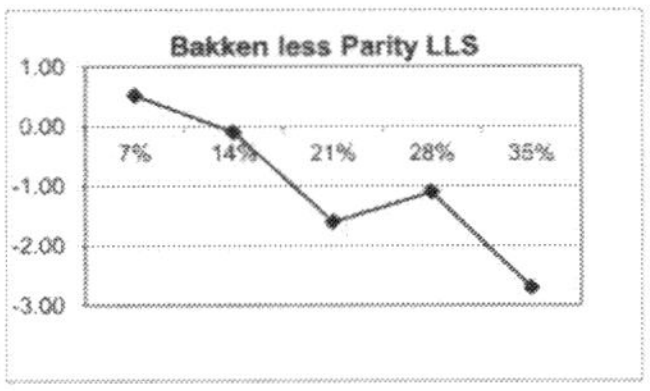

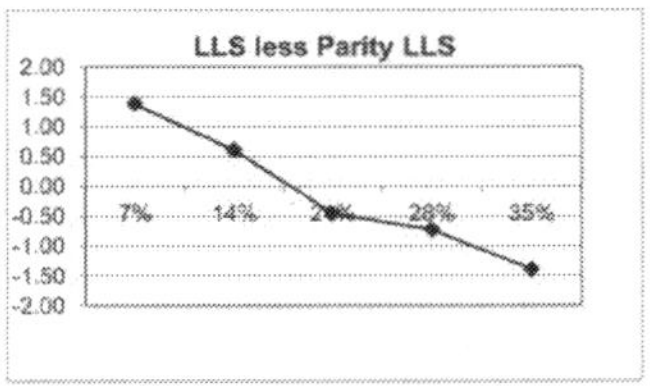

Source: Muse

MUSE STANCIL

AIChE

Gulf Coast Pricing Implications

- **By 2013/14, rising on and offshore Gulf Coast, Eagle Ford, and West Texas crude supply is projected to displace essentially all waterborne light sweet crude imports from the Gulf Coast**
  - Requiring light sweet crudes in the Southern Supply Region to displace light sour and, potentially, medium sour waterborne imports
  - Or find a way to other markets by water
- **Gulf Coast domestic light sweet crudes are already generally trading at a discount to an import parity value**
- **This dynamic is projected to persist throughout the forecast period, absent a sizable reaction by the Gulf Coast refiners to add more capacity to process light sweet crude**
  - The latter is possible, and there already has been some capacity additions announced that are targeted at processing more Eagle Ford production

MUSE STANCIL

North American Crude Oil Market Implications

- **The North American crude oil markets will remain in flux, particularly for the light crude oil segment, as the rapidly evolving supply picture promises to continue to reshape the crude oil distribution patterns within North America**
- **For the Canadian light oil producer, high capacity access to the eastern and western markets will be increasingly important**
  - Rail will continue to play a vital role in transporting crude oil within North America, particularly from the Williston Basin and Western Canada
- **Light crude on the Gulf Coast is expected to price under import parity for some time to come**
- **However, the refiner response is something of a wild card for the light crude oil market – North American refiners will continue to re-configure their refineries to respond to changes in crude supply patterns**
- **An equally difficult forecasting challenge is assessing the market response of the existing heavy crude oil suppliers to the U.S. Gulf Coast to the arrival of the Canadians**
  - A resurgence of Venezuelan and Mexican heavy crude oil supply would be interesting

MUSE STANCIL

-15-

Muse, Stancil & Co.

- **Economic and technical expertise combined with an in-depth understanding of the commercial aspects of the global downstream energy industry**
  - Without exception, our work effort is customized to address each client's specific issues
- **Independent energy consulting firm founded in 1984: www.musestancil.com**
  - Offices in Dallas, Houston, and London

Business Strategy
E&P
Refining
Midstream
Biofuels

Market Analysis
Crude Oil
Natural Gas
NGLs
Biofuels

Asset Valuation and Performance Optimization
Refining
Midstream
Biofuels
PIMS Licensee since 1986

Project Development
Refining
Midstream
Biofuels

Transaction Due Diligence
Refining
Midstream
Biofuels

Expert Services
International Arbitration
Commercial Litigation
Royalty and Contract Auditing
Property Loss and Business Interruption Insurance

- **Muse also owns 50 percent of OSIA, which develops and licenses software that interfaces with LP models and other refining and petrochemical planning tools**

MUSE STANCIL

-16-

TYPICAL MUSE ENGAGEMENTS

AIChE

- **Strategic Plan Development**
  - Specializing in a scenario-oriented, quantitative analysis of the oil markets
  - Incorporating Muse's extensive knowledge of refining economics and objectives
  - Also utilizing proprietary, highly detailed LP-based distribution models of the North American crude and product markets
- **Downstream Integration Support**
  - Identification of potential processing partners and acquisition targets
  - Downstream asset valuation
  - Joint venture operating agreement negotiating assistance
- **Commercial Support**
  - Valuation of pipeline throughput commitments
  - Field upgrader technical and economic assessment
- **Litigation Support**
  - Crude oil and bitumen valuation for royalty administration purposes
  - Expert testimony at NEB, FERC, etc.
  - Commercial disputes

MUSE STANCIL

-17-

MUSE STANCIL

REAL WORLD EXPERTISE

15455 Dallas Parkway
Suite 350
Addison, TX 75001-4690
Phone: 214-954-4455
Fax: 214-954-1521

Two Allen Center
Suite 1600, 1200 Smith Street
Houston, TX 77002
Phone: 713-890-1182
Fax: 214-954-1521

48/54 Moorgate
London EC2R 6EJ
United Kingdom
Phone: 44-0-207-374-8994
Fax: 44-0-207- 374-8995

Level 58 Republic Plaza
9 Raffles Place
Singapore 048619
Phone: 65-6832-1341
Fax: 65-6832-1491

Solutions for FCC Refiners in the Shale Oil Era

Chad Huovie
UOP LLC, A Honeywell Company

AIChE Southwest Process Technology Conference
October 3-4, 2013
Galveston, Texas

UOP
A Honeywell Company

UOP 6166-1

Agenda

UOP
A Honeywell Company

- Market Overview
- Practical Testing of Shale Crudes in FCC Pilot Plants
- Case Study: Impact of Shale on Refinery Operations
- Shale Processing Challenges & Solutions at the FCC

UOP 6166-2

Access to Cheap Energy & Crude Oils Providing Lifeline to US Refinery Industry

UOP A Honeywell Company

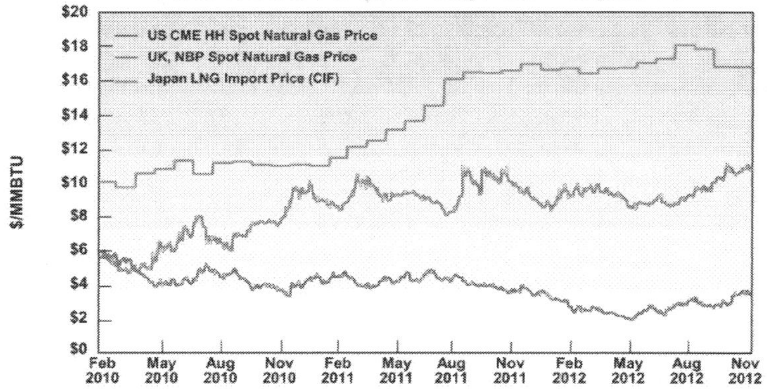

Source: Bloomberg, November 2012

- **US refiners capitalizing on low cost supply of natural gas from shale sources**
- **Enabling well-positioned refiners to become a low cost supplier of transportation fuels**

UOP 6166-3

Surge in US Shale Crude Production is Fueling Significant Growth Prospects

UOP A Honeywell Company

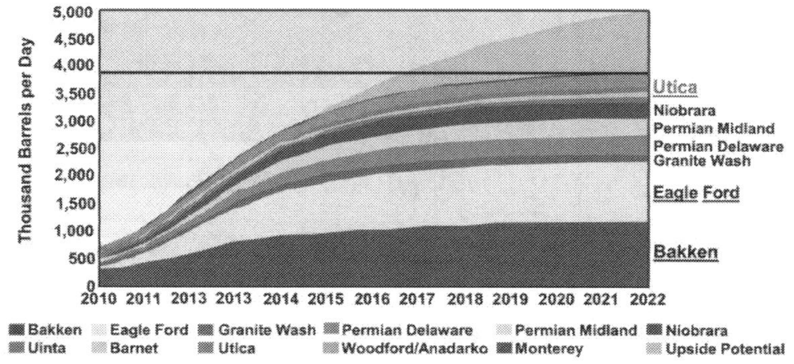

Source: Citi investment Research and Analysis
Source: David Hughes, AGU presentation, December 2012

- **Extension of hydraulic fracturing enabling economic capture of condensates / crude hydrocarbons for refining purposes**
- **Long term forecasts call for 2 - 6 million barrels of production addition from shale sources**

UOP 6166-4

Anatomy of Shale Crude

UOP
A Honeywell Company

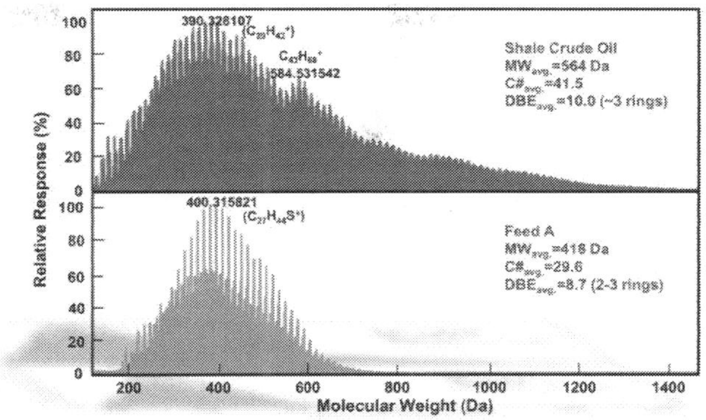

	Feed A	Feed B	Feed C	Feed C-1	Feed C-2
	ME Gasoil	*HT US Gasoil*	*Utica Shale Crude*	*Utica 650-*	*Utica 650+*
API	21	27.8	43.4	51.1	30.1
Sulfur (wppm)	23639	633	290	129	470
Nitrogen (wppm)	1257	422	42	1.2	94
Conradson Carbon (Wt-%)	0.27	0.16	0.4	<0.1	0.5
UOP K	11.74	12.07	12.43	12.21	12.70

UOP 6166-5

Pilot Plant Testing & New Characterization Methods Improve Prediction of Effects

UOP
A Honeywell Company

	Feed A	Feed B	Feed C-2
H (Wt-%), derived from UOP K	12.07	12.99	14.02
H (Wt-%), measured UOP Method 997-13	12.03	12.99	13.55

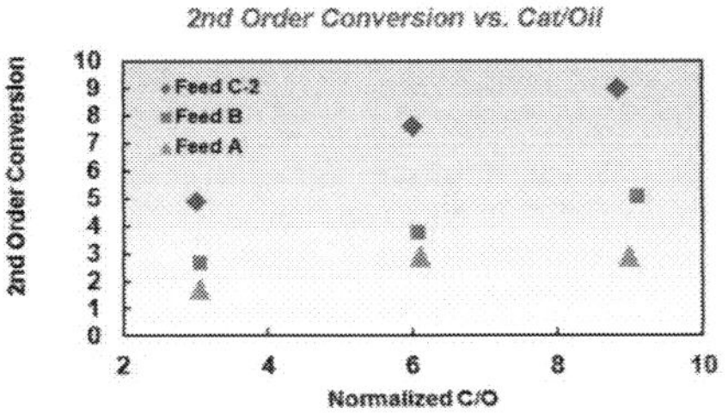

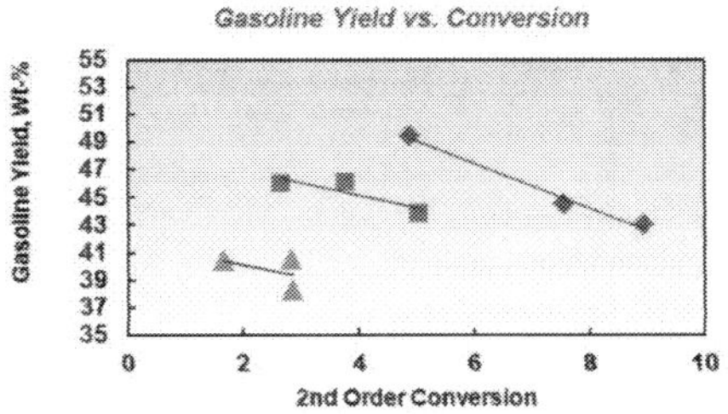

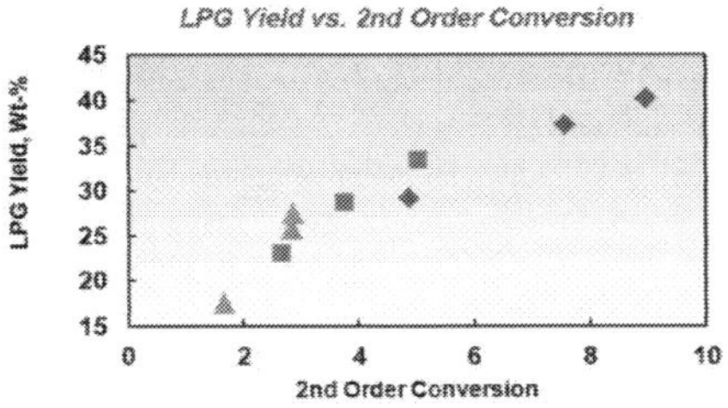

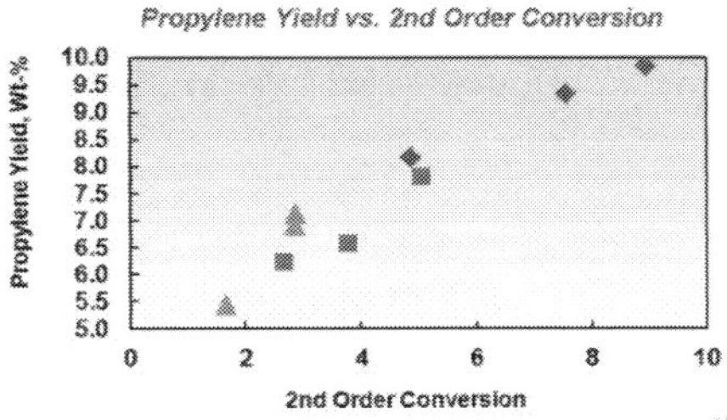

UOP 6166-6

Pilot Plant Work Validates Expected Impact on FCC Operations

UOP
A Honeywell Company

Yield	*Shift Direction*	*Expected Shift*
Feed Hydrogen, Wt-%	⇧	0.5 – 1
Conversion, Wt-%	⇧	6 – 10
Gasoline Yield, Wt-%	⇧	6 – 9
Clarified Oil, Wt-%	⇩	3 – 5
Coke Yield, Wt-%	⇩	0.2 – 0.5
LPG Yield, Wt-%	⇧	4 – 6
Gasoline octane, RONC	⇩	1.5
Delta Coke	⇩	0.08 – 0.16

UOP 6156-7

Refinery Economic Model Capabilities Enhanced with Pilot Plant Data

UOP
A Honeywell Company

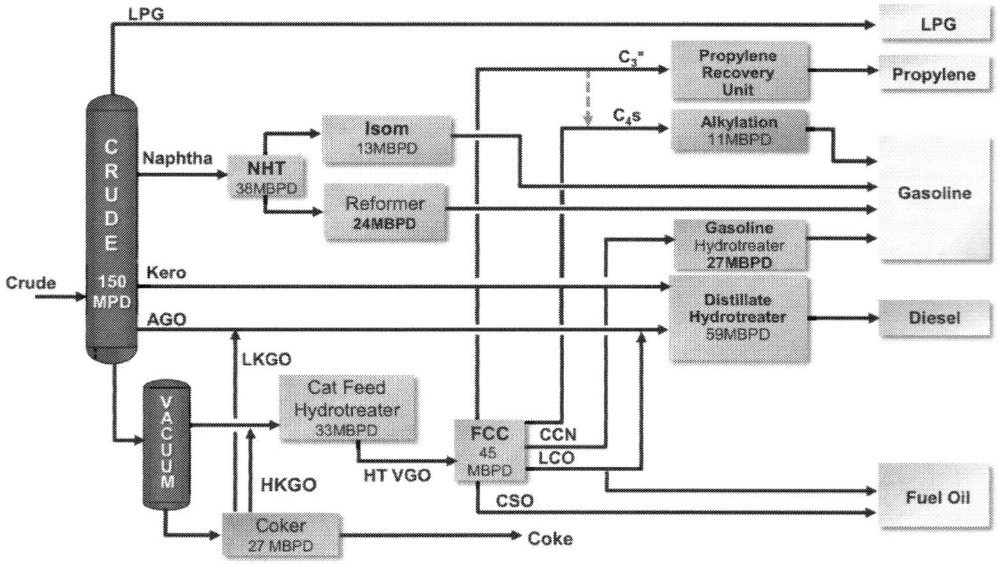

Base Crude (LV%): Arab Medium (50%), Maya (25%), Bonny Light (25%)

UOP 6156-8

Shift to Price Advantaged WCS Provides Significant Reduction in Feed Costs

UOP
A Honeywell Company

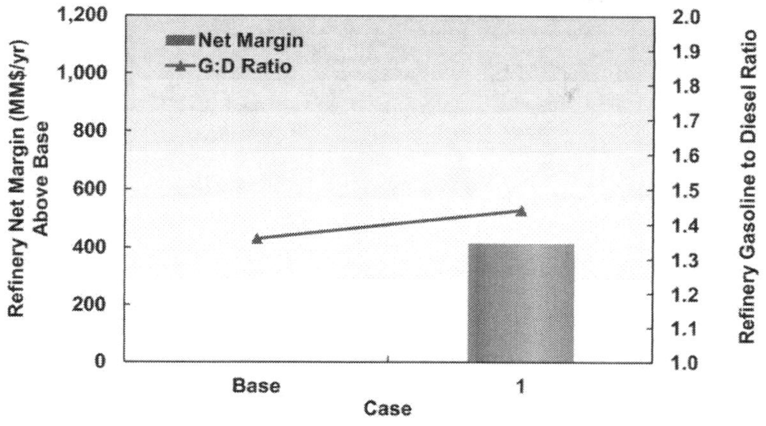

- Replace imported Maya with lower cost WCS
- Minor impact on refinery production rates

UOP 6156-9

Swap of Imported Light with Shale Derived Crude Establishes First Constraint in Operations

UOP
A Honeywell Company

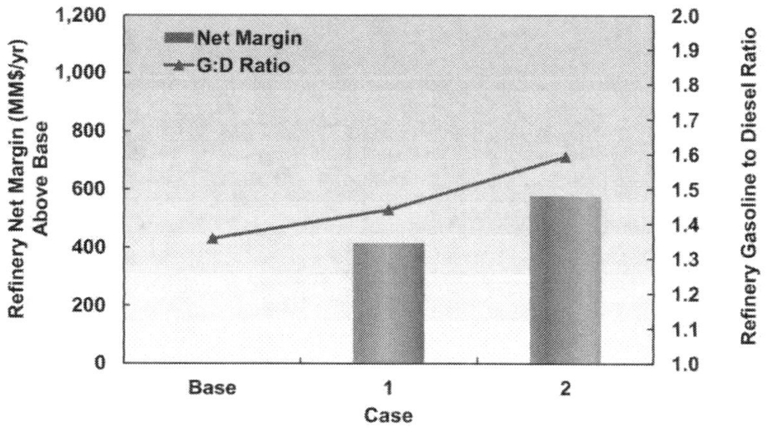

- Imported Bonny Light replaced with Eagle Ford
- +33% in CDU LPG rates
- FCC & downstream unit capacity shifts
- Alky and FCC gasoline treating capacities > 20% above base

UOP 6156-10

Significant Impact +/- on Refinery Operations Anticipated with Import Elimination Scenario

UOP
A Honeywell Company

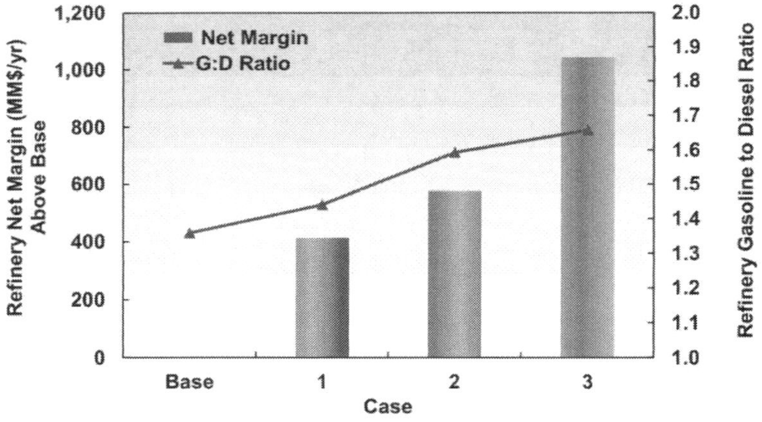

- Waterborne imports are eliminated with NA crudes
- 75% Eagle Ford, 25% WCS
- Represents lowest crude cost of cases modeled
- Highest G:D
- Coker unit at 65% of capacity
- Crude LPG 86% above base

UOP 6165-11

Significant Shift in Capacities Predicted

UOP
A Honeywell Company

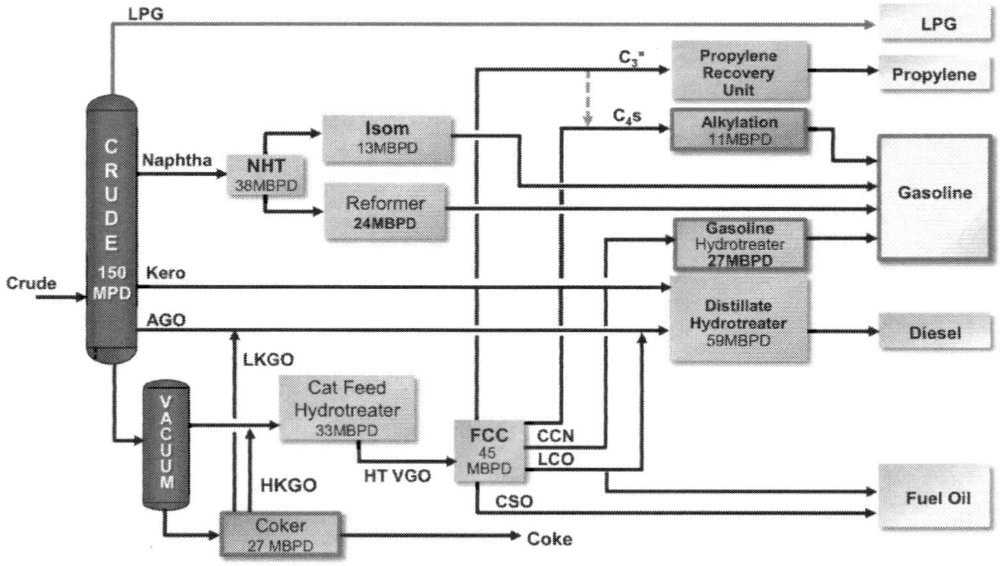

Units that are operating +/- 15% of Base are Highlighted

UOP 6165-12

100% Shale Crude Processing

UOP A Honeywell Company

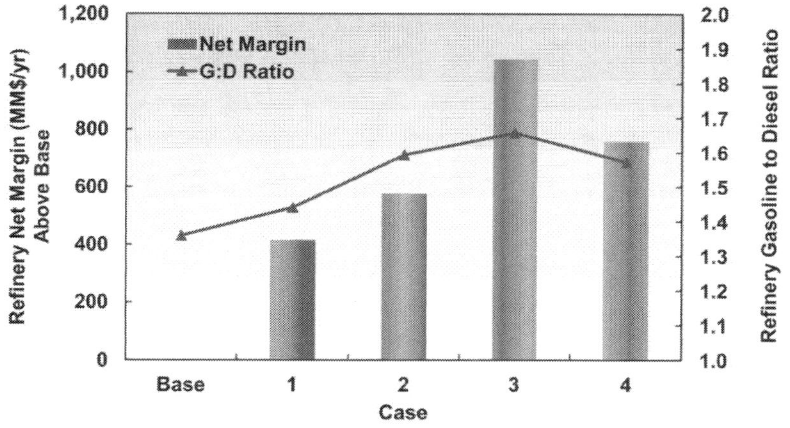

- **Minimal residue content drives low capacity at VDU & Coker**
- **LPG and naphtha complex strained**
- **Consider VDU and Coker shutdown and route AR direct to FCC**

UOP 6186-13

Impacts of Shale Crude Processing on Refinery Balances / Operations

UOP A Honeywell Company

- **Overall refinery G:D ratio is likely to increase**
- **Lower contaminants will decrease hydrotreating severity and overall H_2 requirements**
- **Light Ends / Naphtha**
  - Crude LPG production increases significantly requiring modifications to crude tower LPG recovery section
  - Higher naphtha yields increase demands on the naphtha complex
  - Higher paraffinic naphtha content leads to higher coke, lower yields and poorer activity in CCR Platforming unit
- **Distillate**
  - Diesel pool cetane and cloud point increases
- **Fuel Oil**
  - Higher API gravity and lower residue content will decrease feed to the vacuum and coker units

UOP 6186-14

Shale Crude Processing Challenges at the FCC

UOP
A Honeywell Company

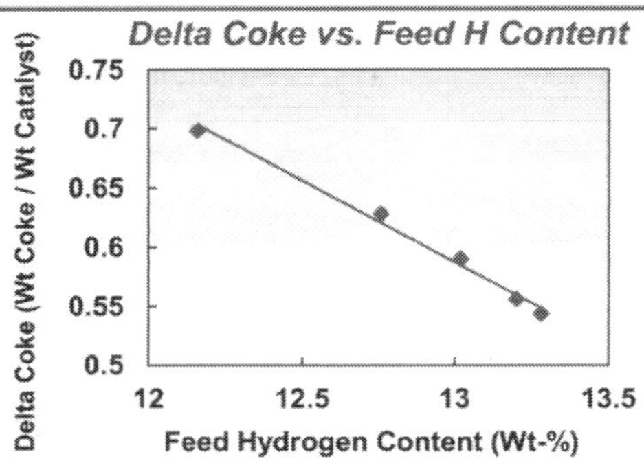

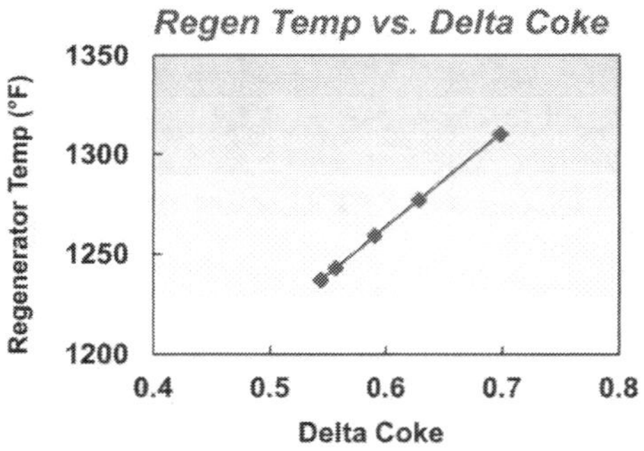

- **FCC feed quality improves yields and conversion**
  - Higher gasoline yield with lower octane
  - Increased LPG yield impacts on WGC
  - Lower delta coke
- **Reduction in regenerator temperature**
  - Coke combustion kinetics negatively impacted
  - Possible higher emissions of NO_x and CO

UOP 6165-15

Exclusive UOP RxCat™ Technology Increases Regenerator Temperature, Conversion & Margins

UOP
A Honeywell Company

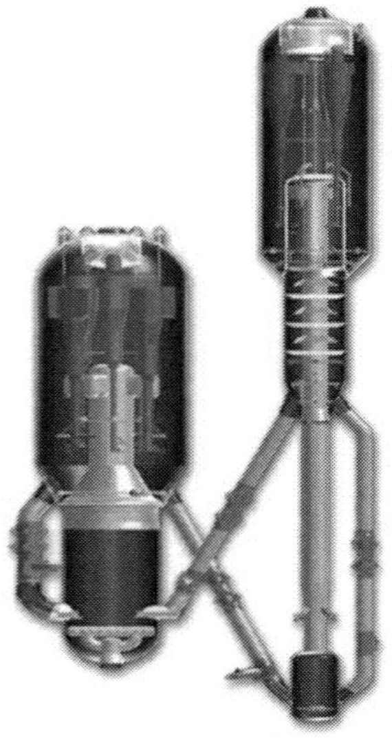

	Case 1 *Base*	*Case 2* *DFAH*	*Case 3* *MCB Recycle*	*Case4* *Reduced Stripping Steam*	*Case 5* *RxCat*
Regenerator Temp (°F)	1260	1340	1328	1290	1350
Delta Conversion (Wt-%)	Base	(2.5)	(1.8)	(1.0)	2.1

UOP 6165-16

RxCat™ Technology Solves Regenerator Temperature Problem in Commercial Unit

UOP A Honeywell Company

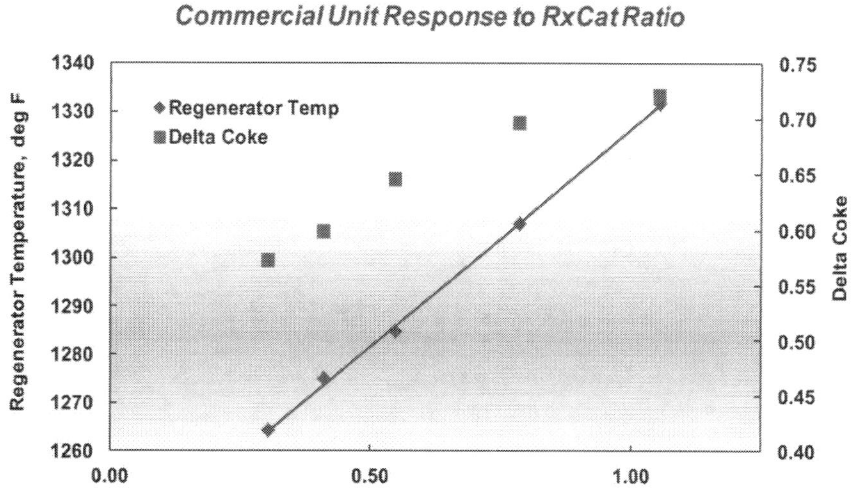

RxCat Ratio, wt/wt	0.0*	0.5	1.0
Delta Coke, wt/wt	0.52	0.62	0.72
Regen Temp, F	1237	1282	1326

UOP 5166-17

Positioning FCC-based Refinery to Produce Additional High Quality Distillate

UOP A Honeywell Company

Solution	Refinery Distillate Production Increase
Oligomerize co-produced FCC LPG olefins to diesel	4 %
Feed system upgrade to remove distillate from FCC charge stock	3 %

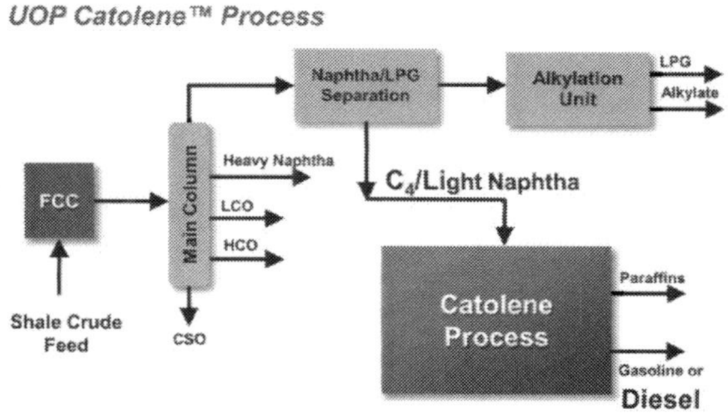

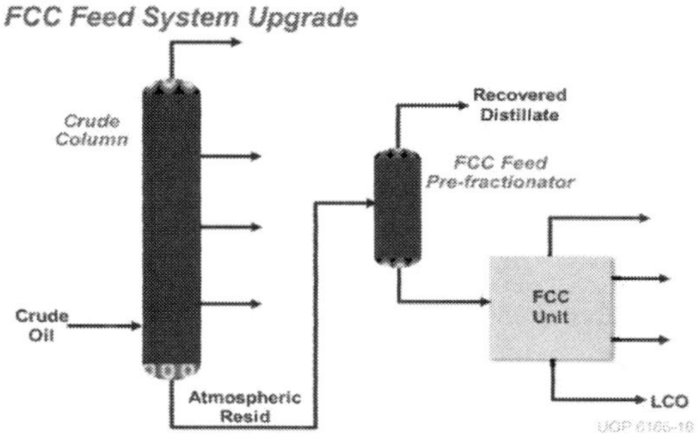

UOP 5166-18

Conclusions

UOP
A Honeywell Company

- **Shale oil crude processing in FCC-based complexes materializing in U.S. requiring solutions to address operational & product challenges**
- **Accurate assessment of refinery impacts requires development of dynamic LP submodels**
  - Validated against pilot plant data
  - Utilizing latest feed characterization methods
- **Enhanced distillate production from FCC-based complexes possible with new technology and design changes**
- **UOP RxCat technology can solve the FCC heat balance issue and provide additional flexibility in an era of shale crude processing**

UOP 6166-19

UOP
A Honeywell Company

Q&A

UOP 6166-20

AIChE

A Simple & Effective Approach for Determining Optimum Refinery Size

By: Fouad A. Aly

Petroleum Refining Technology Session
5th AIChE Southwest Process Technology Conference
(October 4, 2013)

Ability to Influence Project Costs, Schedule, and Economics

2

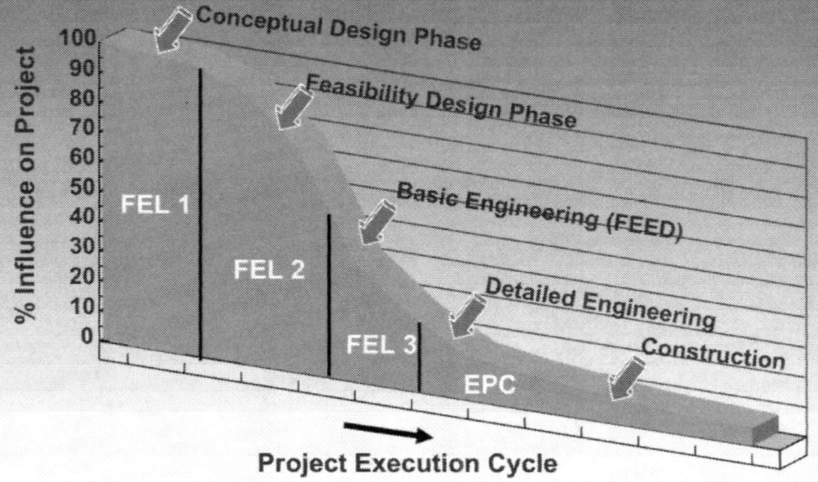

Refinery Configuration

3

Current State-of-the-Art Practice

AIChE

Petroleum Refining Technology Session
5th AIChE Southwest Process Technology Conference
(October 4, 2013)

KBR

Refinery Configuration

4

- **Current State-of-the-Art Practice:**
  - Utilize LP \ NLP to Optimize:
    - Crude Oil Supply (Multiple Sources)
    - Possible Processing Routes
    - Possible Products
    - Blending: On-spec Products
    - Physics-based Model – Mass Balance
    - Objective Function: Profit

AIChE

Petroleum Refining Technology Session
5th AIChE Southwest Process Technology Conference
(October 4, 2013)

KBR

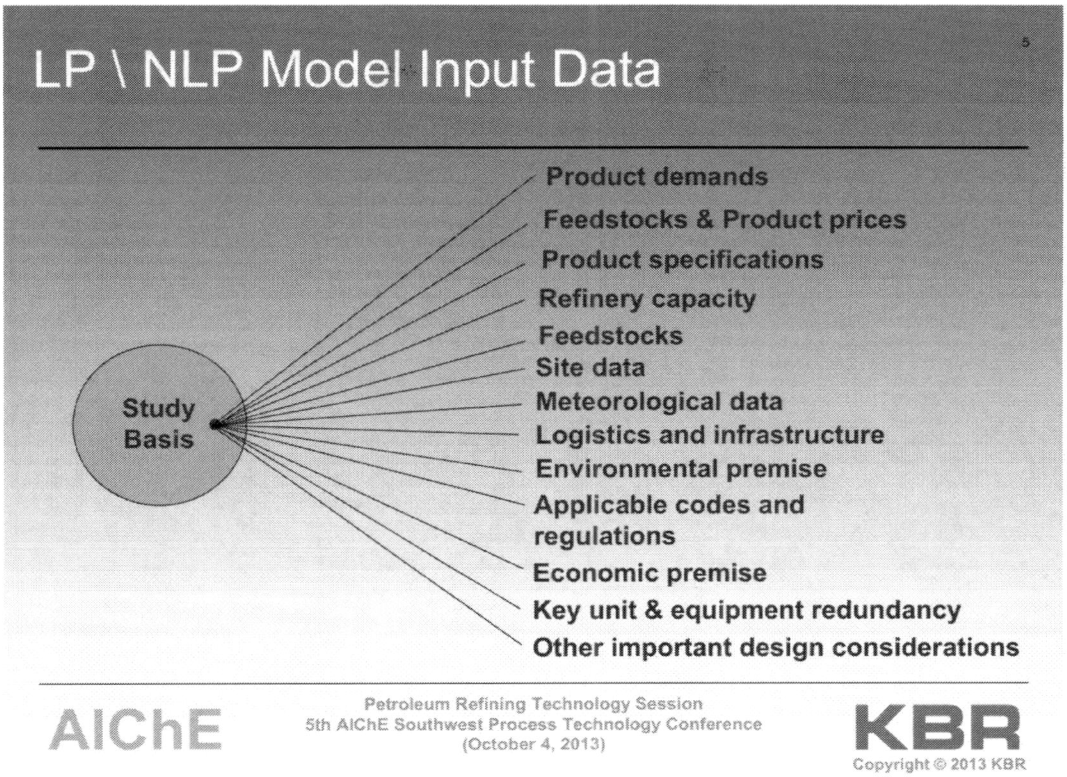
LP \ NLP Model Input Data
5
Study Basis
Product demands
Feedstocks & Product prices
Product specifications
Refinery capacity
Feedstocks
Site data
Meteorological data
Logistics and infrastructure
Environmental premise
Applicable codes and regulations
Economic premise
Key unit & equipment redundancy
Other important design considerations
AIChE
Petroleum Refining Technology Session
5th AIChE Southwest Process Technology Conference
(October 4, 2013)
KBR
Copyright © 2013 KBR

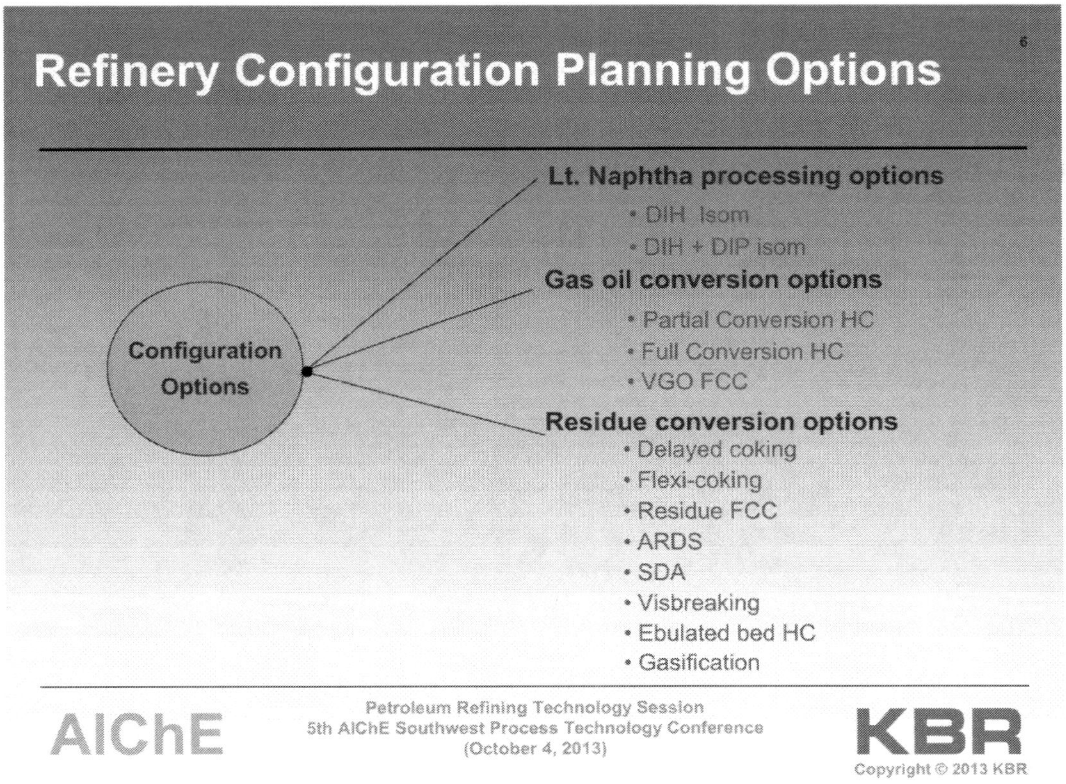
Refinery Configuration Planning Options
6
Configuration Options
Lt. Naphtha processing options
• DIH Isom
• DIH + DIP isom
Gas oil conversion options
• Partial Conversion HC
• Full Conversion HC
• VGO FCC
Residue conversion options
• Delayed coking
• Flexi-coking
• Residue FCC
• ARDS
• SDA
• Visbreaking
• Ebulated bed HC
• Gasification
AIChE
Petroleum Refining Technology Session
5th AIChE Southwest Process Technology Conference
(October 4, 2013)
KBR
Copyright © 2013 KBR

Technologies & Schemes to Consider

7

Technology Area	Scheme B1	Scheme B2	Scheme B4	Scheme A1	Scheme C1	Scheme C2	Scheme C3	Scheme C4
Distillation	CDU/ VDU – All Cases							
Naphtha Processing	NHTU/ CCR/ Isomerization – All Cases							
Mid Barrel Diesel	ULSD Diesel – All Cases							
Gas Oil Prim Conv	PCHC	PCHC		ARDS/ FCHC	PCHC	PCHC	PCHC	PCHC
Gas Oil Sec Conv	FCC	FCC	FCHC	ARDS + RFCC	FCC	FCC	FCC	FCC
Resid Processing	Delayed Coker	SDA + Coking	Delayed Coker	ARDS + RFCC	Flexicoker	Visbreaker	EBHC + Coking	Delayed Coker
C3/C4 Processing	Alkylation + C4 Isom	Alkylation + C4 Isom		Alkylation + C4 Isom	Alkylation + C4 Isom	Alkylation + C4 Isom	Alkylation + C4 Isom	Alkylation + C4 Isom
FCC Gasoline HDS		Selective		Selective				
Mercaptan Treating	LPG/ Jet	LPG/ Jet/ Gasoline	LPG/ Jet	LPG/ Jet/ Gasoline	LPG/ Jet	LPG/ Jet	LPG/ Jet	LPG/ Jet
Sulfur Management	SRU, TGTU, SWS and Amine Treating – All Cases							
Hydrogen Plant	SMR using Sat Gas, LPG and / or Light Naphtha – All Cases							
Gasification								Pet Coke

AIChE

Petroleum Refining Technology Session
5th AIChE Southwest Process Technology Conference
(October 4, 2013)

KBR

Refinery Configuration B-4

8

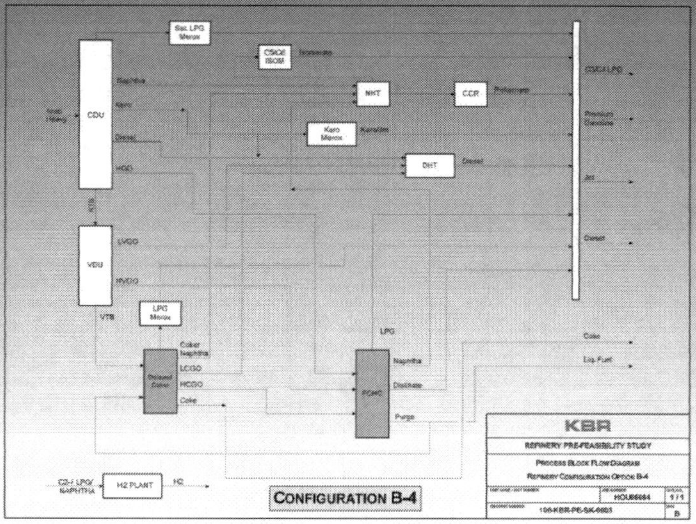

AIChE

Petroleum Refining Technology Session
5th AIChE Southwest Process Technology Conference
(October 4, 2013)

KBR

Typical Partial LP \ NLP Output

CAPACITIES (MBPD)	R-350	R-360	R-370
CRUDE UNIT	400.0	400.0	400.0
VACUUM UNIT	204.6	204.6	247.7
DELAYED COKER	99.3	100.5	101.2
NAPHTHA HT	115.6	86.7	63.8
ISOMERIZATION DIH	0.0	39.5	24.6
ISOM DIP/DIH	51.7	0.0	0.0
CCR REFORMER	62.6	48.4	36.2
JET MEROX	32.9	20.8	23.6
DIESEL HDS/HDAr	129.5	135.6	121.9
HYDROCRACKER (FC)	123.4	124.3	167.5
H2 PLANT MMSCFD	237.5	234.8	309.5
AMINE UNIT MT/D	1,080.7	1,155.1	802.0
SULFUR PLANT MT/D	1,070.0	1,143.7	794.0
GRS POWER-GEN, MW	167.8	189.4	200.3

Refinery Configuration

Improved Practice

LP \ NLP Model Input Data

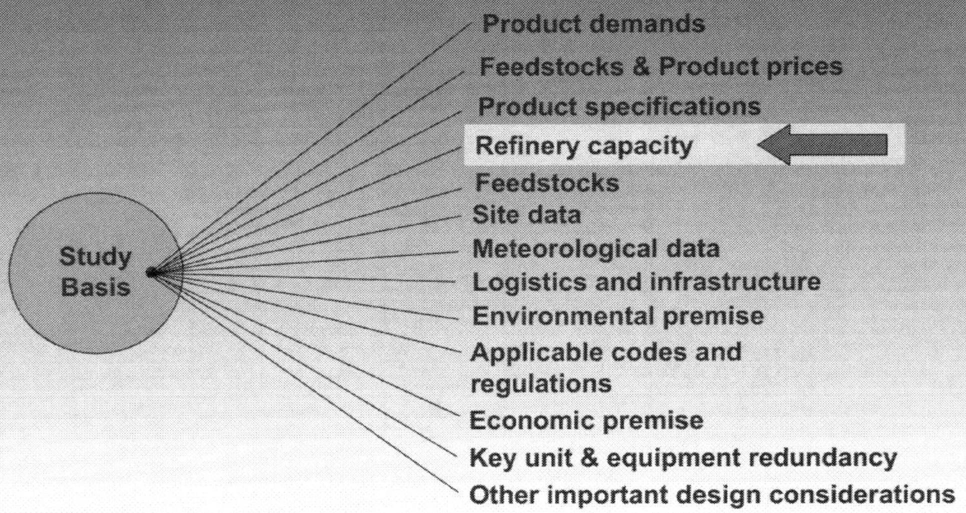

Refinery Total Installed Cost versus Capacity

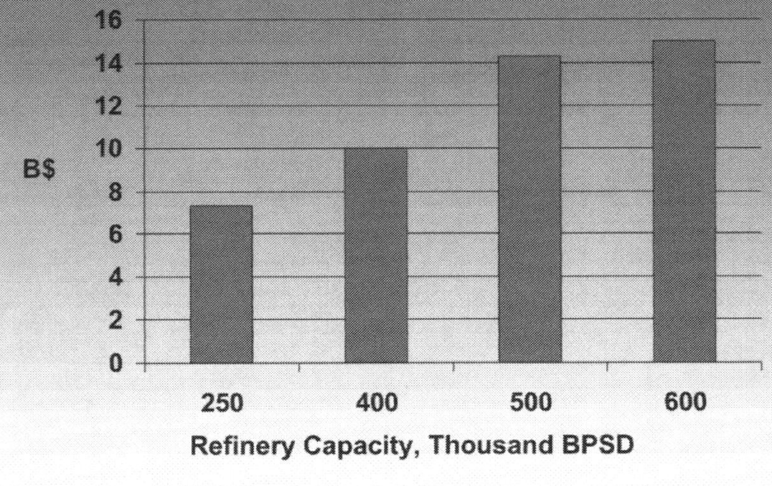

Refinery Configuration

13

- **Improved Practice:**
  - Follow current LP \ NLP based practice plus:
  - In parallel, conduct a <u>Refinery Capacity Constraint analysis</u> as early as possible
    - Identify maximum capacities of individual single-train refinery Units.
    - Identify capacity constraining units for different refinery configurations as a function of refinery capacity.

AIChE

Petroleum Refining Technology Session
5th AIChE Southwest Process Technology Conference
(October 4, 2013)

KBR

What determines maximum single-train unit size?

14

- **One or more major/critical pieces of equipment are:**
  - Supplied by none
  - Supplied by a Single Vendor
  - First of kind
  - One of a kind design
- **Equipment size limitations**
  - Weight (transportation & maintenance)
  - Diameter and length
    - Rail, road, and site clearances and crossings
    - Mechanical design limits
    - Field fabrication (feasibility, desirability)
  - Labor laws and/or company policies
- **Schedule impacts**

AIChE

Petroleum Refining Technology Session
5th AIChE Southwest Process Technology Conference
(October 4, 2013)

KBR

Actual Example

15

400,000-600,000 BPSD REFINERY

Refinery Capacity Constraint Analysis
70% Crude A + 30% Crude B

16

Unit	Capacity Constraint	Identified Limit (BPSD)
DHT	One-Train	76300 - 89100
HCK	One-Train	60300 - 70400
DCU	Two-Drums	32000 - 35000

Refinery Capacity Analysis

Configuration B-4 Transitions

17

Unit	1-2 Train Transition	2 Trains -2 Units Transition
DHT	247-288	494-577
HCK	195-228	391-456

Unit	2-4 Drum Transition	4-6 Drum Transition	6-8 Drum Transition	8-10 Drum Transition	2 Units
DCU	129-141	258-282	387-423	516-564	645-705

All Numbers Reflect Total Refinery Capacity when running design Crude

AIChE

Petroleum Refining Technology Session
5th AIChE Southwest Process Technology Conference
(October 4, 2013)

KBR

Refinery Capacity Constraints

Configuration B4

18

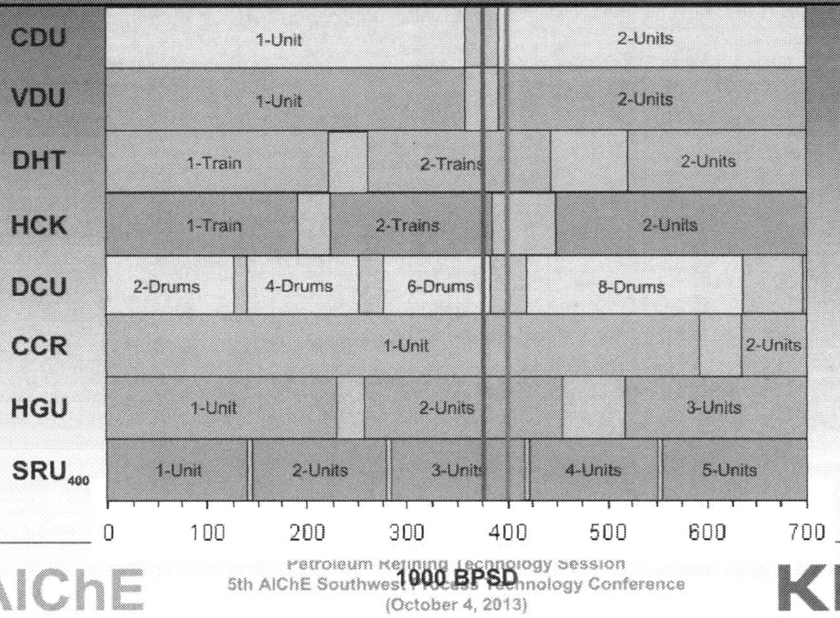

AIChE

Petroleum Refining Technology Session
5th AIChE Southwest Process Technology Conference
(October 4, 2013)

KBR

400 MBPSD Case

19

- Crude unit capacity exceeds maximum recommended single unit capacity range of (357-385 MBPSD)
- Coker unit capacity falls within the transition range from 4-drum to 6-drum unit capacity (equivalent to a refinery throughput of 391-415 MBPSD)
- Sulfur plant capacity is very close to the max available from 3X400 t/d (equivalent to a refinery throughput of 410 MBPSD)

CONCLUSION: a refinery capacity of about **380** MBPSD would be more robust & flexible than the 400 MBPSD target

AIChE

Petroleum Refining Technology Session
5th AIChE Southwest Process Technology Conference
(October 4, 2013)

KBR

600 MBPSD Case

20

- Two, or more, of most process units (ISBL) are required
- Coker, Hydrogen Plant, and Sulfur recovery are likely candidates for significant optimization

Conclusion: ISBL units should be based on **2X300** MBPSD

AIChE

Petroleum Refining Technology Session
5th AIChE Southwest Process Technology Conference
(October 4, 2013)

KBR

Capacity Constraint Analysis Benefits & Conclusions

21

Identified Additional Capital Savings:

- Range from 5% to 12%
- Average 6.2% (~300 MM$ for a 5B$ Refinery)

Approach Also Applicable to:

- Refinery (Revamp or Expansion) Level
- Unit Level
- Equipment Level

AIChE

Petroleum Refining Technology Session
5th AIChE Southwest Process Technology Conference
(October 4, 2013)

KBR

22

Thank You!

Q & A

AIChE

Petroleum Refining Technology Session
5th AIChE Southwest Process Technology Conference
(October 4, 2013)

KBR

ALBEMARLE®

Optimizing FCC – Pretreat Unit Operations To Achieve Tier 3 Gasoline Sulfur Targets

2013 AIChE Southwest Process Technology Conference

Peter-Paul Langerak, Albemarle Corporation

ALBEMARLE®

Contents

1 Background on Tier 3 Gasoline Regulations

2 Implications for Refinery FCC-PT Unit Objectives

3 Case Study Simulations

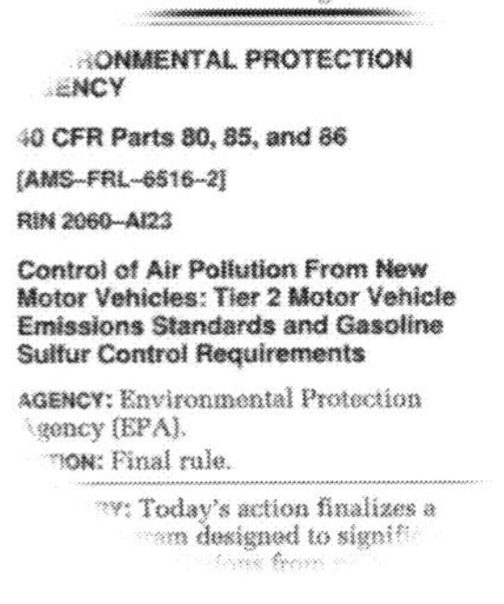

ALBEMARLE®

2013 AICHE SW PROCESS TECHNOLOGY CONFERENCE

1

Background on Tier 3 Gasoline Regulations

ALBEMARLE®

2013 AICHE SW PROCESS TECHNOLOGY CONFERENCE

Outlook for Tier 3 Gasoline Regulations

- Tier 2 Gasoline
  - S ≤ 30 wppm
- Tier 3 Gasoline
  - S ≤10 wppm
- Timing of implementation announced as 2017

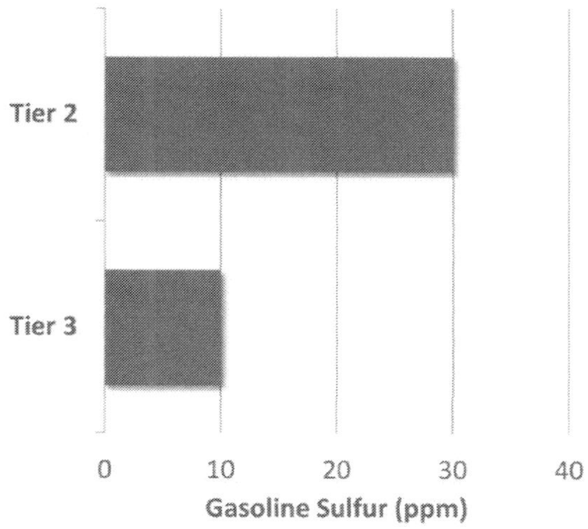

ALBEMARLE®

2013 AICHE SW PROCESS TECHNOLOGY CONFERENCE

FCC-PT Improves FCC Product Quality and Operations

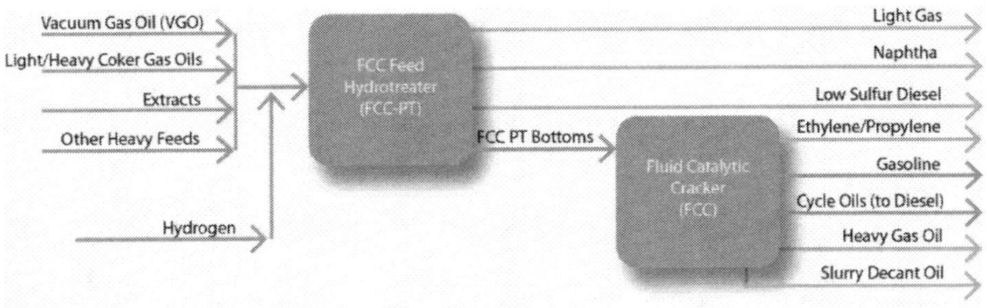

- FCC Feed Hydrotreater generates feed to the FCC to improve the quality by removing contaminants and adding H_2
  - Generates some products directly; e.g., low S diesel
- Fluid Catalytic Cracker uses acid-catalyzed cracking to break heavy hydrocarbons down to gasoline and light cycle oils
  - Unit performs better on low contaminants, C/H ratio and N feed

FCC-PT Operations Span Wide Range of Conditions

- Feed is heavy with high S, N and C/H (i.e. high aromatic) contents, metal contaminants (Ni, V, As, Si etc.), high CCR (0-4 wt.%) and particulates (FeS, coke fines)

Feed		Blend of VGO with lube extracts, coker stocks, resid, DAOs
LHSV	hr^{-1}	0.5-1.2
ppH_2	psia	600-1500
H_2/Oil	SCFB	1400-3000
WABT	°F	640-790
HDS Target	%	≥ 80
HDN Target	%	≥ 50

- Different from ULSD, HC-PT (>99% HDS/HDN)

Hydrotreating FCC Feed Significantly Reduces FCC Gasoline Sulfur Content

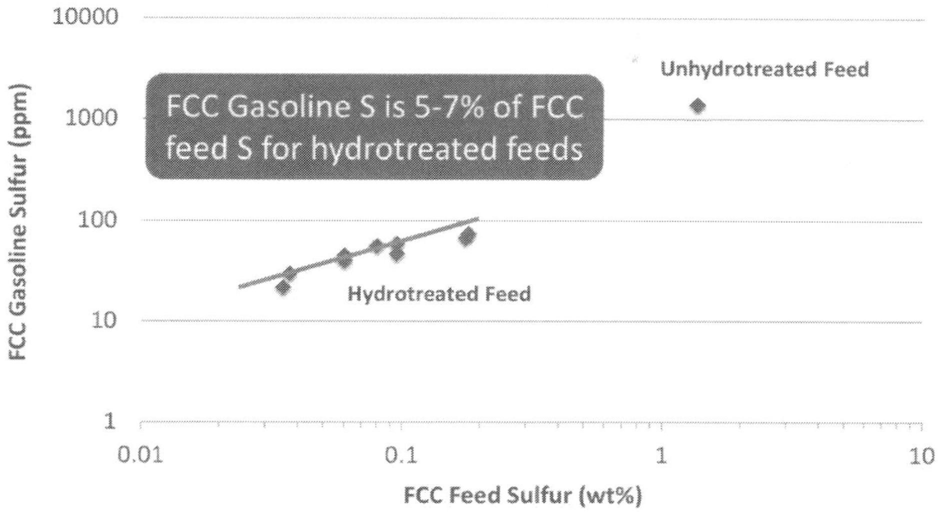

2

Implications for Refinery FCC-PT Unit Objectives

Impacts on FCC-PT and FCCU Products

- Tier 2
  - FCC naphtha typically 60-100 wppm Sulfur
  - FCC-PT HDS Requirement: 90-95%
- Tier 3
  - FCC naphtha will be 20-35 wppm Sulfur
  - FCC-PT HDS Requirement: >97%

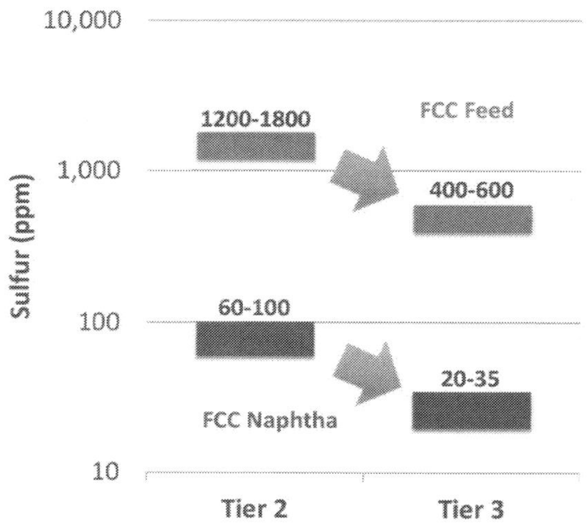

ALBEMARLE® 2013 AICHE SW PROCESS TECHNOLOGY CONFERENCE

Different FCC-PT Operating Strategies

- No single "typical" FCC-PT unit characterization
- FCC-PT operating strategies conducted in 3 modes
- Constant HDS mode with TIR to maintain product S
  - Longest cycle length
  - Roughly 70% of operating units
- Max arosat mode to maximize SOR hydrogenation
  - Typically reduces cycle length by 20-40%
  - Roughly 10-20% of operating units
- Max conversion mode to maximize diesel yield
  - Typically reduces cycle length by ≥50%
  - Roughly 10-20% of operating units

ALBEMARLE® 2013 AICHE SW PROCESS TECHNOLOGY CONFERENCE

Case Studies Illustrate Impacts of Tier 3 Regulations

- Process model simulations used to compare impact on
  - Cycle length
  - H_2 consumption
- Case 1
  - Low P
  - Constant HDS mode
- Case 2
  - Moderate P
  - Constant HDS mode
- Case 3
  - High P
  - Max Arosat mode

3

Case Study Simulations

Simulation Study Inputs

Key Inputs / Case	Case 1	Case 2	Case 3
ppH_2 inlet, psia	650	900	1400
LHSV, 1/hr	0.65	0.80	1.00
H_2 treat rate, SCFB	1600	2400	3000
Feed type	SR VGO	SR VGO + HCGO	SR VGO + HCGO
Feed S, wt%	1.8	2.4	2.8
Feed N, wppm	1400	2000	2400
Tier 2 target	1500 ppm S (92% HDS)	1500 ppm S (94% HDS)	Max arosat, S <1500 ppm
Tier 3 target	500 ppm S (97% HDS)	500 ppm S (98% HDS)	Max arosat, S <500 ppm

Case 1 Simulation

- Low P, constant HDS mode with SR VGO feed
  - 1500 ppm S target for Tier 2 (92% HDS)
  - 500 ppm S target for Tier 3 (97% HDS)
- NiCoMo catalyst optimal at these conditions
  - Catalyst with higher HDS / HDN RVA used for Tier 3 estimate
- Results of simulation
  - 40% shorter cycle length (even with more active catalyst) for Tier 3
  - SOR hydrogen consumption
    - Tier 2: 300 SCFB
    - Tier 3: 420 SCFB

Case 1: Cycle Length Reduced by ~40%

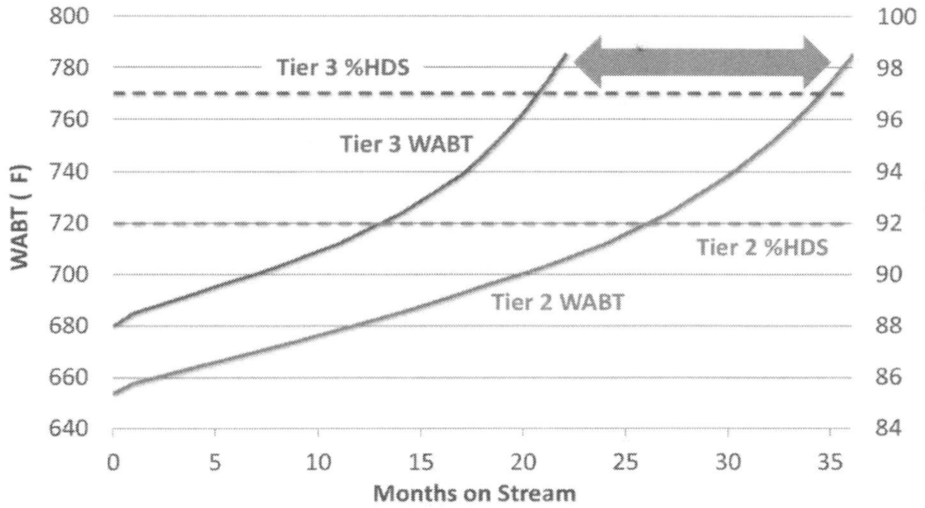

Case 2 Simulation

- Moderate P, constant HDS mode with SR VGO + HCGO
  - 1500 ppm S target for Tier 2 (94% HDS)
  - 500 ppm S target for Tier 3 (98% HDS)
- NiCoMo/NiMo STAX®
  - Catalyst with higher HDS / HDN RVA used for Tier 3 estimate
- Results of simulation
  - 35% shorter cycle length (even with more active catalyst) for Tier 3
  - SOR hydrogen consumption
    - Tier 2: 470 SCFB
    - Tier 3: 600 SCFB

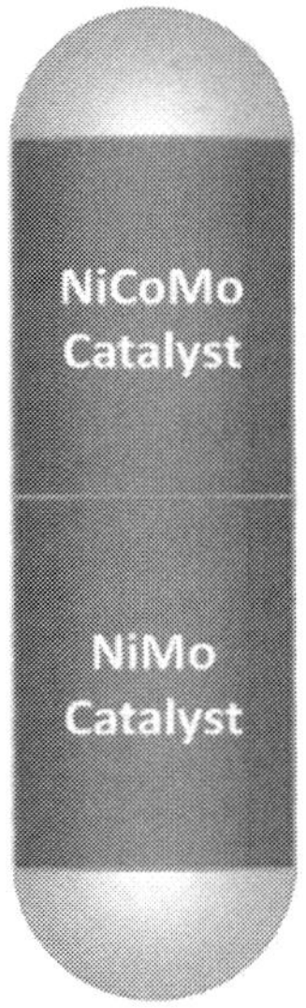

Case 2: Cycle Length Reduced by ~35%

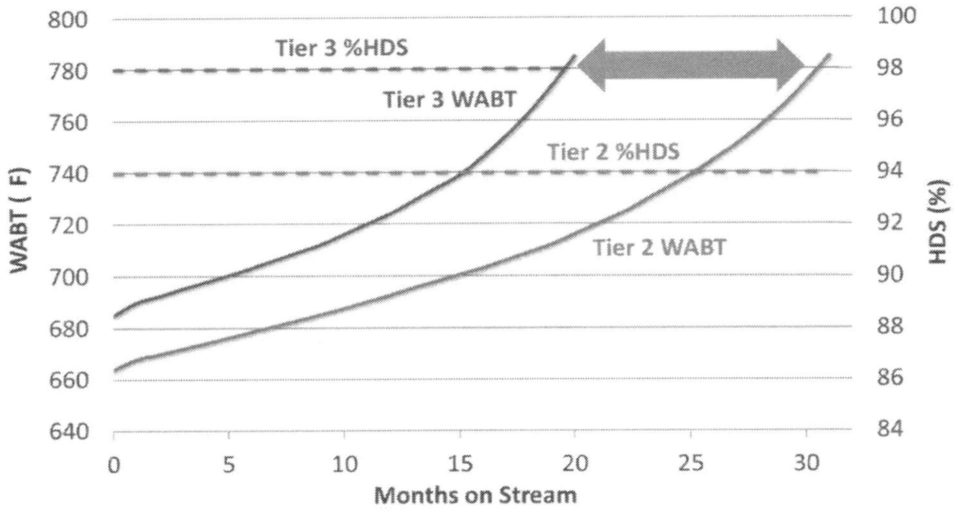

ALBEMARLE®

2013 AICHE SW PROCESS TECHNOLOGY CONFERENCE

Case 3 Simulation

- High P, max Arosat mode with SR VGO + HCGO feed
  - <1500 ppm S target for Tier 2 (>94% HDS)
  - <500 ppm S target for Tier 3 (>98% HDS)
- High RVA NiMo
  - Tier 3 WABT >> Max Arosat T throughout cycle
- Results of simulation
  - >50% shorter cycle length
  - SOR hydrogen consumption
    - Tier 2: 700 SCFB
    - Tier 3: 640 SCFB

ALBEMARLE®

2013 AICHE SW PROCESS TECHNOLOGY CONFERENCE

Case 3: Cycle Length Reduced by >50%

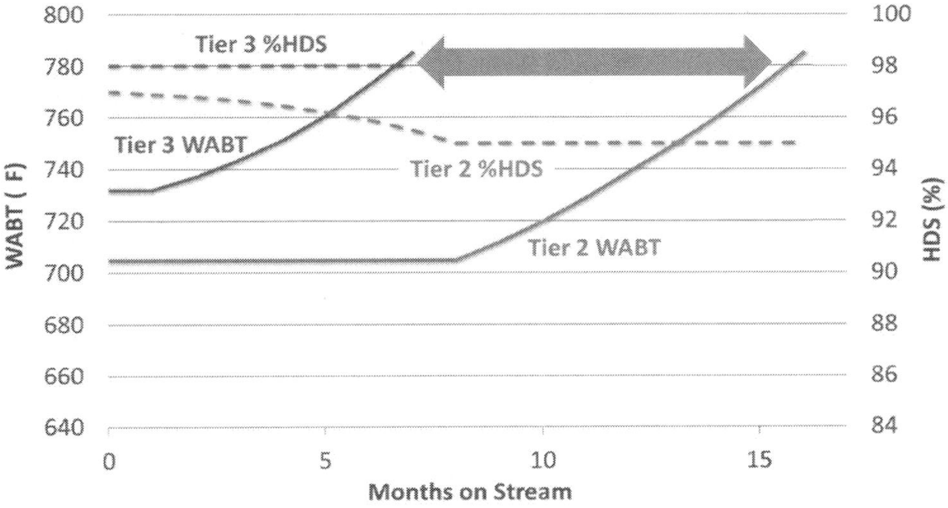

ALBEMARLE®

2013 AICHE SW PROCESS TECHNOLOGY CONFERENCE

Conclusions

- Tier 3 ULSG regulations will have significant impacts on FCC-PT operations
- Specific impacts will be highly individualized
- Essentially all refiners will have shorter cycles and more SOR H_2 consumption
  - Expect 20-40% cycle length impact for constant HDS mode
  - ≥50% cycle length reduction for max arosat mode
- Will increase annualized catalyst costs and complicate integration of FCC-PT / FCCU turnarounds
  - Higher activity catalysts can extend cycles
  - Easier feeds could help mitigate impacts
- Switching to MHC mode could help solve dieselization needs

ALBEMARLE®

2013 AICHE SW PROCESS TECHNOLOGY CONFERENCE

Refinery Challenges from Processing Shale Oils

AIChE Southwest Regional Conference, October 2013

Presented by: Bruce Wright, Baker Hughes Incorporated

BAKER HUGHES

Unconventional Shale Oil

- Production from hydraulic fracturing of shale formations
  - Some areas produce natural gas
  - Other areas produce liquids
- Liquids called Shale Oil or Tight Oil
  - Now 10% of US production
  - Estimated 15% by 2015
- Largest production currently from
  - Eagle Ford (south central Texas)
  - Bakken (North Dakota, south central Canada)
  - Marcellus (northeast U.S.) = Utica fields
- New production wells coming on line every week
- New shale fields being discovered rapidly

BAKER HUGHES

Shale Plays

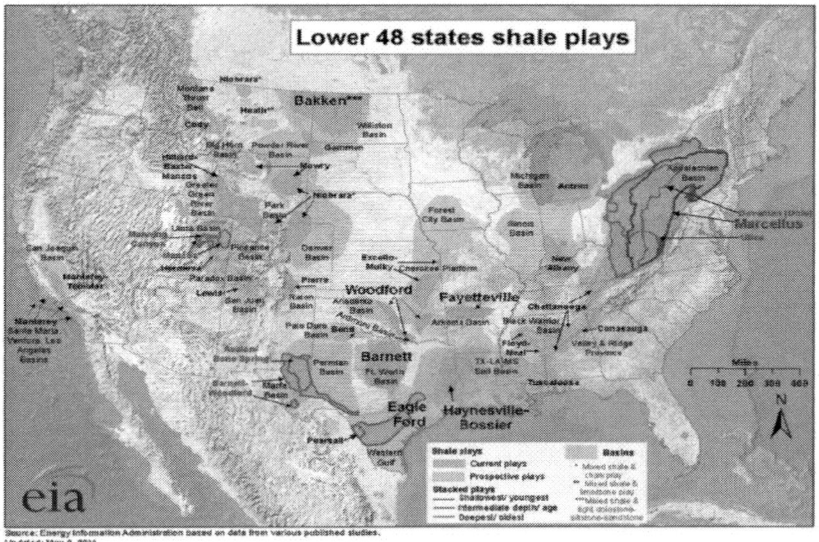

Source: Energy Information Administration based on data from various published studies.
Updated: May 9, 2011

BAKER HUGHES

Shale Oils: The Good, the Bad and the Ugly

- Good
  - Shale oils are low S and low TAN (sweet crudes)
  - Shale oils are paraffinic (good cracking stock)
  - Shale oils can be attractively priced
- Bad
  - Shale oil quality is variable (e.g., Eagle Ford gravity 36 to 55 °API)
  - Shale oils can contain substantial solids (up to 295 ptb measured)
  - Distillate product quality can require additives to meet ASTM specs
- Ugly
  - Waxes: fouling and emulsion issues
  - Light Paraffins: unstable blends with asphaltenic crudes
  - Paraffinic: paraffins crack, can contribute to furnace coking
  - No heavy ends: plants must blend with heavy crude to fill bottoms units
  - Tramp MEA:HCl salts possible

BAKER HUGHES

Eagle Ford Crude Samples

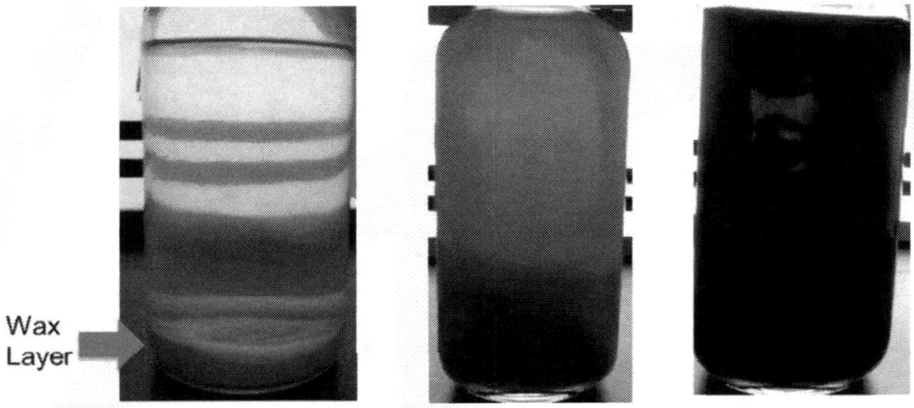

Analytical Breakdown

Value	Yellow	Red	Black
API	55.0	44.6	52.3
TAN	<0.05	0.07	<0.05
Sulfur	<0.20%	<0.2%	<0.2%
Na	1.0 ppm	1.6 ppm	1.6 ppm
K	0.3 ppm	0.4 ppm	0.5 ppm
Mg	3.4 ppm	2.9 ppm	3.0 ppm
Ca	2.6 ppm	2.8 ppm	3.8 ppm
Asphaltene	0%	0%	0.1%
Resin	0.5%	3.2%	1.6%
Filterable solids	176 PTB	295 PTB	225 PTB
P	5.3 ppm	4.6 ppm	5.0 ppm

Paraffin Analysis Bottom of Sample

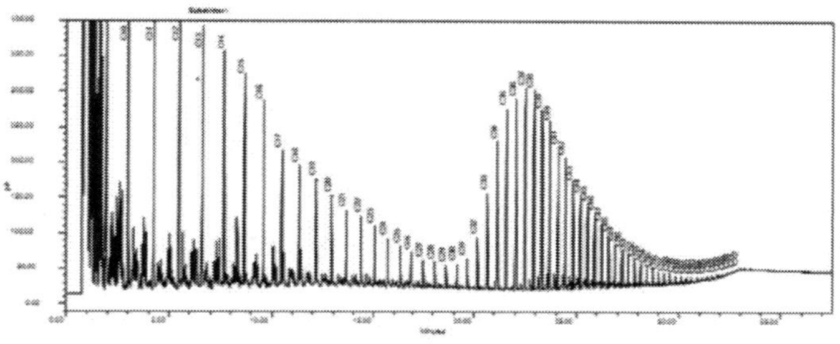

Considerable high molecular weight wax in bottom
High MW = High melting point (> 100 °C)

BAKER HUGHES

Refining Challenge: Paraffinic Oils

- Potential impacts
  - Tank farm
    - Sludge/emulsion accumulation, periodic release to crude unit
  - Preheat train fouling
    - Cold preheat sees wax, solids buildup
  - Desalters
    - Emulsions stabilized by high melting waxes, solids
- Options to manage problems due to paraffins
  - Tankage
    - Pretreatment to break emulsions, disperse waxes
  - Preheat
    - Wax and solids dispersants to reduce preheat fouling
  - Desalters
    - Tank pretreatment, wax dispersants can benefit desalters
    - Specially formulated demulsifier products

BAKER HUGHES

Wax Impact on US Crude Unit

- Concern
  - Waxy sludge in tankage causing desalter upsets
  - High solids/Fe releasing, impacting downstream FCCU
  - Tank had up to 5 feet of sludge/emulsion buildup
  - FCCU catalyst makeup > $500K for one event
- Supporting data
  - GC of tank sludge and desalter interface confirm waxes
- Initial strategy – weekly demulsifier "slug" feeds to relieve desalter emulsions
  - High dosage for 4 hours to break emulsions
  - Inefficient because of difficulty anticipating need/timing

BAKER HUGHES

Wax Impact on US Crude Unit

- Long term strategy: multi-faceted
  - Changed demulsifier product
    - Better product for wax-stabilized emulsions
  - Reduced desalter level
    - Maximize oil residence to break emulsions
    - Level controlled to avoid oil in brine
  - Instituted continuous mudwashing practice
    - Aids in removing solids with brine
  - Instituted tank pretreatment
    - Break emulsions in tankage, improve crude quality to desalter

BAKER HUGHES

Wax Impact on US Crude Unit

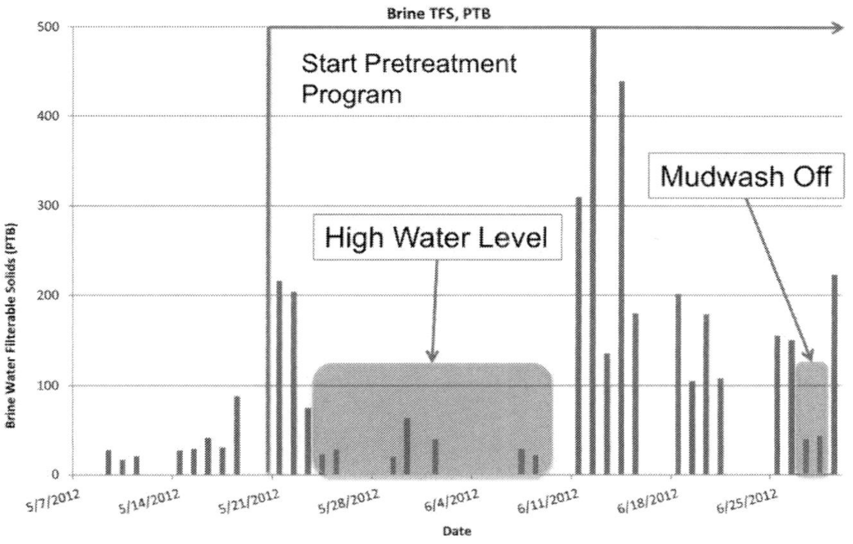

Wax Impact on US Crude Unit

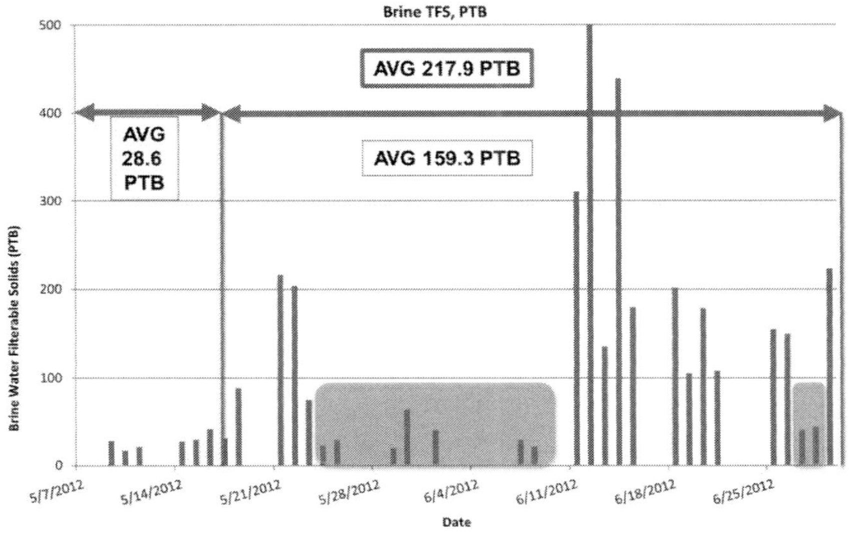

Wax Impact on US Crude Unit

- Results
  - Solids released to brine increased almost 8X
  - Fe in desalted crude reduced 50%
  - Weekly "slug" treatments no longer required
    - Program costs neutral – pretreatment costs = Slug treatment costs
  - FCCU catalyst makeup rates returned to pre-shale oil rates

BAKER HUGHES

Refining Challenge: Asphaltene Instability

- Potential impacts
  - Light paraffins in shale can destabilize asphaltenes in blends
    - Tank fouling/emulsion stabilization
    - Cold preheat fouling
    - Desalter emulsion stability
    - Hot preheat and heater fouling
- Options to manage asphaltene instability
  - Blend and monitor with ASIT ™ technology
    - Asphaltene Stability Index Test
  - Addition of asphaltene stabilizers/antifoulants
    - Crude tank pretreatment
    - Preheat train injection

BAKER HUGHES

ASIT is a trademark of Baker Hughes Incorporated

ASIT Asphaltene Stability Index Test

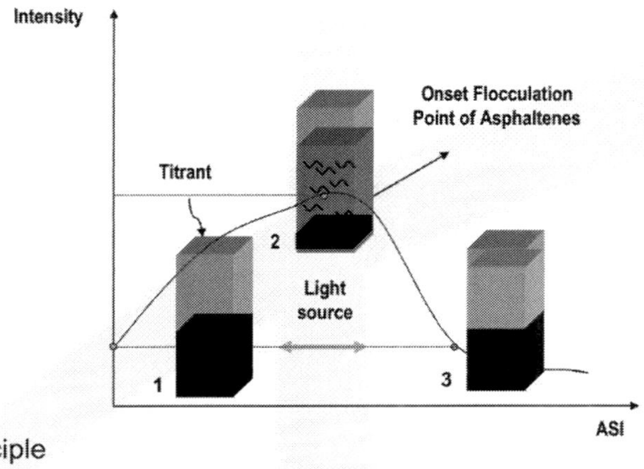

Principle

Shale Oil Destabilizes Asphaltenes

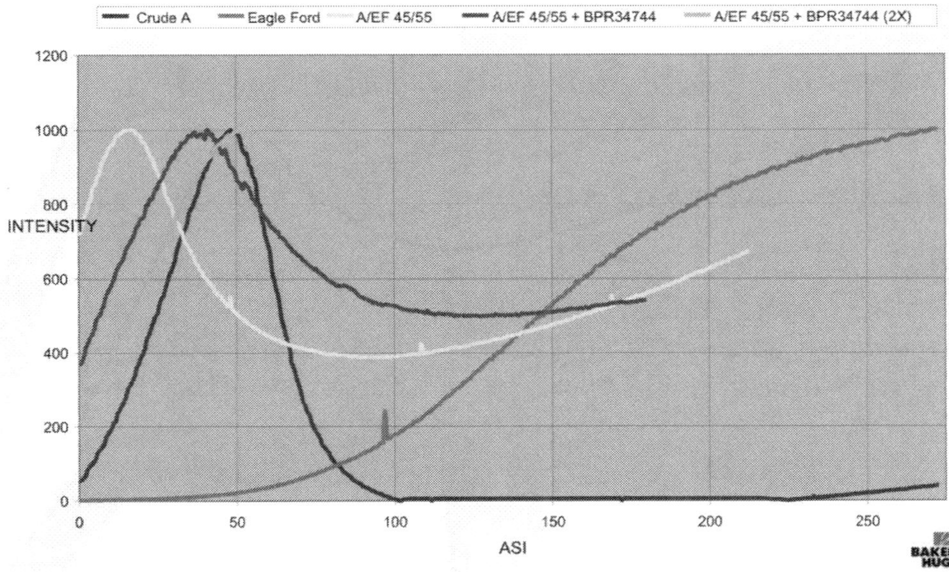

Blending Shale Oils with Conventional Crudes

Hot Preheat Exchanger Train and Furnace Fouling

- Causes
  - Asphaltene deposition
    - May be coupled with paraffins
  - Inorganics imbedded in the organic phase
  - Coke formation because of rapid asphaltene destabilization
- Solutions
  - Stabilize asphaltenes
  - Regularly monitor asphaltene stability (ASIT technology)
  - Add antifoulants designed for asphaltenes and paraffins
  - Monitor exchangers and furnace operations vigilantly

BAKER HUGHES

Asphaltene Fouling – US Refinery

- Concern
  - Crude unit processing blend of heavy, asphaltenic crude with shale oil
  - Severe, rapid fouling of preheat train (over period of 2 months)
  - Heater over-firing led to unplanned shutdown
    - Heater tube temperatures approached tube melting point
  - **Cost of shutdown $45 MM**
- Supporting data
  - Lab fouling tests showed severe fouling
  - Deposit analyses show 70-90% Organics, H/C 1-1.2 (asphaltenes)
  - ASIT data suggested severe asphaltene instability of blends being run

BAKER HUGHES

Shale Oil Blend Resulted in Rapid Fouling

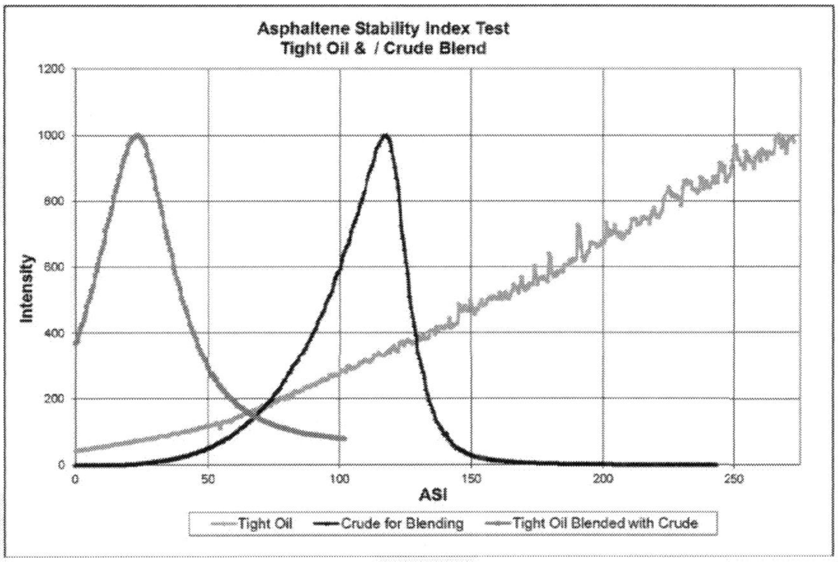

BAKER HUGHES

Asphaltene Fouling – US Refinery

Hot Train Exchanger Deposit Analyses

Crude Unit Deposit Analyses - Results in Wt%										
Sample Description	C	H	N	O	Cl	Fe	S	H/C Atomic Ratio	Ash	Summary
Exchanger 1 - Crude Side	82	8	1	2	1		6	1.16	1	Asphaltenes
Exchanger 2 - Crude Side	78	7	1	4	1	1	8	1.07	3	Asphaltenes
Exchanger 3 - Crude Side	81	8	1	2	1		7	1.18	3	Asphaltenes

BAKER HUGHES

Asphaltene Fouling – US Refinery

- Mitigation and Monitoring
  - Eliminated heavy crude with most severe instability in shale oil blends
  - ASIT technology implemented at refinery to manage crude blending
  - Chemical treatment options
    - Asphaltene stabilizer for storage if ASI < 100
    - Preheat antifoulant possible if ASI < 80
  - Tracking key heat transfer parameters
    - Heater inlet temperature
    - Heater firing rates
    - Tube skin temperatures
- Results
  - Rates of fouling under control with ASI > 100 since unit restart

BAKER HUGHES

Additional Fouling Concerns

- High solids can lead to fouling in exchangers and heaters

- Paraffinic hydrocarbons crack more easily
  - Initially forms olefinic fragments
  - Olefins can polymerize and form coke precursors
  - Crude is light, volatile, but polymers/coke precursors non-volatile
  - Evaporation in furnace can strand polymers/coke precursors
  - Act as binder for solids, salt crystals to accumulate

BAKER HUGHES

Furnace Fouling With Shale Oil

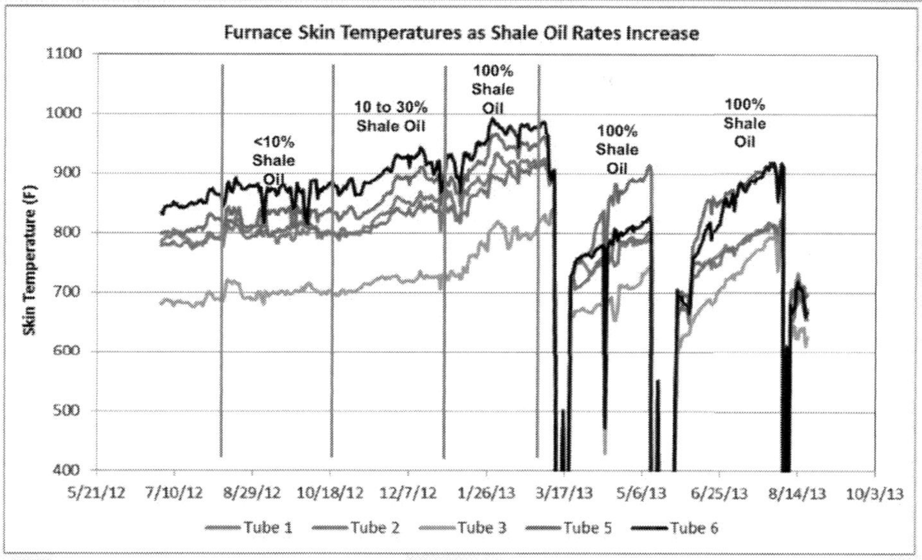

BAKER HUGHES

Furnace Fouling Experience With Shale Oil

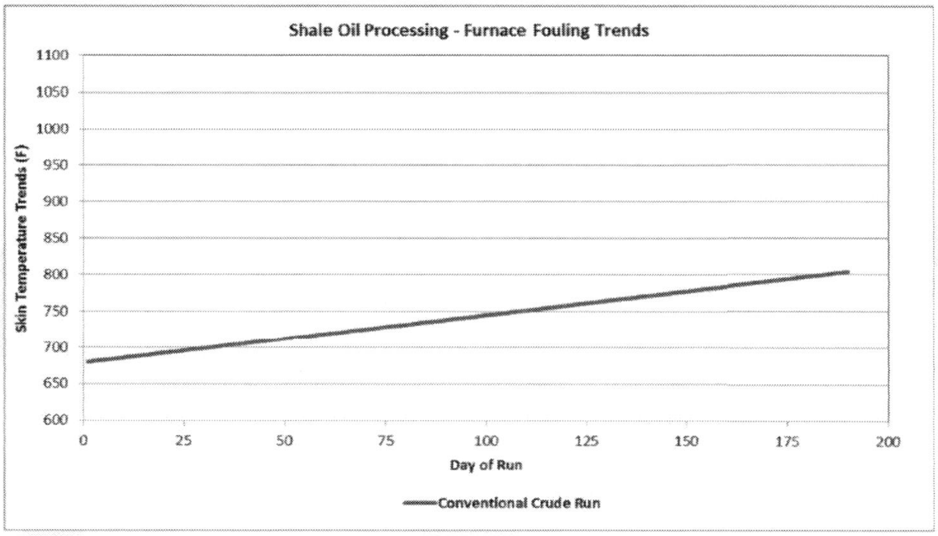

BAKER HUGHES

Furnace Fouling Experience With Shale Oil

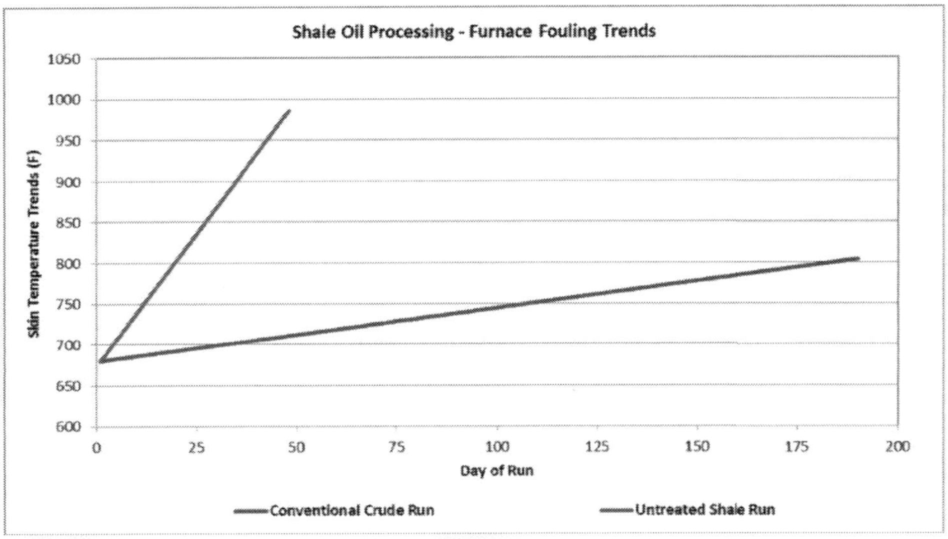

BAKER HUGHES

Furnace Fouling Experience With Shale Oil

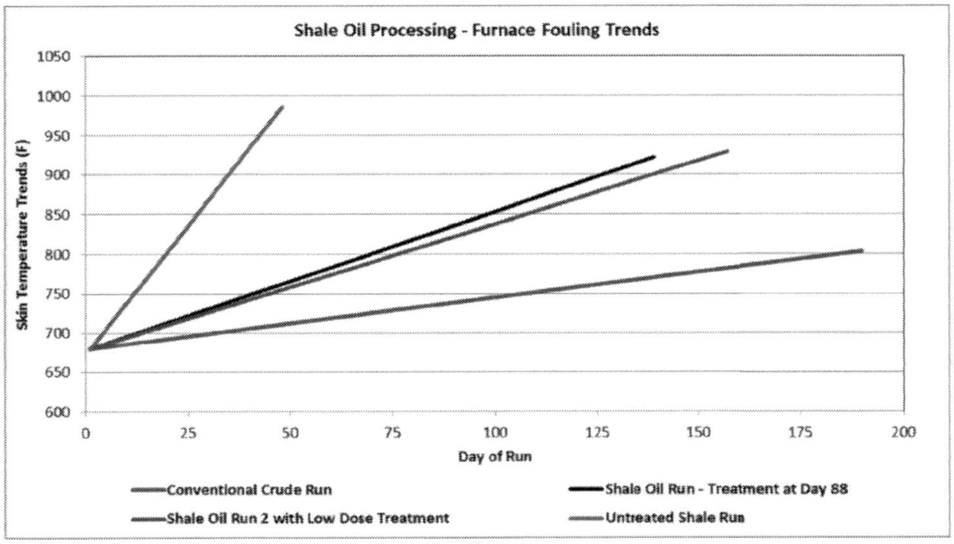

BAKER HUGHES

Furnace Fouling Experience With Shale Oil

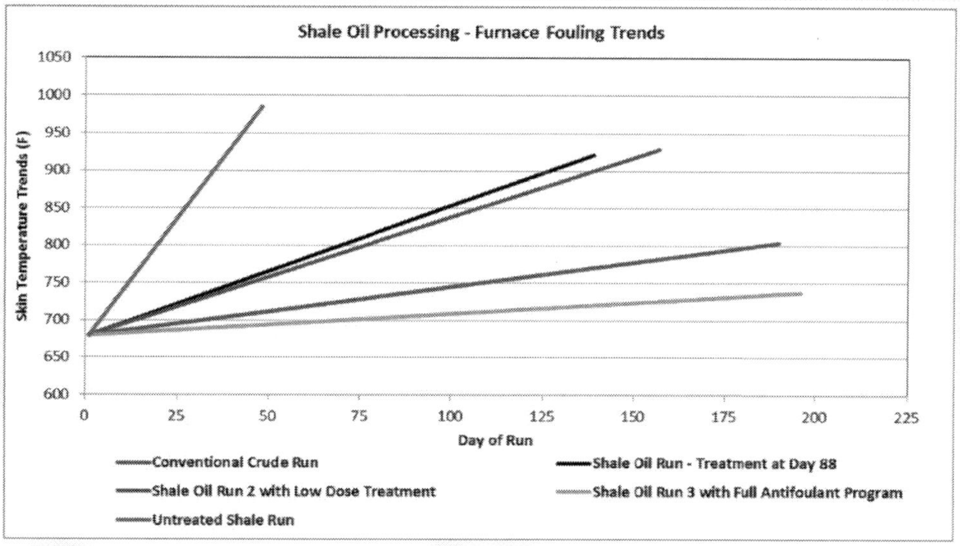

BAKER HUGHES

Refining Challenge: Tramp Amines

- Potential impacts
  - Many shale oils treated for H_2S (road, rail, barge transport)
  - H_2S scavengers react to leave amine in crude oil (MEA, MA)
  - Tramp amines distill to OH system, create amine-HCl salts
    - Corrosive to system metallurgy
- Options to manage salt formation potential
  - Set maximum chloride target
  - Change process conditions (e.g., tower top temperature)
  - Desalter acidification to reduce amine loading
  - TOPGUARD™ Ionic Model to define targets

BAKER HUGHES

TOPGUARD is a trademark of Baker Hughes Incorporated

Analytical Breakdown – OH Waters

Value	Sample 1	Sample 2	Sample 3	Sample 4
pH	7.4	7.5	6.7	7.4
Neutralizer Rate	0 gpd	0 gpd	0 gpd	0 gpd
Chloride (ppm)	73	94	193	93
NH_3 (ppm)	36	29	55	23
Tramp MEA (ppm)	197	235	250	175
Tramp MA (ppm)	126	122	176	115
Raw Crude Salt (ptb)	9.8	7.5	47.2	6.4
Na (ppm)	3.8	3.5	19.3	3.8
K (ppm)	3.2	2.2	8.9	2.4
Mg (ppm)	1.3	1.6	1.5	1.6

BAKER HUGHES

Amine Partitioning in Desalters

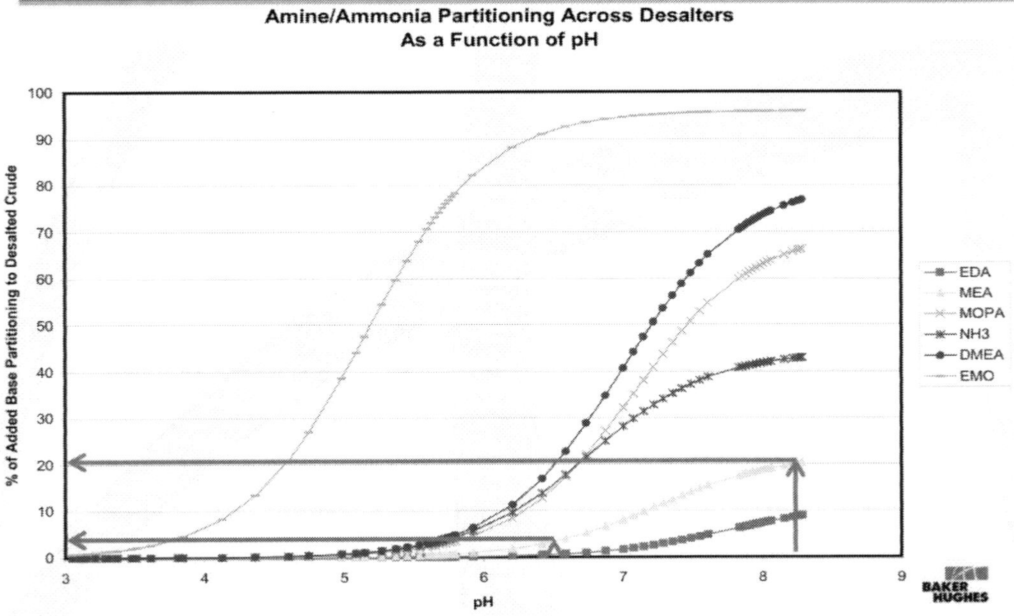

BAKER HUGHES

Amine Contamination From Shale Oil

- Concern
  - Tramp MEA from H_2S Scavenger in crude oil
  - Concentrations sufficient to render neutralizer unnecessary
  - Overhead bundle life reduced to < 5 months (previously 2+ years)
  - Bundle replacements costs increase 5X
  - Unit must shut down if 2 banks fail simultaneously: Cost $4.5MM/day
- Supporting data
  - Bundle tube temperatures 239 to 252 °F
  - Overhead neutralizer usage dropped to 0 when MEA levels rose
  - Initial MEA levels < 10 ppm
    - Salt formation temperatures 230 to 240 °F
  - MEA levels rose to between 20 and 70 ppm
    - Salt formation temperatures between 245 and 260 °F

BAKER HUGHES

Impact of Tramp MEA on OH Water

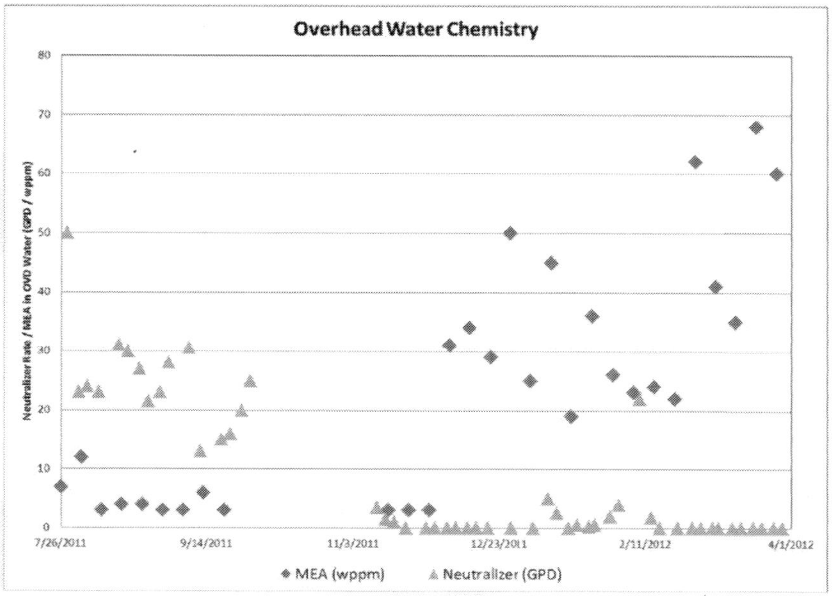

BAKER HUGHES

Salt Risk with MEA Concentration

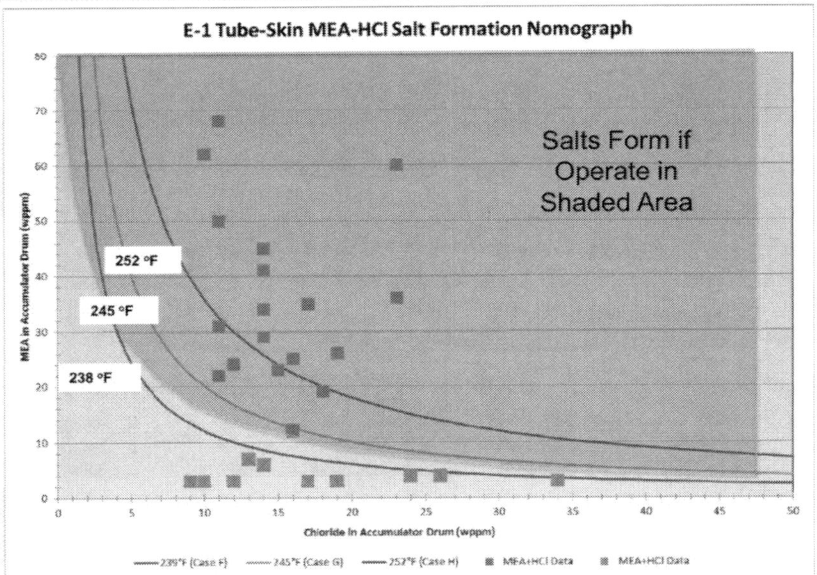

Exchanger Max Allowable Salt Temperature

BAKER HUGHES

Amine Contamination From Shale Oil

- Mitigation
  - Caustic to maintain maximum chloride target of 15 ppm
  - Desalter acidification to reduce MEA in overhead system to < 20 ppm
  - Set hot drum temperature to 265 °F minimum target
  - Note: this strategy aims to return operations to the 2+ year bundle life conditions
  - Monitoring with TOPGUARD corrosion risk monitor (CRM)
- Results
  - TOPGUARD CRM data show system remaining clear of salt formation since implementation of new strategy

BAKER HUGHES

Product Quality Impacts

Shale Derived Naphtha

- Test Results
  - Corrosion fail
  - Water Tolerance pass
  - Oxidative Stability pass
  - Existent Gums pass
- Mitigation
  - Corrosion pass rating easily achieved using corrosion inhibitors

Shale Derived Diesel

- Test Results
  - NACE Corrosion — marginal fail (NACE Corrosion)
  - Thermal Stability — pass
  - Water Tolerance — pass
  - Cold Flow — Summer pass, Winter fail
  - Filter plugging — pass
  - Lubricity — poor on WSD and static dissipation
  - Conductivity — poor
- Mitigation
  - Corrosion inhibitors can raise NACE Corrosion rating to B+
  - Pour point depressants can easily improve Cold Flow
  - Specialty blends bring lubricity and conductivity into spec

BAKER HUGHES

Shale Derived Gas Oil

- Factors
  - Paraffinic – makes for good FCC feed
  - High pour can cause handling issues
    - Intermediate storage
    - Transfer lines
- Mitigation: Easily treated with pour point additive

BAKER HUGHES

Shale Derived Resid/Bunker Fuel

- Factors
  - Paraffinic, high pour point
    - Intermediate storage and transfer line concerns
  - Paraffins can lead to sedimentation in heavy fuel blends
- Mitigation:
  - Transport/storage: Treatable with pour point additive
  - Sedimentation: Measurable with ASIT technology, treatable with asphaltene stabilizers

BAKER HUGHES

Refinery Challenges From Processing Shale Oils

Thanks for your attention

BAKER HUGHES

Process Intensification Strategies for Sustainable Process Development– A "Dow" Perspective

Deshpande, K. T., The Dow Chemical Company
Pendergast, J., The Dow Chemical Company
Dixit, R., The Dow Chemical Company
Originally presented on: 10/4/2013 13:00:00 - 13:30:00

As the world's population rises and new economies emerge, society requires novel solutions to meet its most basic needs, including energy, water, housing, food, health, and transportation. However, limited natural resources entail their efficient use while simultaneously enhancing the quality of life of current and future generations. In this respect, Process Intensification is becoming an immensely important tool to achieve overall sustainability of the chemical process industry and the reduction of our energy footprint through controlled plant size. Numerous chemical companies including Dow, BASF, DSM, DuPont, and Ehrfeld are addressing process sustainability through concerted process intensification undertaking. This talk will focus on Dow's efforts towards sustainable processing through process intensification. In particular, case studies involving process intensification and their impact on the overall plant material and energy balance will be discussed.

Tridiagonal Solutions

compute | innovate | develop

Improving Process Performance of Ethylene Crackers using CFD

Damodaran Vedapuri, Ph.D.
Tridiagonal Solutions Inc.

Presented at the AIChE's 5th Southwest Process Technology Conference

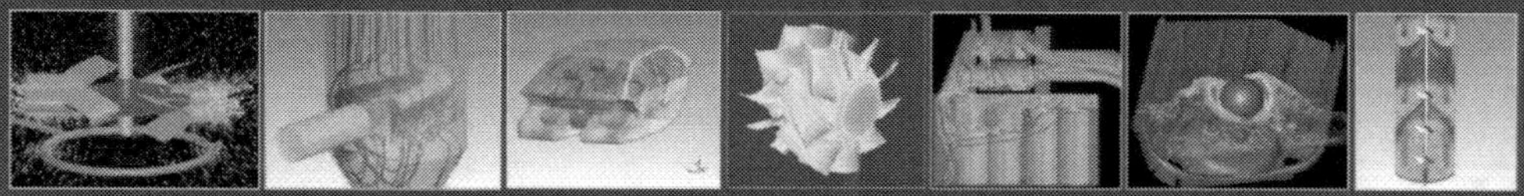

www.tridiagonal.com
info@tridiagonal.com

Outline

Tridiagonal Solutions

compute | innovate | develop

- Introduction
- Improvements in Process Modeling of Crackers
- Firebox – Design Improvements using CFD
- Coupling Firebox Combustion and Cracking Reactions
- Reactor Tube Analysis
  - Straight Pipe, Finned Tube and Swirl Tube

Introduction

Tridiagonal Solutions
compute | innovate | develop

- Furnace technology: Area of active research
  - High-energy consumption, capital and maintenance cost of the current cracking furnace are a driving force to develop improved conversion routes
- Cracking is almost exclusively carried out in tubular coils in fired heaters
- Different configuration of vertical tubes
  - Single or double row, Single Pass, U Type, W type, Hybrid Coil
- Desired characteristics
  - Improved life, maximize yield
- Firebox & Reactor Tube optimization

Firebox Development

Tridiagonal Solutions
compute | innovate | develop

- Process Enhancements
  - Non-uniform heat flux on reactor tubes
  - High thermal stresses, lower tube life
  - High coke formation, frequent shutdowns
  - Pollutant formations
  - Retrofit of existing designs
- New Designs
  - Changes in feedstock
  - Mix fuel combustion (Natural gas + Hydrogen)
  - Low-NOx burners
  - Higher radiant efficiency
  - Flue gas recirculation
  - Burner orientation, location

Improvements in Modeling in Cracking Process

Tridiagonal Solutions
compute | innovate | develop

Firebox

- Approximate representation of flame shape
- Assume a constant heat flux on the tubes based on firing rates

Image source http://www.essentialchemicalindustry.org/processes/cracking-isomerisation-and-reforming.html

Reactor Tube

- 1D calculations for cracking chemistry
- Heat input from radiation
- Failed to consider heat flux variations
- Approximate calculations for heat transfer

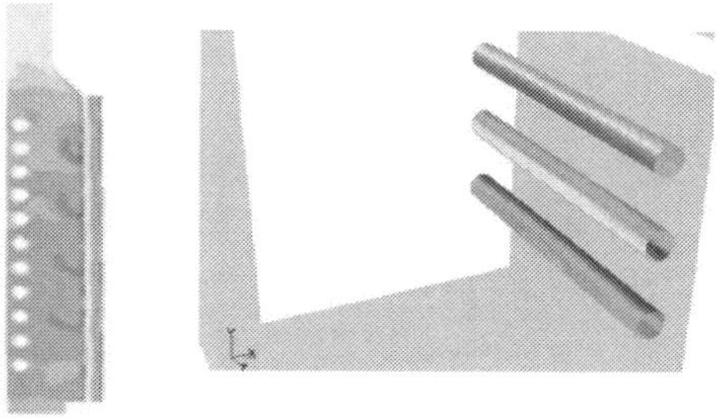

Improvements in Modeling in Cracking Process

Tridiagonal Solutions
compute | innovate | develop

CFD Framework:

- Detailed representation of burners
- Understand flow patterns, flame stabilization
- Flame shape prediction
- Study effect of flue gas recirculation
- Design new concepts
- Computational costs limited study of multi burners

Improvements in Modeling in Cracking Process

Tridiagonal Solutions
compute | innovate | develop

- Multi-burner Simulation
  - More compute power
  - Study of Flame-Flame Interaction
  - Prediction of deviation from ideal flame behavior
  - Radiation calculation from combined flame front

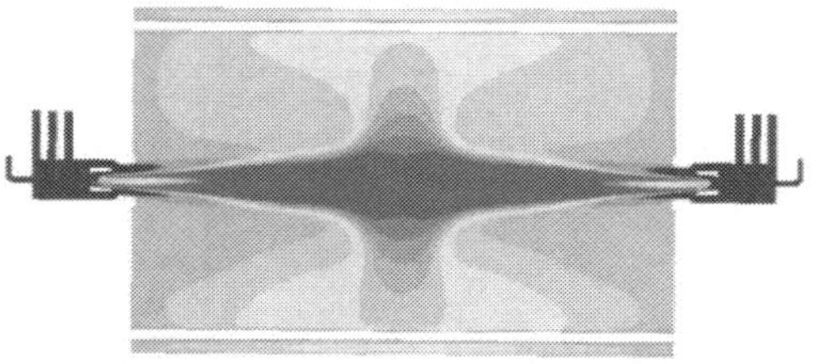

Opposed Burner
Flame Interaction

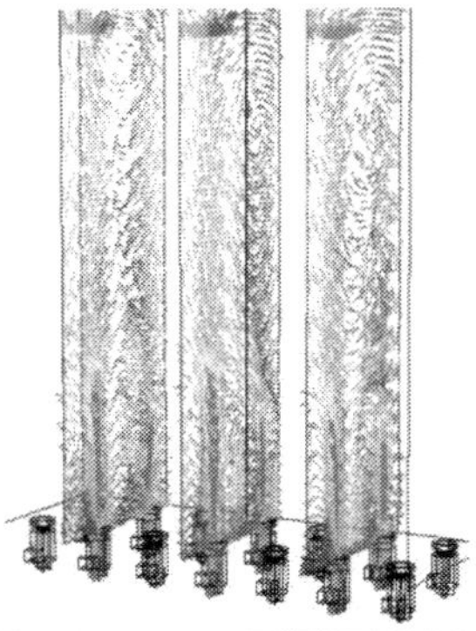

Tridiagonal Solutions
compute | innovate | develop

Fire Box Improvement: A Case Study

Firebox Improvements using CFD

Tridiagonal Solutions
compute | innovate | develop

- Existing Design: Tall and broad flame shape
- High thermal stresses on radiant tubes
- Non-uniform heat flux
- CFD modeling to predict temperature distribution in firebox
- Multiple burner domain divided into periodic zones

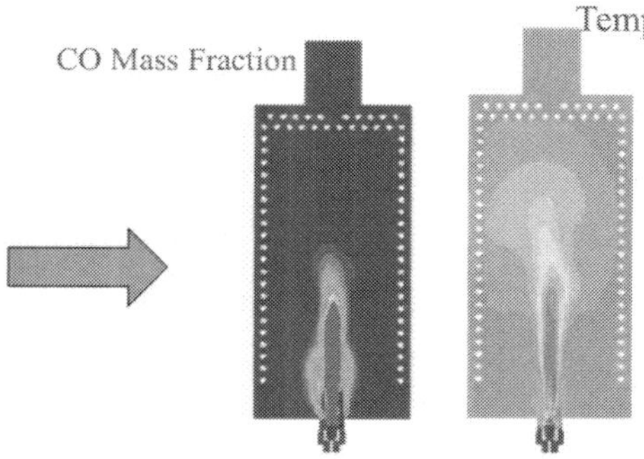

Firebox Improvements using CFD

Tridiagonal Solutions
compute | innovate | develop

- Burner Design modified to
  - Change primary & secondary air distribution through burner
  - Fuel distribution in the combustion zone
  - Fuel tip orientation

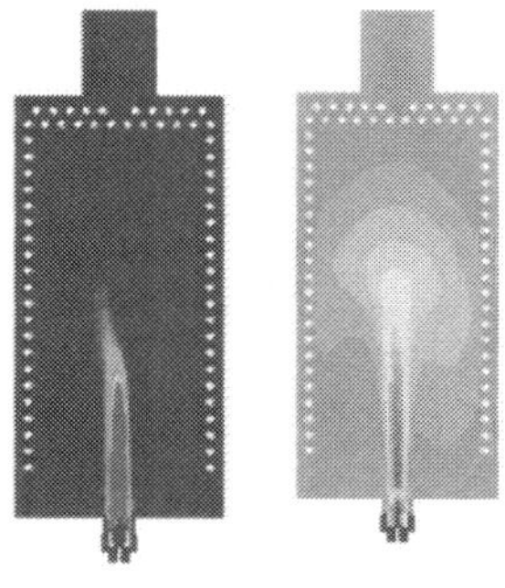

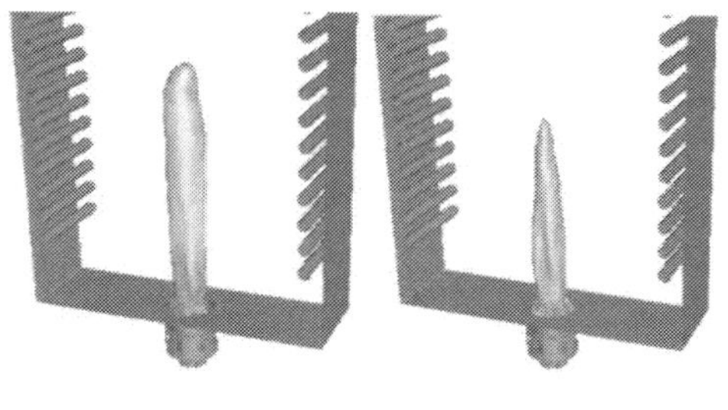

- Flame for modified design is lean and short, consistent observation for different firing rates
- Heat flux on the tubes are more uniform

Firebox Improvement Using CFD

Tridiagonal Solutions
compute | innovate | develop

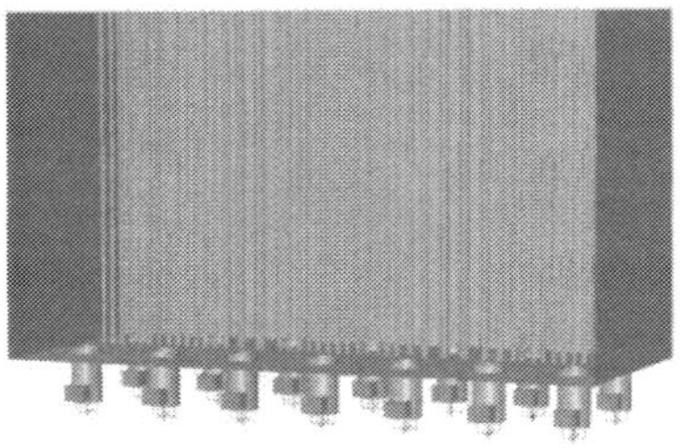

- 35 floor burners
- CFD study to predict & minimize high temperature spots
- Flow recirculation near end burners resulted in flame deflection towards tubes
- Burner orientation modified to direct flame away from tubes

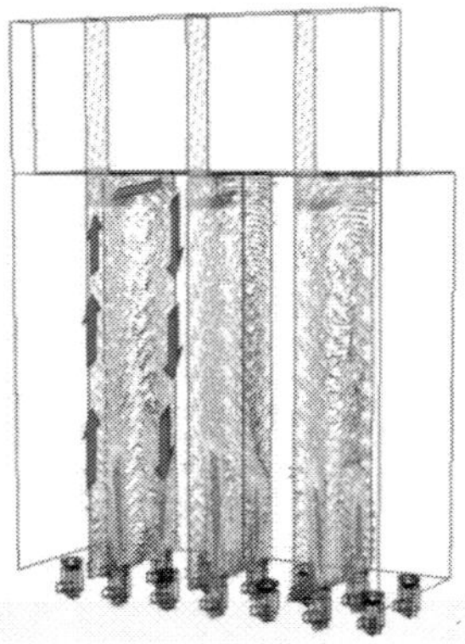

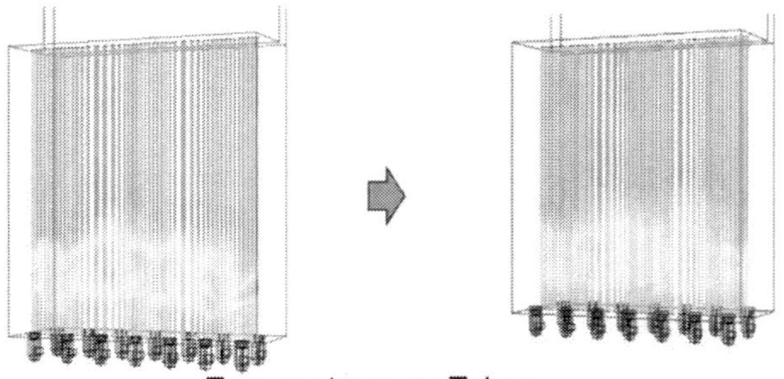

Temperatures on Tubes

Tridiagonal Solutions
compute | innovate | develop

Coupling Firebox Combustion with Cracking Reactions

Burner-Reactor Tube Interaction

Tridiagonal Solutions
compute | innovate | develop

Kinetic Modeling advantage & limitations:

- Improved prediction of yield
- Limited to 1D
- Wall heat flux approximations
- Iteratively coupled with CFD models

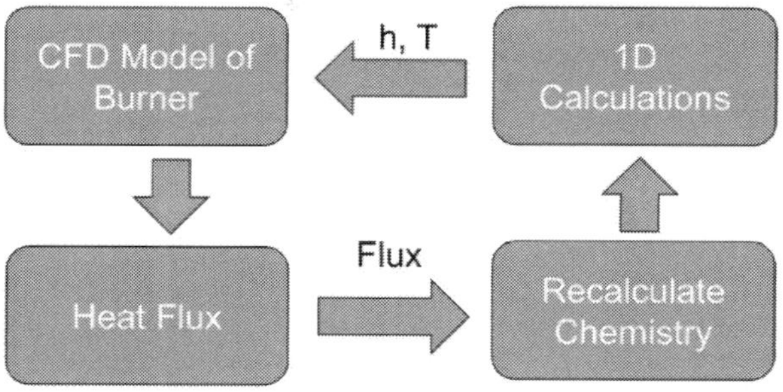

Limitations can be overcome by

- Solving combined Reactor Tube & Firebox
- Different Reactor designs can be studied for process improvement

Complete Furnace Simulation

Tridiagonal Solutions
compute | innovate | develop

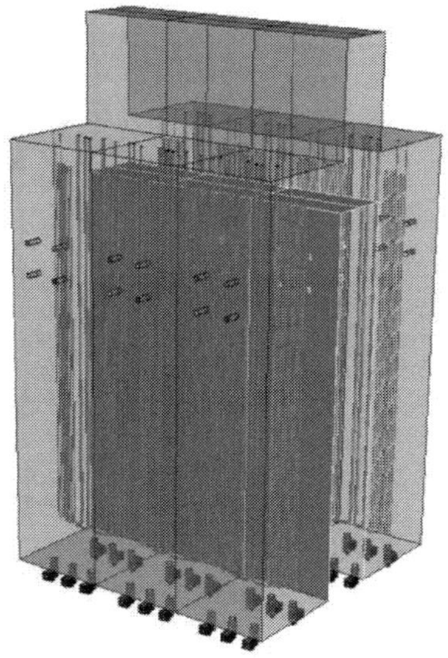

- Furnace Box:
  - Furnace Height: 13.7 m
  - 36 bottom burners, 48 side burners
  - Fuel: Primarily Methane
- Reactor Coils:
  - Number of reactor coils: 24
  - Tube Diameter: 2"
  - Feed Rate: 40 ton/hr
  - Inlet steam : 60%
  - Feedstock Temperature: 883K
- No assumption on heat profile of reactor tubes

Cracking Reactions

Tridiagonal Solutions
compute | innovate | develop

- Highly endothermic reactions
- Require temperatures in the range of 750 to 900 degrees C
- Coke formation reactions not considered
- Reaction Mechanism: 22 reactions model by Kumar and Kunzru (1985)

No.	Reaction equation[a]	Activation energy (KJ/mol)	pre-exponential factor (s^{-1})
1	Naphtha $\rightarrow$ $0.58H_2 + 0.6CH_4 + 0.82C_2H_4 + 0.1C_2H_6 + 0.45C_3H_6 + 0.02C_3H_8 + 0.035C_4H_{10} + 0.2C_4H_8 + 0.07C_4H_6 + 0.09C_{5+}$	220.1	6.565×10^{11}
2	$C_2H_6 \rightleftharpoons C_2H_4 + H_2$	273.0	4.652×10^{13}
3	$C_3H_6 \rightleftharpoons C_2H_2 + CH_4$	273.5	7.284×10^{12}
4	$C_2H_2 + C_2H_4 \rightarrow C_4H_6$	172.7	$(1.026 \times 10^{15})^b$
5	$2C_2H_6 \rightarrow C_3H_8 + CH_4$	273.2	3.75×10^{12}
6	$C_2H_4 + C_2H_6 \rightarrow C_3H_6 + CH_4$	253.0	$(7.083 \times 10^{16})^b$
7	$C_3H_8 \rightleftharpoons C_3H_6 + H_2$	214.7	5.888×10^{10}
8	$C_3H_8 \rightarrow C_2H_4 + CH_4$	211.9	4.692×10^{10}
9	$C_3H_8 + C_2H_4 \rightarrow C_2H_6 + C_3H_6$	247.3	$(2.536 \times 10^{16})^b$
10	$2C_3H_6 \rightarrow 3C_2H_4$	268.7	7.386×10^{12}
11	$2C_3H_6 \rightarrow 0.3C_nH_{2n-6} + 0.14C_5^+ + 3CH_4$	238.2	2.424×10^{11}
12	$C_3H_6 + C_2H_6 \rightarrow 1\text{-}C_4H_8 + CH_4$	251.3	$(1.0 \times 10^{17})^b$
13	$n\text{-}C_4H_{10} \rightarrow C_3H_6 + CH_4$	249.7	1.0×10^{12}
14	$n\text{-}C_4H_{10} \rightarrow 2C_2H_4 + H_2$	295.9	7.0×10^{14}
15	$n\text{-}C_4H_{10} \rightarrow C_2H_4 + C_2H_6$	256.7	4.099×10^{12}
16	$n\text{-}C_4H_{10} \rightleftharpoons 1\text{-}C_4H_8 + H_2$	261.1	1.637×10^{12}
17	$1\text{-}C_4H_8 \rightarrow 0.41C_nH_{2n-6} + 0.19C_5^+$	212.4	2.075×10^{11}
18	$1\text{-}C_4H_8 \rightleftharpoons H_2 + C_4H_6$	209.3	1.0×10^{10}
19	$C_2H_4 + C_4H_6 \rightarrow B + 2H_2$	144.7	$(8.385 \times 10^{12})^b$
20	$C_4H_6 + C_3H_6 \rightarrow T + 2H_2$	149.2	$(9.74 \times 10^{11})^b$
21	$C_4H_6 + 1\text{-}C_4H_8 \rightarrow EB + 2H_2$	242.7	$(6.4 \times 10^{17})^b$
22	$C_4H_6 + C_4H_6 \rightarrow ST + 2H_2$	124.6	$(1.51 \times 10^{12})^b$

Reactor Tube Details

Tridiagonal Solutions
compute | innovate | develop

Feedstock Composition (weight%)	
n-Paraffins	46
i-Paraffins	32
Olefins	15
Aromatics	7

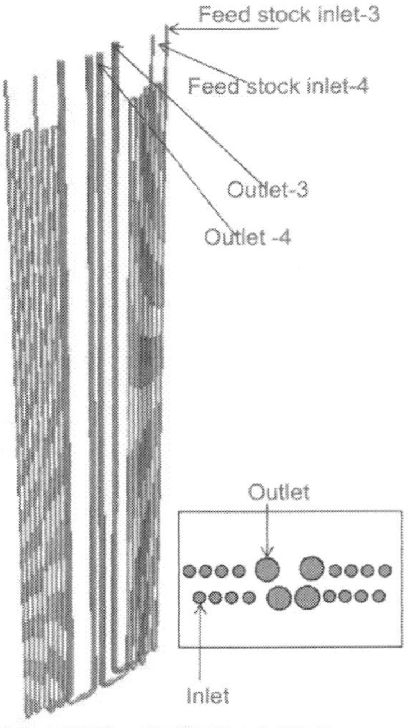

- Modeling Details:
  - Combustion Model: Eddy dissipation with Finite Rate
  - Radiation: Discrete Ordinate
  - Wall emissivity: 0.75
  - Wall Conduction: Through 1D heat transfer (thickness & conductivity)

Flow Field

Tridiagonal Solutions
compute | innovate | develop

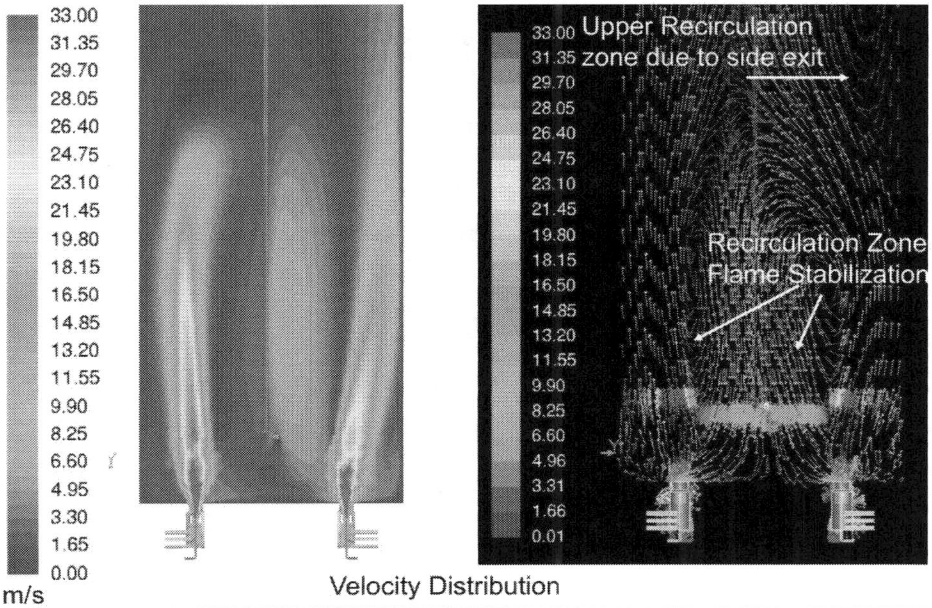

Velocity Distribution

Flow Field

Tridiagonal Solutions
compute | innovate | develop

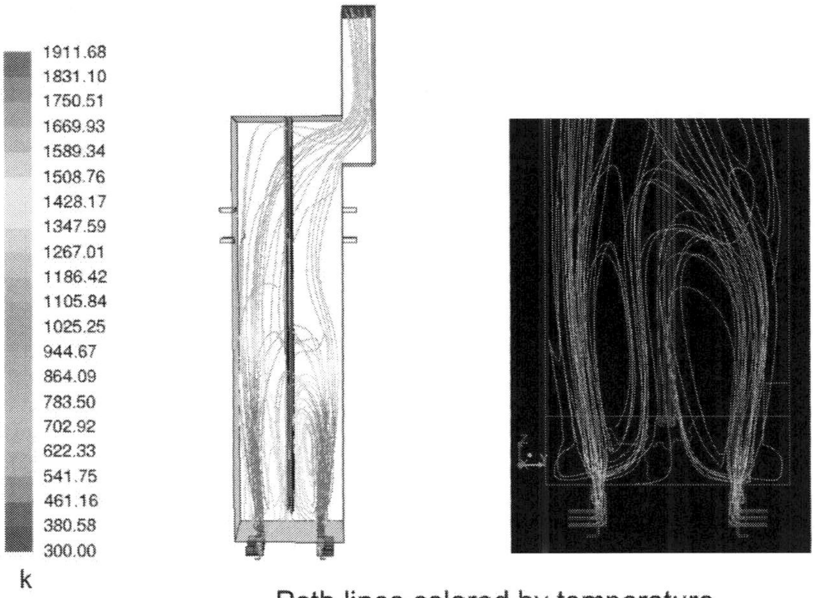

Path lines colored by temperature

Temperature Field

Tridiagonal Solutions
compute | innovate | develop

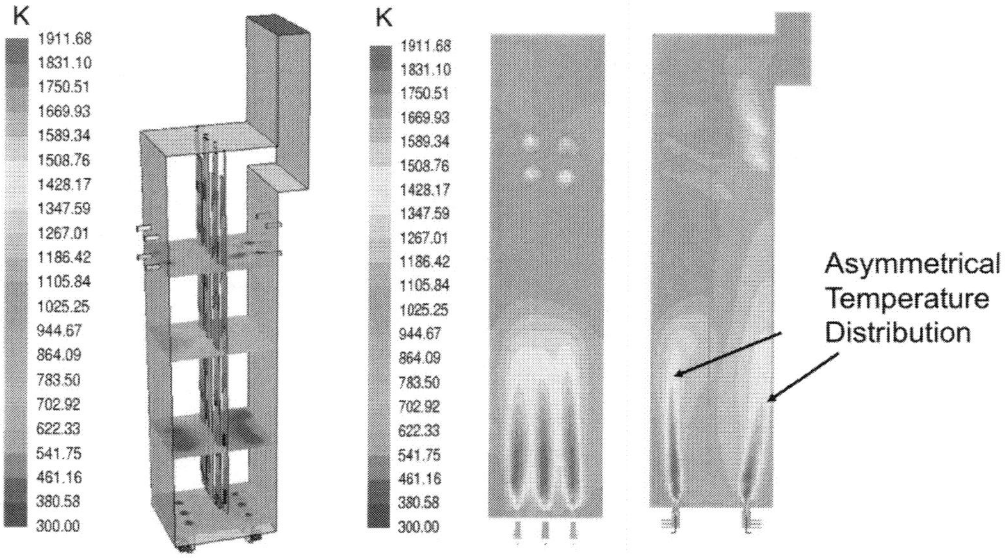

CO Distribution

Tridiagonal Solutions
compute | innovate | develop

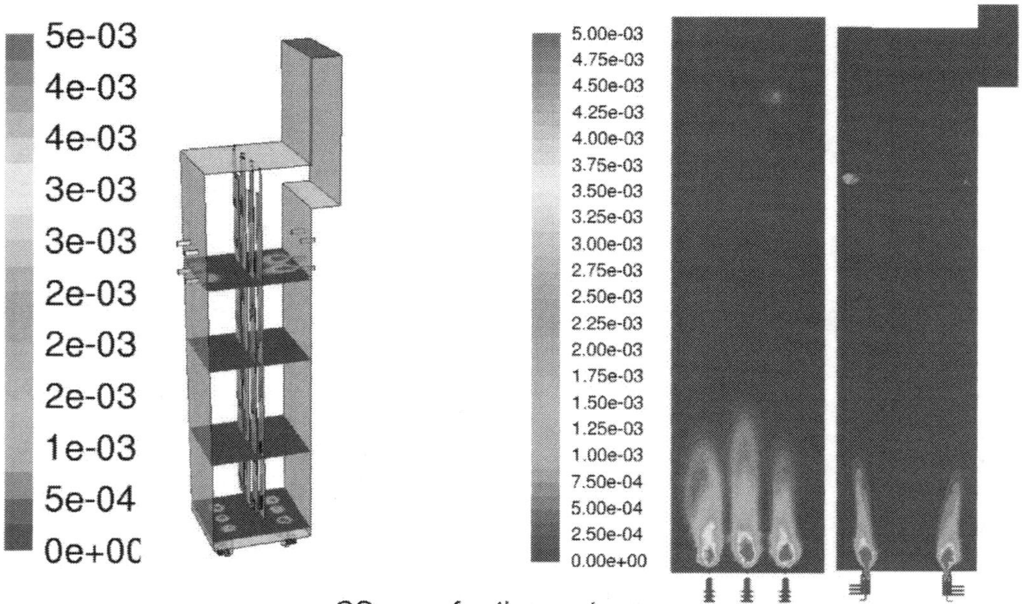

CO mass fraction contours

Oxygen Mass Fraction

Tridiagonal Solutions
compute | innovate | develop

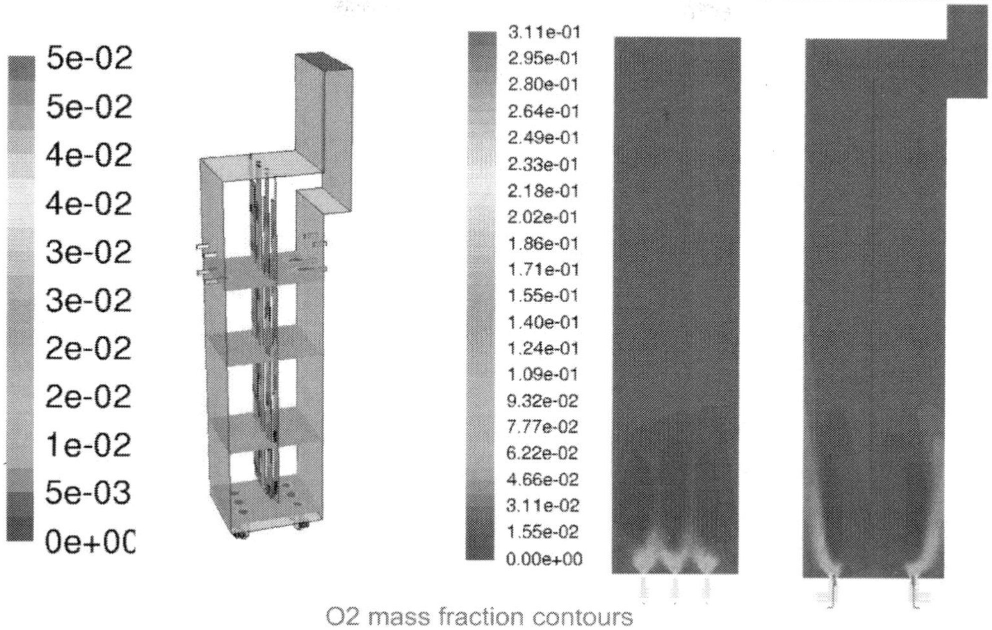

O2 mass fraction contours

Reactor Tube

Tridiagonal Solutions
compute | innovate | develop

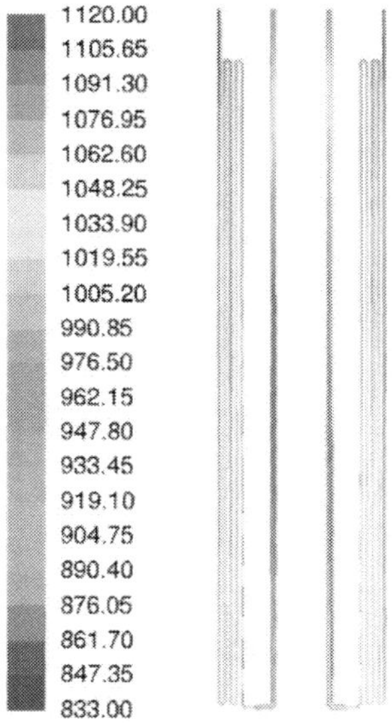

Temperature Contours (k)

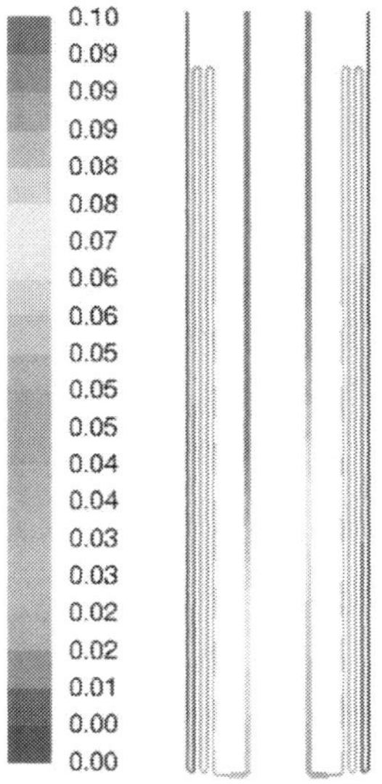

Mass fraction of C2H4

Heat Flux Variation on the reactor pipes

Reactor Tube

Tridiagonal Solutions
compute | innovate | develop

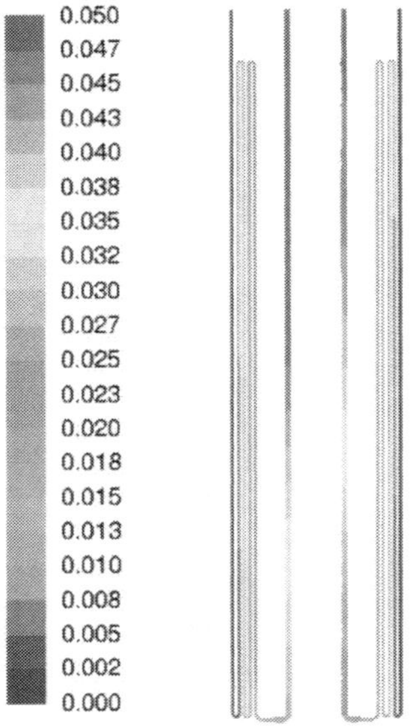

Mass fraction of C3H6

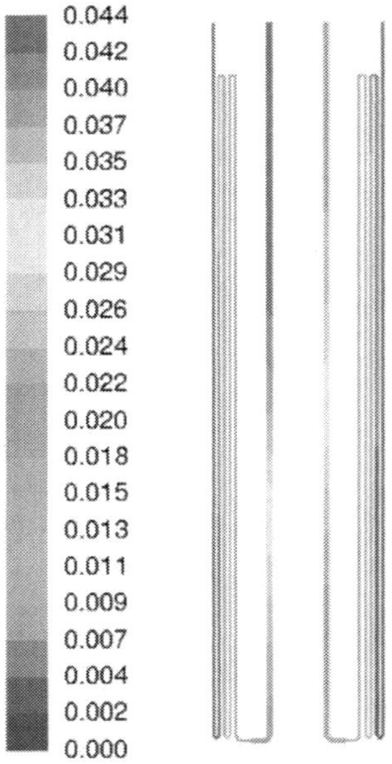

Mass fraction of CH4

Complete Furnace Simulation

Tridiagonal Solutions
compute | innovate | develop

- Yield for different elements

Element	% Yield (CFD)	% Yield (Experimental)
C2H4	24.5	28
C3H6	13.2	15
CH4	11	14

- CFD study slightly under predicted the yield. This can be attributed to
  - Approximations in firebox
  - Limited reaction chemistry for reactor tube side
  - Symmetrical cut section

- Mesh independence study

Mesh	% C2H4 Yield
Coarse	22
Medium	24.5

Effect of Tube Design on Cracker Performance

Reactor Design Modeling

- Complete Furnace Simulation provides mechanism to study different reactor tube designs
- Three designs studied
  - Straight Tube
    - Simple design, Easy to manufacture
    - High coke formation
  - Finned Tube
    - Increased surface area → Increased yield
  - Swirl Tube
    - Increased heat transfer coefficient
    - Uniform temperature distribution

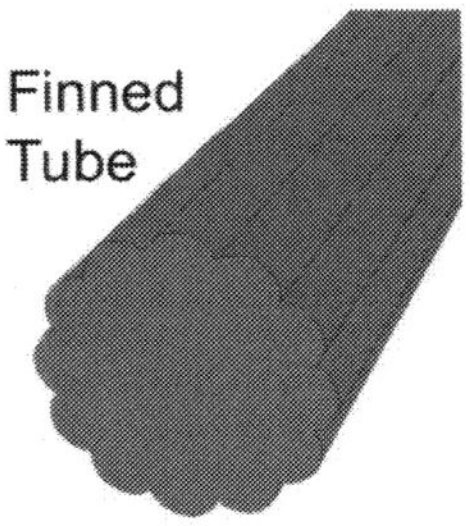

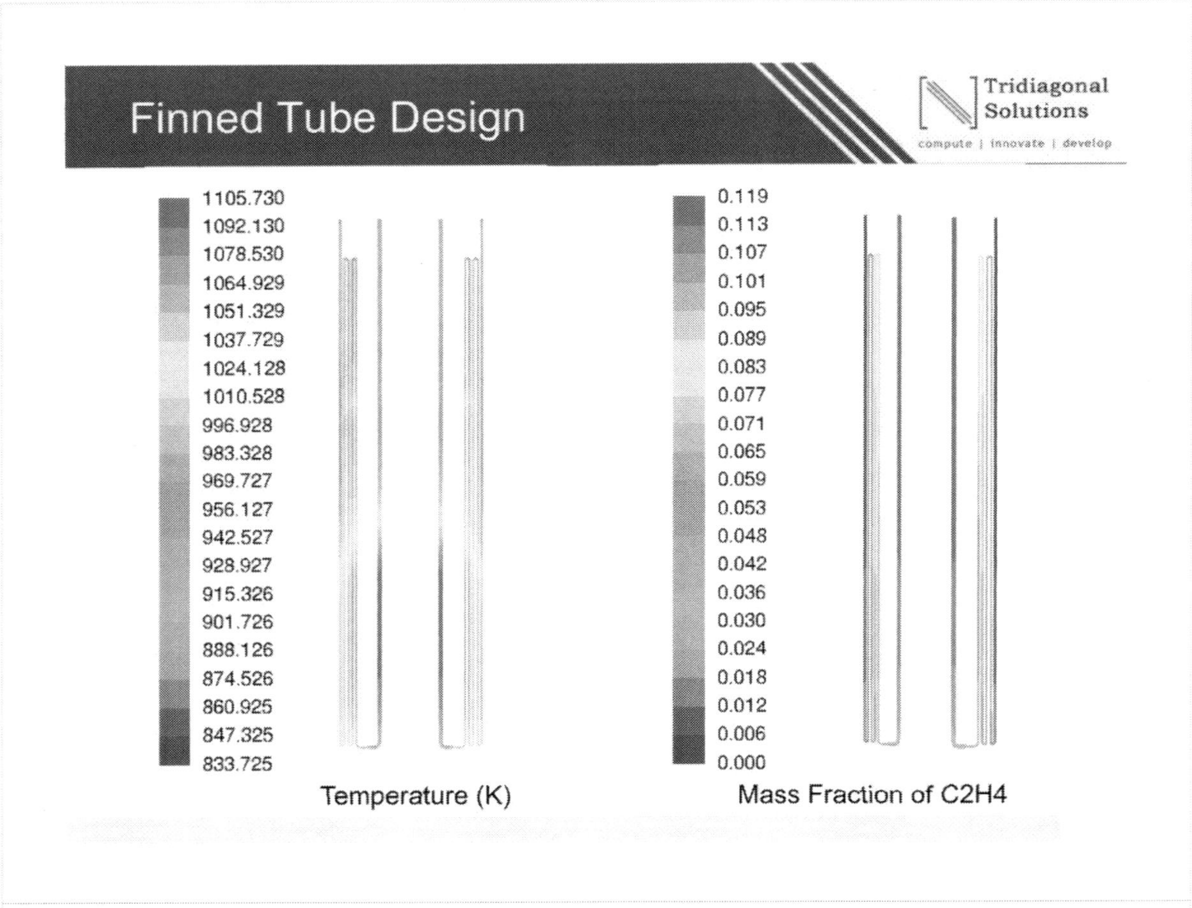
Finned Tube Design
Tridiagonal Solutions
compute | innovate | develop
1105.730
1092.130
1078.530
1064.929
1051.329
1037.729
1024.128
1010.528
996.928
983.328
969.727
956.127
942.527
928.927
915.326
901.726
888.126
874.526
860.925
847.325
833.725
Temperature (K)
0.119
0.113
0.107
0.101
0.095
0.089
0.083
0.077
0.071
0.065
0.059
0.053
0.048
0.042
0.036
0.030
0.024
0.018
0.012
0.006
0.000
Mass Fraction of C2H4

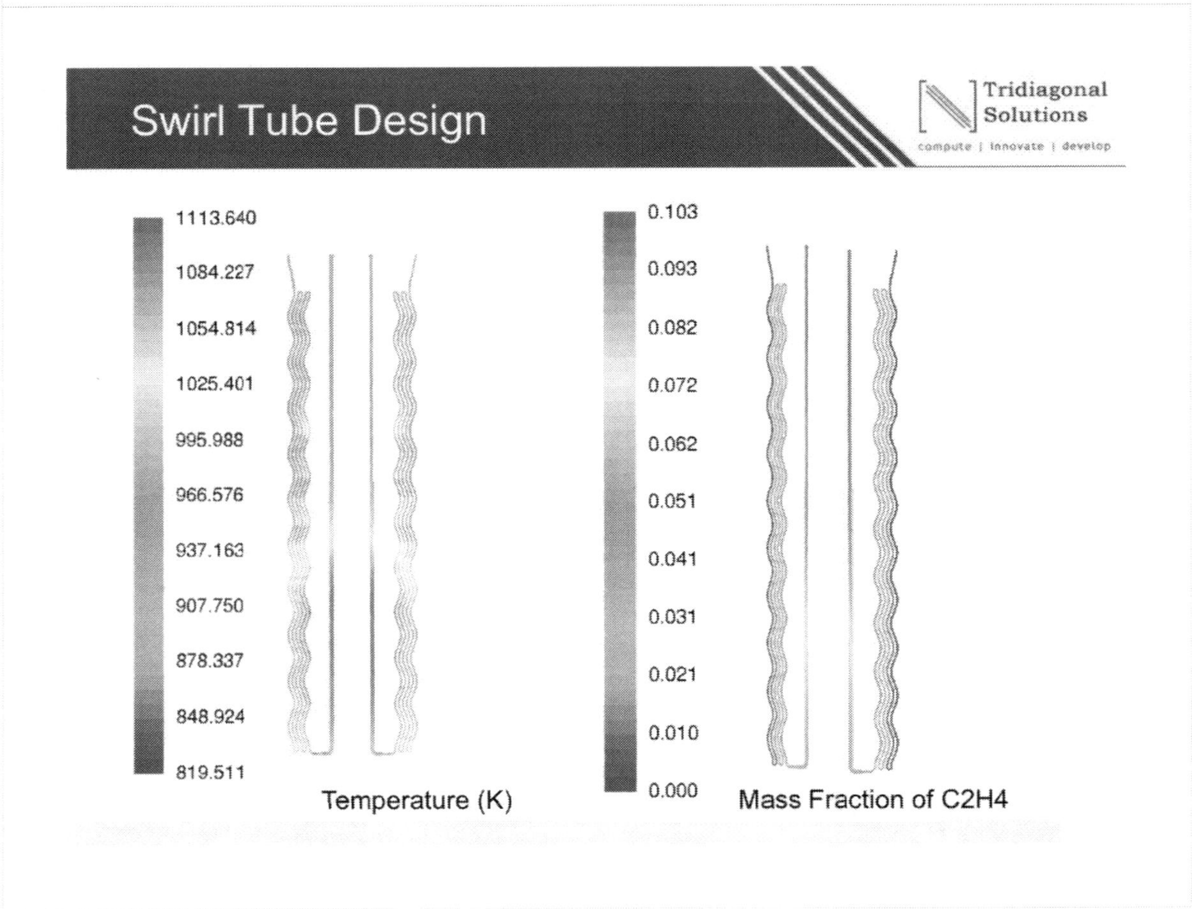
Swirl Tube Design
Tridiagonal Solutions
compute | innovate | develop
1113.640
1084.227
1054.814
1025.401
995.988
966.576
937.163
907.750
878.337
848.924
819.511
Temperature (K)
0.103
0.093
0.082
0.072
0.062
0.051
0.041
0.031
0.021
0.010
0.000
Mass Fraction of C2H4

Velocity Distribution

Tridiagonal Solutions
compute | innovate | develop

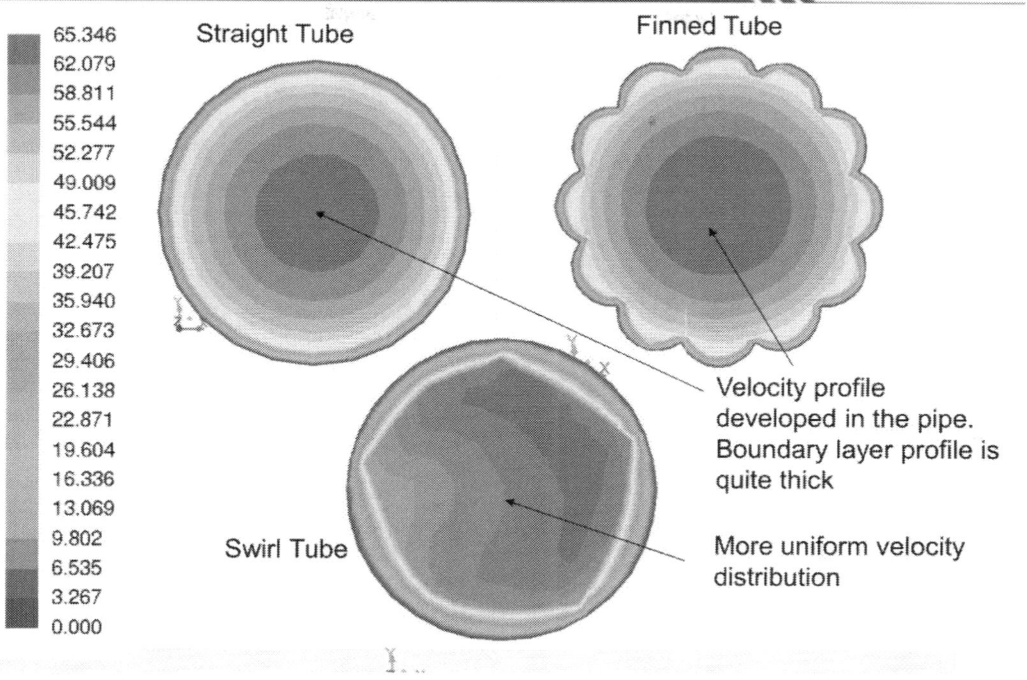

Velocity Distribution

Tridiagonal Solutions
compute | innovate | develop

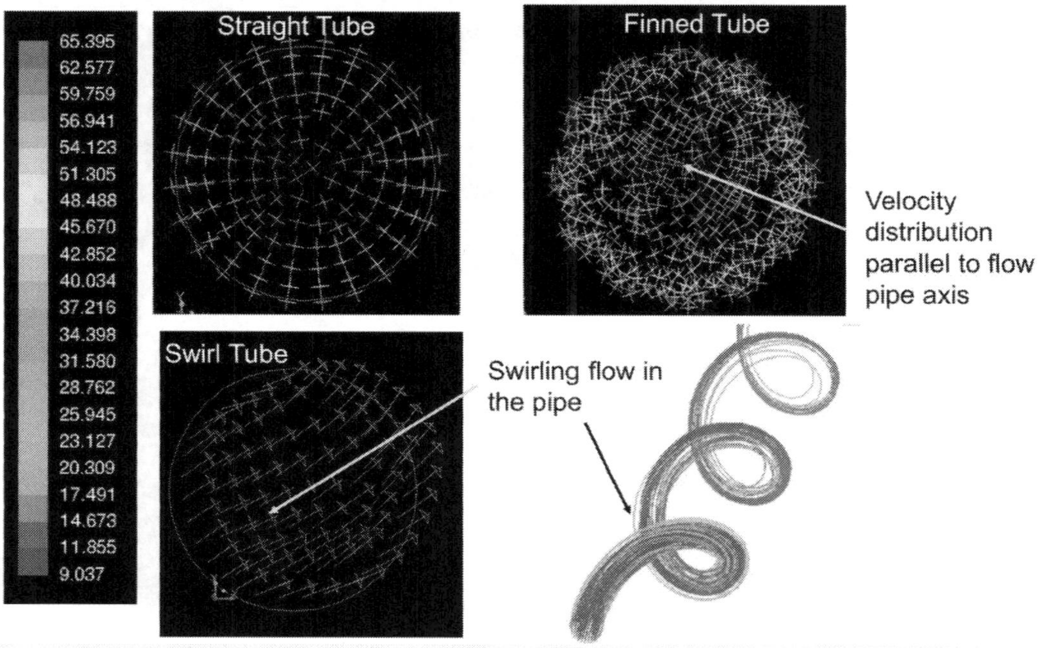

Temperature Distribution

Tridiagonal Solutions
compute | innovate | develop

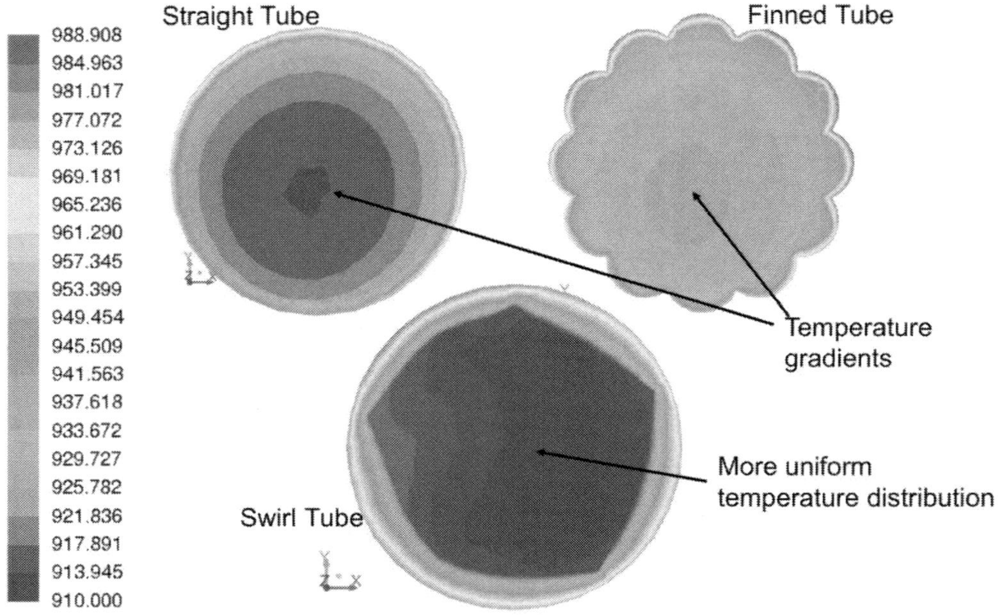

C2H4 Distribution

Tridiagonal Solutions
compute | innovate | develop

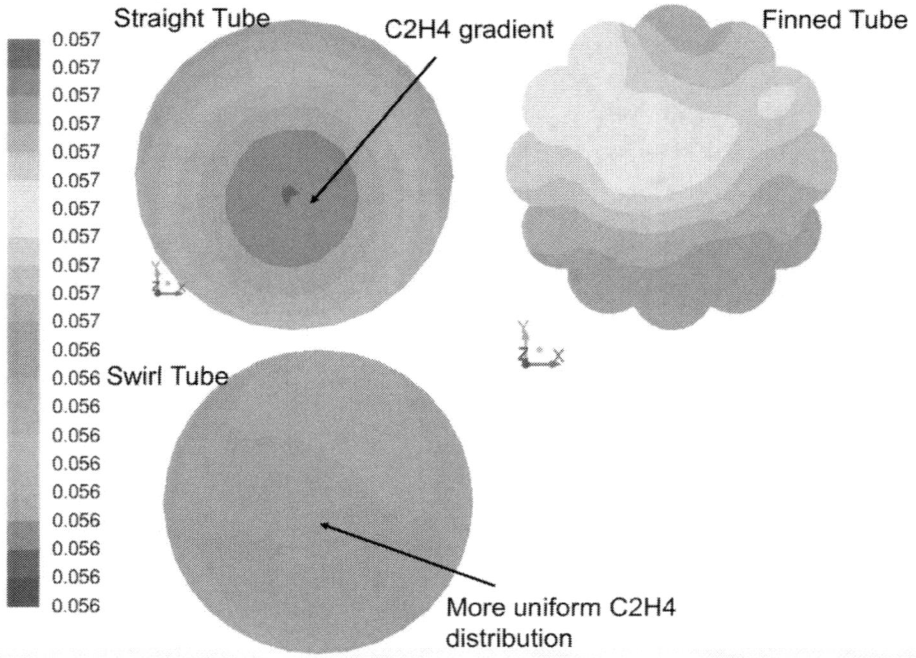

Summary

Reactor Design	C_2H_4 % Yield
Straight	24.5
Finned	27.5
Swirl	26

- Modeling has evolved from simple heat flux calculation to reactor design
- Firebox designs can be improved to meet process needs
- Complete Furnace Simulation (Firebox + Reactor) :
  - Computationally Intensive
  - Predict accurate heat flux
  - Consider variation of heat flux & temperature in the reactor tube
  - Provide mechanism to study complex reactor tube shapes and improve process performance

Adapting Vinyl Acetate Monomer Plant for Manufacturing Allyl Acetate and Allyl Alcohol

Padwal, D.,

Originally presented on: 10/4/2013 14:30:00 - 15:00:00

An economically optimal process configuration was developed for manufacturing of Allyl acetate and Allyl alcohol by adapting Vinyl Acetate Monomer plant. Vinyl acetate monomer is manufactured by Acetoxylation of Ethylene in vapor phase reaction while Allyl acetate is manufactured by Acetoxylation of propylene. Water is formed as by-product in the reaction. Hydrolysis of Allyl acetate produces Allyl alcohol and acetic acid is formed as by-product. Allyl acetate, Allyl alcohol and light end impurities form several binary and ternary low boiling homogeneous and heterogeneous azeotropes with water, which makes purification of Ally acetate and Allyl alcohol a complex process.

A rigorous simulation model was built in Aspen Plus for Allyl acetate manufacturing by adapting process configuration of Vinyl Acetate Monomer plant. Suitable process modification were done in Feed system, Acetoxylation reactor effluent section, Recycle gas section and Purification section considering the difference in feedstock and water content in azeotrope of Allyl acetate-Water. Different process configurations were evaluated for purification of Allyl acetate by azeotropic distillation, to minimize the energy consumption and maximize product recovery.

For Allyl alcohol purification, various different processes were evaluated using advanced process simulation and azeotropic distillation route was found to be the most suitable one. The energy consumption was observed to be sensitive to water content in reaction product and at high water conent, the manufacturing of Allyl alcohol becomes economically non-viable. The trade-off between hydrolysis conversion with different molar ratio of Water to Allyl acetate and the operating cost of purification provided insights for optimizing reaction parameters and determining optimal process configuration for purification.

This paper describes process development approach using advanced process simulation, combined with process knowledge, to determine economically optimum and energy efficient process configuration for manufacturing.

Design of Batch Scheduling

Chemstations, Inc.

Author: Aaron Herrick
Presenter: Wendy Young Reed

Chemstations™

www.chemstations.com

Agenda

- Current practices in design of batch scheduling
- Improvements made possible with
  - Rigorous modeling methodology
  - Microsoft Excel™ to drive a schedule in a simulator
- Example using CITRON (new process)
- Example using WINTEK (existing optimization)
- Summary & suggested procedure

Chemstations™

2

Current practices in design of batch scheduling

- Methods
  - Best guess & experience
  - Lab scale & pilot scale testing
  - Use a basic simulation
- Pros
  - Fast
  - Fairly easy
  - Gives a basic time and heat duty analysis
- Cons
  - Suboptimal schedule times
  - Might not account for equipment/process limitations
  - Difficult to do safety analysis until after process is running

Chemstations™ 3

CITRON (current practice)

Recover 120 kgs of 99.9999% pure Citronellal from 170 kgs essential oil using a 500L pot still with a 10ft packed column.

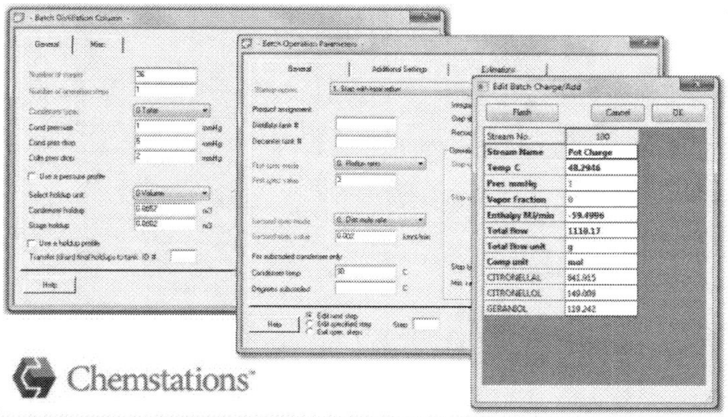

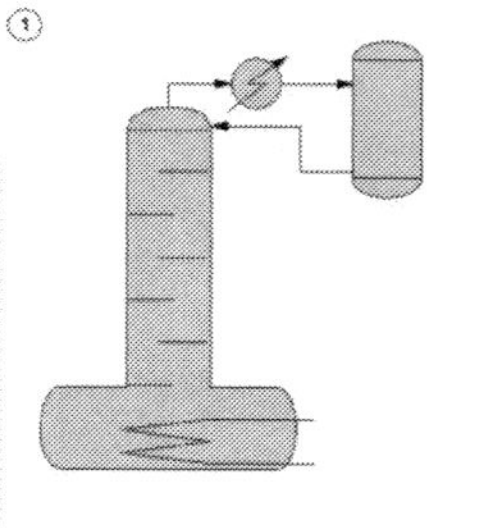

Chemstations™ 4

CITRON (current practice)

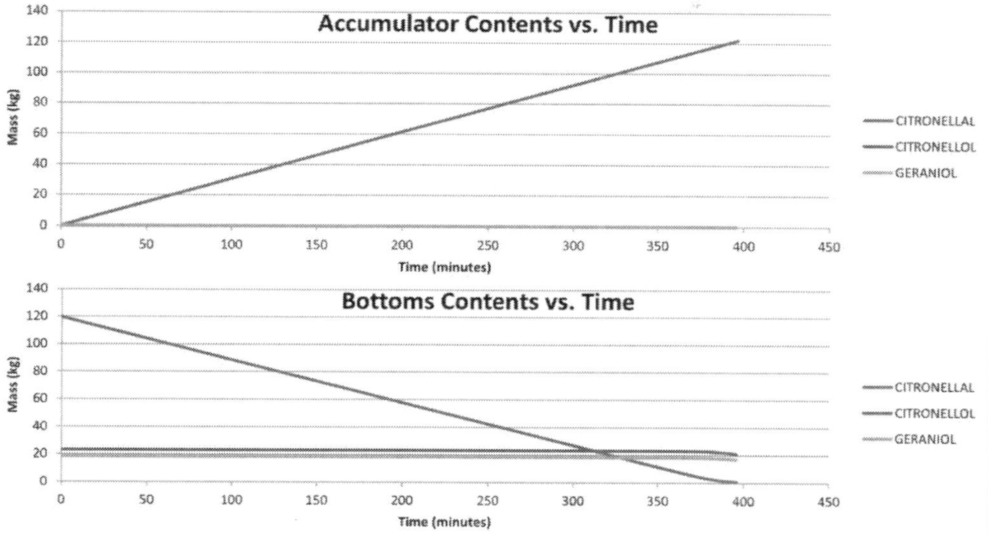

Chemstations 5

CITRON (current practice)

What have we learned from the model?

- Heat duty/flow/time relationship
- Basic operation steps/timing
- Limited equipment sizes and specifications:
  - Heating requirements
  - Condenser requirements
  - Column dimensions

Chemstations 6

CITRON Dynamic (advanced practice)

- Expand our previous example to include
  - Rigorous heat exchanger geometry and performance
  - Event sequencing using DATAMAP to Microsoft Excel
  - Utilities modeling
  - Column metal heat transfer
  - Heat duty / cooling water on control
  - Dry column startup to total reflux
  - Detailed engineering, e.g., nitrogen sweep on vessel, insulation, etc.

Chemstations 7

CITRON Dynamic

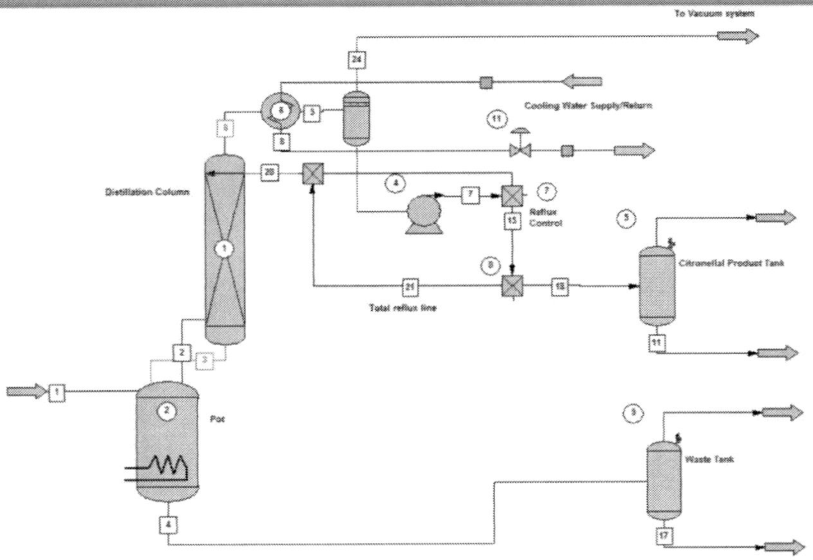

Chemstations 8

Schedule of Events

- A schedule of events is made in Excel and must be connected to CHEMCAD:

	A	B	C	D	E	F	G	H	I
6	Schedule of Events								
7	Time(min)	CW Valve mode	CW Valve Opening	Total reflux	Product Tank	Waste Tank			Notes
8	0	4	0	1	0	0	36	0	Startup
9	21	4	0	1	0	0	29	0	Lower heat duty
10	25	4	5	1	0	0	29	0	Open CW Valve
11	50	4	5	0	1	0	29	0	Switch off Total reflux
12	473	4	5	[illegible]	[illegible]	[illegible]	0		Heater off, drain to tank
13	540	4	0	[illegible]	[illegible]	[illegible]	0		CW off

Schedule of heat duty to pot

Chemstations™ 9

CHEMCAD to Excel connection

CHEMCAD sends time information to Excel cell A2 via datamap

Excel uses time, schedule, and VLOOKUP() function to determine values of process variables

CHEMCAD collects the current time step data from row 2 of the Excel sheet and enters it into the flowsheet via datamap

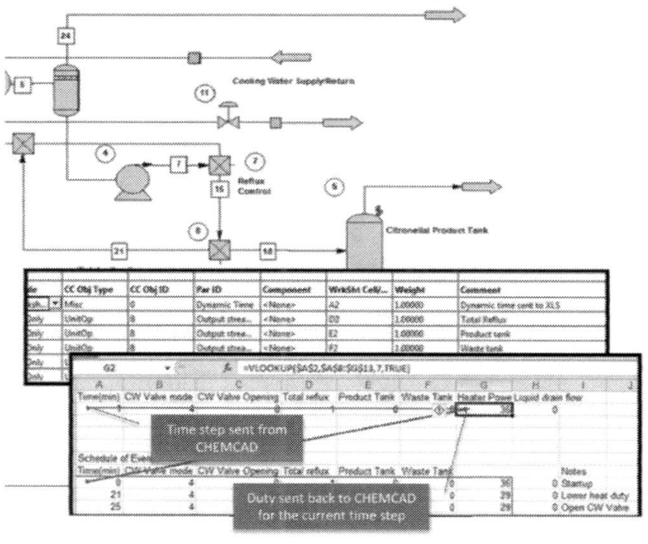

Chemstations™ 10

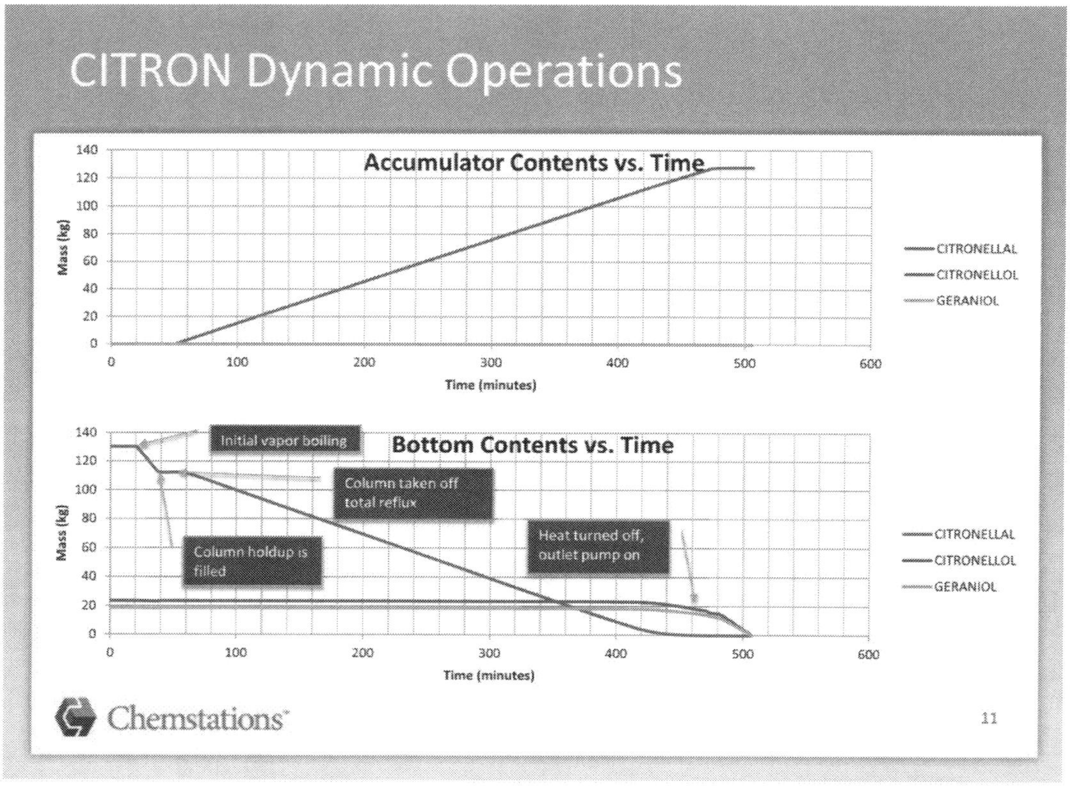
CITRON Dynamic Operations
Accumulator Contents vs. Time
Mass (kg)
Time (minutes)
CITRONELLAL
CITRONELLOL
GERANIOL
Bottom Contents vs. Time
Initial vapor boiling
Column taken off total reflux
Column holdup is filled
Heat turned off, outlet pump on
Chemstations
11

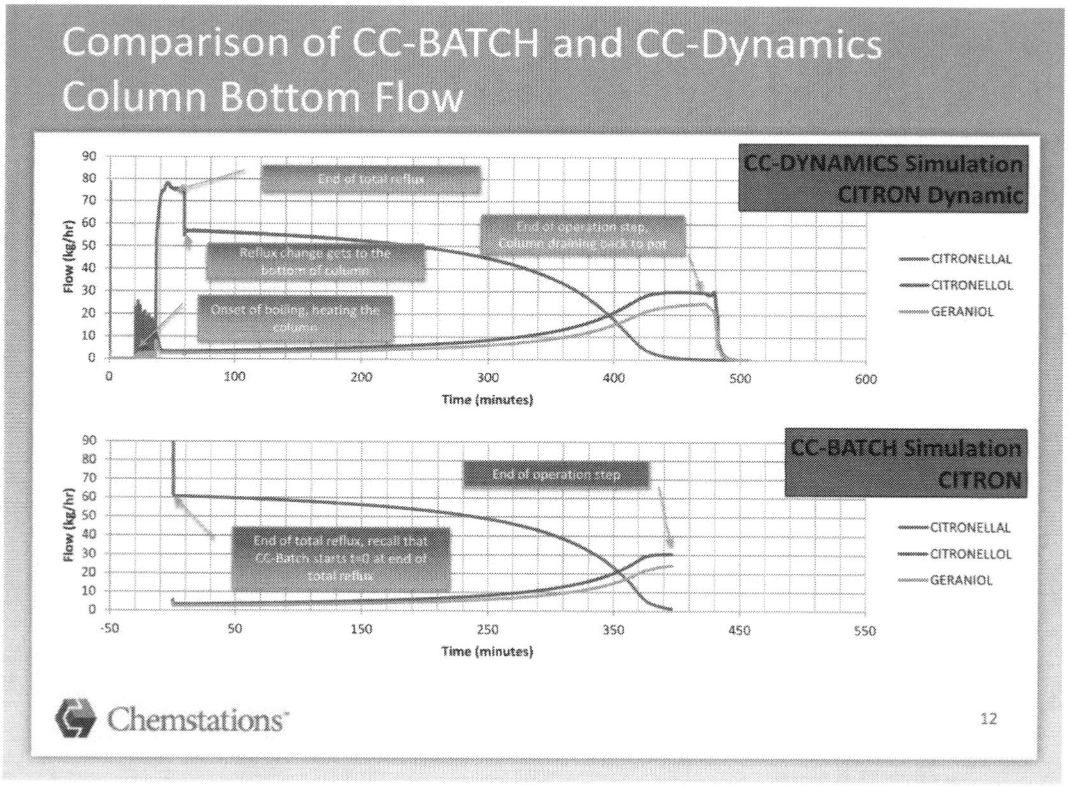
Comparison of CC-BATCH and CC-Dynamics Column Bottom Flow
CC-DYNAMICS Simulation CITRON Dynamic
Flow (kg/hr)
Time (minutes)
End of total reflux
End of operation step, Column draining back to pot
Reflux change gets to the bottom of column
Onset of boiling, heating the column
CITRONELLAL
CITRONELLOL
GERANIOL
CC-BATCH Simulation CITRON
End of operation step
End of total reflux, recall that CC-Batch starts t=0 at end of total reflux
Chemstations
12

CITRON Dynamic Conclusions

We learned from using dynamics:

- Detailed startup procedure
- Detailed operation steps/Sequence details
- Equipment performance limitations
- Vacuum load
- Utility demands
- Equipment optimization is now possible: checking condenser capacity, vacuum system capacity, column flood %, column insulation requirements, etc.
- Higher fidelity simulation provides higher fidelity economic calculation (campaign time and costs)

Chemstations 13

WINTEK Batch Dehydration

Skid mounted solvent dehydration plant

Stripping water from a solvent stream in a two bed adsorber (mol sieve) system. One bed is active and one bed is regenerating under vacuum. Process was already built and operational before modeling analysis.

Process is scheduled with a defined sequence for opening and closing valves to allow one bed to regenerate while the other adsorbs. Rigorous equipment and piping models and pressure/flow calculation included (allows for reversible flow).

Rigorous simulation allows us to simulate effect of malfunction (RB3 blowdown valve malfunction)

Chemstations 14

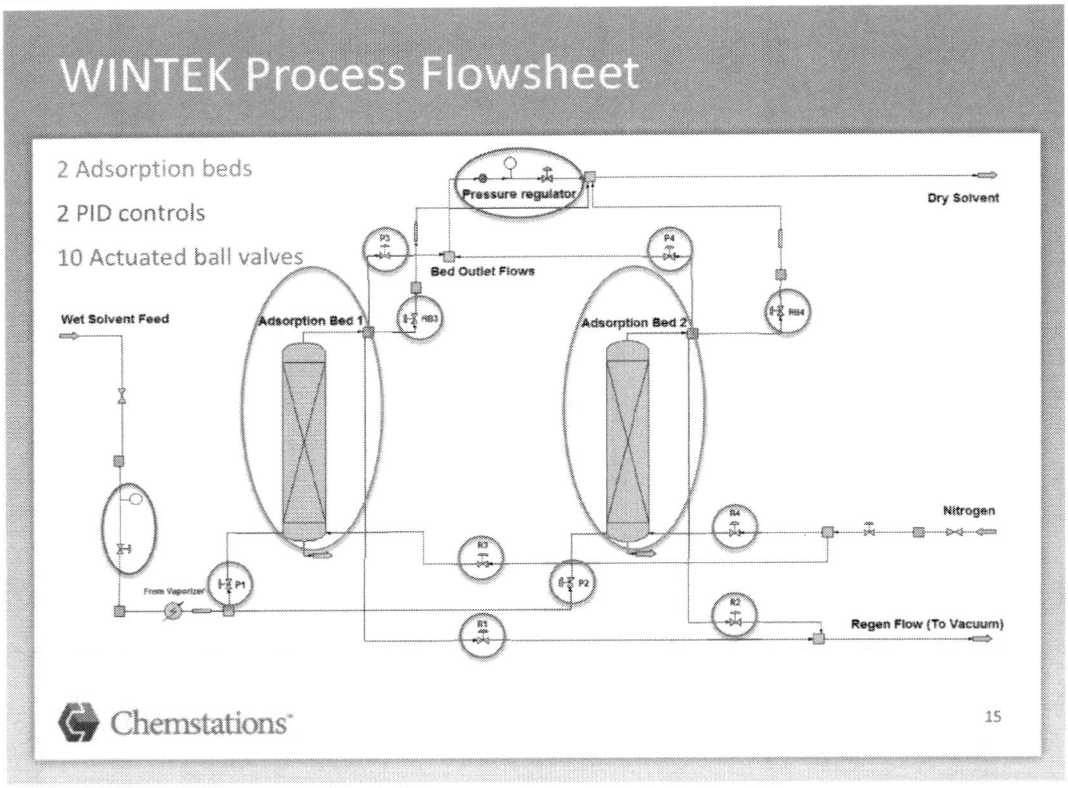
WINTEK Process Flowsheet
2 Adsorption beds
2 PID controls
10 Actuated ball valves
Pressure regulator
Dry Solvent
Bed Outlet Flows
Wet Solvent Feed
Adsorption Bed 1
Adsorption Bed 2
Nitrogen
From Vaporizer
Regen Flow (To Vacuum)
Chemstations
15

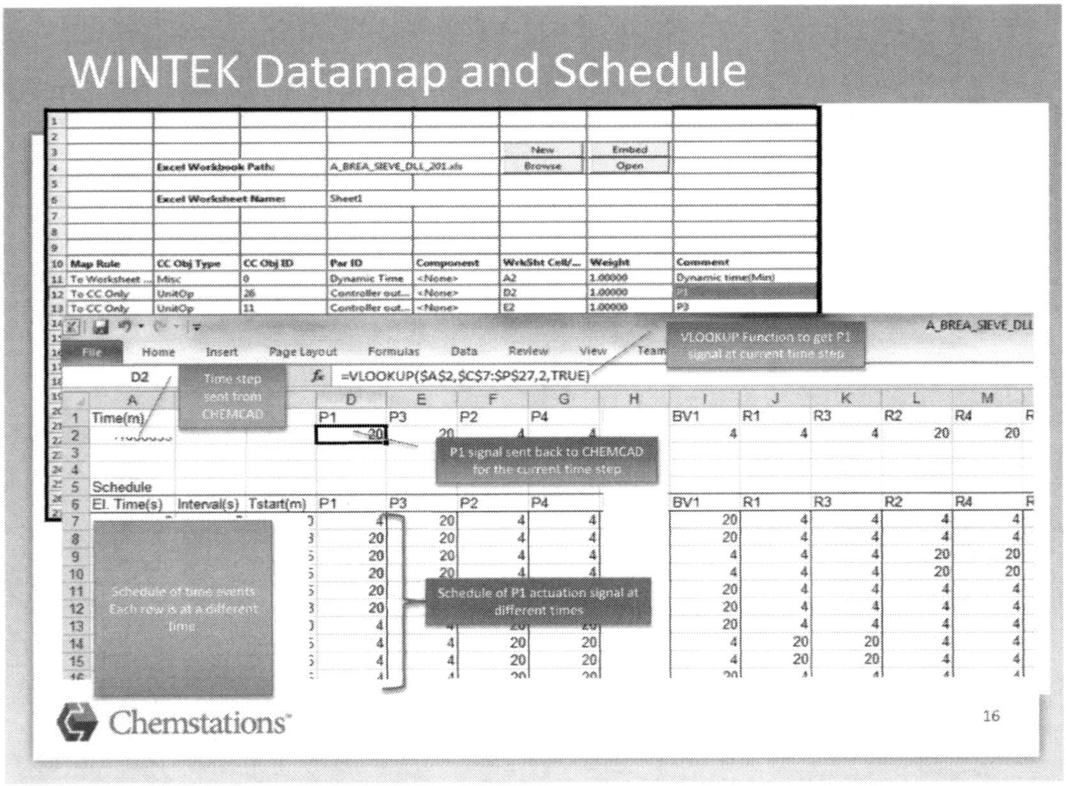
WINTEK Datamap and Schedule
Excel Workbook Path:
A_BREA_SIEVE_DLL_201.xls
Excel Worksheet Name:
Sheet1
=VLOOKUP($A$2,$C$7:$P$27,2,TRUE)
VLOOKUP Function to get P1 signal at current time step
Time step sent from CHEMCAD
P1 signal sent back to CHEMCAD for the current time step
Schedule of time events Each row is at a different time
Schedule of P1 actuation signal at different times
Chemstations
16

WINTEK Process Flowsheet: Malfunction effect

RB3 is simulated blocked, preventing Bed 1 from blowing down properly. Effect is rigorously calculated.

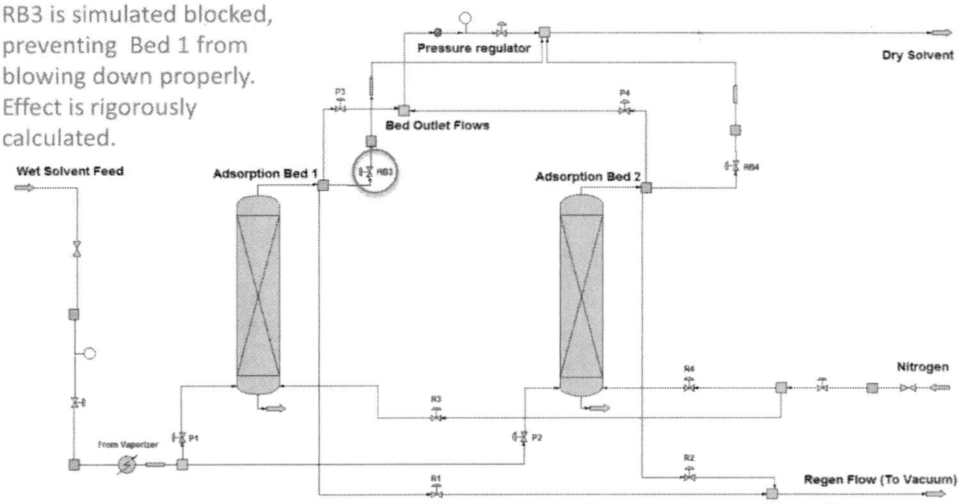

Chemstations

WINTEK Adsorber Beds

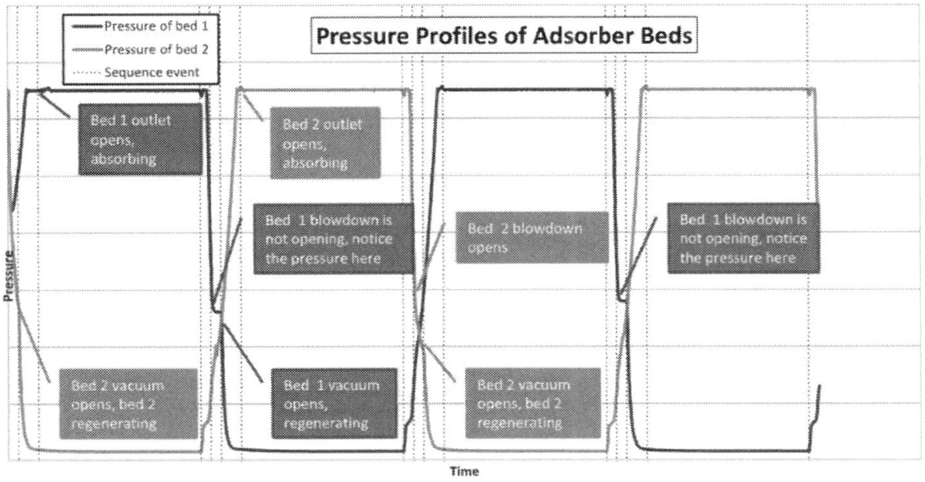

Chemstations

WINTEK Anhydrous Solvent Flowrate

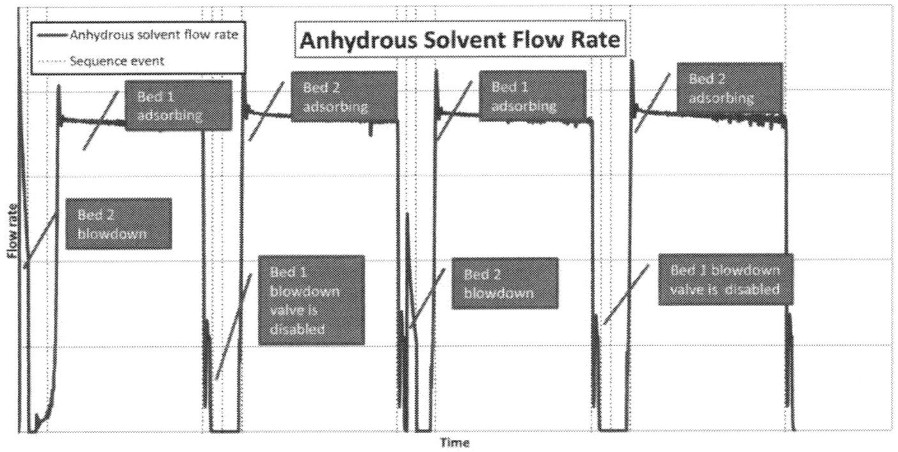

Chemstations™

19

WINTEK Process Flowsheet: reversing flow

P3 and P4 open, flow is forward through one valve, reverse through the other

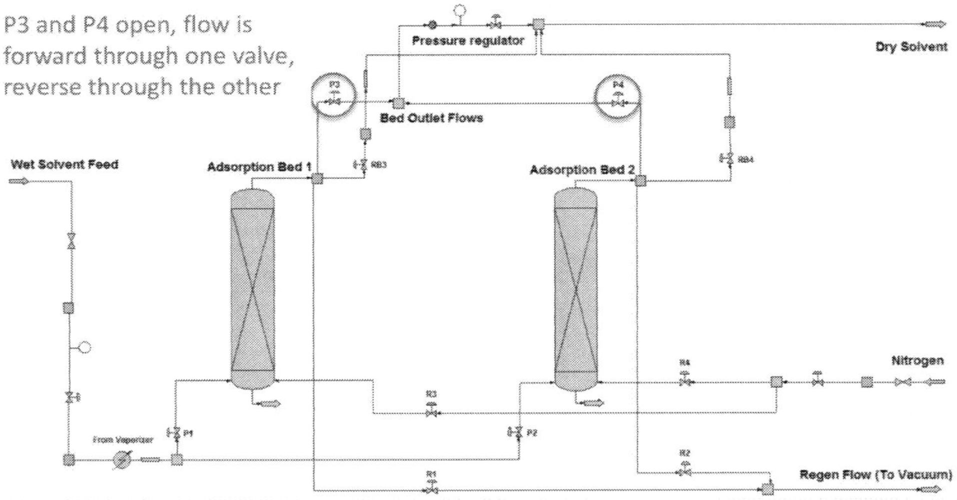

Chemstations™

20

WINTEK Sequence of outlet and regeneration flows out of adsorber beds

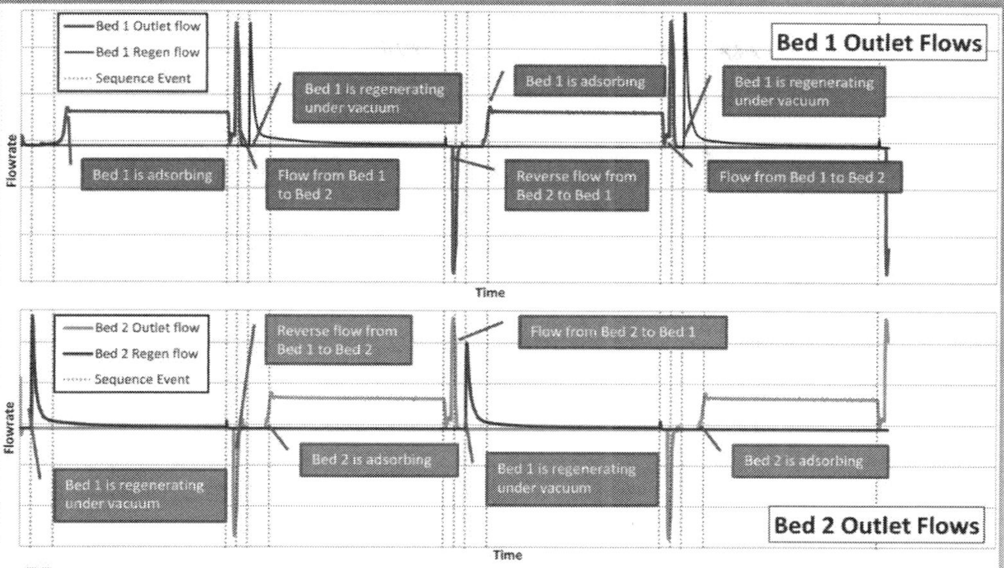

Chemstations

21

Optimization of schedule (future)

- By using an optimization engine (either the built in engine in CHEMCAD, or an external one connected to the Excel spreadsheet) the schedule of valve events can be optimized to maximize on-stream performance (blue area of chart).

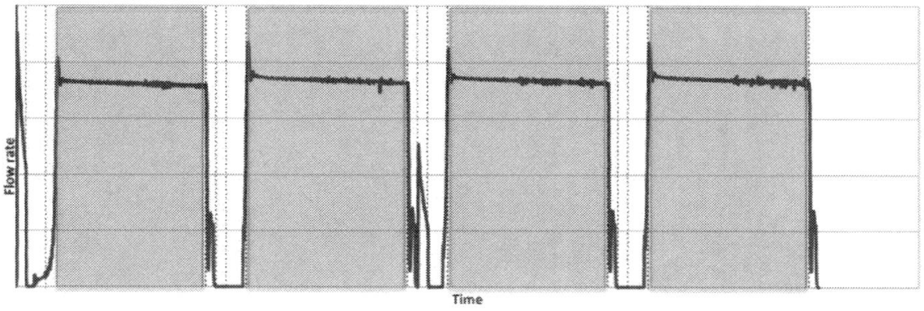

Chemstations

22

WINTEK Conclusions

What did we learn from our model?

- Discrete event scheduler with pressure and flow calculations (including reversible flow) allowed optimization of the process schedule
- Ability to identify and quantify
  - Bottlenecks
  - Sequence timing issues
  - Equipment limitations
  - Malfunction effects (bed 1 blowdown valve sticking)
  - Control loop tuning issues
- Ability to use the model for process scaling: process can be scaled up or down to meet customer requirements using the model before building/assembly
- Ability to optimize schedule of events

Chemstations 23

Suggested Procedure

1. Start with a simplified model (like CC-BATCH) to get heat duties, initial equipment sizes/specifications, and an initial timing/schedule
2. Build a rigorous dynamic model with as much detail as required (but no more than required) to solve the engineering problem (e.g., only add reversible flow if it is a concern)
3. Build an event sequence control scheme using DATAMAP and Excel starting with the information gathered in Step 1 above. Progressively improve the sequence by running the dynamic model with the scheduler

Chemstations 24

Special Thanks

Paul Winter, President and Owner

Wintek Corporation

http://www.wintek-corp.com

230 U.S. Highway 206 STE 401
Flanders, New Jersey 07836

Wintek has been providing equipment solutions to process applications operating at vacuum conditions since 1986, and serves the biodiesel, chemical, environmental, food, pharmaceutical, and plastics industries. Our focus is on process and equipment reliability for your application at a reasonable cost.

Chemstations™ 25

SUSTAINABILITY CHALLENGES AND OPPORTUNITIES IN PROCESS DEVELOPMENT

AIChE Southeast Regional Meeting

4 October 2013
Galveston, TX

Joe Powell, Ph.D.
Shell Chief Scientist -- Chemical Engineering

DISCLAIMER STATEMENT

Resources: Our use of the term "resources" in this presentation includes quantities of oil and gas not yet classified as SEC proved oil and gas reserves or SEC proven mining reserves. Resources are consistent with the Society of Petroleum engineers 2P and 2C definitions.

The companies in which Royal Dutch Shell plc directly and indirectly owns investments are separate entities. In this presentation "Shell", "Shell group" and "Royal Dutch Shell" are sometimes used for convenience where references are made to Royal Dutch Shell plc and its subsidiaries in general. Likewise, the words "we", "us" and "our" are also used to refer to subsidiaries in general or to those who work for them. These expressions are also used where no useful purpose is served by identifying the particular company or companies. "Subsidiaries", "Shell subsidiaries" and "Shell companies" as used in this presentation refer to companies in which Royal Dutch Shell either directly or indirectly has control, by having either a majority of the voting rights or the right to exercise a controlling influence. The companies in which Shell has significant influence but not control are referred to as "associated companies" or "associates" and companies in which Shell has joint control are referred to as "jointly controlled entities". In this presentation, associates and jointly controlled entities are also referred to as "equity-accounted investments". The term "Shell interest" is used for convenience to indicate the direct and/or indirect (for example, through our 23% shareholding in Woodside Petroleum Ltd.) ownership interest held by Shell in a venture, partnership or company, after exclusion of all third-party interest.

This presentation contains forward-looking statements concerning the financial condition, results of operations and businesses of Royal Dutch Shell. All statements other than statements of historical fact are, or may be deemed to be, forward-looking statements. Forward-looking statements are statements of future expectations that are based on management's current expectations and assumptions and involve known and unknown risks and uncertainties that could cause actual results, performance or events to differ materially from those expressed or implied in these statements. Forward-looking statements include, among other things, statements concerning the potential exposure of Royal Dutch Shell to market risks and statements expressing management's expectations, beliefs, estimates, forecasts, projections and assumptions. These forward-looking statements are identified by their use of terms and phrases such as "anticipate", "believe", "could", "estimate", "expect", "intend", "may", "plan", "objectives", "outlook", "probably", "project", "will", "seek", "target", "risks", "goals", "should" and similar terms and phrases. There are a number of factors that could affect the future operations of Royal Dutch Shell and could cause those results to differ materially from those expressed in the forward-looking statements included in this presentation, including (without limitation): (a) price fluctuations in crude oil and natural gas; (b) changes in demand for Shell's products; (c) currency fluctuations; (d) drilling and production results; (e) reserves estimates; (f) loss of market share and industry competition; (g) environmental and physical risks; (h) risks associated with the identification of suitable potential acquisition properties and targets, and successful negotiation and completion of such transactions; (i) the risk of doing business in developing countries and countries subject to international sanctions; (j) legislative, fiscal and regulatory developments including potential litigation and regulatory measures as a result of climate changes; (k) economic and financial market conditions in various countries and regions; (l) political risks, including the risks of expropriation and renegotiation of the terms of contracts with governmental entities, delays or advancements in the approval of projects and delays in the reimbursement for shared costs; and (m) changes in trading conditions. All forward-looking statements contained in this presentation are expressly qualified in their entirety by the cautionary statements contained or referred to in this section. Readers should not place undue reliance on forward-looking statements. Additional factors that may affect future results are contained in Royal Dutch Shell's 20-F for the year ended 31 December, 2011 (available at **www.shell.com/investor** and **www.sec.gov**). These factors also should be considered by the reader. Each forward-looking statement speaks only as of the date of this presentation, October 2013. Neither Royal Dutch Shell nor any of its subsidiaries undertake any obligation to publicly update or revise any forward-looking statement as a result of new information, future events or other information. In light of these risks, results could differ materially from those stated, implied or inferred from the forward-looking statements contained in this presentation. There can be no assurance that dividend payments will match or exceed those set out in this presentation in the future, or that they will be made at all.

We use certain terms in this presentation, such as resources, that the United States Securities and Exchange Commission (SEC) guidelines strictly prohibit us from including in filings with the SEC. U.S. Investors are urged to consider closely the disclosure in our Form 20-F, File No 1-32575, available on the SEC website www.sec.gov. You can also obtain these forms from the SEC by calling 1-800-SEC-0330.

1.0

Opportunity

3

FUTURE ENERGY & SUSTAINABILITY

- Global demand for energy will double by 2050.
- Energy for Mobility is a key driver for prosperity.
- Stakeholder needs: more energy, secure supply, responsible stewardship .
- Result: a Mosiac of energy, fuel and vehicle options!

World population

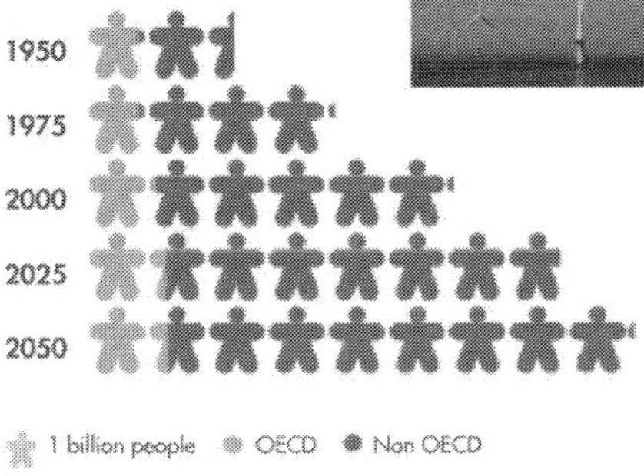

Source: UN Population Division

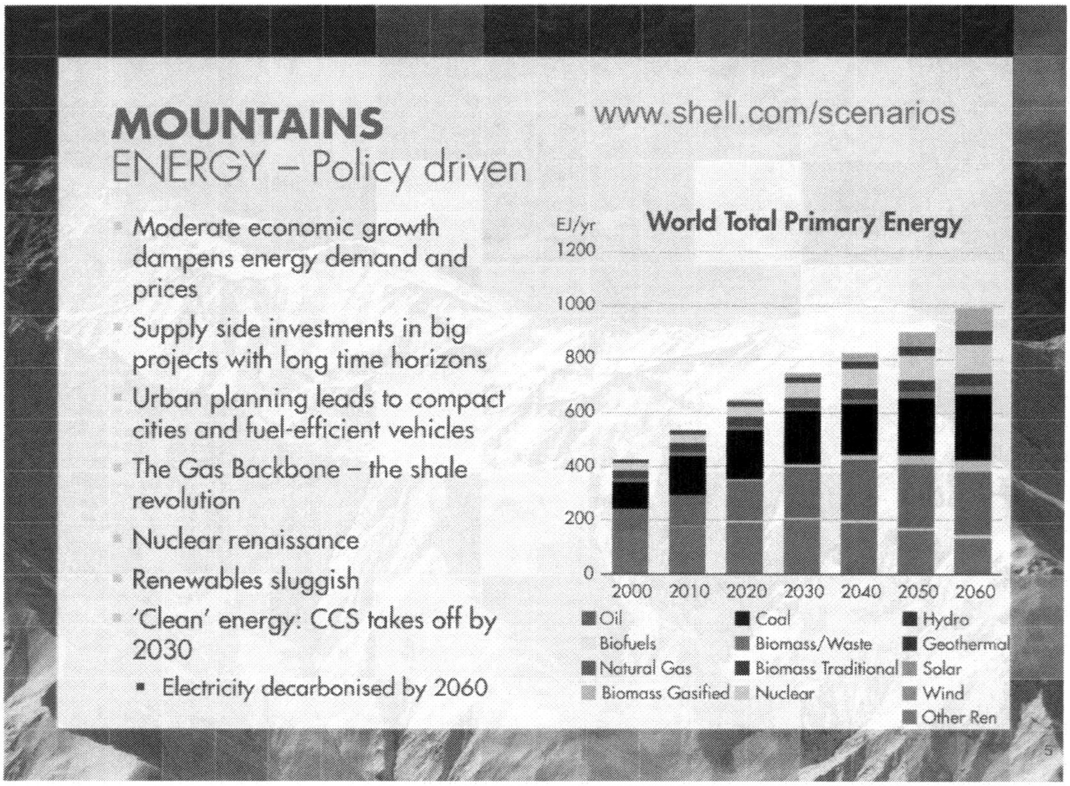
MOUNTAINS
ENERGY – Policy driven
www.shell.com/scenarios
Moderate economic growth dampens energy demand and prices
Supply side investments in big projects with long time horizons
Urban planning leads to compact cities and fuel-efficient vehicles
The Gas Backbone – the shale revolution
Nuclear renaissance
Renewables sluggish
'Clean' energy: CCS takes off by 2030
Electricity decarbonised by 2060
World Total Primary Energy
EJ/yr
1200
1000
800
600
400
200
0
2000 2010 2020 2030 2040 2050 2060
Oil
Coal
Hydro
Biofuels
Biomass/Waste
Geothermal
Natural Gas
Biomass Traditional
Solar
Biomass Gasified
Nuclear
Wind
Other Ren
5

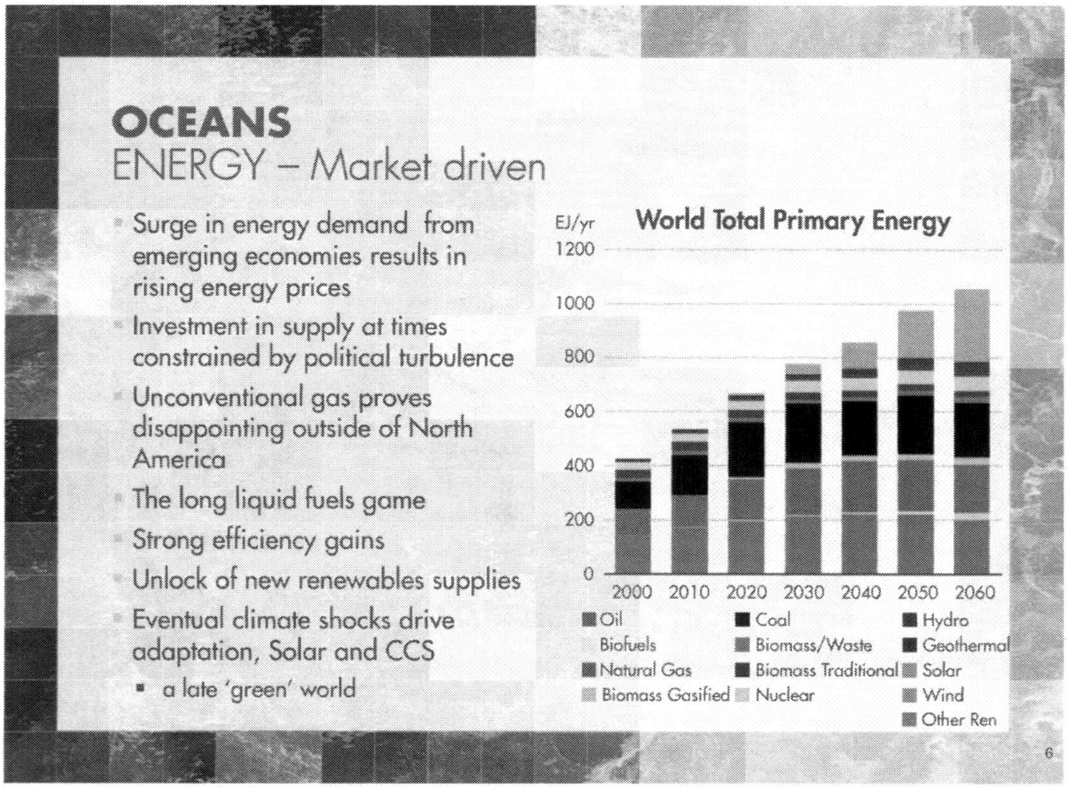
OCEANS
ENERGY – Market driven
Surge in energy demand from emerging economies results in rising energy prices
Investment in supply at times constrained by political turbulence
Unconventional gas proves disappointing outside of North America
The long liquid fuels game
Strong efficiency gains
Unlock of new renewables supplies
Eventual climate shocks drive adaptation, Solar and CCS
a late 'green' world
World Total Primary Energy
EJ/yr
1200
1000
800
600
400
200
0
2000 2010 2020 2030 2040 2050 2060
Oil
Coal
Hydro
Biofuels
Biomass/Waste
Geothermal
Natural Gas
Biomass Traditional
Solar
Biomass Gasified
Nuclear
Wind
Other Ren
6

SHELL ENERGY AND CHEMICALS

WHY BIOFUELS?

- ROAD TRANSPORT = 1/3 TO 1/5 OF CO_2 EMISSIONS.
- BIOFUELS OFFER THE MOST PRACTICAL WAY TO REDUCE ROAD TRANSPORT CO2 EMISSIONS.
- THEY ALSO PROVIDE FOR SECURITY OF SUPPLY.

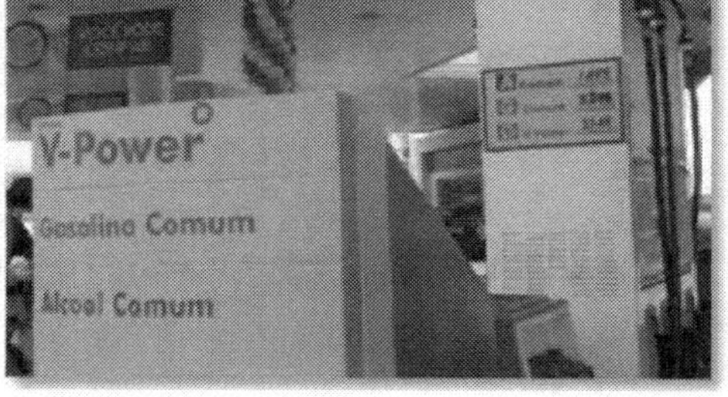

BIOFUELS vs. Energy Efficiency: REDUCING CO_2 EMISSIONS TODAY

- Today's biofuels are the most realistic commercial solution to take CO_2 out of the transport fuels sector over the next twenty years.

Global CO_2 abatement cost curve for the Road Transport sector - Mix Technology World scenario Societal perspective; 2030

raizen

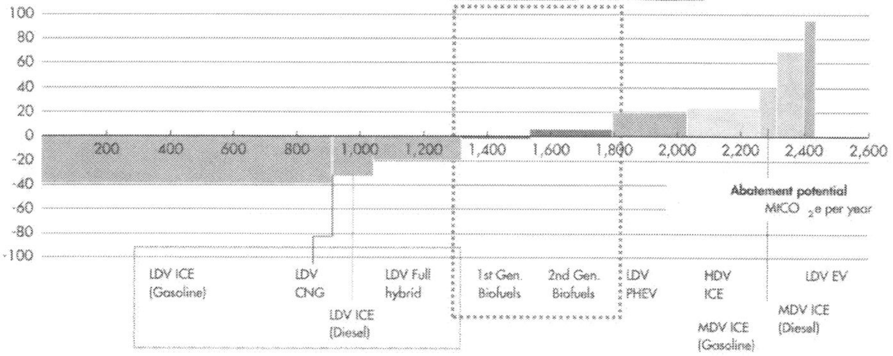

Note: The curve presents an estimate of the maximum potential of all technical GHG abatement measures below €100 per t CO_2e in a penetration scenario if each lever was pursued aggressively. It is not a forecast of what role different abatement measures and technologies will play. Source: Global GHG Abatement Cost Curve v2.0

ENERGY DENSITY: VOLUMETRIC AND MASS

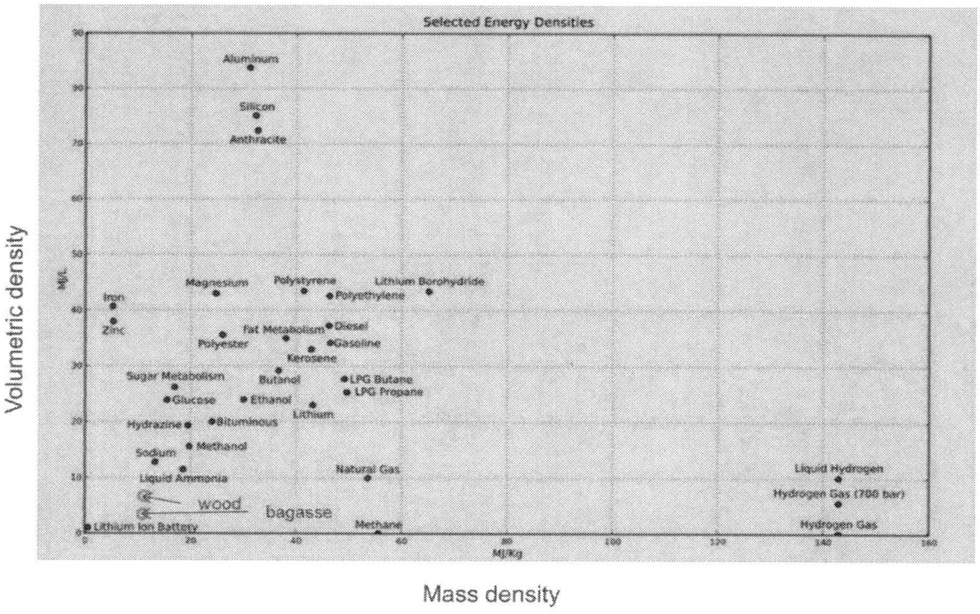

Oxygen content of Biomass

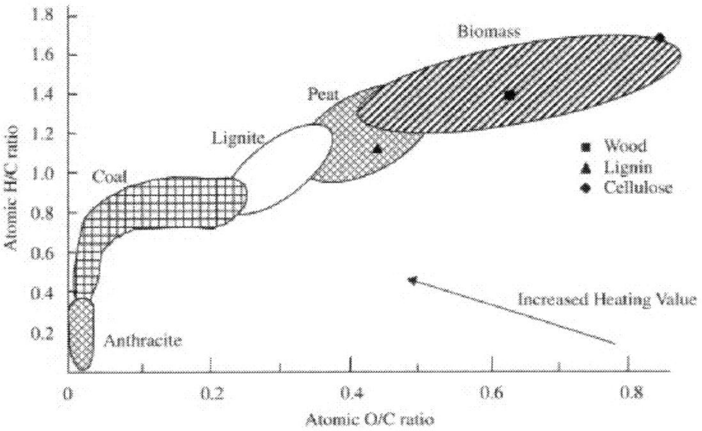

NTNU Combustion Handbook
http://www.handbook.ifrf.net/handbook/cf.html?id=23

CHALLENGES IN BIOFUELS AND BIO-BASED CHEMICALS

Biobased feedstocks = Carbohydrates = $[C(H2O)]_n$

Oxygen is a challenge for thermocatalytic conversion!

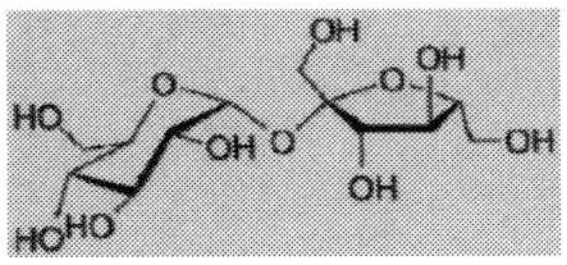

Sucrose is composed of two monosaccharides: D-glucose (left) and D-fructose (right).

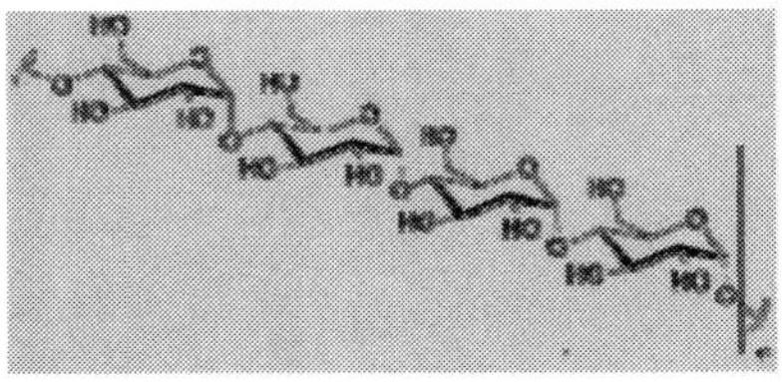

Amylose is a linear polymer of glucose mainly linked with α(1→4) bonds. It can be made of several thousands of glucose units. It is one of the two components of starch, the other being amylopectin.

LIgnin

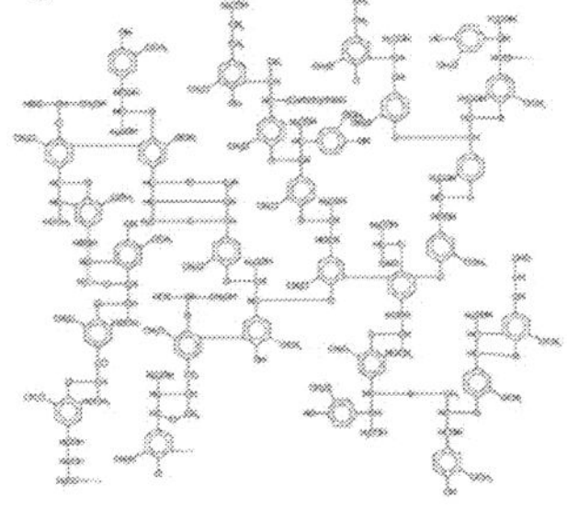

Coal

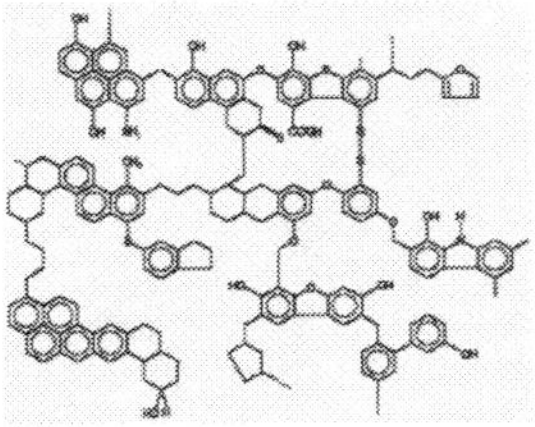

CRUDE OIL	
Element	**Percent range**
Carbon	83 to 87%
Hydrogen	10 to 14%
Nitrogen	0.1 to 2%
Oxygen	0.05 to 1.5%
Sulfur	0.05 to 6.0%
Metals	< 0.1%

CATALYTIC GRAND CHALLENGES: Bio-Based Fuels &Chemicals.

Biomolecular catalysts

- Cost
- Recycle / reuse / *in situ* production?
- Selectivity
- Activity
- Yields (selectivity & conversion vs. inhibition)
- Inhibitor tolerance
- Stability (thermal, chemical, phenotype)

Thermocatalytic

- Water stability
- Oxygen conversion / Hydrogen activation / tar elimination
- Poison tolerance
- Selectivity across complex substrates
- Stability during regeneration

MULTIPLE ROUTES TO CHEMICALS FROM BIO-BASED

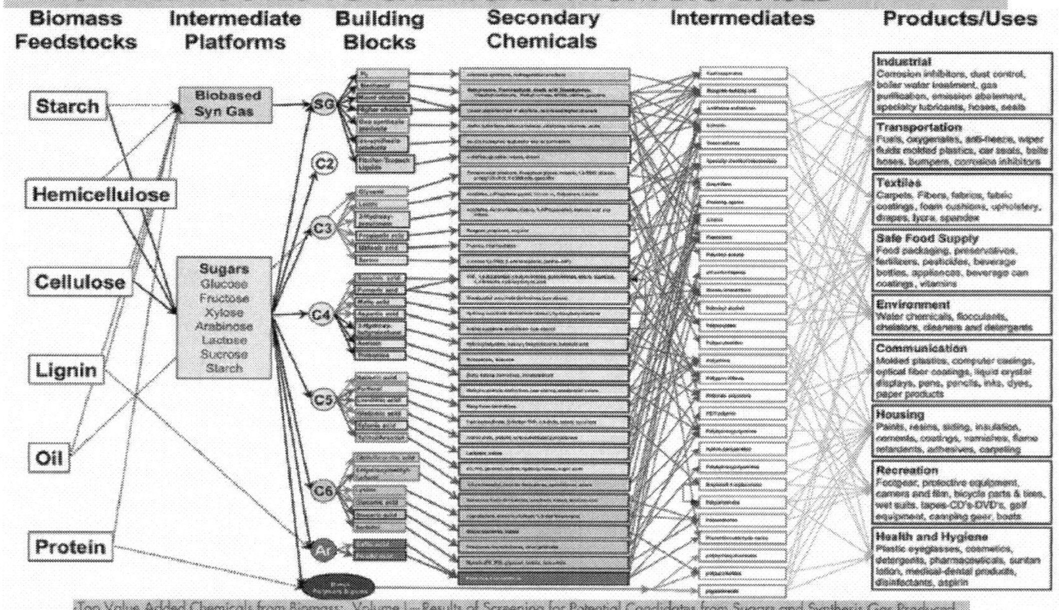

-Top Value Added Chemicals from Biomass: Volume I—Results of Screening for Potential Candidates from Sugars and Synthesis Gas Produced by the Staff at Pacific Northwest National Laboratory (PNNL) National Renewable Energy Laboratory (NREL) Office of Biomass Program (EERE) For the Office of the Biomass Program T. Werpy and G. Petersen, Editors, August 2004.

Advantaged chemicals production derived from fuels feedstocks.

CARBON CAPTURE AND STORAGE

CCS IS A CRITICAL TECHNOLOGY FOR DELIVERING THE CO_2 REDUCTIONS NEEDED BY 2050

WITHOUT CCS THE COST OF MEETING EMISSIONS TARGETS COULD BE SIGNIFICANTLY HIGHER

HUGE GLOBAL GAS RESOURCES

GAS RESOURCES ARE PLENTIFUL, GROWING AND GEOGRAPHICALLY DIVERSE

- Conventional and Tight/Shale/Coal Bed Methane (CBM) gas resources can supply >250 years of current global gas production
- Tight/Shale/CBM gas is transforming the global gas market

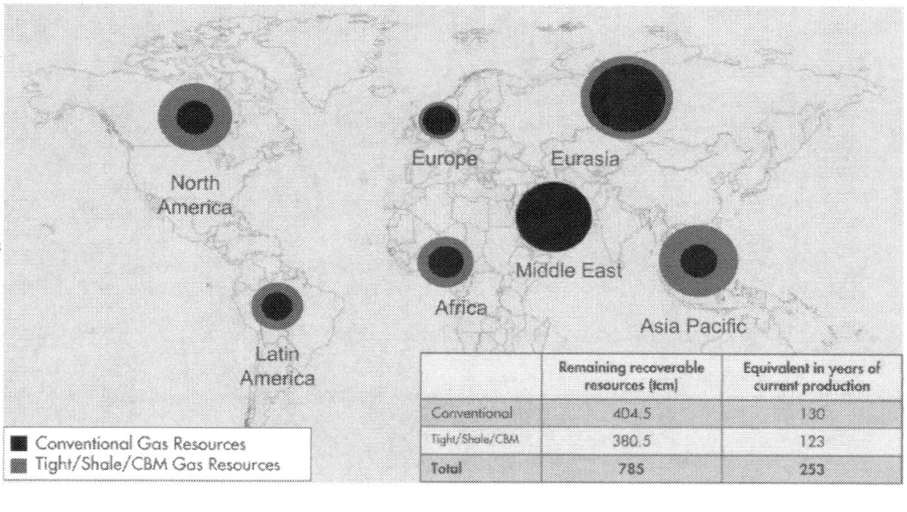

	Remaining recoverable resources (tcm)	Equivalent in years of current production
Conventional	404.5	130
Tight/Shale/CBM	380.5	123
Total	785	253

Source: IEA World Energy Outlook, WoodMackenzie Interpretated

GAS-TO-CHEMICALS

- Natural gas:
  - Affordable
  - Acceptable
  - Abundance
- Advantaged feedstock (ref hydrocarbon liquids) for base chemicals
  - Olefins C2= and C3=
  - Aromatics (BTX)
- Alternative methane activation routes

CHEMICAL PRODUCTS: IMPACT ON REDUCED CONSUMER FOOTPRINT

- **Advantaged chemical products can enable 2:1 to 4:1 footprint reduction relative to energy used to manufacture.**
  - **McKinsey & Company**
  - **International Council of Chemical Associations (ICCA)****

- **Key Shell Chemicals* impacts:**
  - **Building insulation**, where Shell Chemicals* are a leading supplier of raw materials used to make two of the most effective and versatile insulation materials available;
  - **Low-temperature detergents**, which cut the energy used by washing machines by up to half; and
  - **Polyester Textiles**, where Shell Chemicals* are a leading producer of the main component MEG, are less intensive to produce and longer lasting than cotton textiles.

**http://www.icca-chem.org/Home/News-and-press-releases/News-Archive/2009/Global-Chemical-Industry-announces-findings-of-/

*ABOUT SHELL CHEMICALS: With a collective multi-billion dollar turnover and major manufacturing facilities around the world, the companies that make up Shell Chemicals number among the world's largest petrochemical operations. We focus on the production of bulk petrochemicals and their delivery to large industrial customers who, in turn, use them to make many of the essential materials of our modern world.

2.0

Sustainable Partnerships

1

WHY SUSTAINABILITY IS IMPORTANT

BUSINESS CASE

	**Earn licence to operate and grow**	Earn support from communities where we operate, reducing risks and delays. Attract investors, partners, customers and employees.
	**Win market share with better products**	Attract customers by anticipating their demands for affordable energy with lower environmental and social impacts.
	**Drive innovation in technology**	Unlock business opportunities through lower-impact technologies.

IT'S THE RIGHT THING AND THE SMART THING TO DO

2

EARLY INTEGRATION OF SD PRINCIPLES

SOONER RATHER THAN LATER

- Integrating social and environmental considerations early – early in the project life or early in the merger, acquisition or divestment process, or early in the development of new products or technology – is critical to success.

WORKING WITH OTHERS

We work with governments, energy producers, consumers, non-governmental organisationsand local communities to address concerns about our operations and wider environmental and social issues.

Some of our partners include.:

- Global Road Safety Partnership (GRSP)
- Global Alliance for Clean Cookstoves
- The Extractive Industries Transparency Initiative (EITI)
- Global Gas Flaring Reduction Partnership (GGFRP)
- Voluntary Principles on Security and Human Rights
- International Union for Conservation of Nature (IUCN)
- United Nations Partnership for Clean Fuels and Vehicles
- Wetlands International (WI)
- OECD Business and Industry Advisory Council
- International Emissions Trading Association (IETA)
- Earthwatch
- World Business Council for Sustainable Development

MCKINSEY 2030 CO_2 COST ABATEMENT CURVE

Global GHG abatement cost curve beyond business-as-usual – 2030

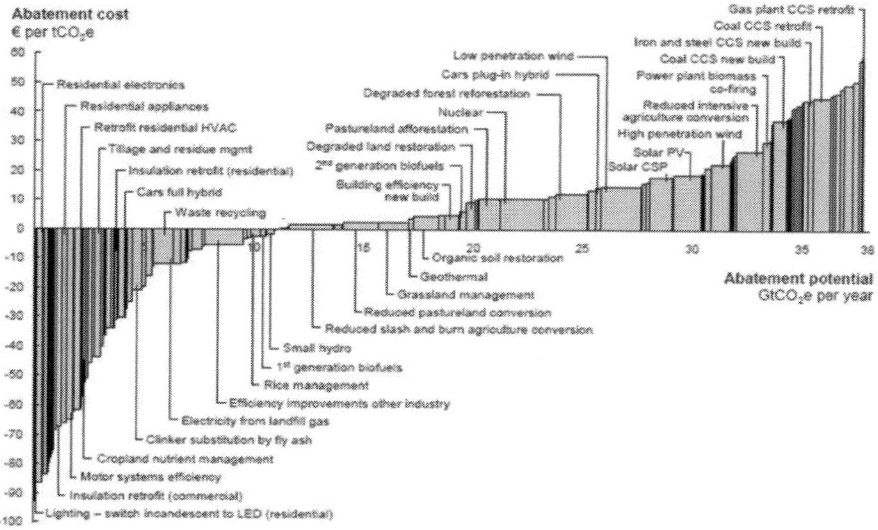

Note: The curve presents an estimate of the maximum potential of all technical GHG abatement measures below €60 per tCO_2e if each lever was pursued aggressively. It is not a forecast of what role different abatement measures and technologies will play.
Source: Global GHG Abatement Cost Curve v2.0

23

Sustainability is more than just CO_2

Water
Land Use
Biodiversity
Managing Chemicals
Social performance
Security / jobs
Other environmental stressors
Safety
Climate and CO_2

24

Shell Gamechangers and Hunters Network

Shell
GameChanger

To become the most **competitive** and **innovative** energy company in the world...

How it works...
When you have a GameChanging idea

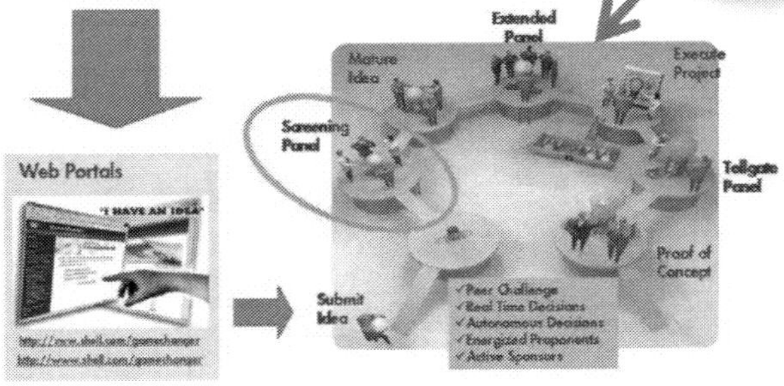

- Hunters are volunteer network to link ideas, opportunities, and challenges.

3.0

Sustainable Process Development

PROCESS DEVELOPMENT: KEY CORE COMPETENCY FOR INDUSTRY

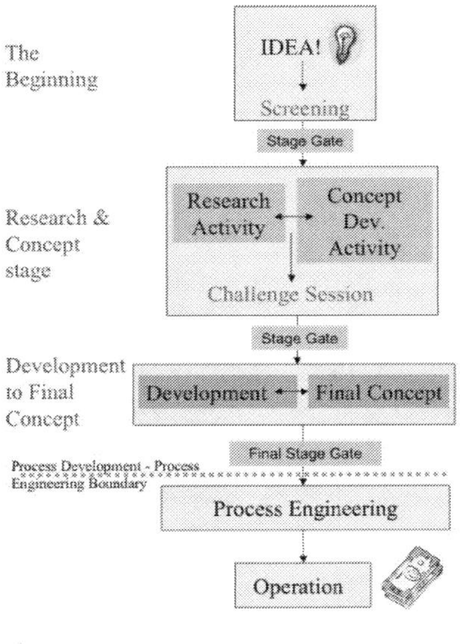

Ability to screen for viability before spending huge $$$

Ability to deliver processes that work

Ability to anticipate business needs

27

TECHNOLOGY, INNOVATION AND DEVELOPMENT

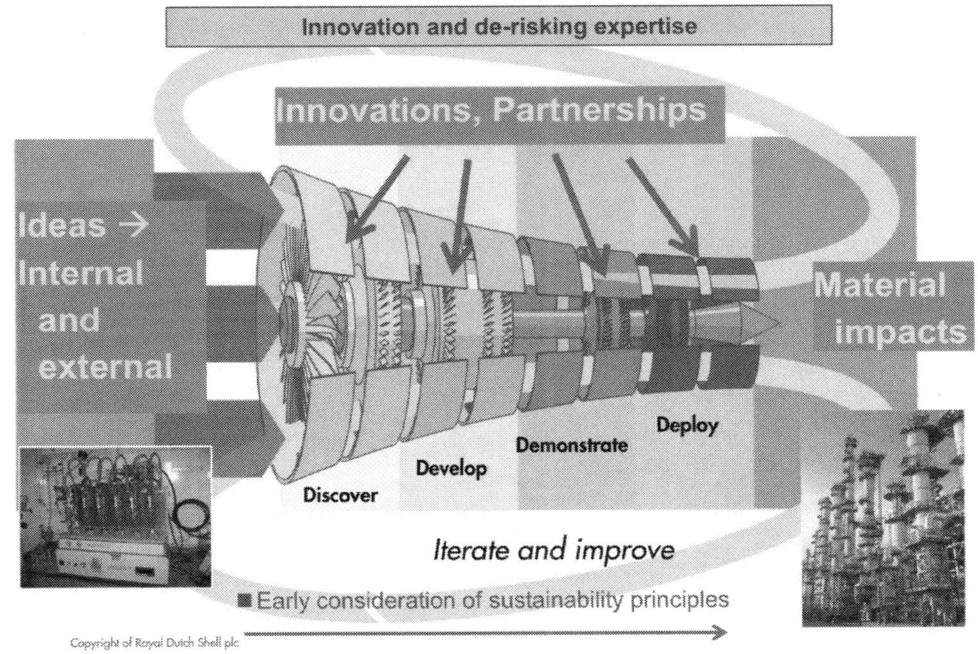

Batch Multi-reactors (P5000)

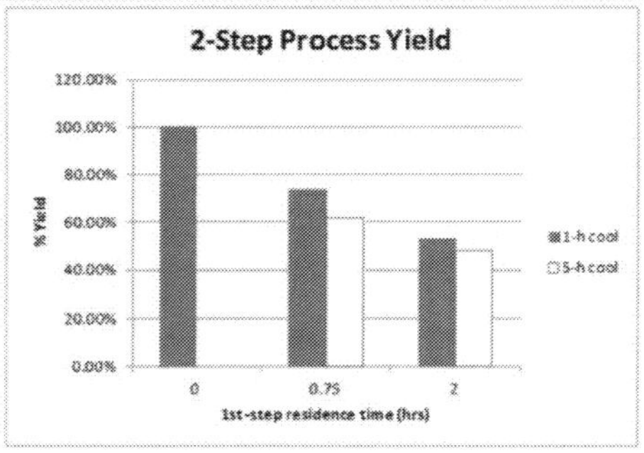

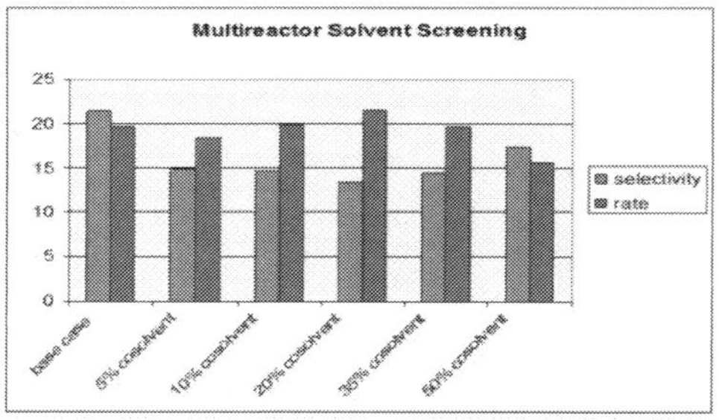

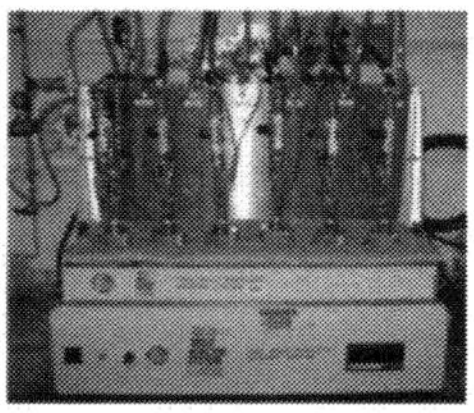

- 1st choice for process screening
- Side-by-side comparison

HOW: *In Situ* Catalyst Characterization

Industrial IR-Reaction Analyzers

- Used for batch, pilot, market demo, and commercial scale units
- Rapid SU troubleshooting

ProcessIR™
Technology for Process Monitoring

ReactIR™ MP Mobile

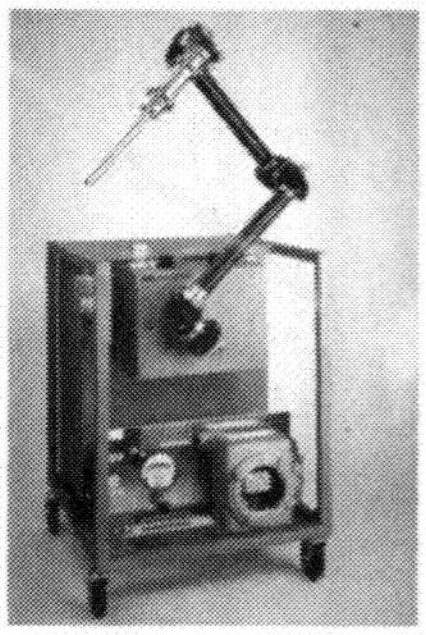

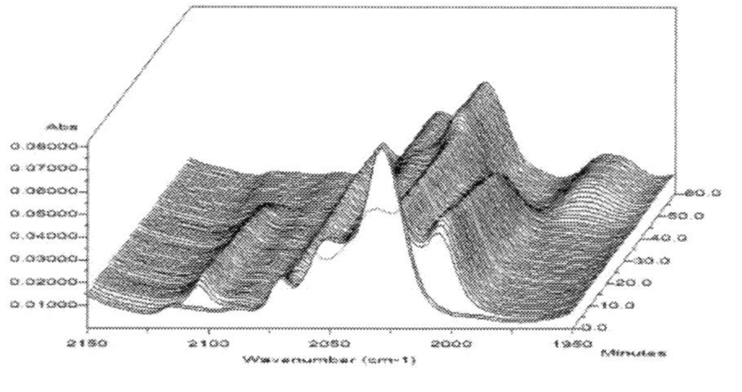

PRW

PILOT PLANT SIZING VS. TIME

year	size	type	industry
1956	4 tons/day	slurry process	fertilizer
1956	100 - 700 g/h	continuous sterilization	bio
1956	55 lb/hr	coal gasification	coal
1956	3000 lb/ batch	wood hydrolysis	wood
1956	300 - 600 gal/batch	fermentation	bio
1956	100 - 300 ml/hr	Midget" FCC	refining
1956	1 - 175 gallon batch	fluorophosphonic	specialty
1956	320 ml/hr	hydrotreating	refining
1956	20- liters catalyst (LHSV=1)	FCC full scale	refining
1956	300 lb/hr	filtration - extraction	agra-chem
2005	2000 - 5000 g/h	new process	petrochem
2007	20 g/hr	new process	petrochem

More potential for scale down 2014 vs. 1956

Compelling reasons remain for larger scale:

- solids handling
- product testing
- distillation / recycle

1956= H. Hoog, W. B. Hawes, and J. J. Leendertse (Royal Dutch Shell) Chemical Engineering Practice (Cremer & Davies), Vol-1 (1956)

31

Development Cost vs. R&D Stage

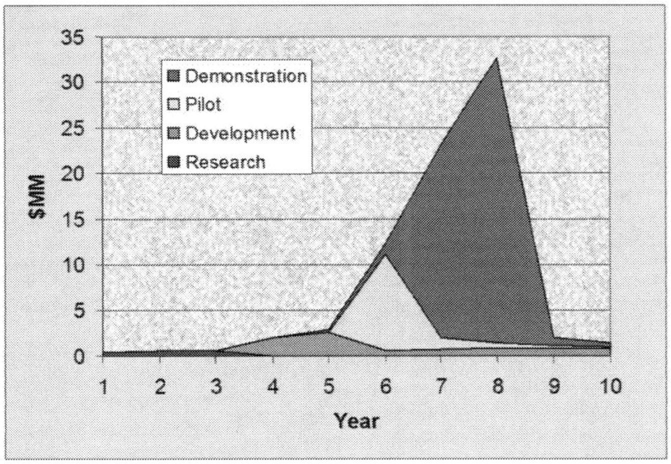

- "Research is cheap; Development is expensive!"
- High cost of demonstration units.

1944 PENICILLIN PILOT PLANT:

Fig. 1. Pilot plant yielding 200 gal./day (1944).

SHELL DEVELOPMENT CO,
SAN FRANCISCO, CA

Countercurrent, pH gradient solvent extraction

THANK YOU!

Joe.Powell@Shell.com

Burns & McDonnell
SINCE 1898

Project Execution using the Stage Gate Process

Noe' H. Saenz, PMP
Burns & McDonnell

Based on 'Front End Planning' Research & Publications from the Construction Industry Institute (CII)

1

Burns & McDonnell
SINCE 1898

Our Mission Today

Share with you this critical process that will improve your company's capital investments / projects -and your career at the same time.

2

Learning Objectives

1. **What is the Stage Gate Process.**
2. **Why this process is Critical for the Success of:**

    - Your Company's Capital Projects.

    - Your Plant's Safety.

    - Your Career.

3. **How it works in Project Front End Planning.**

3

Definition of Capital Project

Long-term investment project that requires large amounts of resources (financial, labor, etc.) to acquire, develop, improve and/or maintain a capital asset (i.e. industrial plants, roads, etc.)

4

Capital Project of a Chemical Plant

Considerations to design & build a chemical plant.

- ☐ Feed Storage
- ☐ Feed Preparation
- ☐ Reactions
- ☐ Product Purification
- ☐ Product Packaging & Storage
- ☐ Recycling, Recovery, and Storage
- ☐ Pollution Control

5

Influences of an Industrial Facility

- Corporate
- Regulatory
- Marketplace
- Location
- Changes
- Trade Groups
- Physical Assets
- Legacy Systems
- Org. Resources
- Governments
- CAPEX

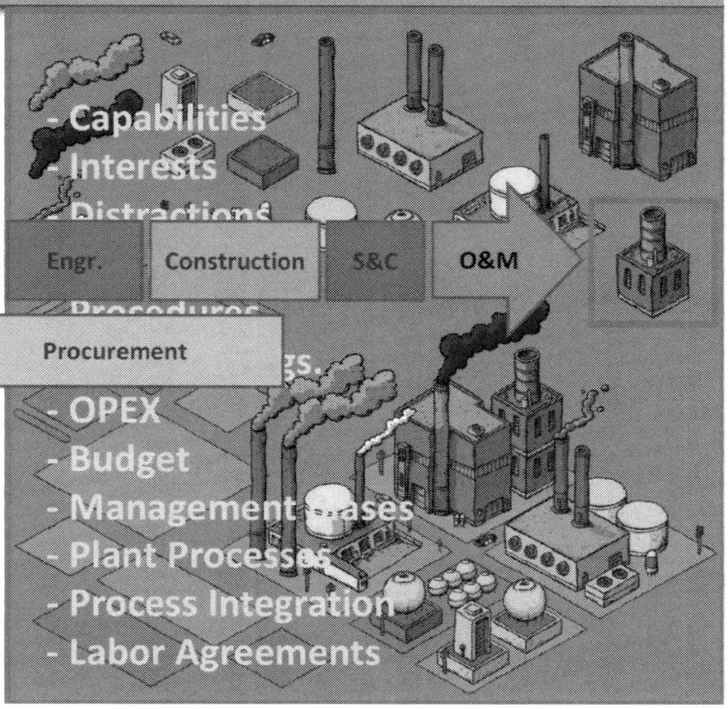

Burns & McDonnell
SINCE 1898

Ability to Influence Project Cost

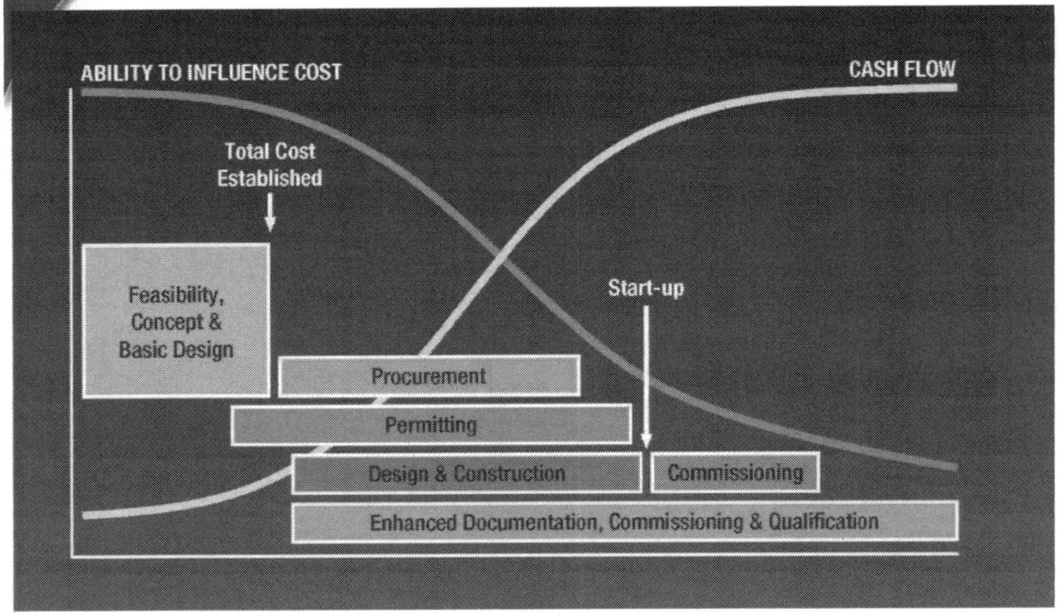

Burns & McDonnell
SINCE 1898

Capital Project Development

Business Opportunity

1. **Front End Planning**
2. **Engineering**
3. **Procurement**
4. **Construction**
5. **Startup & Commissioning**

Operations & Maintenance

Gates
- Feasibility
- Concept / Tech
- Detailed Scope

Performance Assessment

Stage Gate Process in Front End Planning

Developing sufficient information and having the right checks & balances to continuously:

1. Evaluate business feasibility
2. Address risk
3. Obtained alignment
4. Decide to commit resources

to maximize the chance for a successful project.

9

Stage? Gate?

Stage

- A phase with parallel activities
- undertaken by a multi-disciplined team
- gathers information for the next gate

Gate

- A go/no-go decision point with
- a set of required deliverables
- pre defined evaluation criteria
- defined outputs: go/no-go/recycle

Influences of an Industrial Facility

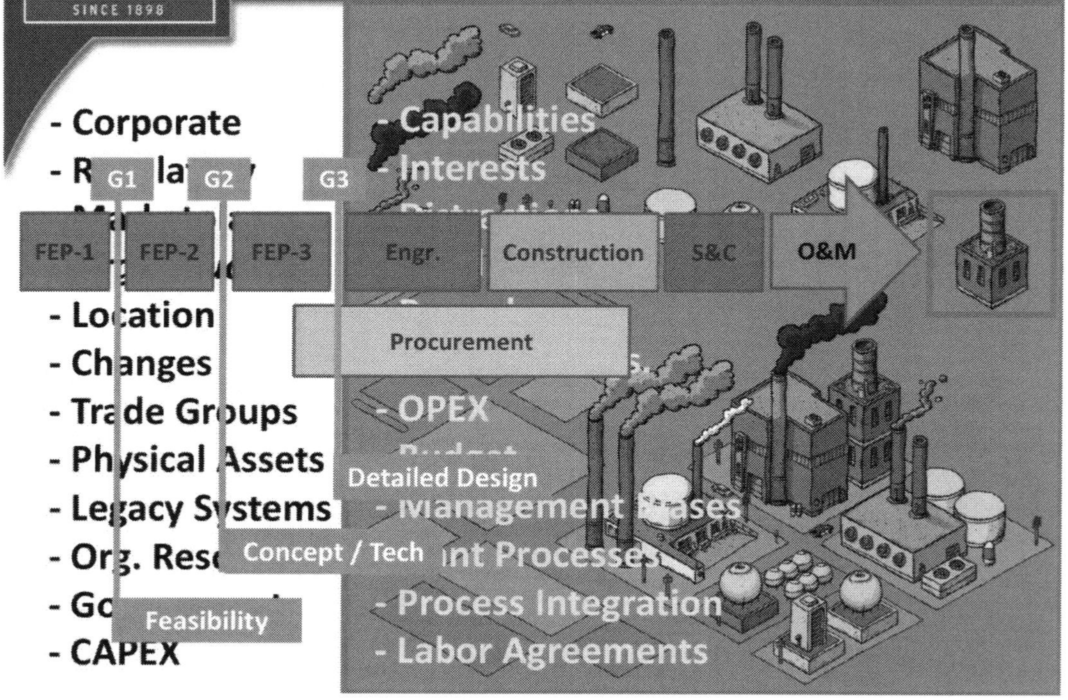

Why Front End Planning with Stage Gates?

In July of 2002, an article appeared in the trade magazine *"Hydrocarbon Engineering"* that asserted the following:

- As much as 80% of costs are committed during the Conceptual (Initiation definition) of a project,
- Poor management of the design phase affects every stage of the project lifecycle.
- 80% of design changes are caused by a lack of data or the wrong data.
- Design Engineers spend between 30% and 50% of their time looking for their "stuff!", and
- Systems that generate and manage data that are not integrated result in duplication and inconsistency.

Why Front End Planning with Stage Gates?

Another set of statistics appeared in 1998 (Standish Group's Chaos research on Information technology Projects):

- 24% of the monitored projects were rated as "Failures",
- 49% were rated as "Challenged" while only
- 27% were rated as "Successful".

Why Front End Planning?

The Construction Industry Institute (CII) states:

- Front End Planning is arguably the single most important process in the development of facility project.
- It creates a strong, early link between the business need, the project strategy, scope, cost and schedule throughout the project phases.

Bottom Line

Stage-Gated Front End Planning

- ✓ Facilitates Communication
- ✓ Helps us to Stop, Think, Evaluate, Prepare
- ✓ Saves Time
- ✓ Saves Money
- ✓ Reduces Change
- ✓ Facilitates Team Alignment
- ✓ Helps Identify & Mitigate Risks
- ✓ Saves Lives

Key to your Company, Key to your Career!

How Front-End Planning Works

Key Factors for Success

1. Defined Front End Planning process
2. Scope definition
3. Existing site conditions definition
4. Contracting strategy
5. Alignment and stakeholder identification
6. Team building

Burns & McDonnell SINCE 1898

FEP @ Burns & McDonnell

100% Employee-Owned EPC Firm

4,300+ Employee-Owners

$2.0 Billion Revenue

Multiple Projects

Multiple Disciplines

Multiple Stakeholders

International
- Canada
- India
- Middle East
- South America

Burns & McDonnell SINCE 1898

Multiple Projects in Multiple Markets

Process & Industrial

Terminals & Pipelines

Energy

Aviation & Federal

Environmental

Infrastructure

Front End Planning at Burns & McDonnell

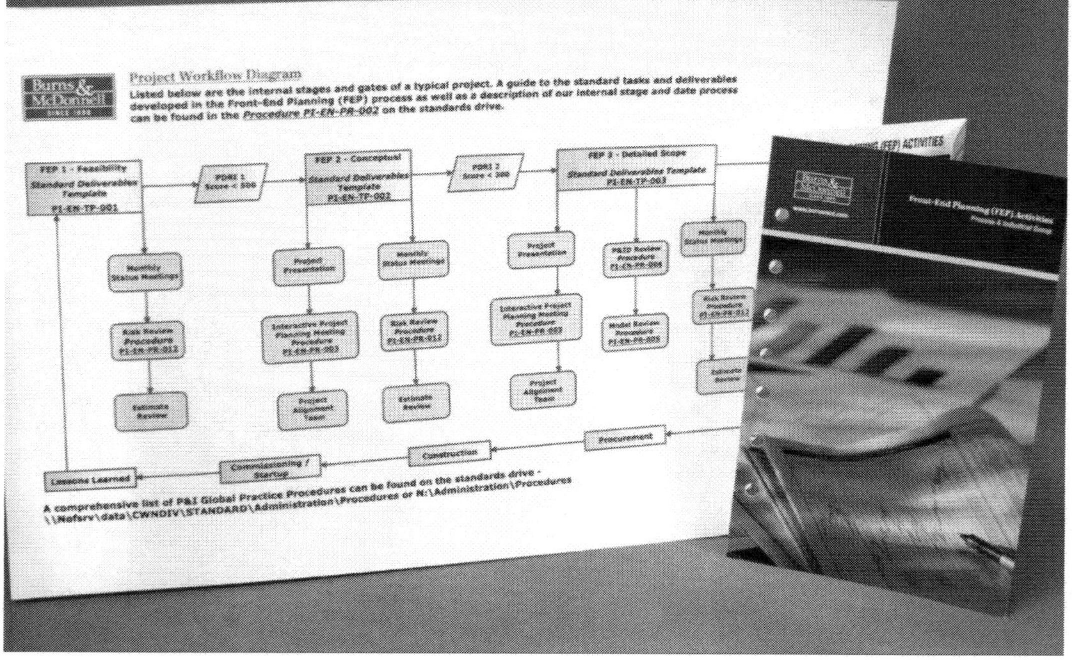

Burns & McDonnell
SINCE 1898

How FEP Works

Key Factors for Success

1. Defined Front End Planning process
2. Scope definition
3. Existing site conditions definition
4. Contracting strategy
5. Alignment and stakeholder identification
6. Team building

1. Defined FEP Process

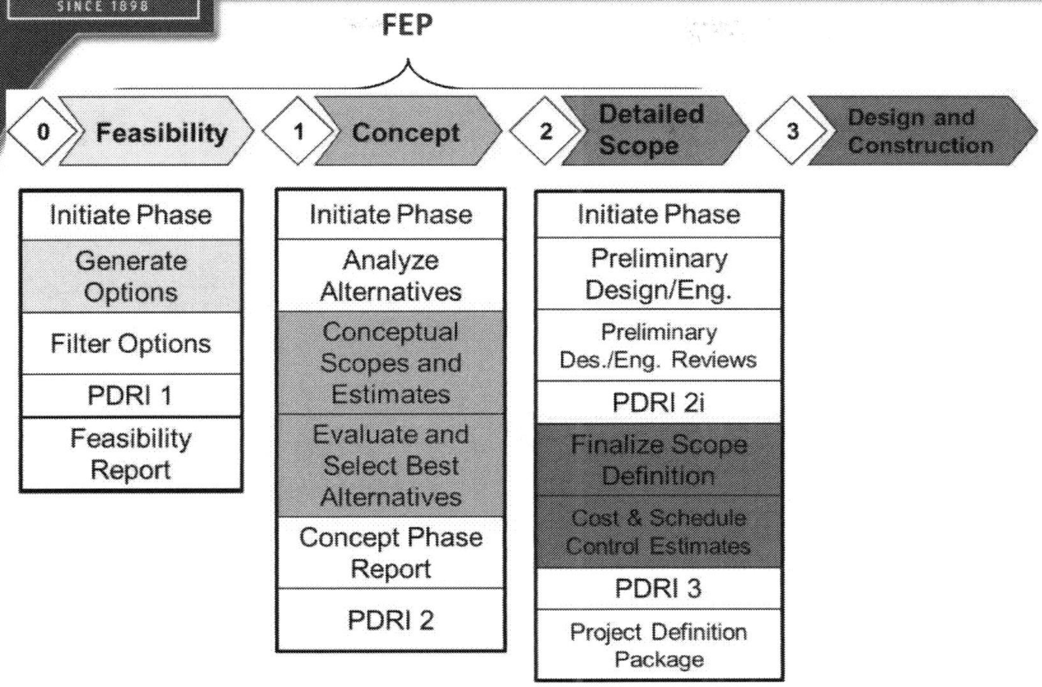

1. Defined FEP Process

Phase	Deliverable Detail	Goals	Options Evaluated
Feasibility	Block Flow	1. Generate Options 2. Order of Magnitude Cost Estimate (+/- 40-50%)	Many
Concept	Process Flow Diagram	1. Evaluate and Pick the Best Option 2. Budgetary Cost Estimate (+/-20-25%)	Few
Detailed Scope	Piping & Instrumentation Diagram	1. Define the Project Scope and Execution Plan 2. Definitive Cost Estimate (+/-10-15%)	One
Design and Construction (EPC)	Piping Isometrics	1. Execute the Project 2. Manage Changes	One

2. Scope Definition

- Options to Capture Opportunity
  - Capital Investment
  - Operations Modification
  - Joint Ventures / Toll Processing
  - Marketing / Trading
- Capital Investment Strategies
  - Manufacturing Objectives
  - Site Selection
  - Technology Selection
  - Project Execution Strategy

3. Existing Conditions Definition

- Underground Conditions
- Integrity (mechanical/structural/electrical)
- Restricted Access
- Competing Projects
- Interface with existing operations

24

Burns & McDonnell
SINCE 1898

4. Contracting Strategy

Design – Bid – Build

Design/Build

Risk Allocation

Cost Plus

Lump Sum

Win-Win

Burns & McDonnell
SINCE 1898

5. Alignment & Stakeholder Identification

5. Alignment & Stakeholder Identification

Key Tool: Project Definition Rating Index (PDRI)

- **An Index** to score along a continuum representing the level of scope definition
- **A Risk Management Tool** to identify and measure risks related to project scope definition

- PDRI 1 = Feasibility
- PDRI 2 = Concept / Technology Selection
- PDRI 2i = Midway through Detailed Design
- PDRI 3 = Toward End of Detailed Design

PDRI

Project Definition Rating Index (PDRI)

I. Basis of Project Decision

1. Business Objectives
2. Manufacturing Objectives
3. Basic Data R&D
4. Project Scope
5. Value Engineering

II. Front End Definition

6. Site Information
7. Process / Mechanical
8. Equipment Scope
9. Civil, Structural & Architectural
10. Infrastructure
11. Instrument & Electrical

III. Execution Approach

12. Procurement Strategy
13. Deliverables
14. Project Control
15. Project Execution Plan

6. Team Building

Identify the Team Needs

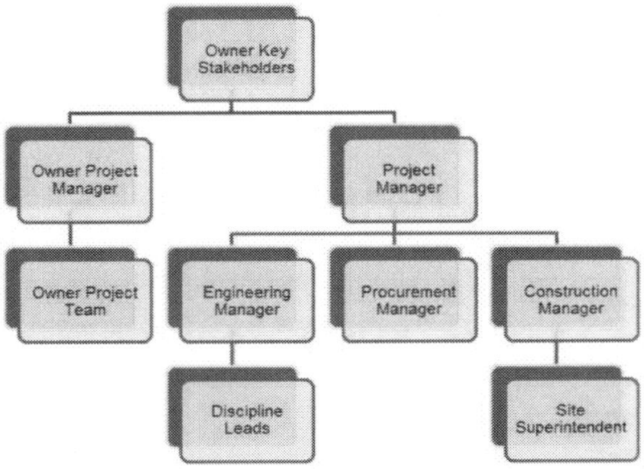

- Written definition of each activity
- Budgeted time for each activity
- Clearly defined roles and responsibilities
- Formal schedule for completion of each activity

Team Accountability

- Team Reporting Structure
- Communication / Reporting Plan
- Required Reviews and Approvals
- Individual Authority and Approval Levels

We have learned

1. What is the Stage Gate Process.
2. Why this process is Critical for the Success of:

   - Your Company's Capital Projects.

   - Your Plant's Safety.

   - Your Career.

3. How it works in Project Front End Planning.

31

Questions?

Thank you!
Noe' H. Saenz, PMP
Burns & McDonnell
nsaenz@burnsmcd.com
SUCCESS
Because you too can own this face of pure accomplishment

SOLVENT DEASPHALTING

The key to unlocking additional profits when processing residues

AIChE Southwest Process Technology Conference
Galveston, TX, October 4, 2013
Joseph Woodson, Senior Process Engineer

FW Technology Group

Centered around delivering solutions

 1

Residue Processing Options

Catalytic
- RFCC
- Residue Hydrotreating
- Residue Hydrocracking

Non-Catalytic
- Visbreaking
- Solvent Deasphalting
- Coking

Residue Processing Options

Hydrogen Addition
- Residue Hydrotreating
- Residue Hydrocracking

Carbon Rejection
- Visbreaking
- Solvent Deasphalting
- Coking
- RFCC

Residue Processing Options

Conversion
- Residue Hydrotreating
- Residue Hydrocracking
- Visbreaking
- Coking
- RFCC

Separation
- Solvent Deasphalting

SDA is unique in how it processes residues

Solvent Deasphalting Process

- Physical separation by molecular type
- Light paraffin solvent
- Products
  - Deasphalted Oil (DAO)
  - Pitch
- "Asphaltenes" are defined by a solvent separation

- Foster Wheeler has been designing and constructing SDA Units since 1940
- SDA technology licensed by UOP/FW

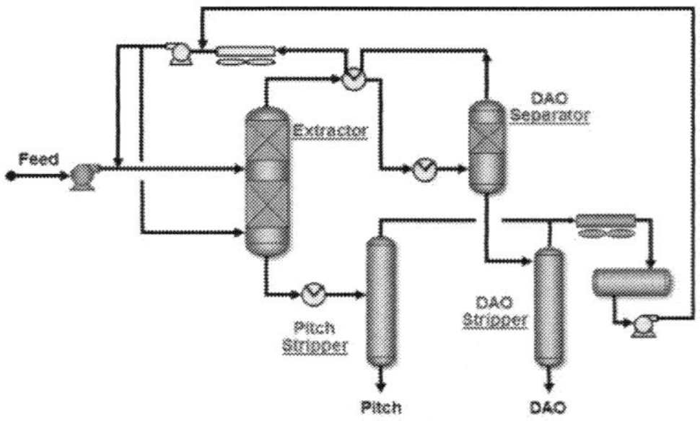

SDA Process

History

- Originally developed for lube oil production
- Early units used propane solvent
- Fuels SDA units became popular once FCC enters
- Relatively low TIC
- Technology improvements
  - Advanced counter-current packed Extractor
  - Optimum solvent-to-oil ratio (SOR)
  - Supercritical solvent recovery
  - Optimized heat integration and efficiency

FOSTER WHEELER

Fuels SDA Capacity in USA

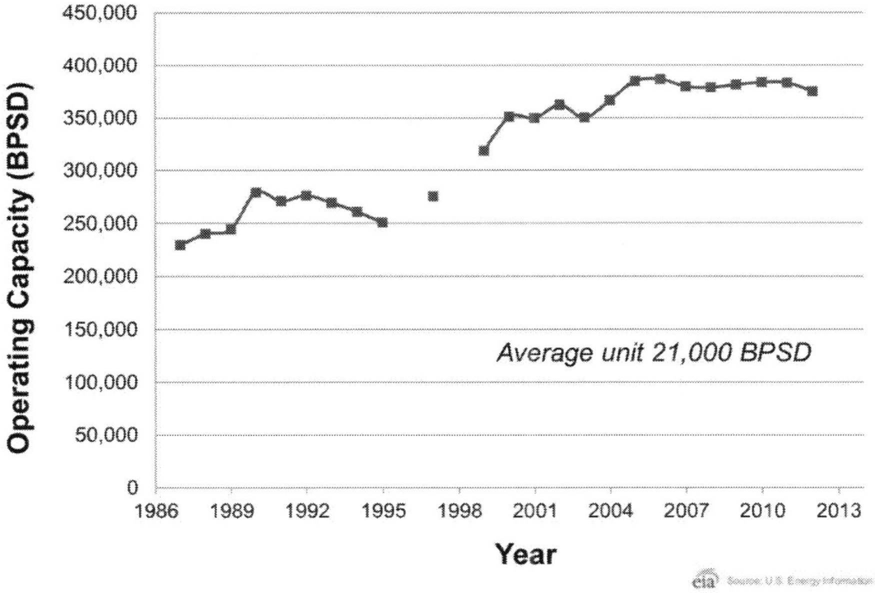

Source: U.S. Energy Information Administration

FOSTER WHEELER

Residue Upgrading Capacity in USA

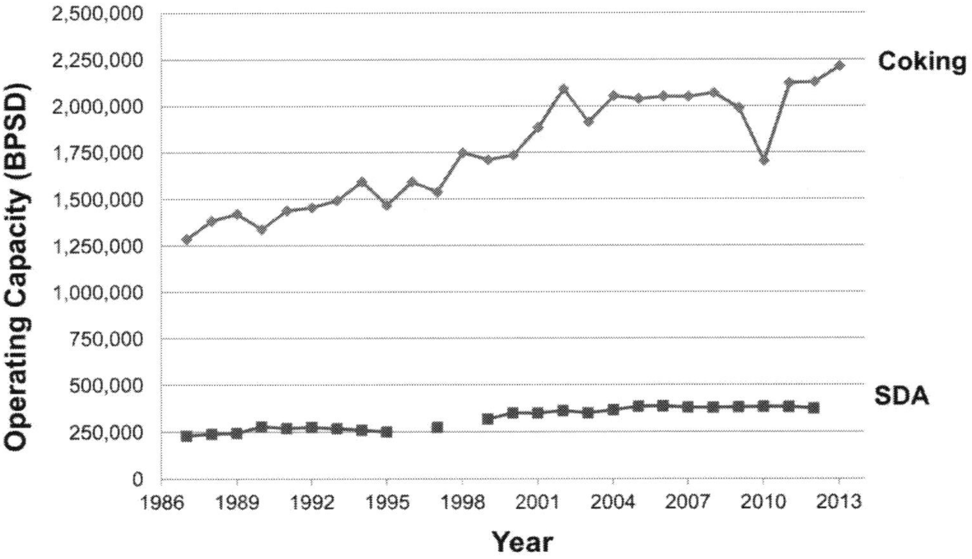

Source: U.S. Energy Information Administration

SDA Process

Traditional Applications

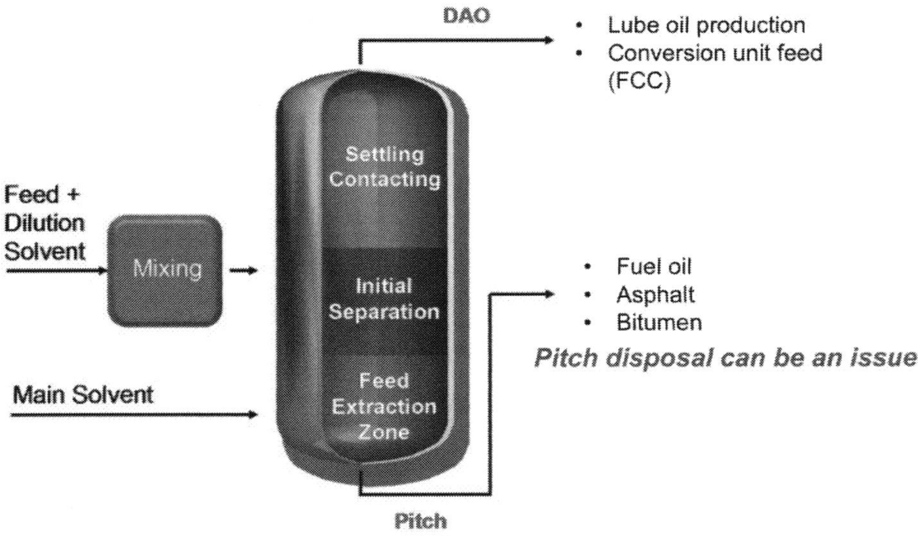

SDA Process

New Applications

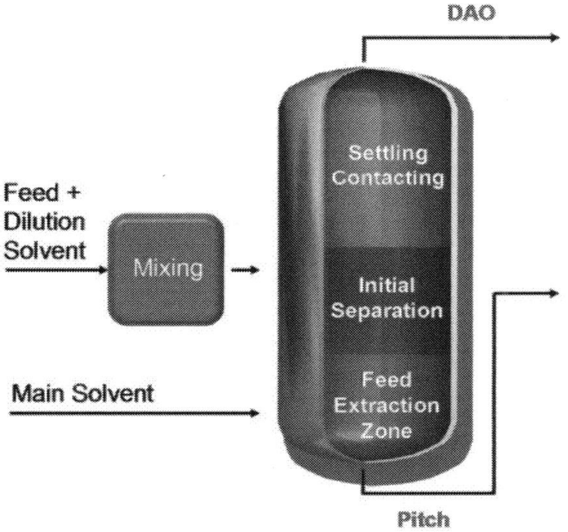

- Conversion unit feed (Hydrocracking or residue conversion)

- Primary residue conversion unit
- Gasification
- Performance asphalts

FOSTER WHEELER

 13

SDA is a bit like David Beckam

FOSTER WHEELER

 14

Conventional Residue Conversion Scheme

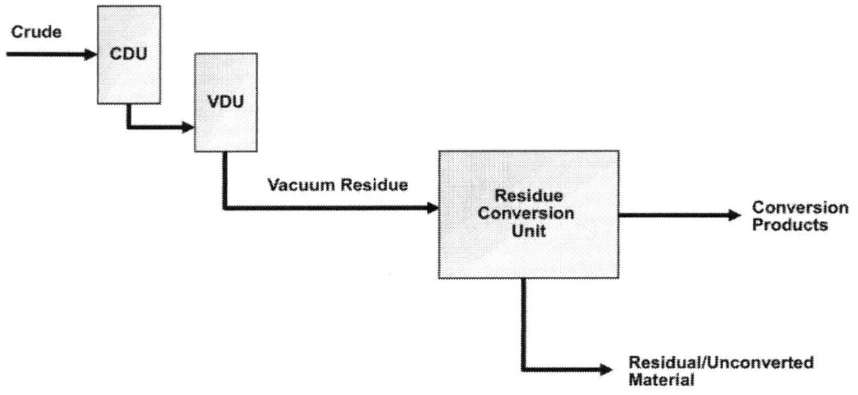

Single process to try and meet all objectives

New Role For Solvent Deasphalting

Residue Conversion Unit Feed Preparation 1

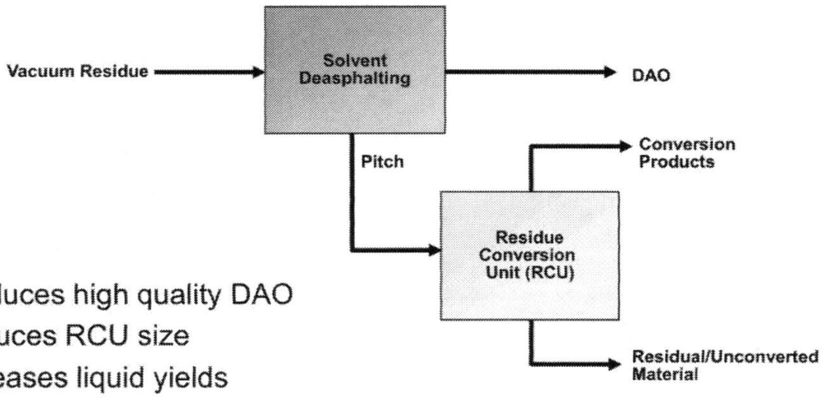

- Produces high quality DAO
- Reduces RCU size
- Increases liquid yields
- Improves project economics

New Role For Solvent Deasphalting

Residue Conversion Unit Feed Preparation 2

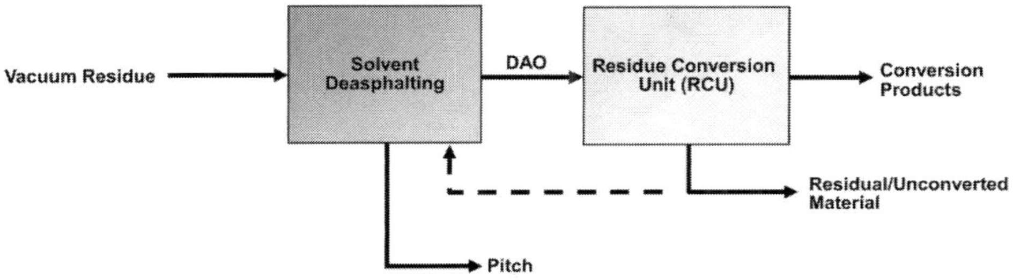

- Reduces RCU size
- Reduces feed impurities to RCU
- Increases conversion
- Improves project economics
- Revamp opportunity

 17

New Role For Solvent Deasphalting

Residue Conversion Unit Recycle

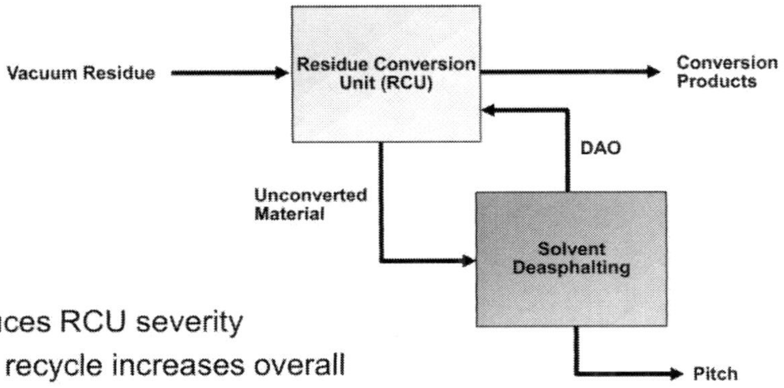

- Reduces RCU severity
- DAO recycle increases overall conversion
- Improves project economics
- Revamp opportunity

 18

Integrating Solvent Deasphalting With Coking

Mastering the Molecules

- SDA selectively separates residue components into those:
  - suitable for catalytic conversion (DAO)
  - suitable for thermal conversion (Pitch)
- Increases yields of desirable products
- Further increases margins
- SDA + Coking overall capital cost similar to Coking

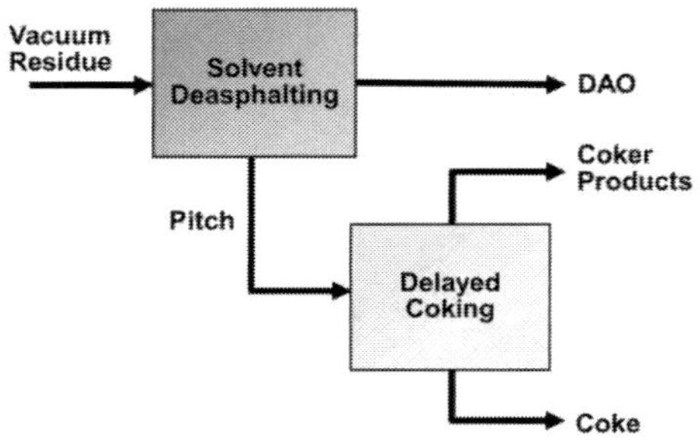

 19

Today's Solvent Deasphalting Requirements

Maximize Refinery Profitability

- High DAO yield
  - Maximize VGO conversion unit (e.g. Hydrocracker) feed
  - Minimize pitch processing unit capacity

- High quality DAO
  - Meet Hydrocracking Unit feed requirements

DAO yield limited by DAO quality

 20

Solvent Deasphalting Selectivity

Rejection of insolubles determines DAO Quality

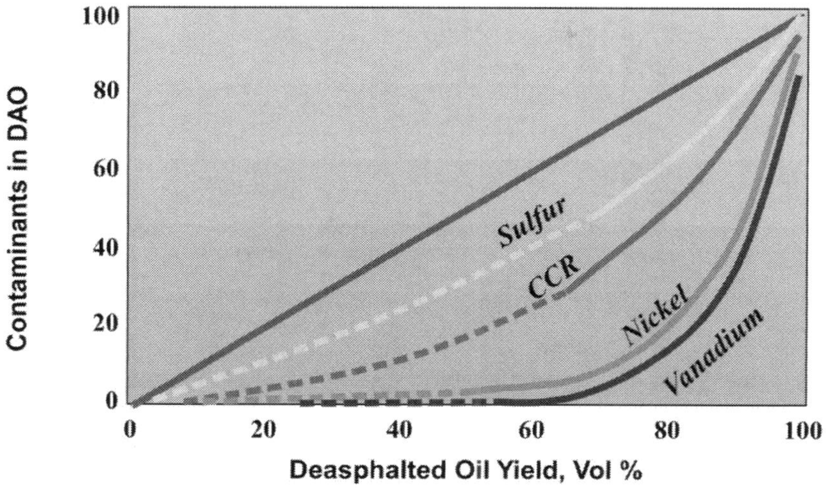

FOSTER WHEELER

Molecular Distribution in SDA Products

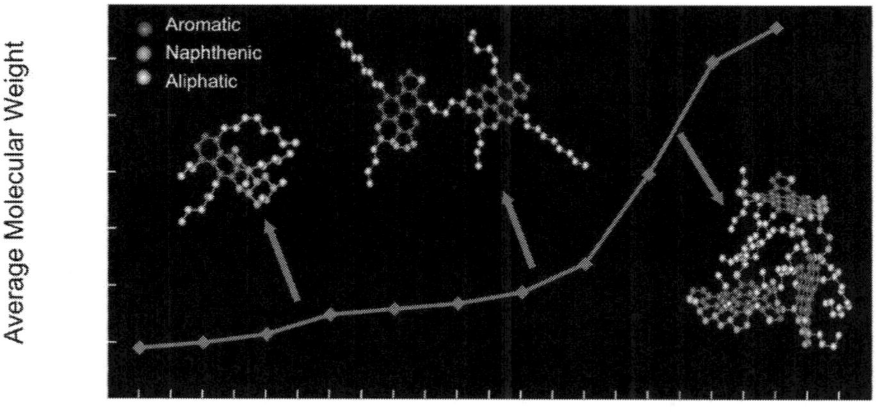

First 50% DAO molecules are suitable for hydrocracking

50-70+% DAO molecules are challenging to hydrocrack

DAO Yield

FOSTER WHEELER

 22

VGO Hydrocracking Feed Comparison

Middle East Sour Vacuum Residue

	Medium Lift DAO	High Lift DAO
VOL-%	50	70
API	16.7	12.9
Sp.Gr.	0.955	0.980
S, wt-%	**2.9**	**3.2**
N, wppm	1310	**1930**
Con Carbon, wt-%	2.9	7.4
C7 insols, wt-%	0.02	**0.04**
Ni +V , wppm	5	15

Color Scheme
Okay
Desirable
Undesirable

High lift DAO is undesirable Hydrocracker feed

Foster Wheeler SDA-RT Process

New Solution for High Yield & High Quality DAO

- Unique 3 product, SDA-based flow scheme
- Resins are hydrotreated and separated again in SDA
- Maximizes high quality DAO
- Reduces pitch
- Integrates very well with Delayed Coking

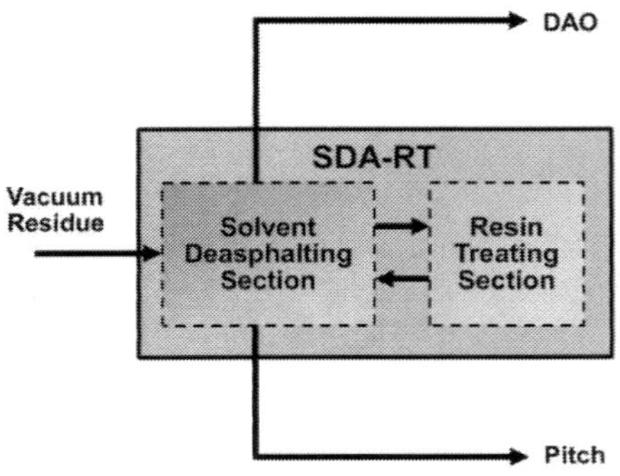

 24

DAO Quality Comparison

Middle East Sour Vacuum Residue

	Medium Lift DAO	High Lift DAO	SDA-RT DAO
VOL-%	50	70	66
API	16.7	12.9	15.9
Sp.Gr.	0.955	0.980	0.960
S, wt-%	**2.9**	**3.2**	2.0
N, wppm	1310	**1930**	1360
Con Carbon, wt-%	2.9	7.4	3.1
C7 insols, wt-%	0.02	**0.04**	0.01
Ni +V , wppm	5	15	4

Color Scheme
Okay
Desirable
Undesirable

SDA-RT maximizes high quality DAO yield

FOSTER WHEELER

SDA-RT & Coking Integration

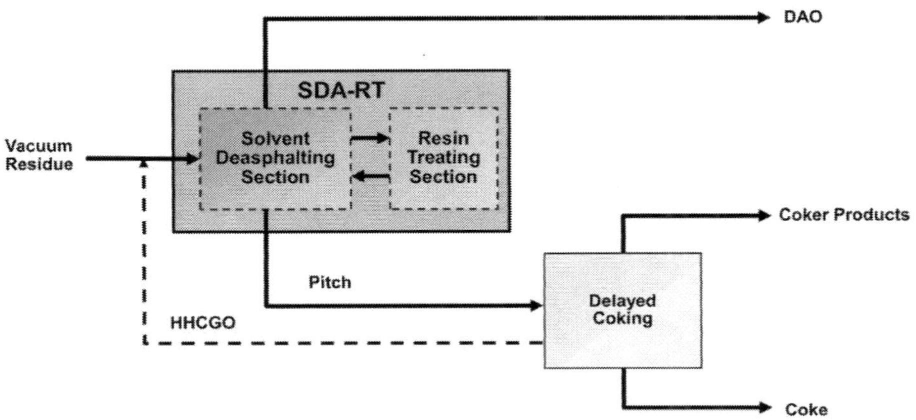

Maximizes catalytic conversion feedstock with conventional residue processing technologies

FOSTER WHEELER

27

Residue Upgrading Case Study

Distillate Hydrocracking Based Refinery

- Basis:
  - 300,000 BPD
    - Middle East sour crude
    - Russian Export Crude
  - 2015 forecast price set
  - Full conversion hydrocracking
- Options:
  - Coking
  - SDA + Coking
  - SDA-RT + Coking
  - EBED HC + Coking
  - Slurry Hydrocracking
  - Coke combustion to power and steam

FOSTER WHEELER

Coke/Pitch Yields

Distillate Hydrocracking Based Refinery

Middle East Sour Crude

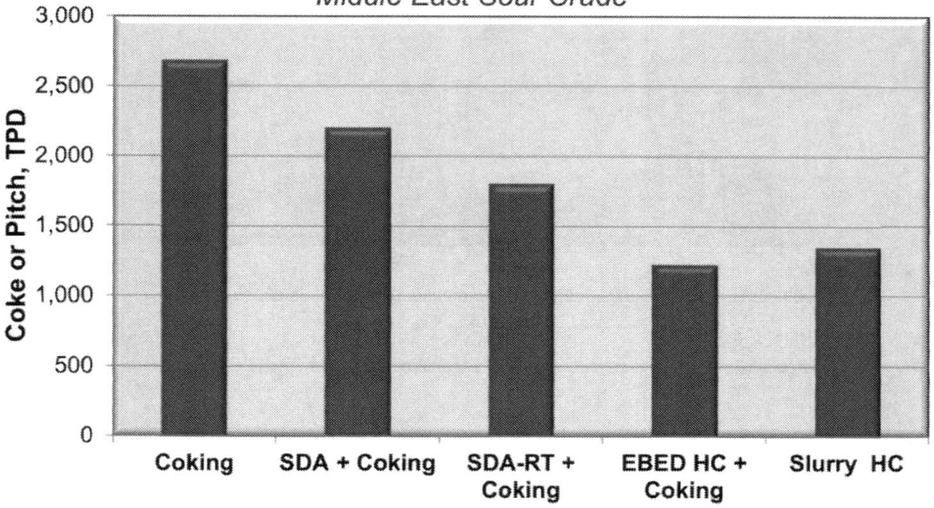

FOSTER WHEELER

Coke/Pitch Yields

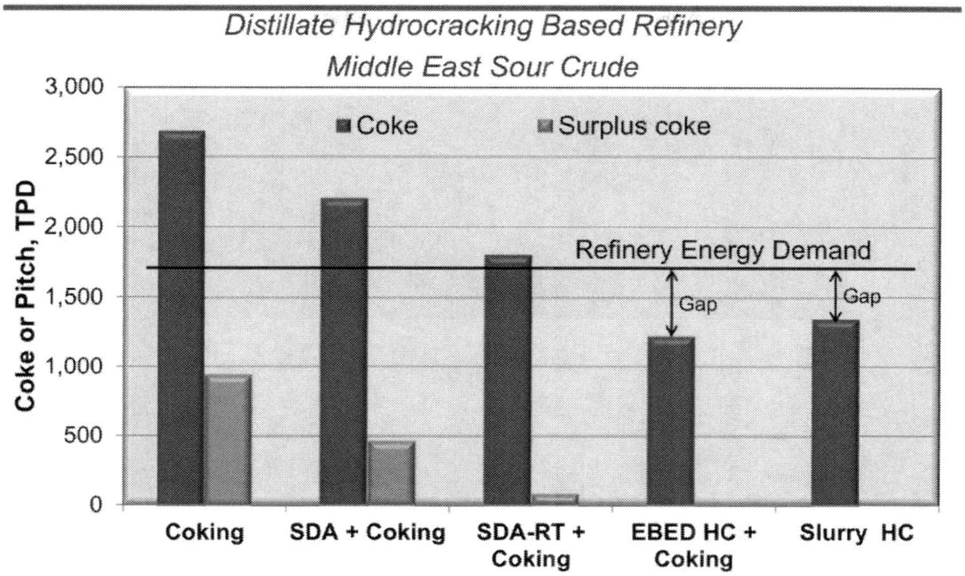

SDA-RT + DCU coke make in balance with energy demand

FOSTER WHEELER

Refinery Product Yields Relative to Coking

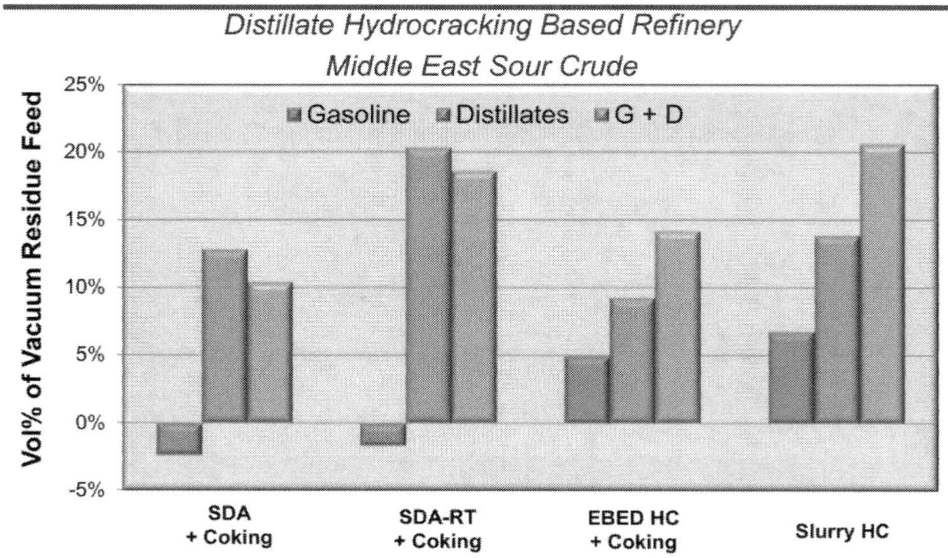

SDA-RT + DCU has highest distillate yield

FOSTER WHEELER

Net Refinery Operating Margins

Distillate Hydrocracking Based Refinery
Middle East Sour Crude

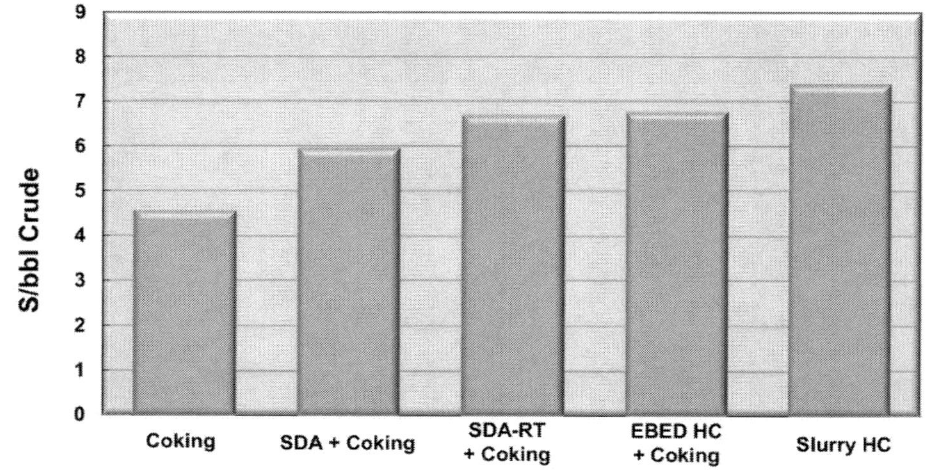

SDA-RT + DCU competes with residue hydrocracking

Net Refinery Operating Margins

Distillate Hydrocracking Based Refinery
Middle East Sour Crude

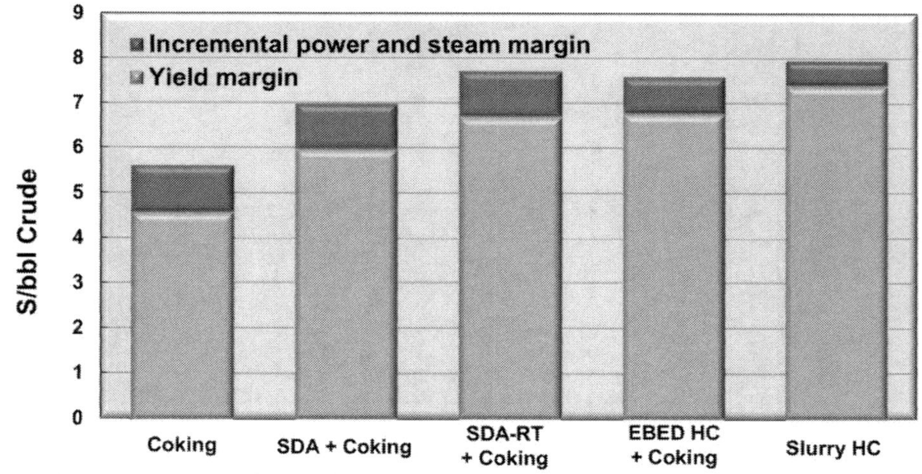

SDA-RT + DCU after coke conversion to energy has margins similar to residue hydrocracking

Refinery Product Yields Relative to Coking

Distillate Hydrocracking Based Refinery

Russian Export Crude

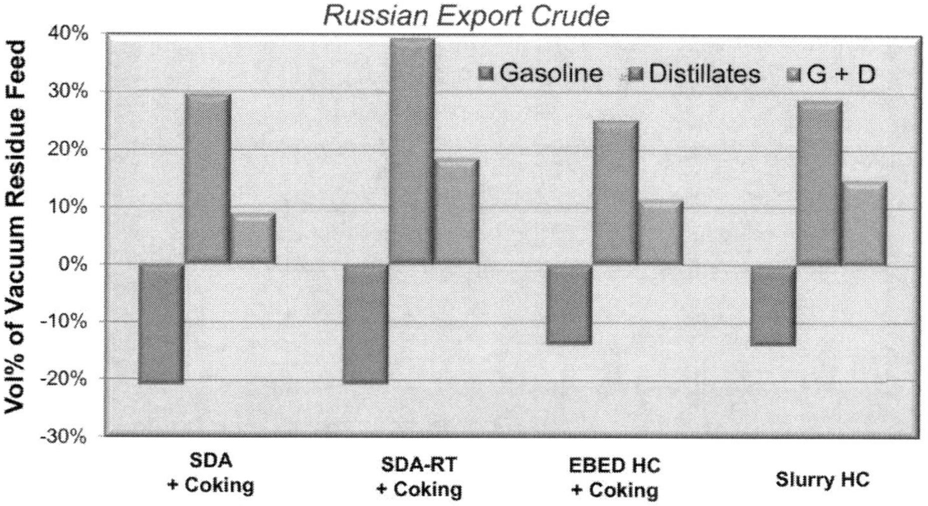

- ***SDA addition has big impact on yields***
- ***SDA-RT + DCU has highest distillate yield***

FOSTER WHEELER

 34

Net Refinery Operating Margins

Distillate Hydrocracking Based Refinery

Russian Export Crude

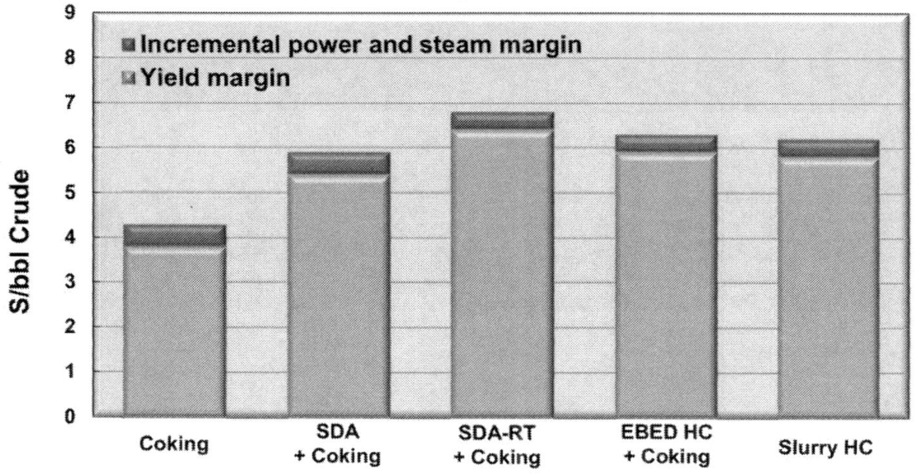

SDA-RT + DCU has highest margins

FOSTER WHEELER

Summary

- Many residue upgrading options available
- Traditional one-step approaches limit:
  - Conversion
  - Yields
- Including separation by SDA
  - Increases overall residue conversion
  - Reduces size of main residue conversion unit
  - Can debottleneck existing residue conversion units
  - Can improve feed quality to residue conversion units
  - Improves refinery margins
- SDA-RT maximizes high quality DAO yield – takes SDA to a new level
- SDA/Coking combinations can compete with other technologies by mastering the molecules

SDA has a key role in any residue conversion scheme

FOSTER WHEELER

 36

FOSTER WHEELER

www.fwc.com

Offshore Process Overview

October 2013

Manan Mehta

KBR

Agenda

- Introduction
- Types of offshore platforms
- Pertinent offshore issues / terms
- Process safety
- Typical Offshore Process
- Typical Offshore Process Utilities & Production Chemicals
- Offshore Platform Challenges
- Future development in offshore
- Rewind…

KBR

Agenda

- Introduction
- Types of offshore platforms
- Pertinent offshore issues / terms
- Process safety
- Typical Offshore Process
- Typical Offshore Process Utilities & Production Chemicals
- Offshore Platform Challenges
- Future development in offshore
- Rewind…

3

KBR

Introduction

- Why an offshore facility?

Source: FPS Nanhai Tiao Zhan, Project Liuhua 11-1, China, from www.offshore-technology.com

4

KBR

Agenda

- Introduction
- Types of offshore platforms
- Pertinent offshore issues / terms
- Process safety
- Typical Offshore Process
- Typical Offshore Process Utilities & Production Chemicals
- Offshore Platform Challenges
- Future development in offshore
- Rewind...

5

KBR

Types of Offshore Platforms

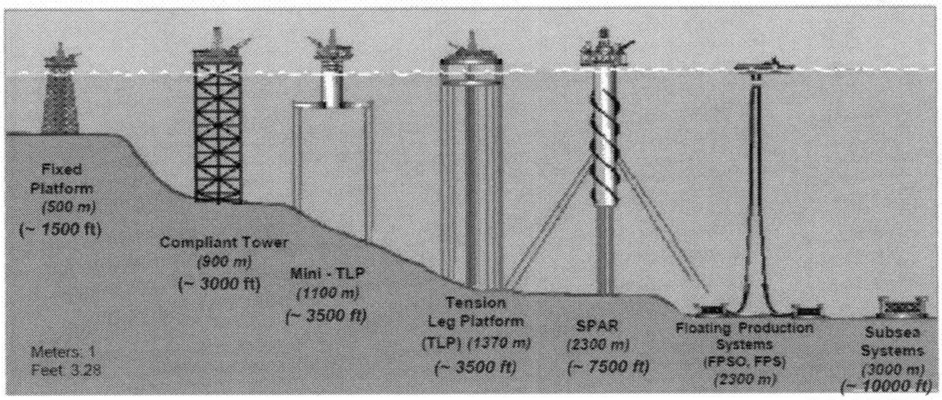

Source: Public Domain

6

KBR

Fixed Platform

Source: The Grane platform, Grane Field, from www.offshore-technology.com

KBR

Fixed Platform Characteristics

Source: www.globalsecurity.org

- Welded steel, tubular framework or jacket to support the topside facilities
- Piles driven into the seafloor secure the jacket.
- Topside facilities consists of 3 or more deck levels.
- Bridge linked jackets tend to favor a separate well head platform, processing platform and accommodation platform due to safety concerns.
- Shallow water applications to a water depths up to about 1500 ft.

KBR

Compliant Tower Platform

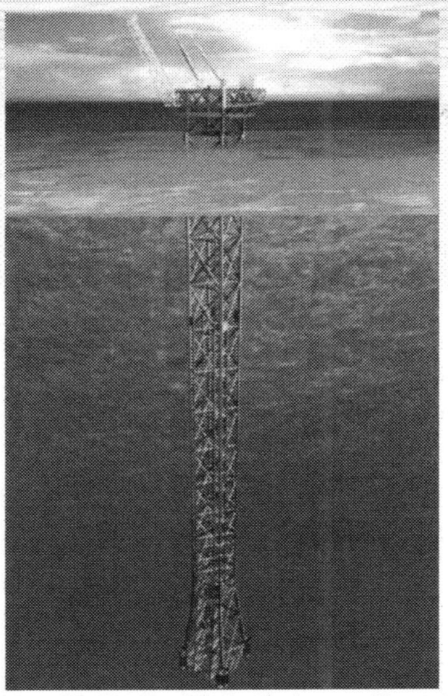

Source: GB 260 platform, Baldplate, USA, from www.offshore-technology.com

Compliant Tower Characteristics

Source: www.globalsecurity.org

- Similar to fixed platforms - steel tubular jacket to support the topside facilities. Topside facilities consist of 3 or more deck levels.
- Jacket structure designed for current and wind in a manner similar to floating structures. Designed to sustain significant lateral deflections and forces.
- Secured to the seafloor with piles – jacket has smaller dimensions than fixed platform
- Used in water depths from 1,500 to 3,000 ft.

Compliant Tower Example

Source: Tombua-Landana Project, Angola, from www.dsme.co.kr

Source: Tombua-Landana Project, Angola, from www.hmc.heerema.com

11

KBR

Tension Leg Platform

Source: Moses TLP, Marco Polo Field, GOM, from www.offshore-technology.com

12

KBR

Tension Leg Platform Characteristics

- Conventional TLP is a 4 column design
- Topside facilities typically modular with two main levels.
- Buoyant platform held in place by a mooring system
- Hull designed to dampen the vertical motions however allows horizontal movements.
- Good application if dry tree wells and drilling rig required.
- Used in water depths from 1500 ft to 7000 ft.

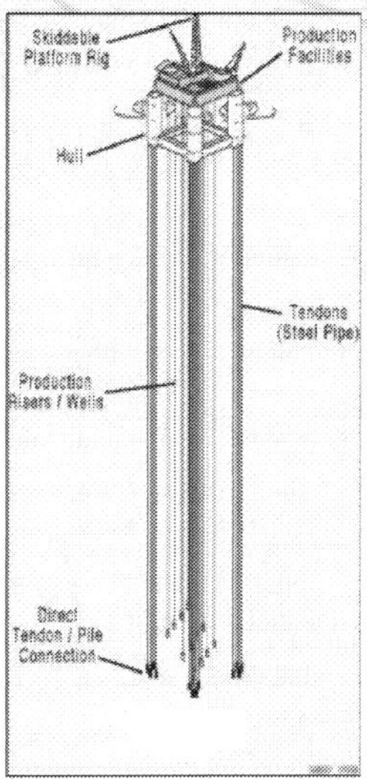

Source: www.globalsecurity.org

KBR

SPAR

Source: www.koreaherald.com

KBR

SPAR Characteristics

- Large diameter single vertical cylinder supporting a deck.
- Topside facilities consists of 3 main deck levels.
- Spar platform hull consists of:
  - Buoyant upper section with compartmentalized annular cylinder
  - Flooded mid-section
  - Compartmentalized keel with fixed ballast
  - Flooded center well through which buoyancy cans and risers are installed
  - ~ 90% of the hull structure is underwater.

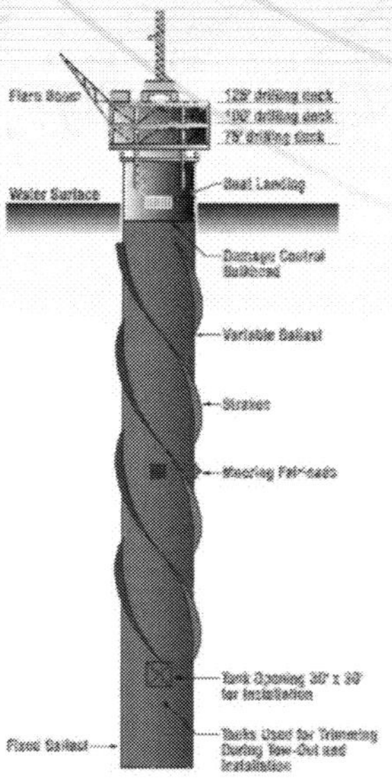

Source: www.globalsecurity.org

15

KBR

SPAR Characteristics

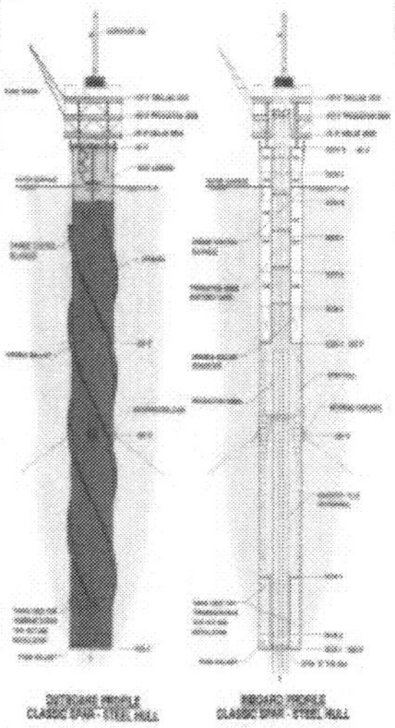

Source: www.globalsecurity.org

- Four major systems are hull, moorings, topsides, and risers.
- Deep-draft hull, produces very favorable riser motion characteristics compared to other floating concepts.
- Low motions and a protected center well also provide an excellent configuration for deepwater operations.

16

KBR

SPAR Example

Source: Constitution / Ticonderoga Field, GOM, from www.offshore-technology.com

Source: Mad Dog, Western Atwater Foldbelt, GOM, from www.offshore-technology.com

17 KBR

Semi-Submersible

Source: West Hercules, South China Sea, from www.offshore-technology.com

18 KBR

Semi-Submersible Characteristics

- Topsides typically two deck levels (modular or integrated topsides).
- Twin hulls (columns and pontoons) of sufficient buoyancy to cause the structure to float, but of weight sufficient to keep the structure upright.
- Can be moved from place to place
- Can be ballasted up or down
- Water depths from 2000 to 10,000 feet.

Source: Public Domain

19

KBR

Semi-Submersible Example

Source: BP Atlantis, GOM, from www.offshore-technology.com

Source: BP Thunder Horse, GOM, from www.gvac.se

20

KBR

FPS / FPSO

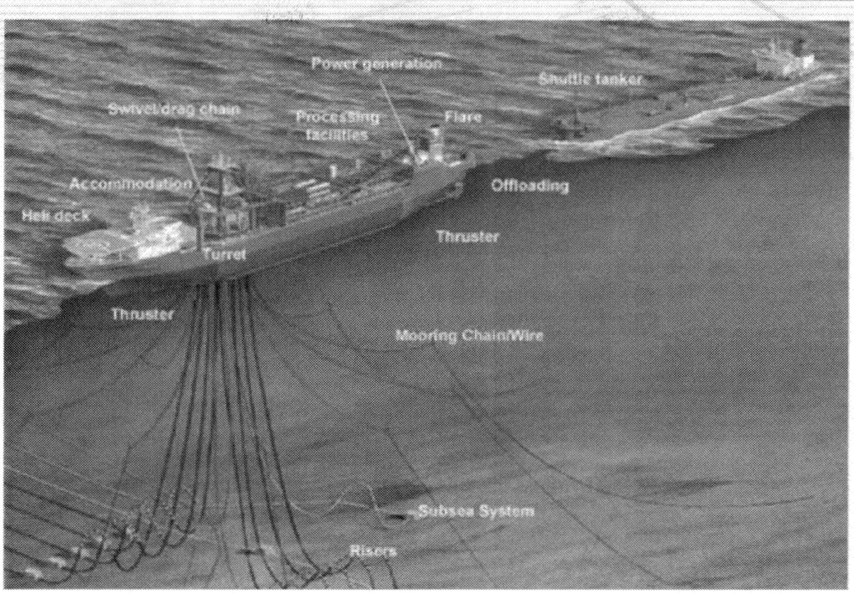

Source: www.teekayoffshore.com

FPS / FPSO Characteristics

- Floating Production, Storage & Offloading System (FPSO).
- Consists of a large tanker type vessel moored to the seafloor.
- FPSO vessel is converted from liquid cargo vessel or new built.

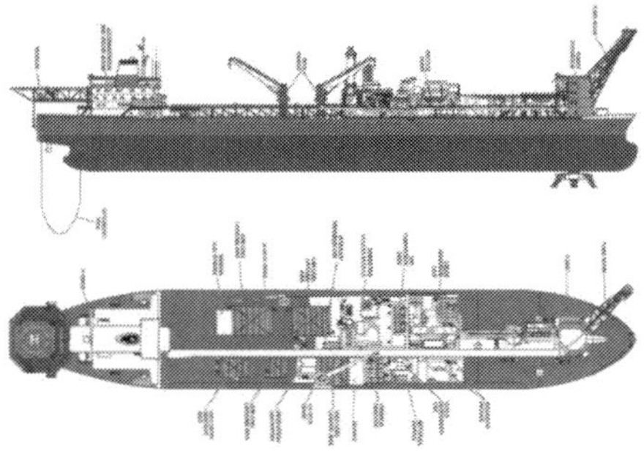

Source: www.marineinsight.com

- Topside facilities normally modular construction.
- Large oil storage capacity is an advantage

FPS / FPSO Characteristics

- FPSO equipped with processing facilities and moored or turret to a location.
- Floating Production Systems are ideal solution for
  - Small and marginal field
  - Applications where field isolated and no established pipeline infrastructure available
  - Deep and Ultra Deep water

Source: FPSO Fluminense, from www.modec.com

23

KBR

FPS / FPSO Example

Source: FPSO Pyrenees Venture, from www.modec.com

24

KBR

FPS / FPSO Example

Source: Belanak LPG, from www.offshore-technology.com

Source: FPSO Aghbami, from www.kbr.com

Source: FPSO Aghbami, from www.statoil.com

25

FPS / FPSO Example

Source: Pazflor FPSO, from www.dsme.co.kr

26

FPS / FPSO Example

Source: Pazflor FPSO, from www.towingline.com

KBR

FLNG – The New Frontier

SHELL PRELUDE FLNG FACILITY

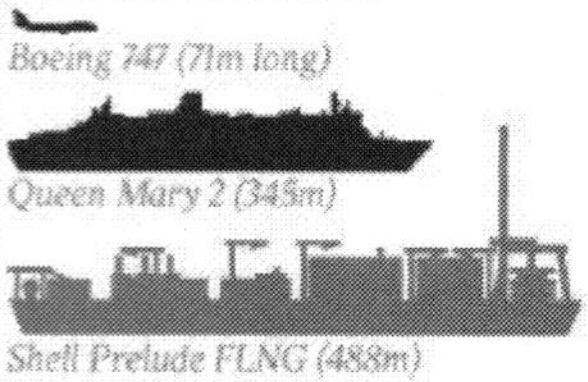

Boeing 747 (71m long)

Queen Mary 2 (345m)

Shell Prelude FLNG (488m)

KEY FACTS

- **The Prelude facility will be 488m long and 74m wide**
- **It will stay moored in water 250m deep for 25 years**
- **First production in 2017 of at least 3.6 million tonnes of LNG per year**
- **It will create 1000 jobs and add $45 billion to the economy**

Source: Prelude FLNG, from www.theaustralian.com.au

KBR

Subsea System

Source: Pazflor Subsea Layout, from www.worldoil.com

KBR

Subsea System

- A Subsea System (SS) ranges from a single subsea well producing to a nearby platform, FPSO, or TLP to multiple wells producing through a manifold and pipeline system to a distant production facility.

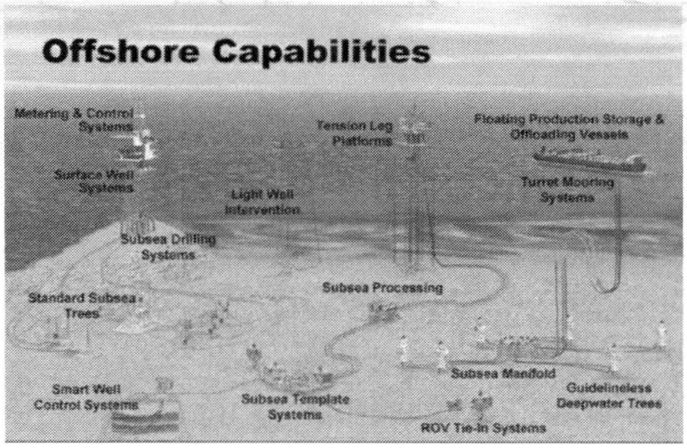

Source: courtesy FMC Technology, from www.sec.gov

- These systems are presently used in water depths greater than 5,000 ft.

KBR

Agenda

- Introduction
- Types of offshore platforms
- Pertinent offshore issues / terms
- Process safety
- Typical Offshore Process
- Typical Offshore Process Utilities & Production Chemicals
- Offshore Platform Challenges
- Future development in offshore
- Rewind…

31

KBR

Pertinent offshore issues / terms

- Water Depth

Definition	U.S. MMS (meters)	North Sea / Atlantic Margin (U.K. / Norway (meters)	Brazil (meters)
Shallow Water	< 400	< 300	< 400
Deepwater	400 – 1,300	300 – 2,000	400 – 1,000
Ultra Deepwater	> 1,300	> 2,000	> 1,000

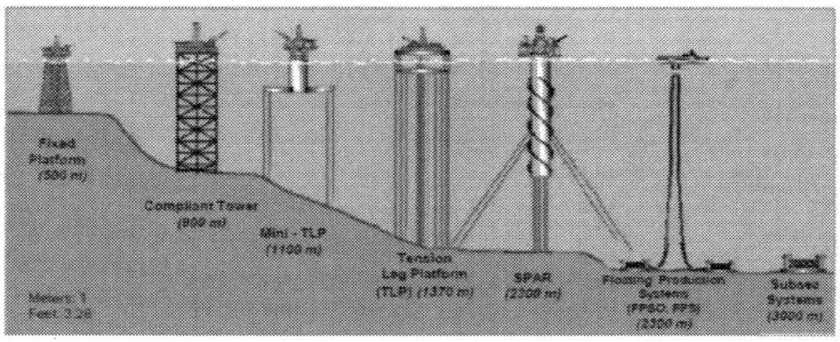

Source: Public Domain

32

KBR

Pertinent offshore issues / terms

- Water Depth

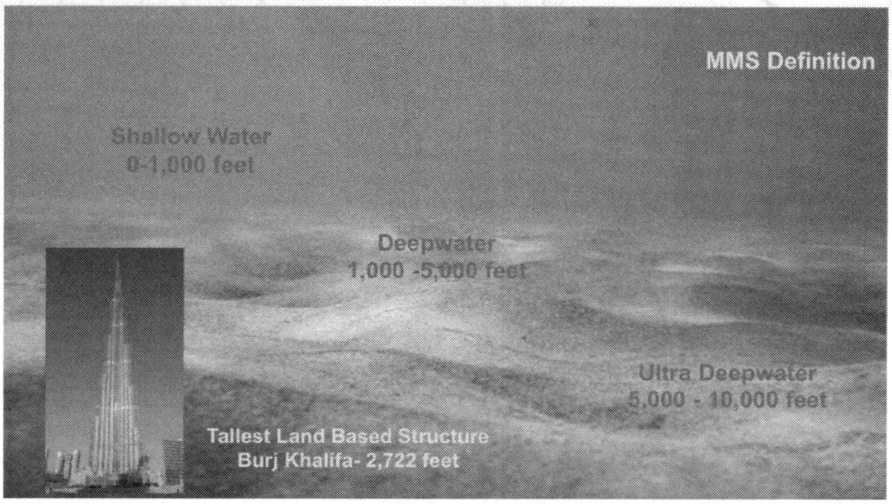

KBR

Pertinent offshore issues / terms

- Flow Assurance
  - Flow assurance = "Cash Flow Assurance"
  - Typical flow assurance issues include:

Hydrates

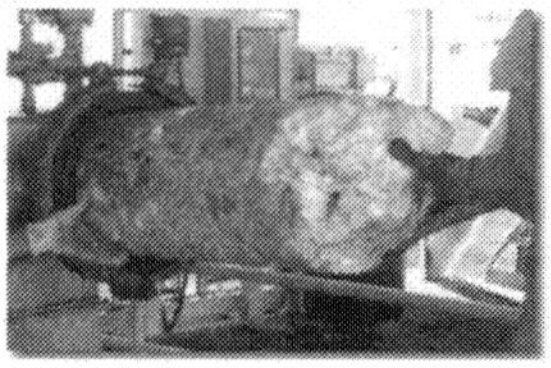

Slugging

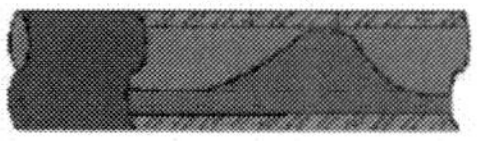

Wax / Asphaltene

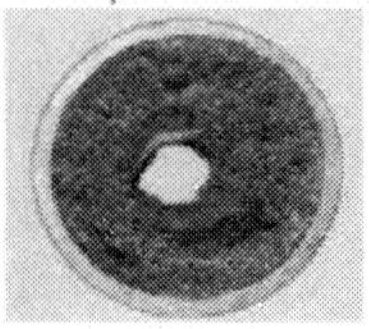

Scaling

Erosion

Corrosion

Source: Public Domain

KBR

Agenda

- Introduction
- Types of offshore platforms
- Pertinent offshore issues / terms
- Process safety
- Typical Offshore Process
- Typical Offshore Process Utilities & Production Chemicals
- Offshore Platform Challenges
- Future development in offshore
- Rewind...

KBR

Process Safety

- API 14C
  - Recommended Practice for Analysis, Design, Installation, and Testing of Basic Surface Safety Systems for Offshore Production Platform
  - Illustrates how system analysis methods can be used to determine safety requirements to protect any process component
  - Safety Analysis Checklist (SAC)

Headers	Pressure vessel
Atmospheric vessels	Fired and exhaust heated component
Pumps	Compressors
Pipelines	Heat exchangers
Flow line segment	Wellhead injection lines

KBR

Process Safety

- Process Safety System for Offshore
  - PFP (Passive Fire Protection) based on Fire Consequence Analysis
  - Confirmed gas detection or fire initiates ESD (Plant Wide Shutdown)
  - Blast or Fire wall
  - Escape, Emergency and Rescue Analysis (EERA)
  - Fire water medium - Seawater
  - Fire water pumps – Caisson type or Sea chest type
  - Additional Safety Regulations to consider (e.g. SOLAS, USCG, IMO, ABS, DNV)

Agenda

- Introduction
- Types of offshore platforms
- Pertinent offshore issues / terms
- Process safety
- Typical Offshore Process
- Typical Offshore Process Utilities & Production Chemicals
- Offshore Platform Challenges
- Future development in offshore
- Rewind...

Typical Oil & Gas Offshore Facility Systems

- Oil / Gas Separation
- Gas compression & processing
- Produced water treatment
- Water treatment & injection
- Living Quarters
- Power generation & utility systems.

Typical Offshore Process

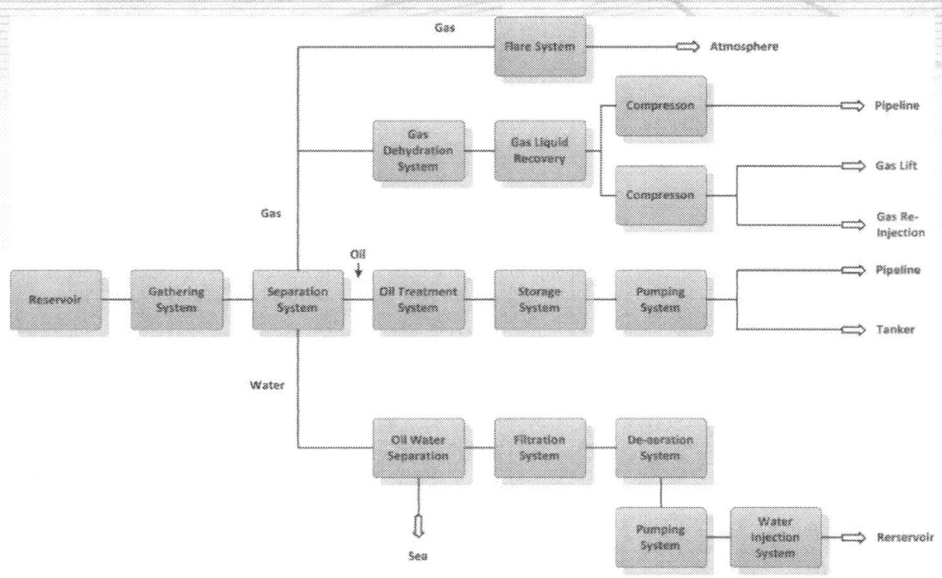

Typical Offshore Process

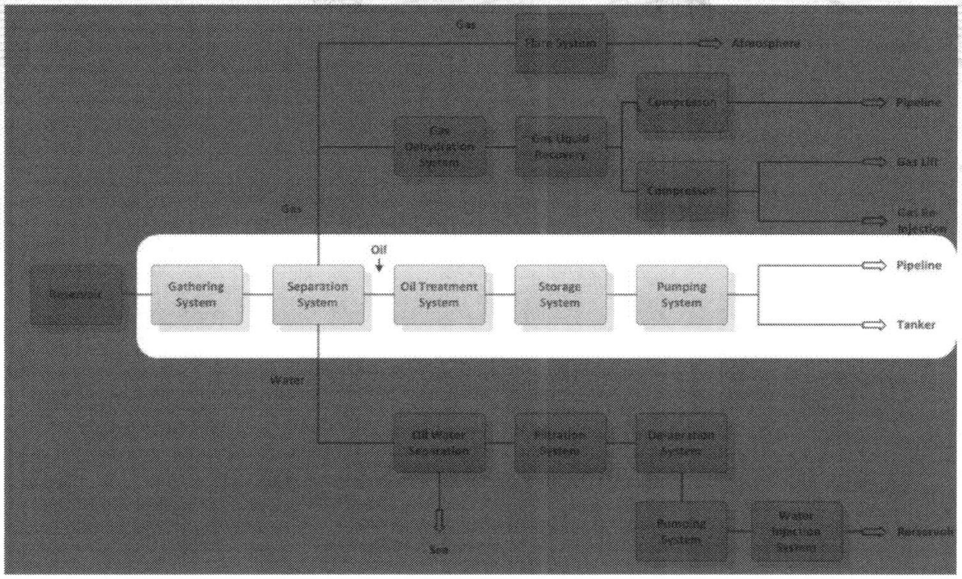

KBR

Typical Offshore Process

- Separation
  - Hydrocarbon flow from the producing wells will be received at production manifolds platform and transferred to separator for separation into oil, gas and water phases.
  - Three phase (oil, gas and water) separator in series.
  - Solids presents in form of sand.
  - Majority of the gas will flash off in first separator.

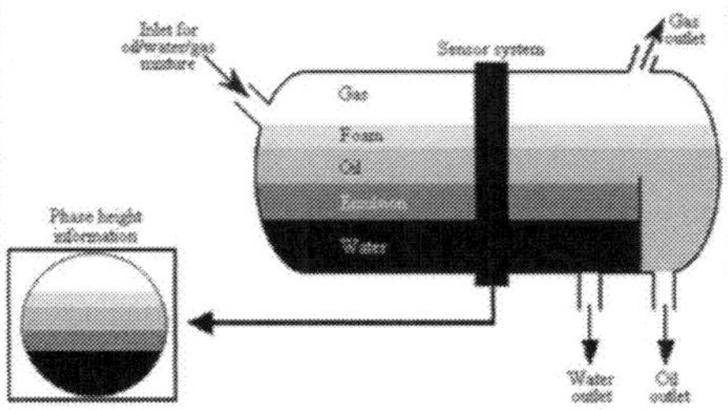

Source: www.oilandgasprocessing.blogspot.com

KBR

Typical Offshore Process

- Separation
  - This gas will be routed to the gas compressor and dehydration system for further processing.
  - The liquid hydrocarbon phase from first separator will be routed for further separation into oil and water phases.
  - Produced oil from the separator will flow to either storage or export oil pipelines via booster pumps.
  - Produced water will be routed to the produced water treatment system and then to the water injection system.
  - Design parameter – Early life, Mid life & Late life

KBR

Typical Offshore Process

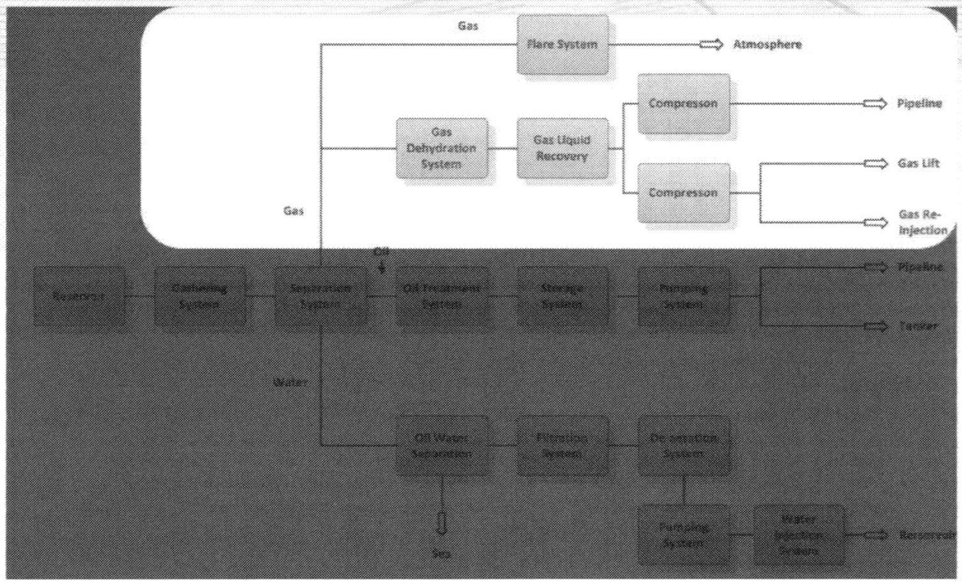

KBR

Typical Offshore Process

- Gas Processing
  - Gas removed from the separator will be compressed and treated prior to export onshore via gas pipeline.
  - Treatment will involve gas cooling and dehydration to remove water.
  - Gas removed from the fluid in the 2nd Separator will be cooled and compressed via flash gas compressor before being co-mingled with 1st Stage gas compressor upstream of the dehydration column.
  - Final dehydration will involve use of glycol to remove any residual moisture to prevent hydrate formation and corrosion within the gas export pipeline.
  - Used glycol will be recovered, treated in a glycol regeneration package and recycled.

KBR

Typical Offshore Process

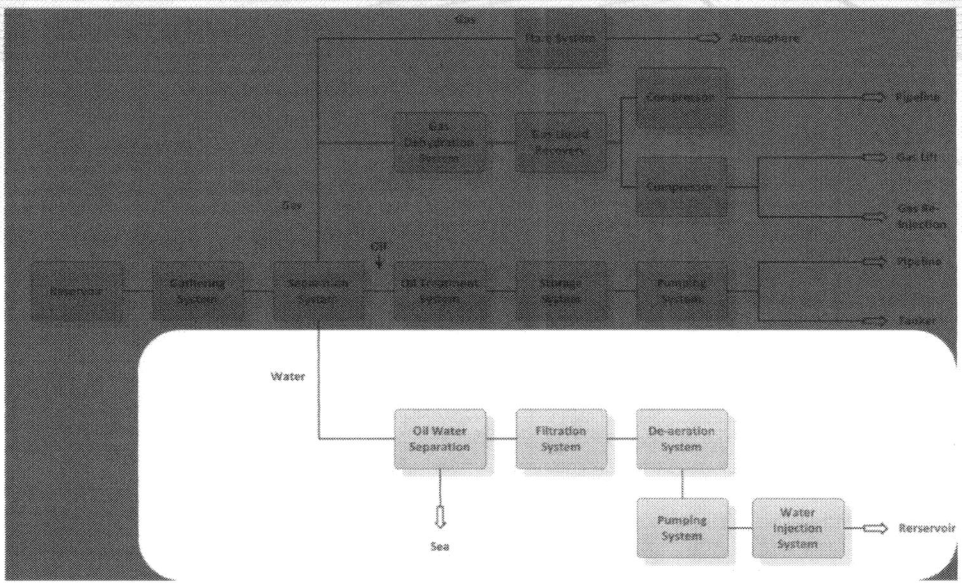

KBR

Typical Offshore Process

- Produced Water System
  - Following treatment, produced water will be sent to the water injection pumps where it will be injected for reservoir pressure maintenance.
  - A produced water treatment package includes solids removal sand cyclone units and de-oiling hydrocyclones, gas flotation, and potentially a polishing unit.

47

KBR

Typical Offshore Process

- Water Injection System
  - Injection water will include produced water and lifted seawater. Seawater will be taken from a depth below the sea surface using lift pumps.
  - Following filtration to remove solids, some seawater will be used the closed loop cooling water system. Filtered seawater required for water injection will be transferred to the water injection system. The injection water treatment system consists of a de-aerator tower where water oxygen levels will be reduced via injection of an oxygen scavenger and other chemicals, and then potentially treated further with a sulfate removal system.

48

KBR

Agenda

- Introduction
- Types of offshore platforms
- Pertinent offshore issues / terms
- Process safety
- Typical Offshore Process
- Typical Offshore Process Utilities & Production Chemicals
- Offshore Platform Challenges
- Future development in offshore
- Rewind...

Typical Offshore Process Utilities

- Power Generation
- WHRU & Heat Medium System
- Fuel Gas & Diesel System
- Flare and Closed Drain System
- Sea Water System
- Cooling Medium System
- Fire Water System
- Fresh Water System
- Instrument Air & Inert Gas System
- Sewage & Drainage System
- Sand Jetting & Separation System
- Chemical Injection System

Typical Offshore Production Chemicals

- Antifoam
- Demulsifier
- Scale Inhibitor
- Reverse Demulsifier
- Corrosion Inhibitor
- Biocide
- Methanol
- MEG / TEG
- Oxygen Scavanger

51

KBR

Agenda

- Introduction
- Types of offshore platforms
- Pertinent offshore issues / terms
- Process safety
- Typical Offshore Process
- Typical Offshore Process Utilities & Production Chemicals
- Offshore Platform Challenges
- Future development in offshore
- Rewind...

52

KBR

Offshore Platform Challenges

- Weight of the topsides affects jacket or hull sizing and should be minimized.
- Space also affects jacket and hull sizing.
- Offshore processing typically minimized.
- New metocean data (wave heights, under water currents) in deep water applications making design of hulls more expensive.
- Designing topside facilities for continually changing reservoir condition (pressure, temperature and composition).

Agenda

- Introduction
- Types of offshore platforms
- Pertinent offshore issues / terms
- Process safety
- Typical Offshore Process
- Typical Offshore Process Utilities & Production Chemicals
- Offshore Platform Challenges
- Future development in offshore
- Rewind...

Future development in offshore

- Arctic Exploration
- Subsea Processing (Separation, Pumping, Metering, etc.)
- Ultra Deep Water
- High Pressure / High Temperature Reservoir
- Heavy Crude Fields
- Bathymetry – Seabed topography (affects moorings of hulls, pipeline routing, etc.)
- Low Energy Reservoir

KBR

Questions?

Thank you!

KBR

Agenda

- Introduction
- Types of offshore platforms
- Pertinent offshore issues / terms
- Process safety
- Typical Offshore Process
- Typical Offshore Process Utilities & Production Chemicals
- Offshore Platform Challenges
- Future development in offshore
- Rewind…

57

KBR

Rewind…

Spars

Source: Genesis, GOM, from www.offshore-technology.com

58

KBR

Rewind…

- Terminology – Water Depth
  - The new reservoir found in GOM at 10,000 ft depth. "Deep water" or "Ultra Deep"?

Ultra Deep

Rewind…

- How many phases on first separator?

Oil, Gas, Water & Sand

TPC Group

Chemical Engineering Calculations with Microsoft Excel ®

Mass Balancing by The Maximum-Likelihood Principle

AIChE Regional Meeting – Young Members Session - Galveston,TX
October 4, 2013

Mark J. Potter, P.E.

Quality Products. Dependable Services.

TPC Group

LEGAL DISCLAIMER *This Report includes forecasts, projections and other predictive statements that represent TPC's assumptions and expectations in light of currently available information. These forecasts, etc., are based on industry trends, markets and other factors, and they involve risks, variables and uncertainties. Consequently, no guarantee is presented or implied as to the accuracy of specific forecasts, projections or predictive statements contained herein.*

2

TPC Group

Microsoft Excel History

» Microsoft's first spreadsheet program, Multiplan, debuted in 1982 on DOS and CP/M operating systems

» First Mac version in 1985
  • Presenter's first experience with Excel in 1987 on Macs

» First Windows version in late 1987

» Excel outpaced Lotus 1-2-3 in sales around 1988

» Excel 5.0 makes Microsoft the dominant spreadsheet seller in 1993

» This presentation based on Excel 2007, WinXP

3

TPC Group

Why mass balance your unit rigorously?

» Discover instrumentation and analytical deficiencies in process data – It is difficult to run a unit properly using misleading metrics!

» Monitor unit performance on a consistent basis
  • Calculate efficiencies, KPIs
  • Rigorous calculation of yields and conversions not swayed by random nor non-random measurement errors

» Provide data for process model building

4

TPC Group

Why Mass Balances Don't Close – The Need To Reconcile

» Random error in perfectly fine analyzers

» Calibration errors/drift

» Zeroing errors in flow meters

» Equipment malfunction or lack of servicing

» Improper sensor installation or design

» Mass balance non-closure usually caused by the last four, but where is the problem? Is there more than one problem? Maybe one which is non-obvious?

5

TPC Group

Example #1 – Flow From Feed Tank

» Mass balancing of a unit indicated poor overall and component balances

» Inconsistent calculated yields

» Hard to reconcile operations with monthly closings

» Maximum Likelihood Principle flagged one process feed as in not only in gross error but wildly fluctuating

6

TPC Group

Example #1 – What's in the Field?

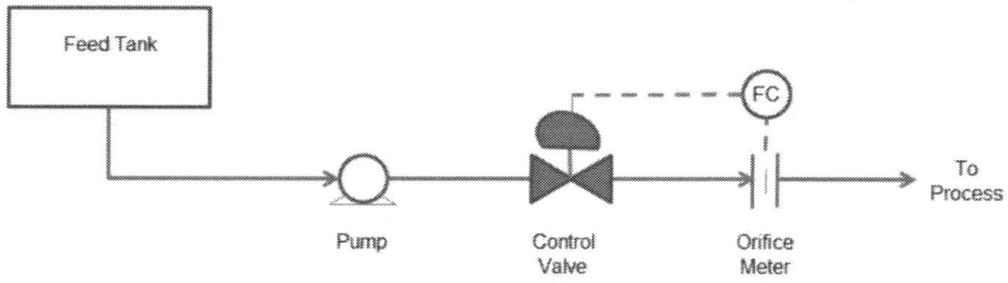

7

TPC Group

Example #1 – What's in the Field?

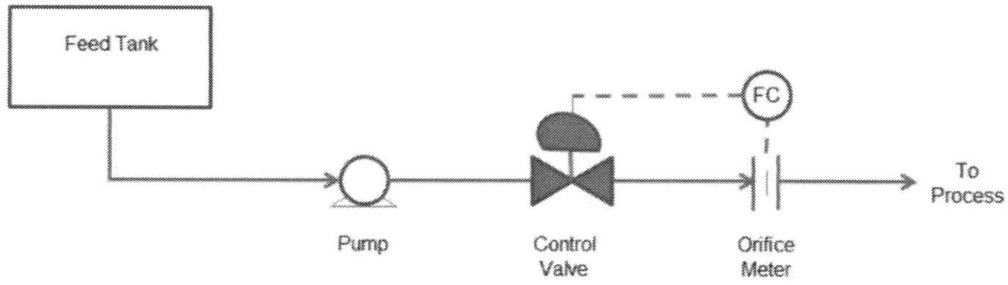

Human Error!

8

TPC Group

Example #2 – Parallel Train Unit

» Attempting to reconcile differences in performance between two parallel trains

» The trains share common feed tanks and a GC to control composition in the of component "B" in the process feed

- No calibration issues with the GC

» MLH principle mass balancing indicated problems with feed composition to one train but not the other.

9

TPC Group

Example #2 – What's In The Field?

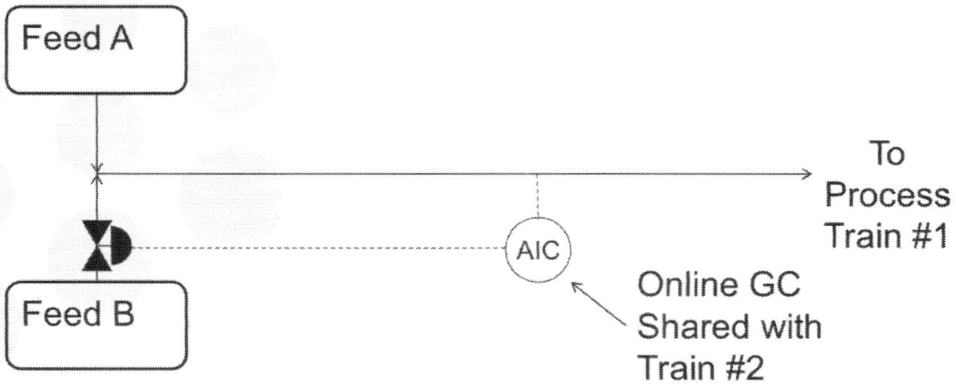

10

TPC Group

Example #2 – What's In The Field?

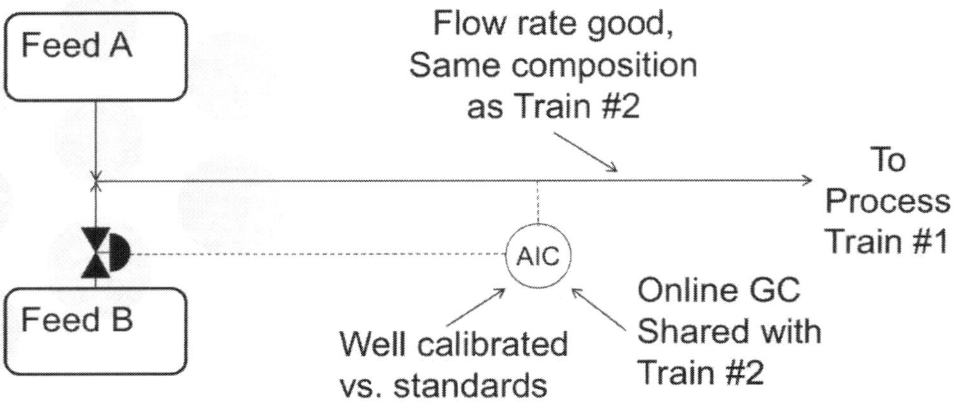

11

TPC Group

Example #2 – What's In The Field?

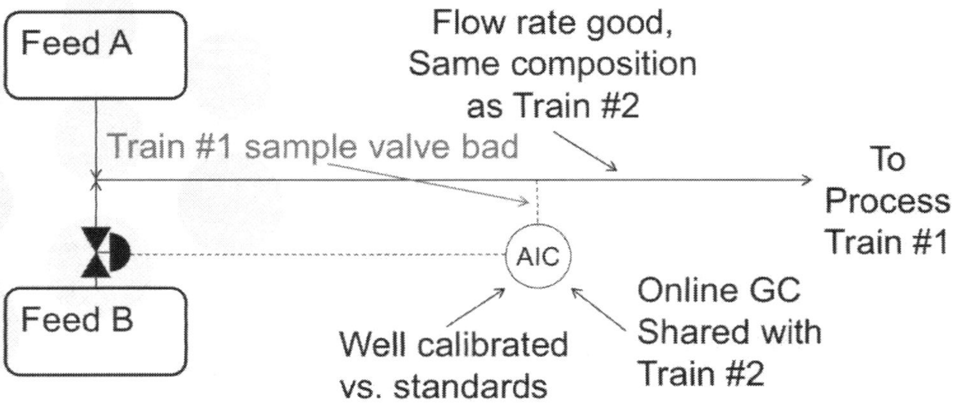

Correctly controlling to the wrong composition !!

12

TPC Group

The Maximum Likelihood Principle

- » **A method of parameter estimation where measured variables are subject to known or estimated error and subject to constraints**
  - **Errors in dependent variables**
  - **Errors in independent variables**
  - **Unmeasured variables**
  - **Mass balance constraints (in = out + generation - depletion)**
  - **Compositional constraints (∑ = 100%)**
  - **Equilibrium constraints (vapor fugacity = liquid fugacity)**
  - **Any other relevant constraints**
- » **Many applications**
  - **Mass and energy balances**
  - **Physical properties regression (vapor pressure, density, heat capacity, etc.)**
  - **Phase equilibrium data regression (VLE, LLE, SLE, etc.)**
  - **Anywhere parameters need to be derived to explain experimental data**

13

TPC Group

The Maximum-Likelihood Principle : Limitation

- » **The maximum-likelihood principle *cannot* turn fundamentally bad data into good data**
  - **At best, the MLH principle will turn bad data into good-looking bad data – this can be dangerous if it falls into the wrong hands**
  - **It *can*, however, isolate what is causing the bad data**

14

TPC Group

The Maximum Likelihood Principle

» **The key is to minimize the following function (Bard, 1974)…..**

Measured Value

Reconciled (estimated actual) value

N = number of data points

$$OBJ = \sum_{i=1}^{i=N} \frac{(x_{m,i} - x_{r,i})^2}{\delta_i^2}$$

Objective Function

Measurement variance ("error")

…… subject to constraints (mass balance, fugacity, etc.)

» **Unmeasured parameters may be varied (with or without constraints) to minimize OBJ**
- **Example – NRTL parameters for VLE fitting**

Bard, Yonathan : *Nonlinear Parameter Estimation*, Academic Press, New York, 1974

TPC Group

The Maximum Likelihood Principle – Mass Balances

» **Measured variables**
- **Total stream flows (convert to mass if not from a coriolis meter)**
- **Stream mass fraction compositions**

» **Parameters – Generally none**
- **Conversion if mass balancing a reactor**

» **Variance – estimation of error in measurement**

» **Constraints**
- **Stream mass fractions must sum to 1.0**
- **The material balance of *each* component must equal 100%**

Minimizing "OBJ" Value Results in Maximum-Likelihood Solution

» **Nonlinear problem**

» **Can solve by linearization, matrix solution techniques (Anderson, 1978)**
  - **FORTRAN code which can be adapted to VBA**
  - **Very fast solution**

Anderson, T. F. : AIChEJ, v24, p20, 1978

FORTRAN code for Maximum-Likelihood solution can be found here

TPC Group

Minimizing "OBJ" Value Results in Maximum-Likelihood Solution

» Nonlinear problem

» Can solve by linearization, matrix solution techniques (Anderson, 1978)
- FORTRAN code which can be adapted to VBA
- Very fast solution

» Use Excel's Solver
- Generalized Reduced Gradient (GRG) algorithm
- Slightly slower
- Often finds better objective function minima than linearization
- Limits to number of variables (~200)
- Third party solvers available with higher number of variables
- Open Office / LibreOffice solvers do not work on this problem

» Process simulators with data reconciliation capability

Anderson, T. F. : AIChEJ, v24, p20, 1978

19

TPC Group

Maximum Likelihood Mass Balancing in Excel

» Enter measured data in an orderly format
- Suggest streams in columns, total flow in top row, compositions beneath
- Sum mass fractions bottom row
- Calculate mass balance of each component is right-most column

» Copy this data to the next set of columns to serve as a holding place for reconciled data (your results)

» Use a third set of columns to enter standard deviations for flows and composition analysis

» A final set of columns to compute the OBJ function

» Use colors to keep the column groups straight in your mind

» Solve, then analyze residuals – important!

20

TPC Group

Enter raw plant data and calculate unreconciled balances

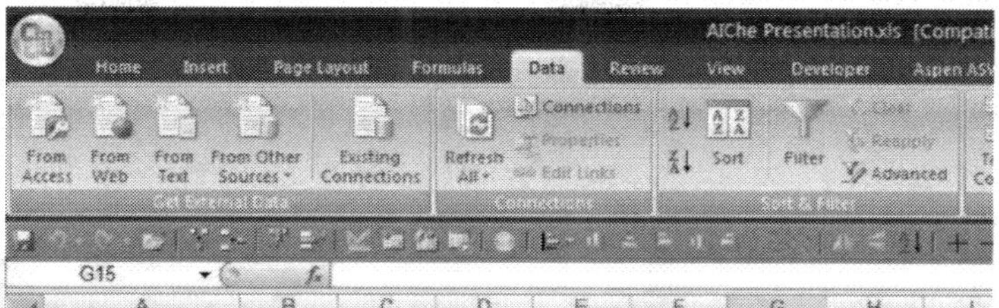

Mass Balance for PolyChickenWire Unit Pre-Fractionator Flash

	Raw Plant Data			
Stream Name	Feed	Vapor	Liquid	Balance
Mass Flow	10123	2811.33	7185	0.987487
Component (m.f.)				
Trimethyl Meatloaf	0.13582	0.491205	0.00001	1.004441
Chicken Wire	0.14398	0.308845	0.0846	1.012766
Ba(Na)2	0.10013	0.200101	0.0553	0.946986
PolyChickenWire	0.60254	0	0.86011	1.013178
Total	0.98247	1.000151	1.00002	

Raw stream mass flows

Component mass fractions

Overall mass balance

Calculated raw component mass balances

TPC Group

Copy the raw data as your initial reconciled data. This is what you will vary.

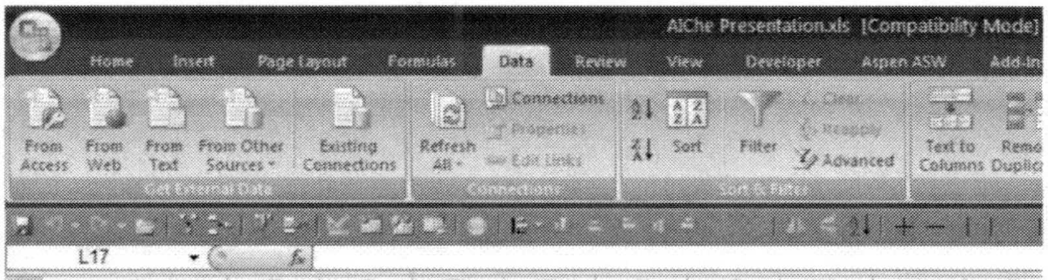

Mass Balance for PolyChickenWire Unit Pre-Fractionator Flash

	Raw Plant Data				Reconciled Data			
Stream Name	Feed	Vapor	Liquid	Balance	Feed	Vapor	Liquid	Balance
Mass Flow	10123	2811.33	7185	0.987487	10123	2811.33	7185	0.987487
Component (m.f.)								
Trimethyl Meatloaf	0.13582	0.491205	0.00001	1.004441	0.13582	0.491205	0.00001	1.004441
Chicken Wire	0.14398	0.308845	0.0846	1.012766	0.14398	0.308845	0.0846	1.012766
Ba(Na)2	0.10013	0.200101	0.0553	0.946986	0.10013	0.200101	0.0553	0.946986
PolyChickenWire	0.60254	0	0.86011	1.013178	0.60254	0	0.86011	1.013178
Total	0.98247	1.000151	1.00002		0.98247	1.000151	1.00002	

TPC Group

	A	B	C	D	E	F	G	H	I	J	K	L
1	Mass Balance for PolyChickenWire Unit Pre-Fractionator Flash											
2												
3		Raw Plant Data				Reconciled Data				Standard Deviations		
4	Stream Name	Feed	Vapor	Liquid	Balance	Feed	Vapor	Liquid	Balance	Feed	Vapor	Liquid
5	Mass Flow	10123	2811.33	7185	0.987487	10123	2811.33	7185	0.987487	506.15	140.5665	359.25
6												
7	Component (m.f.)											
8												
9	Trimethyl Meatloaf	0.13582	0.491205	0.00001	1.004441	0.13582	0.491205	0.00001	1.004441	0.001358	0.004912	0.0001
10	Chicken Wire	0.14398	0.308845	0.0846	1.012766	0.14398	0.308845	0.0846	1.012766	0.00144	0.003088	0.000846
11	Ba(Na)2	0.10013	0.200101	0.0553	0.946986	0.10013	0.200101	0.0553	0.946986	0.001001	0.002001	0.000553
12	PolyChickenWire	0.60254	0	0.86011	1.013178	0.60254	0	0.86011	1.013178	0.006025	0.0001	0.008601
13												
14	Total	0.98247	1.000151	1.00002		0.98247	1.000151	1.00002				
15												

Enter the standard deviations of the measurements.
Approximate values/estimates work fine:

Coriolis meter ~1% or less
Functioning orifice meter ~5%
Poorly functioning orifice meter up to 25%
GC analysis ~1% of value, subject to minimums

TPC Group

AIChe Presentation.xls [Compatibility Mode] - Microsoft Excel

	A	B	C	D	E	F	G	H	I	J	K	L	M	N	O	P
1	Mass Balance for PolyChickenWire Unit Pre-Fractionator Flash															
2																
3		Raw Plant Data				Reconciled Data				Standard Deviations			Objective Function			
4	Stream Name	Feed	Vapor	Liquid	Balance	Feed	Vapor	Liquid	Balance	Feed	Vapor	Liquid	Feed	Vapor	Liquid	
5	Mass Flow	10123	2811.33	7185	0.987487	10123	2811.33	7185	0.987487	506.15	140.5665	[illegible]	0	0	0	
6																
7	Component (m.f.)															
8																
9	Trimethyl Meatloaf	0.13582	0.491205	0.00001	1.004441	0.13582	0.491205	0.00001	1.004441	[illegible]	[illegible]	[illegible]	0	0	0	
10	Chicken Wire	0.14398	0.308845	0.0846	1.012766	0.14398	0.308845	0.0846	1.012766	[illegible]	[illegible]	[illegible]	0	0	0	
11	Ba(Na)2	0.10013	0.200101	0.0553	0.946986	0.10013	0.200101	0.0553	0.946986	[illegible]	[illegible]	[illegible]	0	0	0	
12	PolyChickenWire	0.60254	0	0.86011	1.013178	0.60254	0	0.86011	1.013178	[illegible]	[illegible]	[illegible]	0	0	0	
13															0	Objective function
14	Total	0.98247	1.000151	1.00002		0.98247	1.000151	1.00002								
15																

Calculate the objective function

$$OBJ = \sum_{i=1}^{i=N} \frac{(x_{m,i} - x_{r,i})^2}{\delta_i^2}$$

Value of OBJ, use solver to minimize this cell

TPC Group

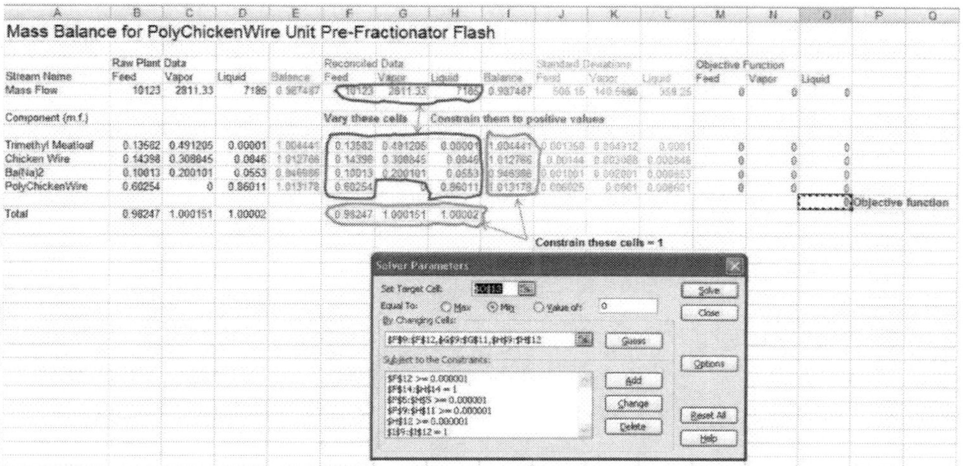

TPC Group

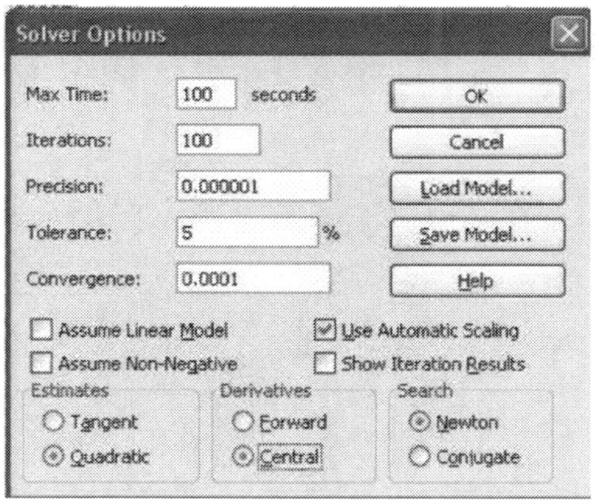

TPC Group

Problem solution :

Reconciled values — Residuals

	A	B	C	D	E	F	G	H	I	J	K	L	M	N	O	P	Q
1	Mass Balance for PolyChickenWire Unit Pre-Fractionator Flash																
2																	
3		Raw Plant Data				Reconciled Data				Standard Deviations			Objective Function				
4	Stream Name	Feed	Vapor	Liquid	Balance	Feed	Vapor	Liquid	Balance	Feed	Vapor	Liquid	Feed	Vapor	Liquid		
5	Mass Flow	10123	2811.33	7185	0.997487	10039.96	2822.122	7217.843	1	501.9962	141.1061	366.9921	0.02736	0.005849	0.008282		
6																	
7	Component (m.f.)																
8																	
9	Trimethyl Meatloaf	0.13582	0.491205	0.00001	1.004441	0.137533	0.48327	6.4E-06	1	0.001358	0.004912	0.0001	1.590341	0.15525	0.0013		
10	Chicken Wire	0.14398	0.308845	0.0846	1.012769	0.146766	0.306966	0.084128	1	0.00444	0.003088	0.000846	3.743236	0.369954	0.310846		
11	Ba(Na)2	0.10013	0.200101	0.0653	0.946986	0.097512	0.203764	0.056968	1	0.001961	0.002001	0.000653	6.8372	3.350921	1.459184		
12	PolyChickenWire	0.60254	0	0.86011	1.013176	0.61819	0	0.859097	1	0.006026	0.0001	0.000601	6.745956	0	0.000612		
13															24.60629	Objective function	
14	Total	0.98247	1.000151	1.00002		1	1	1									
15																	

Constraints satisfied — Objective function value

Analyze Residuals :

Over half the error is in two data values.
Time to have a talk with the GC tech?

	A	B	C	D	E	F
1	Mass Balance for PolyChickenWire Unit Pre-Fract					
2						
3		Objective Function				
4	Stream Name	Feed	Vapor	Liquid		
5	Mass Flow	0.02736	0.005849	0.008282		
6						
7	Component (m.f.)					
8						
9	Trimethyl Meatloaf	1.590341	0.15525	0.0013		
10	Chicken Wire	3.743235	0.369954	0.310846		
11	Ba(Na)2	6.8372	3.350921	1.459184		
12	PolyChickenWire	6.745956	0	0.000612		
13				24.60629	Objective function	

TPC Group

You are now fully equipped to mass balance your units

Happy Reconciling!

AUTHOR INDEX